Stochastic Modelling and Applied Probability

(Formerly:
Applications of Mathematics)

63

Edited by   B. Rozovskiĭ
G. Grimmett

T0143017

For other titles in this series, go to
http://www.springer.com/series/602

G. George Yin • Chao Zhu

Hybrid Switching Diffusions

Properties and Applications

 Springer

G. George Yin
Department of Mathematics
Wayne State University
Detroit, MI 48202
USA
gyin@math.wayne.edu

Chao Zhu
Department of Mathematical Sciences
University of Wisconsin-Milwaukee
Milwaukee, WI 53201
USA
zhu@uwm.edu

Managing Editors
Boris Rozovskiĭ
Division of Applied Mathematics
Brown University
182 George St
Providence, RI 02912
USA
rozovsky@dam.brown.edu

Geoffrey Grimmett
Centre for Mathematical Sciences
University of Cambridge
Wilberforce Road
Cambridge CB3 0WB
UK
g.r.grimmett@statslab.cam.ac.uk

ISSN 0172-4568
ISBN 978-1-4614-2470-3   e-ISBN 978-1-4419-1105-6
DOI 10.1007/978-1-4419-1105-6
Springer New York Dordrecht Heidelberg London

Mathematics Subject Classification (2000): 60J27, 60J60, 93E03, 93E15

Springer is part of Springer Science+Business Media (www.springer.com)

In memory of my sister Kewen Yin, who taught me algebra, calculus, physics, and chemistry during the Cultural Revolution, when I was working in a factory in Beijing and she was over 1000 miles away at a factory in Lanzhou

George Yin

To my parents Yulan Zhong and Changming Zhu and my wife Lijing Sun, with love

Chao Zhu

Contents

Preface

This book encompasses the study of hybrid switching diffusion processes and their applications. The word "hybrid" signifies the coexistence of continuous dynamics and discrete events, which is one of the distinct features of the processes under consideration. Much of the book is concerned with the interactions of the continuous dynamics and the discrete events. Our motivations for studying such processes originate from emerging and existing applications in wireless communications, signal processing, queueing networks, production planning, biological systems, ecosystems, financial engineering, and modeling, analysis, and control and optimization of large-scale systems, under the influence of random environments.

Displaying mixture distributions, switching diffusions may be described by the associated operators or by systems of stochastic differential equations together with the probability transition laws of the switching actions. We either have Markov-modulated switching diffusions or processes with continuous state-dependent switching. The latter turns out to be much more challenging to deal with. Viewing the hybrid diffusions as a number of diffusions joined together by the switching process, they may be seemingly not much different from their diffusion counterpart. Nevertheless, the underlying problems become more difficult to handle, especially when the switching processes depend on continuous states. The difficulty is due to the interaction of the discrete and continuous processes and the tangled and hybrid information pattern.

A salient feature of the book is that the discrete event process is allowed to depend on the continuous dynamics. Apart from the existence and uniqueness of solutions, we treat a number of basic properties such as regularity, the Feller property, the strong Feller property, and continuous and smooth dependence on initial data of the solutions of the associated stochastic differential equations with switching.

A large part of this work is concerned with the stability of switching diffusion processes. Here stability is meant in the broad sense including both weak and strong stability. That is, we focus on both a "neighborhoods of infinity" and neighborhoods of equilibrium points; the stability corresponding to the former is referred to as weak stability, whereas that of the latter is stability in the usual sense. In studying deterministic dynamic systems, researchers used Lagrange stability to depict systems that are ultimately uniformly bounded. Treating stochastic systems, one would still hope to adopt such a notion. Unfortunately, the boundedness excludes many important cases. Thus, we replace this boundedness by a weaker notion known as recurrence (returning to a prescribed compact region in finite time). When the expected returning time is finite, we have so-called positive recurrence. A crucial question is: Under what conditions will the systems be recurrent (resp. positive recurrent). We demonstrate that positive recurrence implies ergodicity and provide criteria for the existence of invariant distributions together with their representation. In addition to studying asymptotic properties of the systems in the neighborhood of infinity, we examine the behavior of the systems at the equilibria. Also considered are invariance principles and stability of differential equations with random switching but without diffusions.

Because the systems are rarely solvable in closed form, numerical methods become a viable alternative. We construct algorithms to approximate solutions of such systems with state-dependent switching, and provide sufficient conditions for convergence for numerical approximations to the invariant measures.

In real-world applications, hybrid systems encountered are often of large scale and complex leading to intensive-computation requirement. Reduction of computational complexity is thus an important issue. To take this into consideration, we consider time-scale separation in hybrid switching-diffusion and hybrid-jump-diffusion models. Several issues including recurrence of processes with switching having multiple-weak-connected ergodic classes, two-time-scale modeling of stochastic volatility, and weak convergence analysis of systems with fast and slow motions involving additional Poisson jumps are studied in details.

This book is written for applied mathematicians, probabilists, systems engineers, control scientists, operations researchers, and financial analysts among others. The results presented in the book are useful to researchers working in stochastic modeling, systems theory, and applications in which continuous dynamics and discrete events are intertwined. The book can be served as a reference for researchers and practitioners in the aforementioned areas. Selected materials from the book may also be used in a graduate-level course on stochastic processes and applications.

This book project could not have been completed without the help and encouragement of many people. We are deeply indebted to Wendell Fleming and Harold Kushner, who introduced the wonderful world of stochastic systems to us. We have been privileged to have the opportunity to work with Rafail Khasminskii on a number of research projects, from whom we have learned a great deal about Markov processes, diffusion processes, and stochastic stability. We express our special appreciation to Vikram Krishnamurthy, Ruihua Liu, Yuanjin Liu, Xuerong Mao, John Moore, Qingshuo Song, Chenggui Yuan, Hanqin Zhang, Qing Zhang, and Xunyu Zhou, who have worked with us on various projects related to switching diffusions. We also thank Eric Key, Jose Luis Menaldi, Richard Stockbridge, and Fubao Xi for many useful discussions. This book, its contents, its presentation, and exposition have benefited a great deal from the comments by Ruihua Liu, Chenggui Yuan, Hanqin Zhang, and by several anonymous reviewers, who read early versions of the drafts and offered many insightful comments. We thank the series editor, Boris Rozovsky, for his encouragement and consideration. Our thanks also go to Springer senior editor, Achi Dosanjh, for her assistance and help, and to the production manager, and the Springer professionals for their work in finalizing the book. During the years of our study, the research was supported in part by the National Science Foundation, and the National Security Agency. Their continuing support is greatly appreciated.

Detroit, Michigan   George Yin
Milwaukee, Wisconsin   Chao Zhu

Conventions

We clarify the numbering system and cross-reference conventions used throughout the book. Equations are numbered consecutively within a chapter. For example, (2.10) indicates the tenth equation in Chapter 2. Corollaries, definitions, examples, lemmas, propositions, remarks, and theorems are numbered sequentially throughout each chapter. For example, Definition 3.1, Theorem 3.2, Corollary 3.3 and so on. Assumptions are marked consecutively within a chapter. For cross reference, an equation is identified by the chapter number and the equation number; similar conventions are used for theorems, remarks, assumptions, and so on.

Throughout the book, we assume that all deterministic processes are Borel measurable and all stochastic processes are measurable with respect to a given filtration. A subscript generally denotes either a finite or an infinite sequence. However, the ε-dependence of a sequence is designated in the superscript. To facilitate reading, we provide a glossary of symbols used in the subsequent chapters.

Glossary of Symbols

A'	transpose of A (either a matrix or a vector)
$\mathrm{Cov}(\xi)$	covariance of a random variable ξ
$\mathbb{C}$	space of complex numbers
$C(D)$	space of real-valued continuous functions defined on D
$C_b^k(D)$	space of real-valued functions with bounded and continuous derivatives up to the order k
C_b^k	$C_b^k(D)$ with $D = \mathbb{R}^r$
C_0^k	C^k-functions with compact support
D^c	complement of a set D
$\overline{D}$	closure of a set D, $\overline{D} = D \cup \partial D$
∂D	boundary of a set D
$D([0,T]; \mathsf{S})$	space of S-valued functions being right continuous and having left-hand limits
$D \subset\subset E$	$D \subset \overline{D} \subset E$ and $\overline{D}$ is compact
$\mathbf{E}_{x,i}$	expectation with $X(0) = x$ and $\alpha(0) = i$
$\mathbf{E}\xi$	expectation of a random variable ξ
$\mathcal{F}$	σ-algebra
$\{\mathcal{F}_t\}$	filtration $\{\mathcal{F}_t, t \geq 0\}$
I	identity matrix of suitable dimension
I_A	indicator function of a set A
K	generic positive constant with convention $K + K = K$ and $KK = K$
$\mathcal{M}$	state space of switching process $\alpha(t)$
$N(x)$	neighborhood of x centered at the origin
$O(y)$	function of y such that $\sup_y |O(y)|/|y| < \infty$

$\mathbf{P}_{x,i}$	probability with $X(0) = x$ and $\alpha(0) = i$
$\mathbf{P}(\xi \in \cdot)$	probability distribution of a random variable ξ
$Qf(\cdot)(i)$	$= \sum_{j \neq i} q_{ij}(f(j) - f(i))$ where $Q = (q_{ij})$
$Q(x)$	x-dependent generator of the switching process
$\mathbb{R}$	space of real numbers
$\mathbb{R}^r$	r-dimensional real Euclidean space
$S(r)$ or S_r	ball centered at the origin with radius r
$X^{x,i}(t)$	$X(t)$ with initial data $X(0) = x$ and $\alpha(0) = i$
a^+	$= \max\{a, 0\}$ for a real number a
a^-	$= \max\{-a, 0\}$ for a real number a
a.s.	almost surely
$\langle a, b \rangle$	inner product of vectors a and b
$a_1 \wedge \cdots \wedge a_l$	$= \min\{a_1, \ldots, a_l\}$ for $a_i \in \mathbb{R}$, $i = 1, \ldots, l$
$a_1 \vee \cdots \vee a_l$	$= \max\{a_1, \ldots, a_l\}$ for $a_i \in \mathbb{R}$, $i = 1, \ldots, l$
$\mathrm{diag}(A^1, \ldots, A^l)$	diagonal matrix of blocks $A^1, \ldots, A^l$
$\exp(Q)$	e^Q for a matrix Q
f_x or $\nabla_x f$	gradient of f with respect to x
f_{xx} or $\nabla_x^2 f$	Hessian of f with respect to x
i.i.d.	independent and identically distributed
$\ln x$ or $\log x$	natural logarithm of x
$m(\cdot)$	Lebesgue measure on $\mathbb{R}$
$o(y)$	a function of y such that $\lim_{y \to 0} o(y)/|y| = 0$
$\mathfrak{p}(dt, dz)$	Poisson random measure with intensity $dt \times m(dz)$
$\mathrm{tr}(A)$	trace of matrix A
w.p.1	with probability one
$\lfloor x \rfloor$	integer part of x
$(\Omega, \mathcal{F}, \mathbf{P})$	probability space
$(\Omega, \mathcal{F}, \{\mathcal{F}_t\}, \mathbf{P})$	filtered probability space
$\alpha(t)$	continuous-time process with right-continuous sample paths
δ^{ij}	$= 1$ if $i = j$; $= 0$ otherwise
ε	positive small parameter
$\lambda_{\max}(A)$	maximal eigenvalue of a symmetric matrix A
$\lambda_{\min}(A)$	minimal eigenvalue of a symmetric matrix A
$\mathbb{1}$	a column vector with all entries being 1
$:=$ or $\overset{\mathrm{def}}{=}$	defined to be equal to
$\square$	end of a proof
$|\cdot|$	norm of an Euclidean space or a function space
$\|\cdot\|$	essential sup-norm

1

Introduction and Motivation

1.1 Introduction

This book focuses on switching diffusion processes involving both continuous dynamics and discrete events. Before proceeding to the detailed
study, we address the following questions. Why should we study such hybrid systems? What are typical examples arising from applications? What
are the main properties we wish to study? This introductory chapter provides motivations of our study, delineates switching diffusions in a simple
way, presents a number of application examples, and gives an outline of the
entire book.

1.2 Motivation

Owing to their wide range of applications, hybrid switching diffusions, also
known as switching diffusion systems, have become more popular recently
and have drawn growing attention, especially in the fields of control and
optimization. Because of the presence of both continuous dynamics and
discrete events, such systems are capable of describing complex systems
and their inherent uncertainty and randomness in the environment. The
formulation provides more opportunity for realistic models, but adds more
difficulties in analyzing the underlying systems. Resurgent efforts have been
devoted to learning more about the processes and their properties. Much
of the study originated from applications arising in control engineering,

G.G. Yin and C. Zhu, *Hybrid Switching Diffusions: Properties and Applications*,
Stochastic Modelling and Applied Probability 63, DOI 10.1007/978-1-4419-1105-6_1,
© Springer Science + Business Media, LLC 2010

manufacturing systems, estimation and filtering, two-time-scale systems, and financial engineering; see [74, 78, 123, 146, 147, 154, 167, 170, 184], among others. In these applications, random-switching processes are used to model demand rate or machine capacity in production planning, to describe the volatility changes over time to capture discrete shifts such as market trends and interest rates and the like in finance and insurance, and to model time-varying parameters for network problems. In reference to recent developments, this book emphasizes the development of basic properties of switching diffusion processes.

The coexistence of continuous dynamics and discrete events and their interactions reflect the salient features of the systems under consideration. Such systems have become increasingly important for formulation, analysis, and optimization in many applications. Perhaps, one of the reasons is that many real-world applications in the new era require sophisticated models, in which the traditional dynamic system setup using continuous dynamics given by differential equations alone is inadequate. In addition, there have been increasing demands for modeling large-scale and complex systems, designing optimal controls, and conducting optimization tasks. In the traditional setting, the design of a feedback controller is based on a plant having fixed parameters, which is inadequate when the actual system differs from the assumed nominal model. As a consequence, the controller is not able to serve its purposes and cannot attenuate disturbances or perturbations. Much effort has been directed to the design of more "robust" controls in recent years. Studies of hybrid systems with Markov regime switching contribute significantly to this end. Various regime-switching models have been proposed and examined. The so-called jump linear systems, widely used in engineering, have been studied in Mariton [123]; controllability and stabilizability of such systems are treated in Ji and Chizeck [78]. Estimation problems are considered in Sworder and Boyd [154]. Manufacturing and production planning under the framework of hierarchical structure are covered in Sethi and Zhang [146], and Sethi, Zhang, and Zhang [147]. Financial engineering applications and the use of hybrid geometric Brownian motion models can be found in [170, 184], among others. To reduce the complexity, effort has been made to deal with hybrid systems with regime switching by means of time-scale separation in Yin and Zhang [176]; see also [167, 185]. Hybrid systems have received increasing attention in recent years. Other than the switching diffusion systems mentioned above, one may find the formulation of a somewhat different setup in Bensoussan and Menaldi [12]. An introductory text on stochastic control for jump diffusions is in Hanson [64], whereas a treatment of stochastic hybrid systems with applications to communication networks is in Hespanha [67]. For references on stochastic control and controlled Markov processes, we refer the reader to Fleming and Rishel [44], Fleming and Soner [45], Krylov [96], and Yong and Zhou [181].

A special feature of switching diffusion processes is: In the systems, continuous dynamics and discrete events are intertwined. For example, one of the early efforts of using such hybrid models for financial applications can be traced back to [5], in which both the appreciation rate and the volatility rate of a stock depend on a continuous-time Markov chain. In the simplest case, a stock market may be considered to have two "modes" or "regimes," up and down, resulting from the state of the underlying economy, the general mood of investors in the market, and so on. The rationale is that in the different modes or regimes, the volatility and return rates are very different. The introduction of hybrid models makes it possible to describe stochastic volatility in a relatively simple manner (simpler than the so-called stochastic volatility models). Another example is concerned with a wireless communication network. Consider the performance analysis of an adaptive linear multiuser detector in a cellular direct-sequence code-division multiple-access wireless network with changing user activity due to an admission or access controller at the base station. Under certain conditions, an associated optimization problem leads to a switching diffusion limit; see [168]. To summarize, a centerpiece in the applications mentioned above is a two-component Markov process: a continuous component and a discrete-event component.

Long-time behavior of such hybrid systems is one of the major concerns in stochastic processes, systems theory, control, and optimization. Nowadays, there is a fairly well-known theory for stability of diffusion processes. The basic setup of the stability study and the foundational work were originated in the work of Khasminskii [83] and Kushner [97]. The origin of the study may be traced back to Kac and Krasovskii [79] for differential equations perturbed by Markov chains. Comprehensive treatment of Markov processes is in Dynkin [38], whereas that of diffusion processes may be found in Gihman and Skorohod [55], Liptser and Shiryayev [110], and Stroock and Varadhan [153] and references therein. Concerning jump processes and Markov chains, we mention the work of Chung [28], Cox and Miller [29], and Doob [33]. Davis's piecewise deterministic viewpoint adds a new twist to Markov models. Ethier and Kurtz [43] examine weak convergence of Markov processes and provide an in-depth study on characterization of the limit process. Chen's book [23] presents an approach from the angle of coupling methods. Recently, stability of diffusion processes with Markovian switching have received much attention (see, e.g., [6, 7, 92, 116, 119, 183, 187, 190] and the references therein). Some of the recent developments in Markov modulated switching diffusions (where the Markov chain and the Brownian motion are independent) are found in Mao and Yuan [120]. Accommodating the many applications, although these systems are more realistically addressing the demands, the nontraditional setup makes the analysis of switching diffusions more difficult. To be able to effectively treat these systems, it is of foremost importance to have a thorough understanding of the underpinning of the systems. This is the main theme of the current work.

1.3   What Is a Switching Diffusion

Let us begin with answering the question: What is a switching diffusion. In this book, we consider a switching diffusion model, in which the discrete events are modeled by a finite-state process. The following diagram provides a visualization of a typical example. Consider a switching diffusion as given in Figure 1.1 that consists of three diffusions sitting on three parallel plans. The discrete event is a three-state jump process. We denote the pair of processes by

$$\text{(continuous process, discrete event)} = (X(t), \alpha(t)).$$

Suppose that initially, the process is at $(X(0), \alpha(0)) = (x, 1)$. The discrete event process sojourns in discrete state 1 for a random duration; during this period, the continuous component evolves according to the diffusion process specified by the drift and diffusion coefficients associated with discrete state 1 until a jump of the discrete component takes place. At random moment τ_1, a jump to discrete state 3 occurs. Then the continuous component evolves according to the diffusion process whose drift and diffusion coefficients are determined by discrete event 3. The process wanders around in the third plan until another random jump time τ_2. At τ_2, the system switches to the second parallel plan and follows another diffusion with different drift and diffusion coefficients and so on.

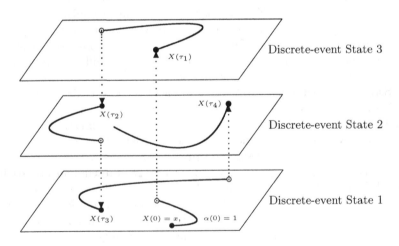

FIGURE 1.1. A "sample path" of regime-switching diffusion $(X(t), \alpha(t))$.

At a first glance, one may feel that the process is not much different from a diffusion because the switching is a finite-state process. Nevertheless, even if the switching is a finite-state Markov chain independent of the Brownian motion (subsequently referred to as Markov-modulated switching diffusions), the switching diffusion is still much harder to handle. The main reason is the coupling and interactions due to the switching process make the analysis much more difficult. For example, when we study recurrence or stability, we have to deal with systems of coupled partial differential equations.

One of the main features of this book is that a large part of it deals with the switching component depending on the continuous component. This is often referred to as a state-dependent switching process in what follows; more precise statements about which will be made later. In this case, the analysis is much more difficult than that of the case in which the switching is independent of the continuous states. As shown in Chapter 2, some of the time-honored characteristics such as continuous dependence of initial data become highly nontrivial; even Feller properties are not straightforward to obtain.

1.4   Examples of Switching Diffusions

To demonstrate the utility of the switching diffusion models, this section provides a number of examples. Some of these examples are revisited in later chapters.

Example 1.1. (Lotka–Volterra model) Concerning ecological population modeling, proposed by Lotka and Volterra, the well-known Lotka–Volterra models have been investigated extensively in the literature. When two or more species live in proximity and share the same basic requirements, they usually compete for resources, food, habitat, or territory. Both deterministic and stochastic models of the Lotka–Volterra systems have been studied extensively.

It has been noted that the growth rates and the carrying capacities are often subject to environmental noise. Moreover, the qualitative changes of the growth rates and the carrying capacities form an essential aspect of the dynamics of the ecosystem. These changes usually cannot be described by the traditional (deterministic or stochastic) Lotka–Volterra models. For instance, the growth rates of some species in the rainy season is much different from those in the dry season. Moreover, the carrying capacities often vary according to the changes in nutrition and food resources. Similarly, the interspecific or intraspecific interactions differ in different environment.

It is natural to consider a stochastic Lotka–Volterra ecosystem in a random environment that can be formulated by use of an additional factor process. Consider the following stochastic differential equation with regime switching

$$
\begin{aligned}
dx(t) = \ & \mathrm{diag}\,(x_1(t), \ldots, x_n(t)) \\
& \times [(b(\alpha(t)) - A(\alpha(t))x(t))dt + \Sigma(\alpha(t)) \circ dw(t)],
\end{aligned}
\tag{1.1}
$$

or equivalently, in componentwise form

$$
\begin{aligned}
dx_i(t) = \ & x_i(t)\Big\{ \Big[b_i(\alpha(t)) - \sum_{j=1}^{n} a_{ij}(\alpha(t))x_j(t) \Big] dt \\
& + \sigma_i(\alpha(t)) \circ dw_i(t) \Big\}, \quad i = 1, \ldots, n,
\end{aligned}
\tag{1.2}
$$

where $w(\cdot) = (w_1(\cdot), \ldots, w_n(\cdot))'$ is an n-dimensional standard Brownian motion, and for $\alpha \in \mathcal{M}$, $b(\alpha) = (b_1(\alpha), \ldots, b_n(\alpha))'$, $A(\alpha) = (a_{ij}(\alpha))$, $\Sigma(\alpha) = \mathrm{diag}(\sigma_1(\alpha), \ldots, \sigma_n(\alpha))$ represent different growth rates, community matrices, and noise intensities in different external environments, respectively, and z' denotes the transpose of z. Assume that $b_i(\alpha) > 0$ for each $\alpha \in \mathcal{M}$ and each $i = 1, \ldots, n$, and the Markov chain $\alpha(\cdot)$ and the Brownian motion $w(\cdot)$ are independent. Without loss of generality, we also assume that the initial conditions $x(0)$ and $\alpha(0)$ are nonrandom. The above formulation is seen to be in the form of Stratonovich. As explained in [84], this form is more suitable for environment modeling. It is well known that (1.2) is equivalent to the following stochastic differential equation in the Itô sense

$$
\begin{aligned}
dx_i(t) = \ & x_i(t)\Big\{ \Big[r_i(\alpha(t)) - \sum_{j=1}^{n} a_{ij}(\alpha(t))x_j(t) \Big] dt \\
& + \sigma_i(\alpha(t))dw_i(t) \Big\}, \quad i = 1, 2, \ldots, n,
\end{aligned}
\tag{1.3}
$$

where $r_i(\alpha(t)) := b_i(\alpha(t)) + \frac{1}{2}\sigma_i^2(\alpha(t))$ for each $i = 1, 2, \ldots, n$.

Regime-switching stochastic Lotka–Volterra models have received much attention lately. For instance, the study of trajectory behavior of Lotka–Volterra competition bistable systems and systems with telegraph noises was considered in [37], stochastic population dynamics under regime switching was treated in [113], the dynamics of a population in a Markovian environment were studied in [151], and the evolution of a system composed of two predator-prey deterministic systems described by Lotka–Volterra equations in a random environment was investigated in [156]. In the absence of regime switching, the system is completely modeled by stochastic time evolution in a fixed environment. The results in a fixed environment correspond to (1.2) or (1.3) in the case when the Markov chain has only one state or the Markov chain always stays in the fixed state (environment). When random environments are considered, the system's qualitative behavior can be drastically different. In a recent paper of Zhu and Yin [189], we considered several asymptotic properties of the models given above.

Example 1.2. Here we consider the problem of balanced realizations, a concept that started gaining popularity in the fields of systems and control in the early 1980s, and has still attracted much attention lately. Balanced realizations have been studied for finite-dimensional linear systems for nearly three decades. The original problem is concerned with linear deterministic systems. In the 1980s, to bridge the gap between minimal realization theory and to treat the problem of finding lower-order approximations, a particular realization was first introduced in the seminal paper by Moore [129], where the problem was studied with principal component analysis of linear systems. The term "balanced" was used because the realizations have a certain symmetry between the input and the output maps characterized by the controllability and observability Grammians. Owing to their importance and their wide range of applications, balanced realizations have attracted much attention. One of the applications areas is model reduction. Asymptotic stability of the reduced-order systems was studied in [129], and error bounds between the reduced-order model and the original system were obtained in [58] in terms of the associated singular values. There have been substantial extensions of the theory to time-varying linear systems. Key existence results concerning balanced realizations were contained in [148] and [157]. Subsequent work can be found in [66, 75, 133, 144] and references therein.

We generalize the ideas to include random switching to represent random perturbations under a stochastic environment. We suppose that for each instant t, the state consists of a pair $(X(t), \alpha(t))$ representing the continuous state component and discrete event component, respectively. Let $\alpha(t)$ be a continuous-time Markov chain with finite state space $\mathcal{M} = \{1, 2, \ldots, m_0\}$ and generator

$$Q = (q_{ij}) \in \mathbb{R}^{m_0 \times m_0} \text{ such that } q_{ij} \geq 0 \text{ for } i \neq j, \text{ and } \sum_{j=1}^{m_0} q_{ij} = 0. \quad (1.4)$$

Our proposed set of equations for balanced realizations includes coupling terms due to the Markov switching of the model. In the trivial case of no switching, the equations are decoupled and are standard balanced realization equations. We loosely carry the term "balanced realization" over to our more general setting. As in the classical theory for the case of no Markovian switching, the condition for balanced realization is represented by a system of algebraic equations. Our system of equations is a set of Riccati equations but with additional couplings between them owing to the discrete events not evident in the classical case. The underlying systems are modulated by a continuous-time Markov chain with state space $\mathcal{M}$. The matrix coefficients of the linear systems depend on the states of the Markov chain. At any given instant t, the Markov chain takes one of the values (e.g., i) in $\mathcal{M}$. Then the system dynamics are determined by the matrix coefficients associated with state i. After a random time, the Markov chain switches to a new state $j \neq i$ and stays there until the next jump. The system dynamics are then determined by the matrix coefficients associated with j. Such random jump linear systems arise frequently in applications where regime changes are utilized to model the random environment. Assume that for each $i \in \mathcal{M}$, $A(i, \cdot)$, $B(i, \cdot)$, and $C(i, \cdot)$ are bounded and continuously differentiable matrix-valued functions with suitable dimensions. Consider the following system

$$
\begin{aligned}
\frac{d}{dt}X(t) &= A(\alpha(t), t)X(t) + B(\alpha(t), t)u(t), \quad X(t_0) = x_0, \\
y(t) &= C(\alpha(t), t)X(t),
\end{aligned}
\tag{1.5}
$$

where the state $X(t) \in \mathbb{R}^{n \times 1}$, input $u(t) \in \mathbb{R}^{r \times 1}$, and output $y(t) \in \mathbb{R}^{m_0}$. Note that because a finite-state Markov chain is used, effectively, (1.5) can be written as

$$
\begin{aligned}
\frac{d}{dt}X(t) &= \sum_{i=1}^{m_0} A(i, t)X(t)I_{\{\alpha(t)=i\}} + \sum_{i=1}^{m_0} B(i, t)u(t)I_{\{\alpha(t)=i\}}, \quad X(t_0) = x_0, \\
y(t) &= \sum_{i=1}^{m_0} C(i, t)X(t)I_{\{\alpha(t)=i\}},
\end{aligned}
$$

where I_S is the indicator function of the set S. In what follows, if a square matrix D is positive definite (resp., nonnegative definite), we often write it as $D > 0$ (resp., $D \geq 0$). For $D_1 \in \mathbb{R}^{\iota \times \ell}$ for some $\iota, \ell \geq 1$, D_1' denotes its transpose. For a suitable function $f(t, i)$, we denote

$$
Qf(t, \cdot)(i) = \sum_{j=1}^{m_0} q_{ij} f(t, j) = \sum_{j \neq i} q_{ij}(f(t, j) - f(t, i)) \quad \text{for each } i \in \mathcal{M}.
$$

For each $t \geq 0$ and $i \in \mathcal{M}$, a realization $(A(i, t), B(i, t), C(i, t))$ is said to be uniformly completely controllable if and only if there is a $\delta > 0$ such that for some positive $L_c(\delta)$ and $U_c(\delta)$,

$$
\infty > U_c(\delta)I \geq G_c(t - \delta, t, i) \geq L_c(\delta)I > 0,
\tag{1.6}
$$

where $G_c(t - \delta, t, i)$ is the controllability Grammian

$$
G_c(t - \delta, t, i) = \int_{t-\delta}^{t} \Phi(t, \lambda, i)B(i, \lambda)B'(i, \lambda)\Phi'(t, \lambda, i)d\lambda,
\tag{1.7}
$$

and $\Phi(t, \lambda, i)$ is the state transition matrix (see [2, p. 349]) of the equation

$$
\frac{dz(t)}{dt} = A(i, t)z(t).
$$

For each $t \geq 0$ and $i \in \mathcal{M}$, a realization $(A(i, t), B(i, t), C(i, t))$ is said to be uniformly completely observable if and only if there is a $\delta > 0$ such that for some positive $L_o(\delta)$ and $U_o(\delta)$,

$$
\infty > U_o(\delta)I \geq G_o(t, t + \delta, i) \geq L_o(\delta)I > 0,
\tag{1.8}
$$

where $G_o(t, t + \delta, i)$ is the observability Grammian

$$G_o(t, t + \delta, i) = \int_t^{t+\delta} \Phi'(\lambda, t, i) C'(i, \lambda) C(i, \lambda) \Phi(\lambda, t, i) d\lambda. \qquad (1.9)$$

System (1.5) is said to have a balanced realization if there are nonsingular coordinate transformations $T(t, i)$ with

$$P(t, i) \stackrel{\text{def}}{=} T'(t, i) T(t, i) > 0$$

such that

$$P(t, i) G_c(t, i) P(t, i) = G_o(t, i) + QP(t, \cdot)(i), \quad i = 1, 2, \dots, m_0. \qquad (1.10)$$

Note that regime switching is one of the main features (fully degenerate regime-switching diffusion with the second order term missing in the system) of such systems compared with the traditional setup. To study the behavior of such systems, it is crucial to have a thorough understanding of the switching diffusion processes. For modeling and analysis of the balanced realizations, we refer the reader to [111] for further reading.

Example 1.3. Continuing with the setup of the hybrid linear system

$$\dot{x}(t) = A(\alpha(t))x(t) + B(\alpha(t))u(t),$$

where $\alpha(t)$ is a continuous-time Markov chain taking values in a finite set $\mathcal{M} = \{1, \dots, m_0\}$, $A(i)$ and $B(i)$ for $i \in \mathcal{M}$ are matrices with compatible dimensions, and $u(\cdot)$ is the control. In lieu of one linear system, we have a number of systems coupled through the Markov chain. Such systems have enjoyed numerous applications in emerging application areas such as financial engineering and wireless communications, as well as in existing applications. A class of important problems concerns the asymptotic behavior of such systems when they are in operation for a long time. Very often, in many engineering problems, one is more interested in whether the system is stable. Much interest lies in finding admissible controls so that the resulting system will be stabilized. There has been continuing interest in dealing with hybrid systems under a Markov switching. In [123], stabilization for robust controls of jump linear quadratic (LQ) control problems was treated. In [78], both controllability and stabilizability of jump linear LQ systems were considered. In [21], adaptive LQG problems with finite-state process parameters were treated. Additional difficulties come when the switching process $\alpha(t)$ cannot be observed, but can only be observed in white noise. That is, we can observe

$$dX(t) = [A(\alpha(t))X(t) + B(\alpha(t))u(t)]dt + dw(t).$$

For such partially observed systems, it is natural to use nonlinear filtering techniques. The associated filter is the well-known Wonham filter [159], which is one of a handful of finite-dimensional filters in existence. Stabilization of linear systems with hidden Markov chains was considered in [22, 36]. In both of these references, averaging criteria were used for the purpose of stabilization under

$$\limsup_{t\to\infty} \mathbf{E}[|X(t)|^2 + |u(t)|^2] < \infty,$$

in [36], whereas stabilization under

$$\limsup_{t\to\infty} \frac{1}{t}\mathbf{E}\int_0^t [|X(s)|^2 + |u(s)|^2]ds < \infty$$

was considered in [22]. A question of considerable practical interest is: Can we design controls so that the resulting system will be stable in the almost sure sense. Using Wonham filters and converting the partially observed system to an equivalent fully observed system, almost sure stabilizing controls were found under the criterion

$$\limsup_{t\to\infty} \frac{1}{t}\log|X(t)| \le 0 \quad \text{almost surely} \tag{1.11}$$

in [13]. The main idea lies in analyzing the sample path properties using a suitable Liapunov function. We refer the reader to the reference given above for further details.

Example 1.4. Let $\{\theta_n\}$ be a discrete-time Markov chain with finite state space

$$\mathcal{M} = \{\bar{\theta}_1, \ldots, \bar{\theta}_{m_0}\} \tag{1.12}$$

and transition probability matrix

$$P^\varepsilon = I + \varepsilon Q, \tag{1.13}$$

where $\varepsilon > 0$ is a small parameter, I is an $m_0 \times m_0$ identity matrix, and $Q = (q_{ij}) \in \mathbb{R}^{m_0 \times m_0}$ is a generator of a continuous-time Markov chain (i.e., Q satisfies $q_{ij} \ge 0$ for $i \ne j$ and $\sum_{j=1}^{m_0} q_{ij} = 0$ for each $i = 1, \ldots, m_0$). For simplicity, suppose that the initial distribution $\mathbf{P}(\theta_0 = \bar{\theta}_i) = p_{0,i}$ is independent of ε for each $i = 1, \ldots, m_0$, where $p_{0,i} \ge 0$ and $\sum_{i=1}^{m_0} p_{0,i} = 1$.

Let $\{X_n\}$ be an S-state conditional Markov chain (conditioned on the parameter process). The state space of $\{X_n\}$ is $\mathcal{S} = \{e_1, \ldots, e_S\}$, where e_i, for $i = 1, \ldots, S$, denotes the ith standard unit vector with the ith component being 1 and rest of the components being 0. For each $\theta \in \mathcal{M}$, $A(\theta) = (a_{ij}(\theta)) \in \mathbb{R}^{S \times S}$, the transition probability matrix of X_n, is defined by

$$a_{ij}(\theta) = \mathbf{P}(X_{n+1} = e_j | X_n = e_i, \theta_n = \theta) = \mathbf{P}(X_1 = e_j | X_0 = e_i, \theta_0 = \theta),$$

where $i, j \in \{1, \ldots, S\}$. Assume that for $\theta \in \mathcal{M}$, $A(\theta)$ is irreducible and aperiodic.

Note that the underlying Markov chain $\{\theta_n\}$ is in fact ε-dependent. We suppress the ε-dependence for notational simplicity. The small parameter ε in (1.13) ensures that the entries of the transition probability matrix are nonnegative, because $p_{ij}^{\varepsilon} = \delta_{ij} + \varepsilon q_{ij} \geq 0$ for $\varepsilon > 0$ small enough, where δ_{ij} denotes the Kronecker δ satisfying $\delta_{ij} = 1$ if $i = j$ and is 0 otherwise. The use of the generator Q makes the row sum of the matrix P be one. Although the true parameter is time varying, it is piecewise constant. Moreover, due to the dominating identity matrix in (1.13), $\{\theta_n\}$ varies slowly in time. The time-varying parameter takes a constant value $\overline{\theta}_i$ for a random duration and jumps to another state $\overline{\theta}_j$ with $j \neq i$ at a random time.

The assumptions on irreducibility and aperiodicity of $A(\theta)$ imply that for each $\theta \in \mathcal{M}$, there exists a unique stationary distribution $\pi(\theta) \in \mathbb{R}^{S \times 1}$ satisfying

$$\pi'(\theta) = \pi'(\theta)A(\theta), \quad \text{and} \quad \pi'(\theta)\mathbb{1}_S = 1,$$

where $\mathbb{1}_\ell \in \mathbb{R}^{\ell \times 1}$ with all entries being equal to 1. We use a stochastic approximation algorithm to track the time-varying distribution $\pi(\theta_n)$ that depends on the underlying Markov chain θ_n.

We use the following adaptive algorithm of least mean squares (LMS) type with constant stepsize in order to construct a sequence of estimates $\{\widehat{\pi}_n\}$ of the time-varying distribution $\pi(\theta_n)$

$$\widehat{\pi}_{n+1} = \widehat{\pi}_n + \mu(X_{n+1} - \widehat{\pi}_n), \tag{1.14}$$

where μ denotes the stepsize. Define $\widetilde{\pi}_n = \widehat{\pi}_n - \mathbf{E}\pi(\theta_n)$. Then (1.14) can be rewritten as

$$\widetilde{\pi}_{n+1} = \widetilde{\pi}_n - \mu\widetilde{\pi}_n + \mu(X_{n+1} - \mathbf{E}\pi(\theta_n)) + \mathbf{E}(\pi(\theta_n) - \pi(\theta_{n+1})). \tag{1.15}$$

Note that $\widehat{\pi}_n$, $\pi(\theta_n)$, and hence $\widetilde{\pi}_n$, are column vectors (i.e., they take values in $\mathbb{R}^{S \times 1}$).

For $0 < T < \infty$, we construct a piecewise constant interpolation of the stochastic approximation iterates $\widehat{\pi}_n$ as

$$\widehat{\pi}^\mu(t) = \widehat{\pi}_n, \quad t \in [\mu n, \mu n + \mu). \tag{1.16}$$

The process $\widehat{\pi}^\mu(\cdot)$ so defined is in $D([0,T];\mathbb{R}^S)$, which is the space of functions defined on $[0,T]$ taking values in $\mathbb{R}^S$ that are right continuous, have left limits, and are endowed with the Skorohod topology.

In one of our recent works [168], using weak convergence methods to carry out the analysis, we have shown that $(\widehat{\pi}^\mu(\cdot), \theta^\mu(\cdot))$ converges weakly to $(\widehat{\pi}(\cdot), \theta(\cdot))$, which is a solution of the following switching ordinary differential equation

$$\frac{d}{dt}\widehat{\pi}(t) = \pi(\theta(t)) - \widehat{\pi}(t), \quad \widehat{\pi}(0) = \widehat{\pi}_0. \tag{1.17}$$

The above switching ODE displays a very different behavior from the trajectories of systems derived from the classical ODE approach for stochastic approximation. It involves a random element because $\theta(t)$ is a continuous-time Markov chain with generator Q. Furthermore, we can show that $\{(\widehat{\pi}_n - \mathbf{E}\pi(\theta_n))/\sqrt{\mu}\}$ is tight for $n \geq n_0$ for some positive integer n_0. In an effort to determine the rate of variation of the tracking error sequence, we define a scaled sequence of the tracking errors $\{v_n\}$ and its continuous-time interpolation $v^\mu(\cdot)$ by

$$v_n = \frac{\widehat{\pi}_n - \mathbf{E}\{\pi(\theta_n)\}}{\sqrt{\mu}}, \quad n \geq n_0, \quad v^\mu(t) = v_n \text{ for } t \in [n\mu, n\mu + \mu). \quad (1.18)$$

We have shown in [168] that $(v^\mu(\cdot), \theta^\mu(\cdot))$ converges weakly to $(v(\cdot), \theta(\cdot))$ satisfying the switching diffusion equation

$$dv(t) = -v(t)dt + \Sigma^{1/2}(\theta(t))dw(t), \quad (1.19)$$

where $w(\cdot)$ is a standard Brownian motion and $\Sigma(\theta)$ is the almost sure limit given by

$$\lim_{\mu \to 0} \frac{1}{n} \sum_{k_1=m}^{n+m-1} \sum_{k=m}^{n+m-1} (X_{k+1}(\theta) - \mathbf{E}X_{k+1}(\theta))(X_{k_1+1}(\theta) - \mathbf{E}X_{k_1+1}(\theta))'$$
$$= \Sigma(\theta) \text{ a.s.}$$
$$(1.20)$$

Note that for each θ, $\Sigma(\theta)$ is an $S \times S$ deterministic matrix and in fact,

$$\frac{1}{n} \sum_{k_1=m}^{n+m-1} \sum_{k=m}^{n+m-1} \mathbf{E}\{(X_{k+1}(\theta) - \mathbf{E}X_{k+1}(\theta))(X_{k_1+1}(\theta) - \mathbf{E}X_{k_1+1}(\theta))'\}$$
$$= \Sigma(\theta) \text{ as } \mu \to 0.$$
$$(1.21)$$

Because of the regime switching, the system is qualitatively different from the existing literature on stochastic approximation methods; see Kushner and Yin [104].

Example 1.5. Suppose that $w(\cdot)$ is a d-dimensional Brownian motion and $\alpha(\cdot)$ is a continuous-time Markov chain generated by Q taking values in $\mathcal{M} = \{1, \ldots, m_0\}$. Define $\mathcal{F}_t = \sigma\{w(s), \alpha(s) : 0 \leq s \leq t\}$, and denote by $L^2_{\mathcal{F}}(0, T; \mathbb{R}^{m_0})$ the set of all $\mathbb{R}^{m_0}$-valued, measurable stochastic processes $f(t)$ adapted to $\{\mathcal{F}_t\}_{t \geq 0}$ satisfying $\mathbf{E}\int_0^T |f(t)|^2 dt < \infty$.

Consider a financial market in which $d + 1$ assets are traded continuously. One of the assets is a bank account whose price $P_0(t)$ is subject to the following stochastic ordinary differential equation with Markovian switching,

$$\begin{cases} dP_0(t) = r(t, \alpha(t))P_0(t)dt, \quad t \in [0, T], \\ P_0(0) = p_0 > 0, \end{cases} \quad (1.22)$$

where $r(t, i) \geq 0$, $i = 1, 2, \ldots, m_0$, are given as the interest rate processes corresponding to different market modes. The other d assets are stocks whose price processes $P_m(t)$, $m = 1, 2, \ldots, d$, satisfy the following system of stochastic differential equations with Markov switching:

$$
\begin{cases}
dP_m(t) = P_m(t) \left\{ b_m(t, \alpha(t))dt + \sum_{n=1}^{d} \sigma_{mn}(t, \alpha(t))dw_n(t) \right\}, & t \in [0, T], \\
P_m(0) = p_m > 0,
\end{cases}
$$

(1.23)

where for each $i = 1, 2, \ldots, m_0$, $b_m(t, i)$ is the appreciation rate process and $\sigma_m(t, i) := (\sigma_{m1}(t, i), \ldots, \sigma_{md}(t, i))$ is the volatility or the dispersion rate process of the mth stock, corresponding to $\alpha(t) = i$.

Define the volatility matrix as

$$
\sigma(t, i) := \begin{pmatrix} \sigma_1(t, i) \\ \vdots \\ \sigma_d(t, i) \end{pmatrix}, \quad \text{for each } i = 1, \ldots, m_0.
$$

(1.24)

We assume that the following non-degeneracy condition

$$
\sigma(t, i)\sigma'(t, i) \geq \delta I, \quad \forall t \in [0, T], \quad \text{and} \quad i = 1, 2, \ldots, m_0,
$$

(1.25)

is satisfied for some $\delta > 0$, and that all the functions $r(t, i)$, $b_m(t, i)$, $\sigma_{mn}(t, i)$ are measurable and uniformly bounded in t. Inequality (1.25) is in the sense of positive definiteness for symmetric matrices; that is, $\sigma(t, i)\sigma'(t, i) - \delta I$ is a positive definite matrix.

Suppose that the initial market mode $\alpha(0) = i_0$. Consider an agent with an initial wealth $x_0 > 0$. Denote by $x(t)$ the total wealth of the agent at time $t \geq 0$. Assuming that the trading of shares takes place continuously and that there are no transaction cost and consumptions, then one has (see, e.g., [181, p.57])

$$
\begin{cases}
dx(t) = \left\{ r(t, \alpha(t))x(t) + \sum_{m=1}^{d} [b_m(t, \alpha(t)) - r(t, \alpha(t))]u_m(t) \right\}dt \\
\qquad + \sum_{n=1}^{d} \sum_{m=1}^{d} \sigma_{mn}(t, \alpha(t))u_m(t)dw_n(t), \\
x(0) = x_0 > 0, \quad \alpha(0) = i_0,
\end{cases}
$$

(1.26)

where $u_m(t)$ is the total market value of the agent's wealth in the mth asset, $m = 0, 1, \ldots, d$, at time t. We call $u(\cdot) = (u_1(\cdot), \ldots, u_d(\cdot))'$ a portfolio of the agent. Note that once $u(\cdot)$ is determined, $u_0(\cdot)$, the asset in the bank account is completed specified inasmuch as

$$u_0(t) = x(t) - \sum_{i=1}^{d} u_i(t).$$

Thus, we need only consider $u(\cdot)$ and ignore $u_0(\cdot)$.

Setting

$$B(t,i) := (b_1(t,i) - r(t,i), \ldots, b_d(t,i) - r(t,i)), \quad i = 1, 2, \ldots, m_0, \quad (1.27)$$

we can rewrite the wealth equation (1.26) as

$$\begin{cases} dx(t) = [r(t,\alpha(t))x(t) + B(t,\alpha(t))u(t)]dt + u'(t)\sigma(t,\alpha(t))dw(t), \\ x(0) = x_0, \quad \alpha(0) = i_0. \end{cases}$$

$$(1.28)$$

A portfolio $u(\cdot)$ is said to be admissible if $u(\cdot) \in L_{\mathcal{F}}^2(0,T;\mathbb{R}^d)$ and the stochastic differential equation (1.28) has a unique solution $x(\cdot)$ corresponding to $u(\cdot)$. In this case, we refer to $(x(\cdot), u(\cdot))$ as an admissible (wealth, portfolio) pair. The agent's objective is to find an admissible portfolio $u(\cdot)$, among all the admissible portfolios whose expected terminal wealth is $\mathbf{E}x(T) = z$ for some given $z \in \mathbb{R}^1$, so that the risk measured by the variance of the terminal wealth

$$\mathbf{Var}\ x(T) \equiv \mathbf{E}[x(T) - \mathbf{E}x(T)]^2 = \mathbf{E}[x(T) - z]^2 \qquad (1.29)$$

is minimized. Finding such a portfolio $u(\cdot)$ is referred to as the mean-variance portfolio selection problem. An interested reader can find further details of the study on this problem in [186].

Example 1.6. This problem arises in the context of insurance and risk theory. Suppose that there is a finite set $\mathcal{M} = \{1, \ldots, m_0\}$, representing the possible regimes (configurations) of the environment. At each $i \in \mathcal{M}$, assume that the premium is payable at rate $c(i)$ continuously. Let $U(t,i)$ be the surplus process given the initial surplus $u > 0$ and initial state i:

$$U(t,i) = u + \int_0^t c(\alpha(s))ds - S(t),$$

where $S(t)$, as in the classical risk model, is a compound Poisson process and $\alpha(t)$ is a continuous-time Markov chain with state space $\mathcal{M}$ representing the random environment. Under suitable conditions, we obtained Lunderberg-type upper bounds and nonexponential upper bounds for the ruin probability, and treated the renewal-type system of equations for ruin probability when the claim sizes were exponentially distributed. To proceed further, we consider a class of jump-diffusions with regime switching to prepare us for treating applications involving more general risk models.

One of the main features of [163] is that there is an additional Markov chain, which enables the underlying surplus to vary in accordance with different regimes. Consider jump-diffusions modulated by a continuous-time Markov chain. Because the dynamic systems are complex, it is of foremost importance to reduce the complexity. Taking into consideration the inherent hierarchy in a complex system [149], and different rates of variations of subsystems and components, we use the two-time-scale method leading to systems in which the fast and slow rates of change are in sharp contrast. Then we proceed to reduce the system complexity by aggregation/decomposition and averaging methods. We demonstrate that under broad conditions, associated with the original systems, there are limit or reduced systems, which are averages with respect to certain invariant measures. Using weak convergence methods [102, 103], we obtain the limit system via martingale problem formulation.

Motivated by risk theory applications [35], we consider a switching jump diffusion model. We formulate the problem in a general way, which allows the consideration of other problems where switching jump diffusions are involved.

Let $\Gamma \subset \mathbb{R}^r - \{0\}$ so that Γ is the range space of the impulsive jumps, $w(\cdot)$ be a real-valued standard Brownian motion, and $N(\cdot, \cdot)$ be a Poisson measure such that $N(t, H)$ counts the number of impulses on $[0, t]$ with values in the set H. Let $f(\cdot, \cdot, \cdot) : [0, T] \times \mathbb{R} \times \mathcal{M} \mapsto \mathbb{R}$, $\sigma(\cdot, \cdot, \cdot) : [0, T] \times \mathbb{R} \times \mathcal{M} \mapsto \mathbb{R}$, $g(\cdot, \cdot, \cdot) : \Gamma \times \mathbb{R} \times \mathcal{M} \mapsto \mathbb{R}$, and $\alpha(\cdot)$ be a continuous-time Markov chain having state space $\mathcal{M}$ and generator $Q(t)$. A brief description of the jump-diffusion process with a modulating Markov chain can be found in the appendix. Consider the following jump-diffusion processes with regime switching

$$
\begin{aligned}
X(t) = x_0 &+ \int_0^t f(s, X(s), \alpha(s))ds + \int_0^t \sigma(s, X(s), \alpha(s))dw(s) \\
&+ \int_0^t \int_\Gamma g(\gamma, X(s^-), \alpha(s^-))N(ds, d\gamma).
\end{aligned}
\tag{1.30}
$$

We assume that $w(\cdot)$, $N(\cdot)$, and $\alpha(\cdot)$ are mutually independent. Note that compared with the traditional jump-diffusion processes, the coefficients involved in (1.30) all depend on an additional switching process–the Markov chain $\alpha(t)$.

In the context of risk theory, $X(t)$ can be considered as the surplus of the insurance company at time t, x_0 is the initial surplus, $f(t, X(t), \alpha(t))$ represents the premium rate (assumed to be ≥ 0), $g(\gamma, X(t), \alpha(t))$ is the amount of the claim if there is one (assumed to be ≤ 0), and the diffusion is used to model additional uncertainty of the claims and/or premium incomes. Similar to the volatility in stock market models, $\sigma(\cdot, \cdot, i)$ represents the amount of oscillations or volatility in an appropriate sense. The model is sufficiently general to cover the traditional as well as the diffusion-perturbed ruin models. It may also be used to represent security price in finance (see [124, Chapter 3]). The process $\alpha(t)$ may be viewed as an environment variable dictating the regime. The use of the Markov chain results from consideration of a general trend of the market environment as well as other economic factors. The economic and/or political environment changes lead to the changes of surplus regimes resulting in markedly different behavior of the system across regimes.

Defining a centered Poisson measure and applying generalized Ito's rule, we can obtain the generator of the jump-diffusion process with regime switching, and formulate a related martingale problem. Instead of a single process, we have to deal with a collection of jump-diffusion processes that are modulated by a continuous-time Markov chain. Suppose that λ is positive such that $\lambda \Delta + o(\Delta)$ represents the probability of a switch of regime in the interval $[t, t + \Delta)$, and $\pi(\cdot)$ is the distribution of the jump. Then the generator of the underlying process can be written as

$$
\begin{aligned}
\mathcal{G}F(t, x, \iota) = {} & \left(\frac{\partial}{\partial t} + \mathcal{L} \right) F(t, x, \iota) \\
& + \int_{\Gamma} \lambda [F(t, x + g(\gamma, x, \iota), \iota) - F(t, x, \iota)] \pi(d\gamma) \qquad (1.31) \\
& + Q(t)F(t, x, \cdot)(\iota), \quad \text{for each } \iota \in \mathcal{M},
\end{aligned}
$$

where

$$
\begin{aligned}
\mathcal{L}F(t, x, \iota) &= \frac{1}{2}\sigma^2(t, x, \iota)\frac{\partial^2}{\partial x^2}F(t, x, \iota) + f(t, x, \iota)\frac{\partial}{\partial x}F(t, x, \iota), \\
Q(t)F(t, x, \cdot)(\iota) &= \sum_{\ell=1}^{m_0} q_{\iota\ell}(t)F(t, x, \ell) = \sum_{\ell \neq \iota} q_{\iota\ell}(t)[F(t, x, \ell) - F(t, x, \iota)].
\end{aligned}
$$

$$(1.32)$$

Of particular interest is the case that the switching process and the diffusion vary at different rates. To reduce the amount of computational effort, we propose a two-time-scale approach. By concentrating on time-scale separations, we treat two cases in Chapter 12. In the first one, the regime switching is significantly faster than the dynamics of the jump diffusions, whereas in the second case, the diffusion vary an order of magnitude faster than the other processes. As shown, averaging plays an essential role in these problems.

Example 1.7. Consider the process $Y(\cdot) = (X(\cdot), \alpha(\cdot))$ that has two components, the diffusion component $X(\cdot)$ and the pure jump component $\alpha(\cdot)$. The state space of the process is $\mathbb{S} \times \mathcal{M}$, where $\mathbb{S}$ is the unit circle and $\mathcal{M} = \{1, \ldots, m_0\}$ is a state space with finitely many elements for the jump process. By identifying the endpoints 0 and 1, let $x \in [0, 1]$ be the coordinates in $\mathbb{S}$. We assume that the generator of the process $(X(t), \alpha(t))$ is of the form (1.31) with the jump part missing (i.e., $\lambda = 0$) and with $(x, i) \in [0, 1] \times \mathcal{M}$.

The probability density $p(x,t) = (p(x,t,1),\ldots,p(x,t,m_0))$ of the process $Y(\cdot)$, with

$$\int_\Gamma p(x,t,i)dx = \mathbf{P}(X(t) \in \Gamma, \alpha(t) = i),$$

satisfies the adjoint equation, namely the system of forward equations

$$\frac{\partial p(x,t,i)}{\partial t} = \mathcal{D}^*p(x,t,i) + \sum_{j=1}^{m_0} p(x,t,j)q_{ji}(x),$$

where for a suitable function $f(x,t,i)$,

$$\begin{aligned}
\mathcal{D}^*f(x,t,i) &= \mathcal{D}^*(x,t,i)f(x,t,i)\\
&= \frac{1}{2}\frac{\partial^2}{\partial x^2}(a_i(x,t)f(x,t,i)) - \frac{\partial}{\partial x}(b_i(x,t)f(x,t,i)),
\end{aligned}$$

$$p(x,0,i) = g_i(x),$$

for $i = 1,\ldots,m_0$, and $g(x) = (g_1(x),\ldots,g_{m_0}(x))$ is the initial distribution for $Y(t)$. The existence and properties of such switching diffusion processes can be found in [55, Section 2.2]. Suppose that all conditions of [46, Theorem 16, p. 82] are satisfied. Then the system of forward equations has a unique solution.

Suppose that $q_{ij}(x) > 0$ for each $i \neq j$. Then the transition density $p(x,t)$ converges exponentially fast to

$$\nu(x) = (\nu_1(x),\ldots,\nu_{m_0}(x)),$$

the density of the stationary distribution; that is,

$$|p(x,t,i) - \nu_i(x)| \leq K \exp(-\gamma t),$$

for some $K > 0$ and $\gamma > 0$. The estimate is the well-known spectrum gap condition. This spectrum gap condition helps us to study related asymptotic properties of the system.

Example 1.8. This example is concerned with a switching diffusion process with a state-dependent switching component. We compare the trajectories of a linear stochastic system with drift and diffusion coefficients given by

$$f(x) = 0.11x \quad \text{and} \quad \sigma(x) = 0.2x \tag{1.33}$$

and another regime-switching linear system with $\alpha(t) \in \{1,2\}$, and the drift and diffusion coefficients given by

$$\begin{aligned}
f(x,1) &= -0.2x, \quad f(x,2) = 0.11x,\\
\sigma(x,1) &= x, \quad \sigma(x,2) = 0.2x,
\end{aligned} \tag{1.34}$$

respectively. For the switching diffusion, the switching process is continuous state-dependent with the $Q(x)$ given by

$$Q(x) = \begin{pmatrix} -5\cos^2 x & 5\cos^2 x \\ 10\cos^2 x & -10\cos^2 x \end{pmatrix}.$$

What happens if the actual model is the one with switching, but we mis-modeled the system so that instead of the switching model, we used a simple linear stochastic differential equation model with drift and diffusion coefficients given by (1.33)? If in fact, the system parameters are really given by (1.34), and if the random environment is included in the model, then we will see a significant departure of the model from that of the true system. This can be seen from the plots in Figure 1.2.

To get further insight from these plots, imagine that we are encountering the following scenarios. Suppose that the above example is for modeling the stock price of a particular equity. Suppose that we modeled the stock price by using the simple geometric Brownian motion model with return rate 11% and the volatility 0.2 as given in (1.33). However, the market is really subject to random environment perturbation. For the bull market, the rates are as before, but corresponding to the bear market, the return rate and the volatility are given by -20% and 1, respectively. The deviation of the plot in Figure 1.2 shows that if one uses a simple linear SDE model, one cannot capture the market behavior.

Example 1.9. This example is concerned with a controlled switching diffusion model. We first give a motivation of the study and begin the discussion on Markov decision processes. Consider a real-valued process $X(\cdot) = \{X(t) : t \geq 0\}$ and a feedback control $u(\cdot) = \{u(t) = u(X(t)) : t \geq 0\}$ such that $u(t) \in \Gamma$ for $t \geq 0$ with Γ being a compact subset of an Euclidean space. Γ denotes the control space. A vast literature is concerned with a certain optimal control problem. That is, one aims to find a feedback $u(\cdot)$ so that an appropriate objective function is minimized.

Extending the idea to switching diffusion processes, we consider the following pair of processes $(X(t), \alpha(t)) \in \mathbb{R}^r \times \mathcal{M}$, where $\mathcal{M} = \{1, \ldots, m_0\}$. Let U be the control space or action space, which is a compact set of an Euclidean space. Suppose that $b(\cdot, \cdot, \cdot) : \mathbb{R}^r \times \mathcal{M} \times U \mapsto \mathbb{R}^r$, and $\sigma(\cdot, \cdot, \cdot) : \mathbb{R}^r \times \mathcal{M} \times U \mapsto \mathbb{R}^{r \times r}$ are appropriate functions satisfying certain regularity conditions, and $Q(x) = (q_{ij}(x)) \in \mathbb{R}^{m_0 \times m_0}$ satisfies that for each x, $q_{ij}(x) \geq 0$ for $i \neq j$, $\sum_{j=1}^{m_0} q_{ij}(x) = 0$ for each $i \in \mathcal{M}$. For each $i \in \mathcal{M}$ and suitable smooth function $h(\cdot, i)$, define an operator

$$\mathcal{L}h(x, i) = \nabla h'(x, i)b(x, i, u) + \frac{1}{2}\mathrm{tr}[\nabla^2 h(x, i)\sigma(x, i, u)\sigma'(x, i, u)]$$
$$+ \sum_{j=1}^{m_0} q_{ij}(x)h(x, j). \tag{1.35}$$

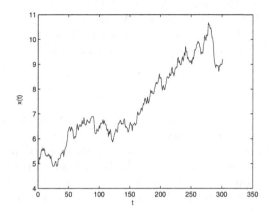

(a) A sample path of a mismodeled system: A linear stochastic differential equation with constant coefficients.

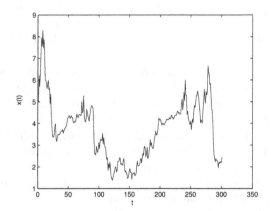

(b) A sample path of the true system: A regime-switching stochastic system of differential equations.

FIGURE 1.2. Comparisons of sample paths for a linear diffusion with that of a regime-switching diffusion.

Then we call $(X(t), \alpha(t))$ a controlled switching diffusion process. Following the notation of Markov decision processes, we may write the process $(X(t), \alpha(t))$ as $(X(t), \alpha(t)) \sim \mathcal{L}(u)$. Note that owing to the feedback controls used, it is natural to require the matrix-valued function $Q(\cdot)$ to be x-dependent. This model is a generalization of the usual Markov decision processes. When the functions $b(x, i, u) \equiv 0$ and $\sigma(x, i, i) \equiv 0$, the model reduces to a Markov decision process. When the matrix-valued function $Q(x) \equiv 0$, the model reduces to controlled diffusions. Typically, our objective is to select the control $u(\cdot)$ so that a control objective function is achieved.

Example 1.10. Originating from statistical mechanics, mean-field models are concerned with many-body systems with interactions. To overcome the difficulty of interactions due to the many bodies, one of the main ideas is to replace all interactions to any one body with an average or effective interaction. This reduces any multibody problem to an effective one-body problem. Although its main motivation and development are in statistical mechanics, such models have also enjoyed recent applications in, for example, graphical models in artificial intelligence.

If the field or particle exhibits many interactions in the original system, the mean field will be more accurate for such a system. The usefulness, the potential impact on many practical scenarios, and the challenges from both physics and mathematics have attracted much attention in recent years. The work of Dawson [31] presents a detailed study on the cooperative behavior of mean fields.

Owing to the rapid progress in technology, more complicated systems are encountered in applications. In response to such challenges, much effort has been devoted to modeling and analysis for more sophisticated systems. Frequently, there are factors that cannot be described by the traditional models. One of the ideas is to bring regime switching into the formulation, so as to deal with the coexistence of continuous dynamics and discrete events. Although one may use a pure jump process to represent the discrete events, due to the interactions of the many bodies, the discrete process in fact, is correlated with the diffusive dynamics.

In Xi and Yin [162], the following system is considered. Suppose that $\alpha(t)$ is a right-continuous random process taking values in a finite state space $\mathcal{M} = \{1, 2, \ldots, m_0\}$. Consider an ℓ-body mean-field model with switching described by the following system of Itô stochastic differential equations. For $i = 1, 2, \ldots, \ell$,

$$
\begin{aligned}
dX_i(t) = & \left[\gamma(\alpha(t))X_i(t) - X_i^3(t) - \beta(\alpha(t))(X_i(t) - \overline{X}(t)) \right] dt \\
& + \sigma_{ii}(X(t), \alpha(t)) dw_i(t),
\end{aligned} \tag{1.36}
$$

where

$$\overline{X}(t) = \frac{1}{\ell} \sum_{j=1}^{\ell} X_j(t),$$

$$X(t) = (X_1(t), X_2(t), \ldots, X_\ell(t))',$$

(1.37)

and $\gamma(i) > 0$ and $\beta(i) > 0$ for $i \in \mathcal{M}$. Moreover, the transition rules of $\alpha(t)$ are specified by

$$\mathbf{P}\{\alpha(t + \Delta) = j | \alpha(t) = j, X(t) = x\}$$
$$= \begin{cases} q_{ij}(x)\Delta + o(\Delta), & \text{if } i \neq j, \\ 1 + q_{ii}(x)\Delta + o(\Delta), & \text{if } i = j, \end{cases}$$

(1.38)

which hold uniformly in $\mathbb{R}^\ell$ as $\Delta \downarrow 0$, for $x = (x_1, x_2, \ldots, x_\ell)' \in \mathbb{R}^\ell$ and $\sigma = (\sigma_{ij}) \in \mathbb{R}^\ell \times \mathbb{R}^\ell$. A number of asymptotic properties were obtained in [162] including regularity, Feller properties, and exponential ergodicity. We refer the interested reader to the aforementioned reference for further details.

As shown in the examples above, all the systems considered involve hybrid switching diffusions. To have a better understanding of each of the problems and the properties of the corresponding processes, it is important that we have a thorough understanding of the switching diffusion process. This is our objective in this book.

1.5   Outline of the Book

This chapter serves as a prelude to the book. It gives motivation and presents examples for the switching-diffusion processes. After this short introduction, the book is divided into four parts.

The first part, including three chapters, presents basic properties such as Feller and strong Feller, recurrence, and ergodicity. Chapter 2 begins with the precise definition of the switching diffusion processes. With a brief review of the existence and uniqueness of solutions of switching diffusions, we deal with basic properties such as regularity, weak continuity, Feller and strong Feller properties and so on. Then continuous and smooth dependence of initial data are presented. The proofs of these results are different from the traditional setup and much more involved due to the coupling resulting from the switching component.

Chapter 3 is concerned with recurrence and positive recurrence. Necessary and sufficient conditions for positive recurrence are given. We show that recurrence and positive recurrence are independent of the open set chosen. Furthermore, we demonstrate that we can work with a fixed discrete component. The approach we are using is based on treatment of elliptic systems of partial differential equations. First, criteria of positive recurrence based on Liapunov functions are given. Then we translate these into conditions on the coefficients of the switching diffusions.

Chapter 4 focuses on ergodicity. It is shown that for a positive recurrent switching diffusion, there exists a unique invariant measure. Not only are the existence and uniqueness proved, but the form of the invariance measure is given. It reveals the salient feature of the underlying process owing to the switching. The ergodic measures obtained enable one to carry out control and optimization tasks with replacement of the instantaneous measures by the ergodic measures.

Part II of the book is devoted to numerical solutions of switching diffusions. As their diffusion counterpart, very often closed-form solutions of nonlinear switching diffusions are hard to come by. Thus, numerical approximation becomes a viable and often the only alternative. Developing numerical results is important. Therefore, Chapters 5 and 6 are devoted to numerical approximations. Chapter 5 presents numerical algorithms of the Euler-Maruyama type. Although decreasing stepsize algorithms are often considered in the literature, we illustrate that constant stepsize algorithms work just as well. Such a fact has been well recognized by researchers working in the fields of systems theory, control, and signal processing among others. Using weak convergence methods, we establish the convergence of the numerical approximations.

Chapter 6 switches gear and concentrates on numerical approximation of the erogdic measures. Sometimes, people refer to convergence to the stationary distribution as stability in distribution. In this language, the subject matter of this chapter is: Under what conditions, will the numerical algorithm be stable in distribution. Again, weak convergence methods are the key for us to reach the conclusion.

Containing three chapters, Part III focuses on stability. Chapters 7 and 8 proceed with the stability analysis. The approach is based on Liapunov function methods. Chapter 7 studies stability of switching diffusions. The notion is in the sense of stability in probability. First, definitions of stability and instability are given. Then criteria are presented based on Liapunov functions. In addition, more verifiable conditions on the coefficients of the systems are presented. Necessary and sufficient conditions for pth-moment stability of linear (in the continuous component) systems and linear approximations are provided. It is noted that contrary to common practice, very often nonquadratic Liapunov functions are easier to use.

Chapter 8 takes up the stability study for fully degenerated systems. We treat the case that the diffusion matrix becomes 0, or equivalently, we have switching ordinary differential equations. Based on Liapunov methods, we obtain sufficient conditions for stability. Similar to the nondegenerate switching diffusions, the results on stability and instability have an "eigenvalue" gap. That is, the stability criteria are in terms of the largest eigenvalue of a certain matrix and the instability is in terms of the smallest eigenvalue of the same matrix. To close up the gap, we introduce a logarithm transformation leading to a necessary and sufficient condition in terms of a one-dimensional function.

Chapter 9 proceeds with the investigation of the invariance principle. In studying deterministic systems represented by ordinary differential equations, the concept of the invariance principle in the sense of LaSalle has come into being (see [106] and [107]). It enables one to obtain far-reaching results compared to the purely equilibrium point analysis alone. Building on this idea, the invariance principle for stochastic counterpart for diffusion processes was considered by Kushner in [98], whereas that for stochastic differential delay equations was treated in [117]. In this chapter, focusing on switching diffusions, we obtain their invariance properties using both a sample path approach and a measure theoretic approach.

The last part of the book concentrates on two-time-scale modeling and applications; it contains three chapters. Up to this point, it has been assumed the switching component is irreducible. Roughly, that means all of the discrete states belong to the same class. One natural question is: What happens if not all states belong to the same recurrent class. Chapter 10 is concerned with such an issue. Using the methods of perturbed Liapunov functions, this problem is answered. We divide the state space of the switching component into several irreducible groups. Within each irreducible group the switching component moves frequently, whereas from one group to another, it moves relatively infrequently. We distinguish the fast and slow motions by introducing a small parameter $\varepsilon > 0$. Aggregating the states in each recurrent group into one superstate, as $\varepsilon \to 0$, we obtain a limit system, whose drift and diffusion coefficients are averaged out with respect to the invariant measure of the switching part. We show that if the limit process is recurrent so is the original process for sufficiently small $\varepsilon > 0$.

Chapter 11 is concerned with an application to financial engineering. The equity asset is assumed to follow a Markov regime-switching diffusion, in which both the return rates and the volatility depend on a continuous-time Markov chain. This model is an alternative to the so-called stochastic volatility model of Hull and White [70] which is well known nowadays. The motivation of our study is similar to the fast mean reverting models treated in [49], in which Fouque, Papanicolaou, and Sircar assumed the volatility to be a fast-varying diffusion process. It was shown that the Black–Scholes pricing formula is a first approximation to the stochastic volatility model when fast mean reversion is observed. We treat an alternative model with regime switching, and reach a similar conclusion.

Chapter 12 considers a slightly more general model with an additional jump component. The key point here is the utilization of two-time scales. The motivation stems from reduction of complexity. Two different situations are considered. In the first case, the switching component changes an order of magnitude faster than the continuous component. In the second case, the diffusion is fast-varying. In both cases, we obtain appropriate limits using a weak convergence approach.

Finally, for convenience, an appendix including a number of mathematical preliminaries is placed at the end of the book. It serves as a quick reference. Topics discussed here including Markov chains, martingales, Gaussian processes, diffusions, jump diffusions, and weak convergence methods. Although detailed developments are often omitted, appropriate references are provided for the reader to facilitate further reading.

Part I

Basic Properties, Recurrence, Ergodicity

2

Switching Diffusion

2.1 Introduction

This chapter provides an introduction to switching diffusions. First the
definition of switching diffusion is given. Then with a short review of the
existence and uniqueness of the solution of associated stochastic differential
equations, weak continuity, Feller, and strong Feller properties are established. Also given here are the definition of regularity and criteria ensuring
such regularity. Moreover, smooth dependence on initial data is presented.

The rest of the chapter is arranged as follows. After this short introductory section, Section 2.2 presents the general setup for switching processes.
Section 2.3 is concerned with regularity. Section 2.4 deals with weak continuity of the pair of process $(X(t), \alpha(t))$. Section 2.5 proceeds with Feller
properties. Section 2.6 goes one step further to obtain strong Feller properties. Section 2.7 presents smooth dependence properties of solutions of
the switching diffusions. Section 2.8 gives remarks on how nonhomogeneous
cases in which both the drift and diffusion coefficients depend explicitly on
time t can be handled. Finally, Section 2.9 provides additional notes and
remarks.

2.2 Switching Diffusions

We work with a probability space $(\Omega, \mathcal{F}, \mathbf{P})$ throughout this book. A family
of σ-algebras $\{\mathcal{F}_t\}$, for $t \geq 0$ or $t = 1, 2, \ldots$, or simply $\mathcal{F}_t$, is termed a

G.G. Yin and C. Zhu, *Hybrid Switching Diffusions: Properties and Applications*,
Stochastic Modelling and Applied Probability 63, DOI 10.1007/978-1-4419-1105-6_2,
© Springer Science + Business Media, LLC 2010

filtration if $\mathcal{F}_s \subset \mathcal{F}_t$ for $s \leq t$. We say that $\mathcal{F}_t$ is complete if it contains all null sets and that the filtration $\{\mathcal{F}_t\}$ satisfies the usual condition if $\mathcal{F}_0$ is complete. A probability space $(\Omega, \mathcal{F}, \mathbf{P})$ together with a filtration $\{\mathcal{F}_t\}$ is said to be a filtered probability space, denoted by $(\Omega, \mathcal{F}, \{\mathcal{F}_t\}, \mathbf{P})$.

Suppose that $\alpha(\cdot)$ is a stochastic process with right-continuous sample paths (or a pure jump process), finite-state space $\mathcal{M} = \{1, \ldots, m_0\}$, and x-dependent generator $Q(x)$ so that for a suitable function $f(\cdot, \cdot)$,

$$Q(x)f(x, \cdot)(\imath) = \sum_{j \in \mathcal{M}} q_{\imath j}(x)(f(x, \jmath) - f(x, \imath)), \quad \text{for each } \imath \in \mathcal{M}. \quad (2.1)$$

Let $w(\cdot)$ be an $\mathbb{R}^d$-valued standard Brownian motion defined in the filtered probability space $(\Omega, \mathcal{F}, \{\mathcal{F}_t\}, \mathbf{P})$. Suppose that $b(\cdot, \cdot) : \mathbb{R}^r \times \mathcal{M} \mapsto \mathbb{R}^r$ and that $\sigma(\cdot, \cdot) : \mathbb{R}^r \times \mathcal{M} \mapsto \mathbb{R}^d$. Then the two-component process $(X(\cdot), \alpha(\cdot))$, satisfying

$$dX(t) = b(X(t), \alpha(t))dt + \sigma(X(t), \alpha(t))dw(t),$$
$$(X(0), \alpha(0)) = (x, \alpha), \quad\quad\quad\quad\quad\quad\quad\quad (2.2)$$

and for $i \neq j$,

$$\mathbf{P}\{\alpha(t + \Delta) = j | \alpha(t) = i, X(s), \alpha(s), s \leq t\} = q_{ij}(X(t))\Delta + o(\Delta), \quad (2.3)$$

is termed a switching diffusion or a regime-switching diffusion. Naturally, for the two-component process $(X(t), \alpha(t))$, we call $X(t)$ the continuous component and $\alpha(t)$ the discrete component, in accordance with their sample path properties.

There is an associated operator defined as follows. For each $\imath \in \mathcal{M}$ and each $f(\cdot, \imath) \in C^2$, where C^2 denotes the class of functions whose partial derivatives with respect to the variable x up to the second-order are continuous, we have

$$\begin{aligned}\mathcal{L}f(x, \imath) &= \nabla f'(x, \imath)b(x, \imath) + \text{tr}(\nabla^2 f(x, \imath)A(x, \imath)) + Q(x)f(x, \cdot)(\imath) \\ &= \sum_{i=1}^{r} b_i(x, \imath)\frac{\partial f(x, \imath)}{\partial x_i} + \frac{1}{2}\sum_{i,j=1}^{r} a_{ij}(x, \imath)\frac{\partial^2 f(x, \imath)}{\partial x_i \partial x_j} \\ &\quad + Q(x)f(x, \cdot)(\imath), \end{aligned} \quad (2.4)$$

where $\nabla f(x, \imath)$ and $\nabla^2 f(x, \imath)$ denote the gradient and Hessian of $f(x, \imath)$ with respect to x, respectively,

$$Q(x)f(x, \cdot)(\imath) = \sum_{j=1}^{m_0} q_{\imath j}f(x, j), \quad \text{and}$$

$$A(x, \imath) = (a_{ij}(x, \imath)) = \sigma(x, \imath)\sigma'(x, \imath) \in \mathbb{R}^{r \times r}.$$

Note that the evolution of the discrete component $\alpha(\cdot)$ can be represented by a stochastic integral with respect to a Poisson random measure (see, e.g., [52, 150]). Indeed, for $x \in \mathbb{R}^r$ and $i, j \in \mathcal{M}$ with $j \neq i$, let $\Delta_{ij}(x)$ be the consecutive (with respect to the lexicographic ordering on $\mathcal{M} \times \mathcal{M}$), left-closed, right-open intervals of the real line, each having length $q_{ij}(x)$. Define a function $h : \mathbb{R}^r \times \mathcal{M} \times \mathbb{R} \mapsto \mathbb{R}$ by

$$h(x, i, z) = \sum_{j=1}^{m_0} (j - i) I_{\{z \in \Delta_{ij}(x)\}}. \tag{2.5}$$

That is, with the partition $\{\Delta_{ij}(x) : i, j \in \mathcal{M}\}$ used and for each $i \in \mathcal{M}$, if $z \in \Delta_{ij}(x)$, $h(x, i, z) = j - i$; otherwise $h(x, i, z) = 0$. Then (2.3) is equivalent to

$$d\alpha(t) = \int_{\mathbb{R}} h(X(t), \alpha(t-), z) \mathfrak{p}(dt, dz), \tag{2.6}$$

where $\mathfrak{p}(dt, dz)$ is a Poisson random measure with intensity $dt \times m(dz)$, and m is the Lebesgue measure on $\mathbb{R}$. The Poisson random measure $\mathfrak{p}(\cdot, \cdot)$ is independent of the Brownian motion $w(\cdot)$.

Similar to the case of diffusions, with the $\mathcal{L}$ defined in (2.4), for each $f(\cdot, \imath) \in C^2$, $\imath \in \mathcal{M}$, a result known as the generalized Itô lemma (see [17, 120] or [150]) reads

$$f(X(t), \alpha(t)) - f(X(0), \alpha(0)) = \int_0^t \mathcal{L}f(X(s), \alpha(s))ds + M_1(t) + M_2(t), \tag{2.7}$$

where

$$M_1(t) = \int_0^t \langle \nabla f(X(s), \alpha(s)), \sigma(X(s), \alpha(s))dw(s) \rangle,$$
$$M_2(t) = \int_0^t \int_{\mathbb{R}} [f(X(s), \alpha(0) + h(X(s), \alpha(s), z)) - f(X(s), \alpha(s))]\mu(ds, dz),$$

and

$$\mu(ds, dz) = \mathfrak{p}(ds, dz) - ds \times m(dz)$$

is a martingale measure.

In view of the generalized Itô formula,

$$M_f(t) = f(X(t), \alpha(t)) - f(X(0), \alpha(0)) - \int_0^t \mathcal{L}f(X(s), \alpha(s))ds \tag{2.8}$$

is a local martingale. If for each $\imath \in \mathcal{M}$, $f(x, \imath) \in C_b^2$ (class of functions possessing bounded and continuous partial derivatives with respect to x of order up to two) or $f(x, \imath) \in C_0^2$ (C^2 functions with compact support), then $M_f(t)$ defined in (2.8) becomes a martingale. Similar to the case of diffusion processes, we can define the corresponding notion of the solution of the martingale problem accordingly.

Another consequence of the generalized Itô formula (2.7) is that the Dynkin formula follows. Indeed, let $f(\cdot, i) \in C^2$ for $i \in \mathcal{M}$, and τ_1, τ_2 be bounded stopping times such that $0 \leq \tau_1 \leq \tau_2$ a.s. If $f(X(t), \alpha(t))$ and $\mathcal{L}f(X(t), \alpha(t))$ and so on are bounded on $t \in [\tau_1, \tau_2]$ with probability 1, then

$$\mathbf{E}f(X(\tau_2), \alpha(\tau_2)) = \mathbf{E}f(X(\tau_1), \alpha(\tau_1)) + \mathbf{E}\int_{\tau_1}^{\tau_2} \mathcal{L}f(X(s), \alpha(s))ds; \quad (2.9)$$

see [120] or [150] for details. In what follows, for convenience (with multi-index notation used, for instance) and emphasis on the x-dependence, we often use $D_x f(x, \alpha)$ and $\nabla f(x, \alpha)$ interchangeably to represent the gradient. In addition, for two vectors x and y with appropriate dimensions, we use $x'y$ and $\langle x, y \rangle$ interchangeably to represent their inner product. To proceed, we present the existence and uniqueness of solutions to a system of stochastic differential equations associated with switching diffusions first.

In what follows, for $Q(x) : \mathbb{R}^r \to \mathbb{R}^{m_0 \times m_0}$, we say it satisfies the q-property, if $Q(x) = (q_{ij}(x))$, for all $x \in \mathbb{R}^r$, $q_{ij}(x)$ is Borel measurable for all $i, j \in \mathcal{M}$ and $x \in \mathbb{R}^r$; $q_{ij}(x)$ is uniformly bounded; $q_{ij}(x) \geq 0$ for $j \neq i$ and $q_{ii}(x) = -\sum_{j \neq i} q_{ij}(x)$ for all $x \in \mathbb{R}^r$.

Note that an alternative definition of the q-property can be devised. The boundedness assumption can be relaxed. However, for our purpose, the current setup seems to be sufficient. The interested reader could find the desired information through [28].

Theorem 2.1. *Let $x \in \mathbb{R}^r$, $\mathcal{M} = \{1, \dots, m_0\}$, and $Q(x) = (q_{ij}(x))$ be an $m_0 \times m_0$ matrix depending on x satisfying the q-property. Consider the two-component process $Y(t) = (X(t), \alpha(t))$ given by (2.2) with initial data (x, α). Suppose that $Q(\cdot) : \mathbb{R}^r \mapsto \mathbb{R}^{m_0 \times m_0}$ is a bounded and continuous function, that the functions $b(\cdot, \cdot)$ and $\sigma(\cdot, \cdot)$ satisfy*

$$|b(x, \alpha)| + |\sigma(x, \alpha)| \leq K(1 + |x|), \quad \alpha \in \mathcal{M}, \quad (2.10)$$

and that for every integer $N \geq 1$, there exists a positive constant M_N such that for all $t \in [0, T], i \in \mathcal{M}$ and all $x, y \in \mathbb{R}^r$ with $|x| \vee |y| \leq M_N$,

$$|b(x, i) - b(y, i)| \vee |\sigma(x, i) - \sigma(y, i)| \leq M_N|x - y|. \quad (2.11)$$

Then there exists a unique solution $(X(t), \alpha(t))$ to the equation (2.2) with given initial data in which the evolution of the jump process is specified by (2.3).

Remark 2.2. For brevity, the detailed proof is omitted. Instead, we make the following remarks. There are a number of possible proofs. For example, the existence can be obtained as in [150, pp. 103-104]. Viewing the switching diffusion as a special case of a jump-diffusion process (see the stochastic integral representation of $\alpha(t)$ in (2.6)), one may prove the existence and uniqueness using [77, Section III.2]. Another possibility is to use a martingale problem formulation together with utilization of truncations and stopping times as in [72, Chapter IV]. In Chapter 5, we present numerical approximation algorithms for solutions of switching diffusions, and show the approximation algorithms converge weakly to the switching diffusion of interest by means of a martingale problem formulation. Then using Lipschitz continuity and the weak convergence, we further obtain the strong convergence of the approximations. As a byproduct, we can obtain the existence and uniqueness of the solution. The verbatim proof can be found in [172]; see Remark 5.10 for further explanations. While most proofs of the uniqueness take two different solutions with the same initial data and show their difference should be 0 by using Lipschitz continuity and Gronwall's inequality, it is possible to consider the difference of the two solutions with different initial data whose difference is arbitrarily small. In this regard, the uniqueness can be derived from Proposition 2.30. Earlier work using such an approach may be found in [130].

Note that (2.10) is the linear growth condition and (2.11) is the local Lipschitz condition. These conditions for the usual diffusion processes (without switching) are used extensively in the literature. Theorem 2.1 provides existence and uniqueness of solutions of (2.2) with (2.3) specified. To proceed, we obtain a moment estimate on $X(t)$. This estimate is used frequently in subsequent development. As its diffusion counter part, the main ingredient of the proof is the use of Gronwall's inequality.

Proposition 2.3. *Assume the conditions of Theorem 2.1. Let $T > 0$ be fixed. Then for any positive constant γ, we have*

$$\mathbb{E}_{x,i}\left[\sup_{t\in[0,T]} |X(t)|^{\gamma}\right] \le C < \infty, \quad (x,i) \in \mathbb{R}^r \times \mathcal{M}, \qquad (2.12)$$

where the constant C satisfies $C = C(x, T, \gamma)$.

Proof. Step 1. Because

$$X(t) = x + \int_0^t b(X(s), \alpha(s))ds + \int_0^t \sigma(X(s), \alpha(s))dw(s),$$

we have for any $p \ge 2$ that

$$|X(t)|^p \le 3^{p-1}\left[|x|^p + \left|\int_0^t b(X(s), \alpha(s))ds\right|^p + \left|\int_0^t \sigma(X(s), \alpha(s))dw(s)\right|^p\right].$$

Using the Hölder inequality and the linear growth condition, detailed computations lead to

$$\mathbf{E}_{x,i}\left[\sup_{0\leq t\leq T_1}\left|\int_0^t b(X(s),\alpha(s))ds\right|^p\right]$$

$$\leq T^{p-1}\int_0^{T_1}|b(X(s),\alpha(s))|^p\,ds$$

$$\leq KT^{p-1}\int_0^{T_1}(1+|X(s)|^p)ds$$

$$\leq c_1(T,p)\int_0^{T_1}\mathbf{E}_{x,i}\left[1+\sup_{0\leq u\leq s}|X(u)|^p\right]ds,$$

where $0\leq T_1\leq T$, and $c_1(T,p)$, $i=1,2$ is a positive constant depending only on T, p, and the linear growth constant in (2.10). Similarly, by the Burkholder–Davis–Gundy inequality (see Lemma A.32; see also [120, p. 70]), the Hölder inequality, and the linear growth condition, we compute

$$\mathbf{E}_{x,i}\left[\sup_{0\leq t\leq T_1}\left|\int_0^t \sigma(X(s),\alpha(s))dw(s)\right|^p\right]$$

$$\leq K\mathbf{E}_{x,i}\left[\int_0^{T_1}|\sigma(X(s),\alpha(s))|^2\,ds\right]^{p/2}$$

$$\leq KT^{p/2-1}\mathbf{E}_{x,i}\int_0^{T_1}|\sigma(X(s),\alpha(s))|^p\,ds$$

$$\leq c_2(T,p)\int_0^{T_1}\mathbf{E}_{x,i}\left[1+\sup_{0\leq u\leq s}|X(u)|^p\right]ds,$$

where $0\leq T_1\leq T$, and $c_2(T,p)$ is a positive constant depending only on T, p, and the linear growth constant in (2.10). Thus we have

$$\mathbf{E}_{x,i}\left[1+\sup_{0\leq t\leq T_1}|X(t)|^p\right]$$

$$\leq c_3(x,p)+c_4(T,p)\int_0^{T_1}\mathbf{E}_{x,i}\left[1+\sup_{0\leq u\leq s}|X(u)|^p\right]ds,$$

where $c_3(x,p)=1+3^{p-1}|x|^p$, and $c_4(T,p)=3^{p-1}[c_1(T,p)+c_2(T,p)]$. Hence the Gronwall inequality (see, e.g., [44]) implies that

$$\mathbf{E}_{x,i}\left[1+\sup_{0\leq t\leq T_1}|X(t)|^p\right]$$

$$\leq c_3(x,p)\exp(c_4(T,p)T_1)$$

$$\leq c_3(x,p)\exp(c_4(T,p)T)$$

$$:=C(x,T,p).$$

Thus we have $\mathbf{E}_{x,i}\left[\sup_{0\leq t\leq T_1}|X(t)|^p\right]\leq C(x,T,p)$. Because $T_1\leq T$ is arbitrary, it follows that

$$\mathbf{E}_{x,i}\left[\sup_{0\leq t\leq T}|X(t)|^p\right]\leq C(x,T,p),\quad p\geq 2.$$

Step 2. Note that $\sup_{0\leq t\leq T}|X(t)|^{\gamma} = \left(\sup_{0\leq t\leq T}|X(s)|\right)^{\gamma}$ for any $\gamma > 0$. Thus we obtain from Hölder's inequality and Step 1 that for any $1 \leq p < 2$,

$$\mathbf{E}_{x,i}\left[\sup_{0\leq t\leq T}|X(t)|^p\right] \leq \left(\mathbf{E}_{x,i}\left[\sup_{0\leq t\leq T}|X(t)|^{2p}\right]\right)^{1/2} \leq C(T,x,p) < \infty.$$

Step 3. Finally, if $0 < p < 1$, note that

$$|X(t)|^p = |X(t)|^p I_{\{|X(t)|\geq 1\}} + |X(t)|^p I_{\{|X(t)|<1\}} \leq 1 + |X(t)|^{1+p}.$$

Hence it follows from Step 2 that

$$\mathbf{E}_{x,i}\left[\sup_{0\leq t\leq T}|X(t)|^p\right] \leq \mathbf{E}_{x,i}\left[1 + \sup_{0\leq t\leq T}|X(t)|^{1+p}\right] \leq C(x,T,p).$$

This completes the proof of the proposition.   □

Proposition 2.4. *The process* $(X(t),\alpha(t))$ *is càdlàg. That is, the sample paths of* $(X(t),\alpha(t))$ *are right continuous and have left limits.*

Proof. It is well known that the sample paths of the discrete component $\alpha(\cdot)$ are right continuous with left limits. Hence it remains to show that the same is true for the continuous component $X(\cdot)$. To this end, let $0 \leq s < t \leq T$ with T being any fixed positive number, and consider

$$X(t) - X(s) = \int_s^t b(X(u),\alpha(u))du + \int_s^t \sigma(X(u),\alpha(u))dw(u).$$

Using Lemma 2.3 and [47, Theorem 4.6.3], detailed computations lead to

$$\mathbf{E}_{x,i}|X(t) - X(s)|^4 \leq C|t-s|^2,$$

where C is a constant dependent on T, the initial condition x, and the linear growth and Lipschitz constant of the coefficients b and σ. Then the desired result follows from Kolmogorov's continuity criterion.   □

Remark 2.5. By considering x-dependent generator $Q(x)$, our model provides a more realistic formulation allowing the switching component depending on the continuous states. This, in turn, allows the coupling and correlation between $X(t)$ and $\alpha(t)$.

2.3 Regularity

It follows from Theorem 2.1 that if the coefficients satisfy the linear growth and the local Lipschitz condition, then the solution $(X(t),\alpha(t))$ of (2.2)–(2.3) is defined for all $t > 0$. Nevertheless, the linear growth condition puts some restrictions on the applicability. For instance, consider the real-valued regime-switching diffusion

$$dX(t) = -X^3(t)dt - \sigma(\alpha(t))dw(t), \qquad (2.13)$$

where $\sigma(i), i \in \mathcal{M}$ are constants. Even though the drift coefficient is only local Lipschitz continuous and does not satisfy the linear growth condition, the regime-switching diffusion will not blow up in finite time because the drift coefficient forces the process to move towards the origin. Hence it is expected that the solution to (2.13) exists for all $t \geq 0$.

Let $(X(t), \alpha(t))$ be a regime-switching diffusion given by (2.2) and (2.3) whose drift and diffusion coefficients satisfy the local Lipschitz condition (2.11). For any $n = 1, 2, \ldots$ and $(x, \alpha) \in \mathbb{R}^r \times \mathcal{M}$, we define β_n to be the first exit time of the process $(X(t), \alpha(t))$ from the ball centered at the origin with radius n, $B(0, n) := \{\xi \in \mathbb{R}^r : |\xi| < n\}$. That is,

$$\beta_n = \beta_n^{x,\alpha} := \inf\{t \geq 0 : |X^{x,\alpha}(t)| \geq n\}. \qquad (2.14)$$

Note that the sequence $\{\beta_n\}$ is monotonically increasing and hence has a (finite or infinite) limit. Denote the limit by β_∞. To proceed, we first give a definition of regularity.

Definition 2.6. *Regularity.* A Markov process $(X(t), \alpha(t))$ with initial data $(X(0), \alpha(0)) = (x, \alpha)$ is said to be *regular*, if

$$\beta_\infty = \lim_{n \to \infty} \beta_n = \infty \text{ a.s.} \qquad (2.15)$$

It is obvious to see that the process $(X^{x,\alpha}(t), \alpha^{x,\alpha}(t))$ is regular if and only if for any $0 < T < \infty$,

$$\mathbf{P}\{\sup_{0 \leq t \leq T} |X^{x,\alpha}(t)| = \infty\} = 0. \qquad (2.16)$$

That is, a process is regular if and only if it does not blow up in finite time. Thus, (2.16) can be used as an alternate definition. Nevertheless, it is handy to use equation (2.15) to delineate the regularity.

In what follows, we take up the regularity issue of a regime-switching diffusion when its coefficients do not satisfy the linear growth condition. The following theorem gives sufficient conditions for regularity, which is based on "local" linear growth and Lipschitz continuity.

Theorem 2.7. *Suppose that for each $i \in \mathcal{M}$, both the drift $b(\cdot, i)$ and the diffusion coefficient $\sigma(\cdot, i)$ satisfy the linear growth and Lipschitz condition in every bounded open set in $\mathbb{R}^r$, and that there is a nonnegative function $V(\cdot, \cdot) : \mathbb{R}^r \times \mathcal{M} \mapsto \mathbb{R}^+$ that is twice continuously differentiable with respect to $x \in \mathbb{R}^r$ for each $i \in \mathcal{M}$ such that there is an $\gamma_0 > 0$ satisfying*

$$\mathcal{L}V(x, i) \leq \gamma_0 V(x, i), \text{ for all } (x, i) \in \mathbb{R}^r \times \mathcal{M},$$

$$V_R := \inf_{|x| \geq R, \ i \in \mathcal{M}} V(x, i) \to \infty \quad as \quad R \to \infty. \qquad (2.17)$$

Then the process $(X(t), \alpha(t))$ is regular.

Proof. As argued earlier, it suffices to prove (2.15). Suppose on the contrary that (2.15) were false, then there would exist some $T > 0$ and $\varepsilon > 0$ such that $\mathbf{P}_{x,i}\{\beta_\infty \leq T\} > \varepsilon$. Therefore we could find some $n_1 \in \mathbb{N}$ such that

$$\mathbf{P}_{x,i}\{\beta_n \leq T\} > \varepsilon, \quad \text{for all } n \geq n_1. \tag{2.18}$$

Define

$$S(x, i, t) = V(x, i) \exp(-\gamma_0 t), \quad (x, i) \in \mathbb{R}^r \times \mathcal{M}, \text{ and } t \geq 0.$$

Then it satisfies $[(\partial/\partial t) + \mathcal{L}]S(x, i, t) \leq 0$. By virtue of Dynkin's formula, we have

$$\mathbf{E}_{x,i}\left[V(X(\beta_n \wedge T), \alpha(\beta_n \wedge T)) \exp(-\gamma_0(\beta_n \wedge T))\right] - V(x, i)$$
$$= \mathbf{E}_{x,i} \int_0^{\beta_n \wedge T} \left(\frac{\partial}{\partial t} + \mathcal{L}\right) S(X(u), \alpha(u), u) du \leq 0.$$

Hence we have

$$V(x, i) \geq \mathbf{E}_{x,i}\left[V(X(\beta_n \wedge T), \alpha(\beta_n \wedge T)) \exp(-\gamma_0(\beta_n \wedge T))\right].$$

Note that $\beta_n \wedge T \leq T$ and V is nonnegative. Thus we have

$$V(x, i) \exp\{\gamma_0 T\} \geq \mathbf{E}_{x,i}\left[V(X(\beta_n \wedge T), \alpha(\beta_n \wedge T))\right]$$
$$\geq \mathbf{E}_{x,i}\left[V(X(\beta_n), \alpha(\beta_n))I_{\{\beta_n \leq T\}}\right].$$

Furthermore, by the definition of β_n and (2.17), we have

$$V(x, i) \exp\{\gamma_0 T\} \geq V_n \mathbf{P}_{x,i}\{\beta_n \leq T\} > \varepsilon V_n \to \infty, \quad \text{as } n \to \infty.$$

This is a contradiction. Thus we must have $\lim_{n \to \infty} \beta_n = \infty$ a.s. This completes the proof. $\square$

Similar to the proof of [83, Theorem 3.4.2], we can also prove the following theorem.

Theorem 2.8. *Suppose that for each $i \in \mathcal{M}$, both the drift $b(\cdot, i)$ and the diffusion coefficient $\sigma(\cdot, i)$ satisfy the linear growth and Lipschitz condition in every bounded open set in $\mathbb{R}^r$, and that there is a nonnegative and bounded function $V(\cdot, \cdot) : \mathbb{R}^r \times \mathcal{M} \mapsto \mathbb{R}^+$ that is not identically zero, that for each $i \in \mathcal{M}$, $V(x, i)$ is twice continuously differentiable with respect to $x \in \mathbb{R}^r$ such that there is a $\gamma_1 > 0$ satisfying*

$$\mathcal{L}V(x, i) \geq \gamma_1 V(x, i), \quad \text{for all } (x, i) \in \mathbb{R}^r \times \mathcal{M},$$

*Then the process $(X(t), \alpha(t))$ is not regular. In particular, for any $\varepsilon > 0$,
we have*

$$\mathbf{P}_{x_0, \ell} \{ \beta_\infty < \kappa + \varepsilon \} > 0,$$

where $\beta_\infty := \lim_{n \to \infty} \beta_n$, $(x_0, \ell) \in \mathbb{R}^r \times \mathcal{M}$ satisfying $V(x_0, \ell) > 0$, and

$$\kappa = \frac{1}{\gamma_1} \log \left(\frac{\sup\limits_{(x,i) \in \mathbb{R}^r \times \mathcal{M}} V(x,i)}{V(x_0, \ell)} \right).$$

Remark 2.9. If the coefficients of (2.2)–(2.3) satisfy both the local Lipschitz condition (2.11) and linear growth condition (2.10), then by The-
orem 2.1, there is a unique solution $(X^{x,\alpha}(t), \alpha^{x,\alpha}(t))$ to (2.2)–(2.3) for
all $t \geq 0$. The regime-switching diffusion $(X^{x,\alpha}(t), \alpha^{x,\alpha}(t))$ is thus regular. Alternatively, we can use Theorem 2.7 to verify this. In fact, detailed
computations show that the function

$$V(x,i) = (|x|^2 + 1)^{r/2}, \quad (x,i) \in \mathbb{R}^r \times \mathcal{M}$$

satisfies all conditions in Theorem 2.7. Thus the desired assertion follows.

In particular, we can verify that $V(\cdot, i)$ is twice continuously differentiable
for each $i \in \mathcal{M}$ and that $\mathcal{L}V(x,i) \leq \gamma_0 V(x,i)$ for some positive constant
γ_0 and all $(x,i) \in \mathbb{R}^r \times \mathcal{M}$. Thus for any $t > 0$, a similar argument as in
the proof of Theorem 2.7 leads to

$$\mathbf{P}_{x,i} \{ \beta_n < t \} \leq \frac{V(x,i) e^{\gamma_0 t}}{V_n},$$

where $V_n = (1 + n^2)^{r/2}$. Therefore for any $t > 0$ and $\varepsilon > 0$, there exists an
$N \in \mathbb{N}$ such that

$$\mathbf{P}_{x,i} \{ \beta_n < t \} < \varepsilon, \quad \text{for all } n \geq N, \tag{2.19}$$

uniformly for x in any compact set $F \subset \mathbb{R}^r$ and $i \in \mathcal{M}$.

Example 2.10. Consider a real-valued regime-switching diffusion

$$dX(t) = X(t) \left[(b(\alpha(t)) - a(\alpha(t))X^2(t))dt + \sigma(\alpha(t))dw(t) \right], \tag{2.20}$$

where $w(\cdot)$ is a one-dimensional standard Brownian motion, and $\alpha(\cdot) \in
\mathcal{M} = \{1, 2, \ldots, m_0\}$ is a jump process with appropriate generator $Q(x)$.
Clearly, the coefficients of (2.20) do not satisfy the linear growth condition
if not all $a(i) = 0$ for $i \in \mathcal{M}$. We claim that if $a(i) > 0$ for each $i \in \mathcal{M}$,
then (2.20) is regular. To see this, we apply Theorem 2.7 and consider

$$V(x,i) := |x|^2, \quad (x,i) \in \mathbb{R} \times \mathcal{M}.$$

Clearly, V is nonnegative and satisfies

$$\lim_{|x| \to \infty} V(x,i) = \infty, \quad \text{for each } i \in \mathcal{M}.$$

Thus by Theorem 2.7, it remains to verify that for all $(x, i) \in \mathbb{R} \times \mathcal{M}$, we have $\mathcal{L}V(x, i) \leq KV(x, i)$ for some positive constant K. To this end, we compute

$$\mathcal{L}V(x, i) = 2x \cdot x[b(i) - a(i)x^2] + \frac{1}{2} 2 \cdot x^2 \sigma^2(i)$$
$$= (2b(i) + \sigma^2(i))x^2 - 2a(i)x^4$$
$$\leq KV(x, i),$$

where $K = \max\{2b(i) + \sigma^2(i), i = 1, \ldots, m_0\}$. Note in the above, we used the assumption that $a(i) > 0$ for each $i \in \mathcal{M}$. Hence it follows from Theorem 2.7 that (2.20) is regular.

For demonstration, we plot a sample path of (2.20) with its coefficients specified as follows. The jump process $\alpha(\cdot) \in \mathcal{M} = \{1, 2\}$ is generated by

$$Q(x) = \begin{pmatrix} -3 - \sin x \cos x & 3 + \sin x \cos x \\ 2 & -2 \end{pmatrix},$$

and

$$b(1) = 3, \quad a(1) = 2, \quad \sigma(1) = 1$$
$$b(2) = 10, \quad a(1) = 1, \quad \sigma(1) = -1.$$

Figure 2.1 plots a sample path of (2.20) with initial condition $(x, \alpha) = (2.5, 1)$.

Example 2.11. (Lotka–Volterra model cont.) We continue the discussion initiated in Example 1.1. Regarding the concept of regularity, we can show that if the ecosystem is self-regulating or competitive, then the population will not explode in finite time almost surely. Recall that the system is competitive if all the values in the community matrix $A(\alpha)$ are nonnegative, that is, $a_{ij}(\alpha) \geq 0$ for all $\alpha \in \mathcal{M} = \{1, \ldots, m\}$ and $i, j = 1, 2, \ldots, n$. The competition among the same species is assumed to be strictly positive so that each species grows "logistically" in any environment. To interpret it in another way, members of the same species compete with one another. Take, for instance, bee colonies in a field. They will compete for food strongly with the colonies located near them. We assume that the self-regulating competition within the same species is strictly positive, that is, for each $\alpha \in \mathcal{M} = \{1, 2, \ldots, m\}$ and $i, j = 1, 2, \ldots, n$ with $j \neq i$, $a_{ii}(\alpha) > 0$ and $a_{ij}(\alpha) \geq 0$. It was proved in [189], for any initial conditions $(x(0), \alpha(0)) = (x_0, \alpha) \in \mathbb{R}_+^n \times \mathcal{M}$, where

$$\mathbb{R}_+^n = \{(x_1, \ldots, x_n) : x_i > 0, i = 1, \ldots, n\},$$

there is a unique solution $x(t)$ to (1.1) on $t \geq 0$, and the solution will remain in $\mathbb{R}_+^n$ almost surely; that is, $x(t) \in \mathbb{R}_+^n$ for any $t \geq 0$ with probability 1.

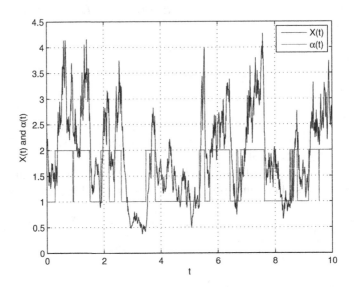

FIGURE 2.1. A sample path of switching diffusion (2.20) with initial condition $(x, \alpha) = (2.5, 1)$.

2.4 Weak Continuity

Recall that a stochastic process $Y(t)$ with right continuous sample paths is said to be *continuous in probability* at t if for any $\eta > 0$,

$$\lim_{\Delta \to 0} \mathbf{P}(|Y(t + \Delta) - Y(t)| \geq \eta) = 0. \qquad (2.21)$$

It is *continuous in mean square* at t if

$$\lim_{\Delta \to 0} \mathbf{E}|Y(t + \Delta) - Y(t)|^2 = 0. \qquad (2.22)$$

The process $Y(t)$ is said to be continuous in probability in the interval $[0, T]$ (or in short continuous in probability if $[0, T]$ is clearly understood), if it is continuous in probability at every $t \in [0, T]$. Likewise it is continuous in mean square if it is continuous in mean square at every $t \in [0, T]$. We proceed to obtain a continuity result for the two-component switching diffusion processes $Y(t) = (X(t), \alpha(t))$. The results are presented in two parts. The first part concentrates on Markovian switching diffusions, whereas the second one is concerned with state-dependent diffusions.

Theorem 2.12. *Suppose that the conditions of Theorem 2.1 are satisfied with the modification $Q(x) = Q$ that generates a Markov chain independent* *of the Brownian motion. Then the process $Y(t) = (X(t), \alpha(t))$ is continuous in probability and also continuous in mean square.*

Proof. We show that for any $\eta > 0$,

$$\mathbf{P}(|Y(t + \Delta) - Y(t)| \geq \eta) \to 0 \quad \text{as} \quad \Delta \to 0, \quad \text{and}$$
$$\mathbf{E}|Y(t + \Delta) - Y(t)|^2 \to 0 \quad \text{as} \quad \Delta \to 0. \tag{2.23}$$

We proceed first to establish the mean square convergence above. Note that

$$Y(t + \Delta) - Y(t) = (X(t + \Delta), \alpha(t + \Delta)) - (X(t + \Delta), \alpha(t))$$
$$+ (X(t + \Delta), \alpha(t)) - (X(t), \alpha(t)). \tag{2.24}$$

We divide the rest of the proof into several steps.

Step 1: First we recognize that in view of (2.24),

$$\mathbf{E}|Y(t + \Delta) - Y(t)|^2 \leq 2\big[\mathbf{E}|\alpha(t + \Delta) - \alpha(t)|^2$$
$$+ \mathbf{E}|X(t + \Delta) - X(t)|^2\big]. \tag{2.25}$$

Thus to estimate the difference of the second moment, it suffices to consider the two marginal estimates separately. We do this in the next two steps.

Step 2: We claim that for any $t \geq 0$ and $\Delta \geq 0$,

$$\mathbf{E}|X(t + \Delta) - X(t)|^2 \leq K\Delta. \tag{2.26}$$

This estimate, in fact, is a modification of the standard estimates for stochastic differential equations. It mainly uses the linear growth and Lipschitz conditions of the drift and diffusion coefficients and Proposition 2.3. We thus omit the details.

Step 3: Note that for any $t \geq 0$,

$$\alpha(t) = \sum_{i=1}^{m_0} i I_{\{\alpha(t)=i\}} = \chi(t)(1, \ldots, m_0)',$$

where

$$\chi(t) = (\chi_1(t), \ldots, \chi_{m_0}(t)) = (I_{\{\alpha(t)=1\}}, \ldots, I_{\{\alpha(t)=m_0\}}) \in \mathbb{R}^{1 \times m_0}, \tag{2.27}$$

and $(1, \ldots, m_0)' \in \mathbb{R}^{m_0}$ is a column vector. Because the Markov chain $\alpha(t)$ is independent of the Brownian motion $w(\cdot)$ (Q is a constant matrix), it is well known that

$$\chi(t + \Delta) - \chi(t) - \int_t^{t+\Delta} \chi(s) Q \, ds$$

is a martingale; see Lemma A.5 in this book or [176, Lemma 2.4]. It follows that

$$\mathbf{E}_t \left[\chi(t + \Delta) - \chi(t) - \int_t^{t+\Delta} \chi(s) Q ds \right] = 0,$$

where $\mathbf{E}_t$ denotes the conditional expectation on the σ-algebra

$$\mathcal{F}_t = \{ (X(u), \alpha(u)) : u \leq t \}.$$

It then follows that

$$\left| \int_t^{t+\Delta} \chi(s) Q ds \right| = O(\Delta) \text{ a.s.} \tag{2.28}$$

Thus, we obtain

$$\mathbf{E}_t \chi(t + \Delta) = \chi(t) + O(\Delta) \text{ a.s.} \tag{2.29}$$

In view of this structure, we see that

$$\chi(t + \Delta) - \chi(t) = (\chi_1(t + \Delta) - \chi_1(t), \ldots, \chi_{m_0}(t + \Delta) - \chi_{m_0}(t))$$

with $\chi_i(\cdot)$ given by (2.27). This together with (2.29) implies

$$\begin{aligned}
\mathbf{E}_t &[\chi_i(t + \Delta) - \chi_i(t)]^2 \\
&= \mathbf{E}_t [I_{\{\alpha(t+\Delta)=i\}} - I_{\{\alpha(t)=i\}}]^2 \\
&= \left[\mathbf{E}_t I_{\{\alpha(t+\Delta)=i\}} - 2 I_{\{\alpha(t)=i\}} \mathbf{E}_t I_{\{\alpha(t+\Delta)=i\}} + I_{\{\alpha(t)=i\}} \right] \\
&= O(\Delta) \text{ a.s.}
\end{aligned} \tag{2.30}$$

Step 4: Next we consider

$$\begin{aligned}
\mathbf{E}&[\alpha(t + \Delta) - \alpha(t)]^2 \\
&= \mathbf{E}|[\chi(t + \Delta) - \chi(t)](1, \ldots, m_0)'|^2 \\
&\leq K \mathbf{E}|\chi(t + \Delta) - \chi(t)|^2 \\
&\leq K \sum_{i=1}^{m_0} \mathbf{E}\mathbf{E}_t[\chi_i(t + \Delta) - \chi_i(t)]^2 \\
&\leq K \Delta \to 0 \text{ as } \Delta \to 0.
\end{aligned} \tag{2.31}$$

From the next to the last line to the last line above, we have used (2.30). By combining (2.26) and (2.31), we obtain that (2.25) leads to

$$\mathbf{E}|Y(t + \Delta) - Y(t)|^2 \to 0 \text{ as } \Delta \to 0.$$

The mean square continuity has been established. Then the desired continuity in probability follows from Tchebyshev's inequality. □

We next generalize the above result and allow the switching process to be x-dependent. The result is presented next.

Theorem 2.13. *Suppose that the conditions of Theorem 2.1 are satisfied. Then the process $Y(t) = (X(t), \alpha(t))$ is continuous in probability and continuous in mean square.*

Proof. Step 1 and Step 2 are the same as before. We only point out the main difference as compared to Theorem 2.12.

Consider the function $h(x, \alpha) = I_{\{\alpha = i\}}$ for each $i \in \mathcal{M}$. Because h is independent of x, it is readily seen that $\mathcal{L}h(x, \alpha) = Q(x)h(x, \cdot)(\alpha)$. Consequently,

$$\mathbf{E}_t \left[h(X(t + \delta), \alpha(t + \Delta)) - h(X(t), \alpha(t)) - \int_t^{t+\Delta} \mathcal{L}h(X(s), i)ds \right] = 0.$$

However,

$$\mathbf{E}_t \left[h(X(t + \Delta), \alpha(t + \Delta)) - h(X(t), \alpha(t)) - \int_t^{t+\Delta} \mathcal{L}h(X(s), \alpha(s))ds \right]$$

$$= \mathbf{E}_t \left[h(X(t + \Delta), \alpha(t + \Delta)) - h(X(t), \alpha(t)) \right.$$

$$\left. - \int_t^{t+\Delta} Q(X(s))h(X(s), \cdot)(\alpha(s))ds \right]$$

$$= \mathbf{E}_t \left[I_{\{\alpha(t+\Delta)=i\}} - I_{\{\alpha(t)=i\}} - \int_t^{t+\Delta} \sum_{j=1}^{m_0} q_{ij}(X(s))I_{\{\alpha(s)=i\}}ds \right].$$

Because $Q(x)$ is bounded, similar to (2.28), we obtain

$$\left| \int_t^{t+\Delta} \chi(s)Q(X(s))ds \right| = O(\Delta) \quad \text{a.s.} \tag{2.32}$$

With (2.32) at our hands, we proceed with the rest of Step 3 and Step 4 in the proof of Theorem 2.12. The desired result follows. □

2.5 Feller Property

In the theory of Markov processes and their applications, dealing with a Markov process $\xi(t)$ with $\xi(0) = x$, for a suitable function $f(\cdot)$, often one must consider the function $u(t, x) = \mathbf{E}_x f(\xi(t))$. Following [38], the process $\xi(t)$ is said to be *Feller* if $u(t, \cdot)$ is continuous for any $t \geq 0$ and $\lim_{t \downarrow 0} u(t, x) = f(x)$ for any bounded and continuous function f and it is said to be *strong Feller* if $u(t, \cdot)$ is continuous for any $t > 0$ and any bounded and measurable function f. This is a natural condition in physical or social modeling. It indicates that a slight perturbation of the initial data should result in a small perturbation in the subsequent movement. For example, let $X^x(t)$ be a diffusion process of appropriate dimension with $X(0) = x$, we say that $X^x(\cdot)$ satisfies the Feller property or $X^x(\cdot)$ is a Feller process if for any $t \geq 0$ and any bounded and continuous function $g(\cdot)$, $u(t,x) = \mathbf{E}_x g(X(t))$ is continuous with respect to x and $\lim_{t \downarrow 0} u(t,x) = g(x)$; see [38, 57] and many references therein for the pioneering work on Feller properties for the diffusions. For switching-diffusion processes, do the Feller and the strong Feller properties carry over? This section is concerned with Feller property, whereas the next section deals with strong Feller property.

We show that such properties carry over to the switching diffusions, but the proofs are in fact nontrivial. The difficulty stems from the consideration of x-dependent generator $Q(x)$ for the discrete component $\alpha(t)$. The classical arguments for Feller and strong Feller properties of diffusions will not work here. We need to prove that the function $u(t,x,i) = \mathbf{E}_{x,i} f(X(t), \alpha(t))$ is continuous with respect to the initial data (x,i) for any $t \geq 0$ and $\lim_{t \downarrow 0} u(t,x,i) = f(x,i)$ for any bounded and continuous function f. By virtue of Proposition 2.4 and the boundedness and continuity of f, we have $\lim_{t \downarrow 0} u(t,x,i) = \mathbf{E}_{x,i} f(X(0), \alpha(0)) = f(x,i)$. Note also that $u(0,x,i)$ is automatically continuous by the continuity of f. Because $\mathcal{M} = \{1, \ldots, m_0\}$ is a finite set, it is enough to show that $u(t,x,i)$ is continuous with respect to x for any $t > 0$. In this section, we establish the Feller property for switching diffusions under Lipschitz continuity and linear growth conditions. This section is motivated by the recent work [161]. The discussion of the strong Feller property is deferred to Section 2.6.

We first present the following lemma. In lieu of the local Lipschitz condition, we assume a global Lipschitz condition holds henceforth without specific mentioning. Also, we assume for the moment that the discrete component $\alpha(\cdot)$ is generated by a constant Q.

Lemma 2.14. *Assume the conditions of Theorem 2.1 hold with the modification of the local Lipschitz condition replaced by a global Lipschitz condition. Moreover, suppose that $\alpha(\cdot)$ is generated by $Q(x) = Q$ and that $\alpha(\cdot)$ is independent of the Brownian motion $w(\cdot)$. Then for any fixed $T > 0$, we have*

$$\mathbf{E}\left[\sup_{0 \leq t \leq T} \left| X^{\widetilde{x},\alpha}(t) - X^{x,\alpha}(t) \right|^2 \right] \leq C \left| \widetilde{x} - x \right|^2, \qquad (2.33)$$

where C is a constant depending only on T and the global Lipschitz and the linear growth constant K.

Proof. Assume that $\widetilde{x} = x + \Delta x$ and let $0 < t \leq T$. Moreover, for notational simplicity, we denote $\widetilde{X}(t) = X^{\widetilde{x},\alpha}(t)$ and $X(t) = X^{x,\alpha}(t)$. By the assumption of the theorem, $Q(x) = Q$ that is independent of x, we have $\alpha^{\widetilde{x},\alpha}(t) = \alpha^{x,\alpha}(t) = \alpha(t)$. Then we have

$$\widetilde{X}(t) - X(t) = \Delta x + \int_0^t [b(\widetilde{X}(s), \alpha(s)) - b(X(s), \alpha(s))]ds$$
$$+ \int_0^t [\sigma(\widetilde{X}(s), \alpha(s)) - b(X(s), \alpha(s))]dw(s).$$

Thus it follows that

$$\left|\widetilde{X}(t) - X(t)\right|^2 \leq 3|\Delta x|^2 + 3\left|\int_0^t [b(\widetilde{X}(s), \alpha(s)) - b(X(s), \alpha(s))]ds\right|^2$$
$$+ 3\left|\int_0^t [\sigma(\widetilde{X}(s), \widetilde{\alpha}(s)) - b(X(s), \alpha(s))]dw(s)\right|^2.$$

(2.34)

Using the assumption that $b(\cdot, j)$ is Lipschitz continuous for each $j \in \mathcal{M}$, we have

$$\left|b(\widetilde{X}(s), \alpha(s)) - b(X(s), \alpha(s))\right|$$
$$= \left|\sum_{j=1}^{m_0}[b(\widetilde{X}(s), j) - b(X(s), j)]I_{\{\alpha(s)=j\}}\right|$$
$$\leq K\left|\widetilde{X}(s) - X(s)\right|,$$

where in the above and hereafter, K is a generic positive constant not depending on x, $\widetilde{x}$, or t whose exact value may change in different appearances. Consequently, we have from the Hölder inequality that

$$\mathbf{E}\left[\sup_{0 \leq t \leq T_1}\left|\int_0^t [b(\widetilde{X}(s), \alpha(s)) - b(X(s), \alpha(s))]ds\right|^2\right]$$
$$\leq T\mathbf{E}\int_0^{T_1}\left|b(\widetilde{X}(s), \alpha(s)) - b(X(s), \alpha(s))\right|^2 ds \qquad (2.35)$$
$$\leq KT\int_0^{T_1}\mathbf{E}\left|\widetilde{X}(s) - X(s)\right|^2 ds,$$

where $0 \leq T_1 \leq T$. Similarly the Lipschitz continuity yields that

$$\left|\sigma(\widetilde{X}(s), \widetilde{\alpha}(s)) - \sigma(X(s), \widetilde{\alpha}(s))\right| \leq K\left|\widetilde{X}(s) - X(s)\right|,$$

and hence the Burkholder–Davis–Gundy inequality (see Lemma A.32) leads to

$$\mathbf{E}\left[\sup_{0 \leq t \leq T_1}\left|\int_0^t [\sigma(\widetilde{X}(s), \alpha(s)) - \sigma(X(s), \alpha(s))]dw(s)\right|^2\right]$$
$$\leq \mathbf{E}\int_0^{T_1}\left|\sigma(\widetilde{X}(s), \alpha(s)) - \sigma(X(s), \alpha(s))\right|^2 ds \qquad (2.36)$$
$$\leq K\int_0^{T_1}\mathbf{E}\left|\widetilde{X}(s) - X(s)\right|^2 ds.$$

Therefore it follows from (2.34)–(2.36) that

$$
\mathbf{E}\left[\sup_{0 \le t \le T_1} \left|\widetilde{X}(t) - X(t)\right|^2\right]
$$

$$
\le 3\,|\Delta x|^2 + K(T+1)\int_0^{T_1} \mathbf{E}\left|\widetilde{X}(s) - X(s)\right|^2 ds
$$

$$
\le 3\,|\Delta x|^2 + K(T+1)\int_0^{T_1} \mathbf{E}\left[\sup_{0 \le u \le s}\left|\widetilde{X}(u) - X(u)\right|^2\right] ds.
$$

Then Gronwall's inequality implies that

$$
\mathbf{E}\left[\sup_{0 \le t \le T_1}\left|\widetilde{X}(t) - X(t)\right|^2\right] \le 3\,|\Delta x|^2 \exp\left\{K(T+1)T_1\right\}
$$

$$
\le 3\,|\Delta|^2 \exp\left\{K(T+1)T\right\}.
$$

The above inequality is true for any $0 \le T_1 \le T$, hence we have

$$
\mathbf{E}\left[\sup_{0 \le t \le T}\left|\widetilde{X}(t) - X(t)\right|^2\right] \le 3\,|\Delta x|^2 \exp\left\{K(T+1)T\right\} = K\,|\widetilde{x} - x|^2.
$$

This completes the proof.   □

Remark 2.15. An immediate consequence of (2.33) is that we obtain the Feller property for the case when the discrete component $\alpha(\cdot)$ is generated by a constant Q following the argument in [130, Lemma 8.1.4]. But for the general case when the generator $Q(x)$ of $\alpha(\cdot)$ is x-dependent, the proof in [130] is not applicable since (2.33) is not proven yet. New methodology is needed to treat (2.33); see Section 2.7 for more details.

To proceed, in addition to the assumptions of Theorem 2.1, we assume that the coefficients of (2.2)–(2.3) are bounded, continuously differentiable, and satisfy the global Lipschitz condition (all with respect to x). That is, we assume:

(A2.1) For $\alpha, \ell \in \mathcal{M}$ and $i, j = 1, \ldots, r$, $a_{ij}(\cdot, \alpha) \in C^2$, $b(\cdot, \alpha) \in C^1$, and $q_{\alpha\ell}(\cdot) \in C^1$, and for some positive constant K, we have

$$
|b(x, i)| + |\sigma(x, i)| + |q_{ij}(x)| \le K, \quad x \in \mathbb{R}^r, i, j \in \mathcal{M}, \quad (2.37)
$$

and for $\alpha, \ell \in \mathcal{M}$ and $\widetilde{x}, x \in \mathbb{R}^r$, we have

$$
|\sigma(\widetilde{x}, \alpha) - \sigma(x, \alpha)| + |b(\widetilde{x}, \alpha) - b(x, \alpha)| + |q_{\alpha\ell}(\widetilde{x}) - q_{\alpha\ell}(x)|
$$

$$
\le K\,|\widetilde{x} - x|.
$$

$$
(2.38)
$$

Consider the auxiliary process $(Z^{x,\alpha}(t), r^{\alpha}(t)) \in \mathbb{R}^r \times \mathcal{M}$ defined by

$$
\begin{cases}
dZ(t) = b(Z(t), r(t))dt + \sigma(Z(t), r(t))dw(t), \\
\mathbf{P}\left\{r(t+\Delta) = j | r(t) = i, r(s), 0 \le s \le t\right\} = \Delta + o(\Delta), \quad j \ne i,
\end{cases}
\tag{2.39}
$$

with initial conditions $Z(0) = x$ and $r(0) = \alpha$, where $w(t)$ is a d-dimensional standard Brownian motion independent of the Markov chain $r(t)$. The Markov chain $r(t)$ thus can be viewed as one with the generator

$$
\begin{pmatrix}
-(m_0 - 1) & 1 & \cdots & 1 & 1 \\
1 & -(m_0 - 1) & \cdots & 1 & 1 \\
\cdots & \cdots & \cdots & \cdots & \cdots \\
1 & 1 & \cdots & 1 & -(m_0 - 1)
\end{pmatrix}
$$
$$
= -m_0 I_{m_0} + (\mathbf{1}_{m_0}, \mathbf{1}_{m_0}, \ldots, \mathbf{1}_{m_0}),
$$

where I_{m_0} is the $m_0 \times m_0$ identity matrix and $\mathbf{1}_{m_0}$ is a column vector of dimension m_0 with all components being 1.

For any $T > 0$, denote by $\mu_{T,1}(\cdot)$ the measure induced by the process $(X^{x,\alpha}(t), \alpha^{x,\alpha}(t))$ and by $\mu_{T,2}(\cdot)$ the measure induced by the auxiliary process $(Z^{x,\alpha}(t), r^{\alpha}(t))$. Then by virtue of [41, 42], $\mu_{T,1}(\cdot)$ is absolutely continuous with respect to $\mu_{T,2}(\cdot)$. Moreover, the corresponding Radon–Nikodyn derivative has the form

$$
\begin{aligned}
&p_T(Z^{x,\alpha}(\cdot), r^{\alpha}(\cdot)) \\
&= \frac{d\mu_{T,1}}{d\mu_{T,2}}(Z^{x,\alpha}(\cdot), r^{\alpha}(\cdot)) \\
&= \exp\left\{(m_0 - 1)T\right\} \exp\left\{-\int_{\tau_{n(T)}}^{T} q_{r(\tau_{n(T)})}(Z(s))ds\right\} \\
&\quad \times \prod_{i=0}^{n(T)-1}\left[q_{r(\tau_i)r(\tau_{i+1})}(Z(\tau_{i+1}))\exp\left\{-\int_{\tau_i}^{\tau_{i+1}} q_{r(\tau_i)}(Z(s))ds\right\}\right],
\end{aligned}
\tag{2.40}
$$

where for each $i \in \mathcal{M}$, $q_i(x) := -q_{ii}(x) = \sum_{j \ne i} q_{ij}(x)$, τ_i is a sequence of stopping times defined by:

$$
\tau_0 = 0,
$$

and for $i = 0, 1, \ldots,$

$$
\tau_{i+1} := \inf\left\{t > \tau_i : r(t) \ne r(\tau_i)\right\},
$$

and $n(T) = \max\left\{n \in \mathbb{N} : \tau_n \le T\right\}$. Note that if $n(T) = 0$, then

$$
p_T(Z^{x,\alpha}(\cdot), r^{\alpha}(\cdot)) = \exp\left\{-\int_0^T \left[q_{\alpha}(Z(s)) - m_0 + 1\right]ds\right\}.
$$

Concerning the Radon–Nikodyn derivative $p_T(Z^{x,\alpha}(\cdot), r^{\alpha}(\cdot))$, we have the following results.

Lemma 2.16. *Assume the conditions of Theorem 2.1 and (A2.1). Then for any $T > 0$ and $(x, \alpha) \in \mathbb{R}^r \times \mathcal{M}$, we have*

$$\mathbf{E} \left| p_T(Z^{x,\alpha}(\cdot), r^\alpha(\cdot)) \right| \leq K < \infty.$$

Proof. By virtue of (2.37) and (2.40), we have

$$\mathbf{E} \left| p_T(Z^{x,\alpha}(\cdot), r^\alpha(\cdot)) \right| \leq \exp\left\{ (m_0 - 1)T \right\} \mathbf{E} K^{n(T)}.$$

Note that $n(T)$ is a Poisson process with rate $m_0 - 1$. Thus it follows that $\mathbf{E}\left[K^{n(T)}\right] < \infty$ and the desired conclusion follows. $\qquad\square$

Lemma 2.17. *Assume the conditions of Lemma 2.16. Then for any $T > 0$, $\tilde{x}, x \in \mathbb{R}^r$, and $\alpha \in \mathcal{M}$, we have*

$$\mathbf{E} \left| p_T(Z^{\tilde{x},\alpha}(\cdot), r^\alpha(\cdot)) - p_T(Z^{x,\alpha}(\cdot), r^\alpha(\cdot)) \right| \leq K \left| \tilde{x} - x \right|.$$

Proof. Similar to the notations in the proof of Lemma 2.14, we denote $\tilde{Z}(t) = Z^{\tilde{x},\alpha}(t)$, $Z(t) = Z^{x,\alpha}(t)$, and $r(t) = r^\alpha(t)$. Note that for any positive sequences $\{c_k\}_{k=1}^n$ and $\{d_k\}_{k=1}^n$, we have by induction that (see also [41, 42])

$$\left| \prod_{k=1}^n c_k - \prod_{k=1}^n d_k \right| \leq n \left(\max_{k=1,\ldots,n} \{c_k, d_k\} \right)^{n-1} \max_{k=1,\ldots,n} \{|c_k - d_k|\}. \quad (2.41)$$

Applying (2.41) to $p_T(\tilde{Z}(\cdot), r(\cdot))$ and $p_T(Z(\cdot), r(\cdot))$, we have

$$\mathbf{E} \left| p_T(\tilde{Z}(\cdot), r(\cdot)) - p_T(Z(\cdot), r(\cdot)) \right|$$
$$\leq \exp\left\{ (m_0 - 1)T \right\} \mathbf{E} \left[\sum_{n=0}^{\infty} (n+1) K^n I_{\{n(T)=n\}} \right.$$
$$\times \max_{i=0,1,\ldots,n-1} \left\{ \left| e^{-\int_{\tau_n}^{T} q_{r(\tau_n)}(\tilde{Z}(s))ds} - e^{-\int_{\tau_n}^{T} q_{r(\tau_n)}(Z(s))ds} \right|, \right.$$
$$\left| q_{r(\tau_i)r(\tau_{i+1})}(\tilde{Z}(\tau_{i+1})) e^{-\int_{\tau_i}^{\tau_{i+1}} q_{r(\tau_i)}(\tilde{Z}(s))ds} \right.$$
$$\left. \left. - q_{r(\tau_i)r(\tau_{i+1})}(Z(\tau_{i+1})) e^{-\int_{\tau_i}^{\tau_{i+1}} q_{r(\tau_i)}(Z(s))ds} \right| \right\} \right].$$

Note that $\left| e^{-c} - e^{-d} \right| \leq |c - d|$ for any $c, d \geq 0$. Meanwhile, (2.38) implies that

$$|q_i(\tilde{x}) - q_i(x)| \leq K \left| \tilde{x} - x \right|, \quad \tilde{x}, x \in \mathbb{R}^r, \quad i \in \mathcal{M}.$$

Thus it follows that

$$
\begin{aligned}
&\left| \exp\left(-\int_{\tau_n}^{T} q_{r(\tau_n)}(\widetilde{Z}(s)) ds \right) - \exp\left(-\int_{\tau_n}^{T} q_{r(\tau_n)}(Z(s)) ds \right) \right| \\
&\leq \left| \int_{\tau_n}^{T} q_{r(\tau_n)}(\widetilde{Z}(s)) ds - \int_{\tau_n}^{T} q_{r(\tau_n)}(Z(s)) ds \right| \\
&\leq \int_{\tau_n}^{T} \left| q_{r(\tau_n)}(\widetilde{Z}(s)) - q_{r(\tau_n)}(Z(s)) \right| ds \\
&\leq \int_{\tau_n}^{T} K \left| \widetilde{Z}(s) - Z(s) \right| ds \\
&\leq T \sup\left\{ \left| \widetilde{Z}(s) - Z(s) \right| : 0 \leq s \leq T \right\}.
\end{aligned}
$$

Similarly, for any $i = 0, 1, \ldots, n-1$, we have

$$
\begin{aligned}
&\left| q_{r(\tau_i)r(\tau_{i+1})}(\widetilde{Z}(\tau_{i+1})) e^{-\int_{\tau_i}^{\tau_{i+1}} q_{r(\tau_i)}(\widetilde{Z}(s)) ds} \right. \\
&\quad \left. - q_{r(\tau_i)r(\tau_{i+1})}(Z(\tau_{i+1})) e^{-\int_{\tau_i}^{\tau_{i+1}} q_{r(\tau_i)}(Z(s)) ds} \right| \\
&\leq \left| q_{r(\tau_i)r(\tau_{i+1})}(\widetilde{Z}(\tau_{i+1})) \left[e^{-\int_{\tau_i}^{\tau_{i+1}} q_{r(\tau_i)}(\widetilde{Z}(s)) ds} - e^{-\int_{\tau_i}^{\tau_{i+1}} q_{r(\tau_i)}(Z(s)) ds} \right] \right| \\
&\quad + \left| \left[q_{r(\tau_i)r(\tau_{i+1})}(\widetilde{Z}(\tau_{i+1})) - q_{r(\tau_i)r(\tau_{i+1})}(Z(\tau_{i+1})) \right] e^{-\int_{\tau_i}^{\tau_{i+1}} q_{r(\tau_i)}(Z(s)) ds} \right| \\
&\leq K \int_{\tau_i}^{\tau_{i+1}} \left| q_{r(\tau_i)}(\widetilde{Z}(s)) - q_{r(\tau_i)}(Z(s)) \right| ds + K \left| \widetilde{Z}(\tau_{i+1}) - Z(\tau_{i+1}) \right| \\
&\leq K(T+1) \sup\left\{ \left| \widetilde{Z}(s) - Z(s) \right| : 0 \leq s \leq T \right\}.
\end{aligned}
$$

Hence it follows that

$$
\begin{aligned}
&\mathbf{E} \left| p_T(\widetilde{Z}(\cdot), r(\cdot)) - p_T(Z(\cdot), r(\cdot)) \right| \\
&\leq \mathbf{E}\left[\sum_{n=0}^{\infty} \exp\{(m_0 - 1)T\} (n+1) K^n I_{\{n(T)=n\}} \right. \\
&\qquad \left. \times K(T+1) \sup\left\{ \left| \widetilde{Z}(s) - Z(s) \right| : 0 \leq s \leq T \right\} \right] \\
&\leq \sum_{n=0}^{\infty} \exp\{(m_0 - 1)T\} (T+1)(n+1) K^{n+1} \\
&\qquad \times \mathbf{E}^{1/2}\left[I_{\{n(T)=n\}} \right] \mathbf{E}^{1/2}\left[\left| \sup\left\{ \left| \widetilde{Z}(s) - Z(s) \right| : 0 \leq s \leq T \right\} \right|^2 \right].
\end{aligned}
$$

As noted in the proof of Lemma 2.16, $n(T)$ is a Poisson process with rate $m_0 - 1$. Hence

$$
\mathbf{E}\left[I_{\{n(T)=n\}} \right] = \mathbf{P}\{n(T) = n\} = \exp\{-(m_0 - 1)T\} \frac{[(m_0 - 1)T]^n}{n!}.
$$

Also, because the generator of $r(t)$ of the auxiliary process $(Z(t), r(t))$ is a constant matrix, by virtue of Lemma 2.14, we have

$$
\mathbf{E}\left[\left|\sup\left\{\left|\widetilde{Z}(s) - Z(s)\right| : 0 \le s \le T\right\}\right|^2\right]
$$
$$
= \mathbf{E}\left[\sup\left\{\left|\widetilde{Z}(s) - Z(s)\right|^2 : 0 \le s \le T\right\}\right]
$$
$$
\le K\left|\widetilde{x} - x\right|^2.
$$

Therefore, it follows that

$$
\mathbf{E}\left|p_T(\widetilde{Z}(\cdot), r(\cdot)) - p_T(Z(\cdot), r(\cdot))\right| \le \sum_{n=0}^{\infty} K\left|\widetilde{x} - x\right| \frac{(n+1)K^{n/2}}{\sqrt{n!}}
$$
$$
\le K\left|\widetilde{x} - x\right|,
$$

where in the above, we used the fact that

$$
\sum_{n=0}^{\infty} \frac{(n+1)K^{n/2}}{\sqrt{n!}} < \infty,
$$

and as before, K is a generic constant independent of $\widetilde{x}, x, \widetilde{x} - x$, or t. This completes the proof of the lemma.   □

With the preparations above, we prove the Feller property for the switching diffusion process.

Theorem 2.18. *Let $(X^{x,\alpha}(t), \alpha^{x,\alpha}(t))$ be the solution to the system given by (2.2)–(2.3) with $(X(0), \alpha(0)) = (x, \alpha)$. Assume the conditions of Theorem 2.1 hold. Then for any bounded and continuous function $g(\cdot, \cdot) : \mathbb{R}^r \times \mathcal{M} \to \mathbb{R}$, the function $u(x, \alpha) = \mathbf{E}_{x,\alpha} g(X(t), \alpha(t)) = \mathbf{E} g(X^{x,\alpha}(t), \alpha^{x,\alpha}(t))$ is continuous with respect to x.*

Proof. We prove the theorem in two steps. The first step deals with the special case when Assumption (A2.1) is true. The general case is treated in Step 2.

Step 1. Assume (A2.1). Then for any fixed $t > 0$, we have

$$
\mathbf{E}\left[g(X^{x,\alpha}(t), \alpha^{x,\alpha}(t))\right] = \mathbf{E}\left[g(Z^{x,\alpha}(t), r^\alpha(t))p_t(Z^{x,\alpha}(\cdot), r^\alpha(\cdot))\right].
$$

Let $\{x_n\}$ be a sequence of points converging to x. Then Lemma 2.14 implies that

$$
\mathbf{E}\left|Z^{x_n,\alpha}(t) - Z^{x,\alpha}(t)\right|^2 \to 0 \quad \text{as} \quad n \to \infty.
$$

Thus there is a subsequence $\{y_n\}$ of $\{x_n\}$ such that

$$
Z^{y_n,\alpha}(t) \to Z^{x,\alpha}(t) \quad \text{a.s. as} \quad n \to \infty.
$$

By virtue of Lemma 2.17,

$$\mathbf{E}\,|p_t(Z^{y_n,\alpha}(\cdot),r^\alpha(\cdot)) - p_t(Z^{x,\alpha}(\cdot),r^\alpha(\cdot))| \to 0 \quad \text{as} \quad n \to \infty.$$

Hence there is a subsequence $\{z_n\}$ of $\{y_n\}$ such that

$$p_t(Z^{z_n,\alpha}(\cdot),r^\alpha(\cdot)) \to p_t(Z^{x,\alpha}(\cdot),r^\alpha(\cdot)) \quad \text{a.s. as} \quad n \to \infty.$$

Now g is bounded and continuous, it follows from Lemma 2.16 and the dominated convergence theorem that

$$
\begin{aligned}
u(x,\alpha) &= \mathbf{E}\,[g(X^{x,\alpha}(t),\alpha^{x,\alpha}(t))] \\
&= \mathbf{E}\,[g(Z^{x,\alpha}(t),r^\alpha(t))p_t(Z^{x,\alpha}(\cdot),r^\alpha(\cdot))] \\
&= \mathbf{E}\,\Big[\lim_{n\to\infty} g(Z^{z_n,\alpha}(t),r^\alpha(t))p_t(Z^{z_n,\alpha}(\cdot),r^\alpha(\cdot))\Big] \\
&= \lim_{n\to\infty} \mathbf{E}\,[g(Z^{z_n,\alpha}(t),r^\alpha(t))p_t(Z^{z_n,\alpha}(\cdot),r^\alpha(\cdot))] \\
&= \lim_{n\to\infty} \mathbf{E}\,[g(X^{z_n,\alpha}(t),\alpha^{z_n,\alpha}(t))] \\
&= \lim_{n\to\infty} u(y_n,\alpha).
\end{aligned}
$$

Therefore, every sequence $\{x_n\}$ converging to x has a subsequence $\{z_n\}$ such that $u(x,\alpha) \le \liminf_{n\to\infty} u(z_n,\alpha)$. It can be shown that the set $\{x \in \mathbb{R}^r : u(x,\alpha) > \beta\}$ is open for any real number β and each $\alpha \in \mathcal{M}$. That is, $u(\cdot,\alpha)$ is lower semi-continuous for each $\alpha \in \mathcal{M}$ (see [143, p. 37]). Applying the above argument to $-u(x,\alpha) = \mathbf{E}[-g(X^{x,\alpha}(t),\alpha^{x,\alpha}(t))]$, we obtain that

$$u(x,\alpha) \ge \limsup_{n\to\infty} u(z_n,\alpha),$$

and hence $u(\cdot,\alpha)$ is upper semi-continuous for each $\alpha \in \mathcal{M}$. Therefore, $u(x,\alpha)$ is continuous with respect to x.

Step 2. Fix any $t \ge 0$. We construct an N-truncated process in light of the approach in [102]. Let N be any positive integer, and define an N-truncation process $X^N(t)$ so that $X^N(t) = X(t)$ up until the first exit from the N-ball $B(0,N) = \{x \in \mathbb{R}^r : |x| < N\}$. Associated with the N-truncated process $X^N(t)$, we construct an auxiliary operator as follows. For $N = 1,2,\ldots$, let $\phi^N(x)$ be a C^∞ function with range $[0,1]$ satisfying

$$\phi^N(x) = \begin{cases} 1, & \text{if } |x| \le N, \\ 0, & \text{if } |x| \ge N+1. \end{cases}$$

Now for $j,k = 1,2,\ldots,n$ and $(x,i) \in \mathbb{R}^r \times \mathcal{M}$, define

$$
\begin{aligned}
a_{jk}^N(x,i) &:= a_{jk}(x,i)\phi^N(x), \\
b_j^N(x,i) &:= b_j(x,i)\phi^N(x).
\end{aligned}
$$

For any $\varphi(\cdot, i) \in C^2, i \in \mathcal{M}$, define the operator $\mathcal{L}^N$ as

$$
\begin{aligned}
\mathcal{L}^N \varphi(x, i) = \frac{1}{2} \sum_{j,k=1}^{n} a_{jk}^N(x, i) \frac{\partial^2}{\partial x_j \partial x_k} \varphi(x, i) \\
+ \sum_{j=1}^{n} b_j^N(x, i) \frac{\partial}{\partial x_j} \varphi(x, i) + \sum_{j=1}^{m_0} q_{ij}(x) \varphi(x, j).
\end{aligned}
\tag{2.42}
$$

Denote by $\mathbf{P}_{x,i}^N$ the probability measure for which the associated martingale problem [153] has operator $\mathcal{L}^N$ with coefficients $a_{jk}^N(x, i)$, $b_j^N(x, i)$, and $q_{ij}(x)$, and denote by $\mathbf{E}_{x,i}^N$ the corresponding expectation. Then by Step 1, $(X^N(t), \alpha(t))$ is Feller.

As in (2.14), let β_N be the first exit time from the ball $B(0, N)$. Then by virtue of the strong uniqueness result in [172], the probabilities $\mathbf{P}_{x,i}$ and $\mathbf{P}_{x,i}^N$ agree until the moment when the continuous component reaches the boundary $|x| = N$. Hence it follows that for any bounded and Borel measurable function $f(\cdot, \cdot) : \mathbb{R}^r \times \mathcal{M} \mapsto \mathbb{R}$, we have

$$
\mathbf{E}_{x,i} \left[f(X(t), \alpha(t)) I_{\{\beta_N > t\}} \right] = \mathbf{E}_{x,i}^N \left[f(X(t), \alpha(t)) I_{\{\beta_N > t\}} \right].
\tag{2.43}
$$

(Alternatively, one can obtain (2.43) by showing that the solution to the martingale problem with operator $\mathcal{L}$ is unique in the weak sense. The weak uniqueness can be established by using the characteristic function as in the proof of [176, Lemma 7.18].)

Fix $(x_0, \alpha) \in \mathbb{R}^r \times \mathcal{M}$. By virtue of (2.19) and (2.43), it follows that for any $t > 0$ and $\varepsilon > 0$, there exists an $N \in \mathbb{N}$ sufficiently large and a bounded neighborhood $N(x_0)$ of x_0 ($N(x_0) = \{x : |x - x_0| \le N\}$) such that for any $(x, i) \in N(x_0) \times \mathcal{M}$, we have

$$
\begin{aligned}
\left| \mathbf{E}_{x,i} g(X(t), \alpha(t)) - \mathbf{E}_{x,i}^N g(X(t), \alpha(t)) \right| \\
= \left| \mathbf{E}_{x,i} \left[g(X(t), \alpha(t)) I_{\{\beta_N \le t\}} \right] - \mathbf{E}_{x,i}^N \left[g(X(t), \alpha(t)) I_{\{\beta_N \le t\}} \right] \right| \\
\le \|g\| \left[\mathbf{P}_{x,i} \{\beta_N \le t\} + \mathbf{P}_{x,i}^N \{\beta_N \le t\} \right] \\
= 2 \|g\| \left[1 - \mathbf{P}_{x,i}^N \{\beta_N > t\} \right] \\
= 2 \|g\| \left[1 - \mathbf{P}_{x,i} \{\beta_N > t\} \right] < \varepsilon/2,
\end{aligned}
\tag{2.44}
$$

where $\|\cdot\|$ is the essential sup-norm.

Now let $\{x_n\} \subset N(x_0)$ be any sequence converging to x_0. Then it follows from (2.44) and Step 1 that for N sufficiently large,

$$
\begin{aligned}
\left| \mathbf{E} \left[g(X^{x_n,\alpha}(t), \alpha^{x_n,\alpha}(t)) \right] - \mathbf{E} \left[g(X^{x_0,\alpha}(t), \alpha^{x_0,\alpha}(t)) \right] \right| \\
\leq \left| \mathbf{E} \left[g(X^{x_n,\alpha}(t), \alpha^{x_n,\alpha}(t)) \right] - \mathbf{E}^N \left[g(X^{x_n,\alpha}(t), \alpha^{x_n,\alpha}(t)) \right] \right| \\
+ \left| \mathbf{E}^N \left[g(X^{x_n,\alpha}(t), \alpha^{x_n,\alpha}(t)) \right] - \mathbf{E}^N \left[g(X^{x_0,\alpha}(t), \alpha^{x_0,\alpha}(t)) \right] \right| \\
+ \left| \mathbf{E}^N \left[g(X^{x_0,\alpha}(t), \alpha^{x_0,\alpha}(t)) \right] - \mathbf{E} \left[g(X^{x_0,\alpha}(t), \alpha^{x_0,\alpha}(t)) \right] \right| \\
\leq \varepsilon + \left| \mathbf{E}^N \left[g(X^{x_n,\alpha}(t), \alpha^{x_n,\alpha}(t)) \right] - \mathbf{E}^N \left[g(X^{x_0,\alpha}(t), \alpha^{x_0,\alpha}(t)) \right] \right| \\
\to \varepsilon + 0 = \varepsilon.
\end{aligned}
$$

This proves that as $n \to \infty$

$$
\begin{aligned}
u(x_n, \alpha) &= \mathbf{E} \left[g(X^{x_n,\alpha}(t), \alpha^{x_n,\alpha}(t)) \right] \\
&\to \mathbf{E} \left[g(X^{x_0,\alpha}(t), \alpha^{x_0,\alpha}(t)) \right] = u(x_0, \alpha),
\end{aligned}
$$

as desired. $\qquad\qquad\square$

Corollary 2.19. *Assume that the conditions of Theorem 2.18 hold. Then the process $(X(t), \alpha(t))$ is strong Markov.*

Proof. This claim follows from [47, Theorem 2.2.4], Lemma 2.4, and Theorem 2.18. $\qquad\qquad\square$

In Step 2 of the proof of Theorem 2.18, we used a truncation device and defined $X^N(t)$. As a digression, using such a truncation process, we obtain the following result as a by-product. It enables us to extend the existence and uniqueness to more general functions satisfying only local linear growth and Lipschitz conditions.

Proposition 2.20. *Under the conditions of Theorem 2.7, equation (2.2) together with (2.3) has a unique solution a.s.*

Proof. By virtue of Theorem 2.7, $(X(t), \alpha(t))$ is regular and

$$
\mathbf{P}(\beta = \infty) = 1. \tag{2.45}
$$

Define the N-truncation process $X^N(t)$ as in Step 2 of the proof of Theorem 2.18. By virtue of Theorem 2.1, and the local linear growth and Lipschitz conditions, $X^N(t)$ is the unique solution of (2.2)–(2.3) for $t < \beta_N$. The regularity implies that $X^N(t)$ is a solution of (2.2)–(2.3) for all $t \geq 0$. Since $X^N(t)$ coincides with $X(t)$ for $t < \beta_N$,

$$
\mathbf{P} \left(\sup_{0 \leq t < \beta_N} |X^N(t) - X(t)| > 0 \right) = 0.
$$

Using (2.45) and letting $N \to \infty$, the desired result follows. $\qquad\qquad\square$

2.6 Strong Feller Property

Assume the following conditions hold throughout the section.

(A2.2) For α and $i \in \mathcal{M}$ and $j, k = 1, \ldots, r$, the coefficients $b_j(x, i)$, $\sigma_{jk}(x, i)$, and $q_{i\alpha}(x)$ are Hölder continuous with exponent γ ($0 < \gamma \leq 1$).

(A2.3) The matrix-valued $Q(x)$ is irreducible for each $x \in \mathbb{R}^r$.

(A2.4) For each $i \in \mathcal{M}$, $a(x, i) = (a_{jk}(x, i))$ is symmetric and satisfies

$$\langle a(x, i)\xi, \xi \rangle \geq \kappa |\xi|^2, \quad \text{for all } \xi \in \mathbb{R}^r, \tag{2.46}$$

with some positive constant $\kappa \in \mathbb{R}$ for all $x \in \mathbb{R}^r$.

Lemma 2.21. *Assume that $f(\cdot, i) \in C_b(\mathbb{R}^r)$ for $i \in \mathcal{M}$. If for each $i \in \mathcal{M}$, the function $u(\cdot, \cdot, i) \in C^{1,2}(\mathbb{R} \times \mathbb{R}^r)$ (the class of functions whose derivatives with respect to t are continuous and partial derivatives with respect to x up to the second order are continuous) is bounded and satisfies*

$$\begin{cases} \dfrac{\partial u}{\partial t} = \mathcal{L}u, & t > 0, \ (x, i) \in \mathbb{R}^r \times \mathcal{M}, \\ u(0, x, i) = f(x, i), & (x, i) \in \mathbb{R}^r \times \mathcal{M}. \end{cases} \tag{2.47}$$

Then $u(t, x, i) = \mathbf{E}_{x,i} f(X(t), \alpha(t))$.

Proof. For fixed $t > 0$ and $(x, i) \in \mathbb{R}^r \times \mathcal{M}$, we apply generalized Itô's lemma (2.7) (see also [150]) to the function $u(t - r, X(r), \alpha(r))$, $r \leq t$, to obtain

$$\begin{aligned} &u(t - r \wedge \beta_N, X(r \wedge \beta_N), \alpha(r \wedge \beta_N)) \\ &= u(t, x, i) + \int_0^{r \wedge \beta_N} \left[\frac{\partial}{\partial s} + \mathcal{L} \right] u(t - s, X(s), \alpha(s)) ds \\ &+ M_1(r \wedge \beta_N) + M_2(r \wedge \beta_N), \end{aligned} \tag{2.48}$$

where $(x, i) = (X(0), \alpha(0))$, $N \in \mathbb{N}$ satisfying $N > |x|$, β_N is the first exit time from the bounded ball $B(0, N)$ as defined in (2.14), and

$$M_1(r \wedge \beta_N) = \int_0^{r \wedge \beta_N} \langle \nabla u(t - s, X(s), \alpha(s)), \sigma(X(s), \alpha(s)) dw(s) \rangle.$$

$$\begin{aligned} M_2(r \wedge \beta_N) = \int_0^{r \wedge \beta_N} \int_{\mathbb{R}} &[u(t - s, X(s), \alpha(0) + h(X(s), \alpha(s), z)) \\ &- u(t - s, X(s), \alpha(s))] \, \mu(ds, dz), \end{aligned}$$

with $\mu(ds, dz) = \mathfrak{p}(ds, dz) - ds \times m(dz)$ being a martingale measure. By virtue of Proposition 2.3, the linear growth condition of σ, using the standard argument, we can show that $M_1(r \wedge \beta_N)$ is a mean zero martingale.

The boundedness of u immediately implies that $M_2(r \wedge \beta_N)$ is also a martingale with zero mean. Therefore by taking expectations on both sides of (2.48) and noting (2.47), we obtain

$$u(t, x, i) = \mathbf{E}_{x,i} u(t - r \wedge \beta_N, X(r \wedge \beta_N), \alpha(r \wedge \beta_N)).$$

Recall that $\beta_N \to \infty$ a.s. as $N \to \infty$. Thus it follows from the bounded convergence theorem that $u(t, x, i) = \mathbf{E}_{x,i} u(t-r, X(r), \alpha(r))$. Finally by taking $r = t$ and noting that by virtue of (2.47), $u(0, X(t), \alpha(t)) = f(X(t), \alpha(t))$, we have $u(t, x, i) = \mathbf{E}_{x,i} f(X(t), \alpha(t))$.   □

Lemma 2.22. *Assume in addition to (A2.2)–(A2.4) that for $i, \ell \in \mathcal{M}$ and $j, k = 1, 2, \ldots, r$, the coefficients $a_{jk}(x, i)$, $b_j(x, i)$, and $q_{i\ell}(x)$ are bounded. Then the process $(X(t), \alpha(t))$ is strong Feller.*

Proof. By virtue of [39, Theorem 2.1], the backward equation

$$\frac{\partial u}{\partial t} = \mathcal{L}u, \quad t > 0$$

has a unique fundamental solution $p(x, i, t, y, j)$ (as a function of t, x, i), which is positive, jointly continuous in t, x, and y, and satisfies

$$\left| D_x^\theta p(x, i, t, y, j) \right| \le Ct^{(-r+|\theta|)/2} \exp\left\{ \frac{-c|y - x|^2}{t} \right\}, \tag{2.49}$$

where C and c are positive constants not depending on x, y, or t, and $\theta = (\theta_1, \theta_2, \ldots, \theta_r)$ is a multi-index with $|\theta| := \theta_1 + \theta_2 + \cdots + \theta_r \le 2$, and

$$D_x^\theta = \frac{\partial^\theta}{\partial x^\theta} := \frac{\partial^{|\theta|}}{\partial x_1^{\theta_1} \partial x_2^{\theta_2} \cdots \partial x_r^{\theta_r}}.$$

To see that p is the transition probability density of the process $(X(t), \alpha(t))$, consider an arbitrary bounded and continuous function $\phi(x, i)$, $(x, i) \in \mathbb{R}^r \times \mathcal{M}$ and define

$$\Phi(t, x, i) := \sum_{j=1}^{m_0} \int_{\mathbb{R}^r} p(x, i, t, y, j)\phi(y, j)dy.$$

Then Φ satisfies (2.47). Moreover, by virtue of (2.49) and the boundedness of ϕ, Φ is also bounded. Thus it follows from Lemma 2.21 that $\Phi(t, x, i) = \mathbf{E}_{x,i}\phi(X(t), \alpha(t))$ or

$$\mathbf{E}_{x,i}\phi(X(t), \alpha(t)) = \sum_{j=1}^{m_0} \int_{\mathbb{R}^r} p(x, i, t, y, j)\phi(y, j)dy.$$

Hence the fundamental solution p is the transition probability density of $(X^{x,i}(t), \alpha^{x,i}(t))$. Then for each $\iota \in \mathcal{M}$ and for any bounded and measurable function $f(x, i)$, the function

$$x \mapsto \mathbf{E}_{x,i} f(X(t), \alpha(t)) = \sum_{j=1}^{m_0} \int_{\mathbb{R}^r} f(y, j) p(x, i, t, y, j) dy$$

is continuous by the dominated convergence theorem.   □

Using the argument in [38, Theorem 13.1], we obtain the following lemma.

Lemma 2.23. *Assume the process* $(X(t), \alpha(t))$ *is strong Feller. Denote* $U := D \times J \subset \mathbb{R}^r \times \mathcal{M}$, *where* D *is a nonempty bounded open subset of* $\mathbb{R}^r$. *Then for any* $t > 0$ *and every bounded real Borel measurable function* $f(\cdot, \cdot)$ *on* U, *the functions*

$$F(x, i) := \mathbf{E}_{x,i} \left[I_{\{\tau_U > t\}} f(X(t), \alpha(t)) \right],$$

$$G(x, i) := \mathbf{E}_{x,i} \left[f(X(t \wedge \tau_U), \alpha(t \wedge \tau_U)) \right],$$

are continuous in U, *where* τ_U *is the first exit time from* U.

Now we are ready to present the main result of this section.

Theorem 2.24. *Assume* (A2.2)–(A2.4) *hold. Then the process* $(X(t), \alpha(t))$ *possesses the strong Feller property.*

Proof. As in the proof of Theorem 2.18, we denote by $\mathbf{P}_{x,i}^N$ the probability measure for which the associated martingale problem [153] has operator $\mathcal{L}^N$ defined in (2.42) and denote by $\mathbf{E}_{x,i}^N$ the corresponding expectation. Then by Lemma 2.22, $\mathbf{P}_{x,i}^N$ is strong Feller. As in (2.14), let β_N be the first exit time from the ball $B(0, N) = \{x \in \mathbb{R}^r : |x| < N\}$. Then as argued in the proof of Theorem 2.18, for any bounded and Borel measurable function $f(\cdot, \cdot) \in \mathbb{R}^r \times \mathcal{M}$ and $(x, i) \in \mathbb{R}^r \times \mathcal{M}$, we have

$$\mathbf{E}_{x,i} \left[f(X(t), \alpha(t)) I_{\{\beta_N > t\}} \right] = \mathbf{E}_{x,i}^N \left[f(X(t), \alpha(t)) I_{\{\beta_N > t\}} \right].$$

It follows that

$$\left| \mathbf{E}_{x,i} f(X(t), \alpha(t)) - \mathbf{E}_{x,i}^N f(X(t), \alpha(t)) \right|$$

$$= \left| \mathbf{E}_{x,i} \left[f(X(t), \alpha(t)) I_{\{\beta_N \le t\}} \right] - \mathbf{E}_{x,i}^N \left[f(X(t), \alpha(t)) I_{\{\beta_N \le t\}} \right] \right|$$

$$= \left| \int_\Omega f(X(t), \alpha(t)) I_{\{\beta_N \le t\}} \left[\mathbf{P}_{x,i}(d\omega) - \mathbf{P}_{x,i}^N(d\omega) \right] \right| \qquad (2.50)$$

$$\le \|f\| \left[\mathbf{P}_{x,i} \{\beta_N \le t\} + \mathbf{P}_{x,i}^N \{\beta_N \le t\} \right]$$

$$= 2 \|f\| \left[1 - \mathbf{P}_{x,i}^N \{\beta_N > t\} \right],$$

where $\|\cdot\|$ is the essential sup norm. Fix some $(x_0, i) \in \mathbb{R}^r \times \mathcal{M}$. Because the process $(X(t), \alpha(t))$ is regular, by (2.15), $\beta_N \to \infty$ a.s. $\mathbf{P}_{x_0,i}$ as $N \to \infty$. Therefore for any positive number $\varepsilon > 0$, we can choose some N sufficiently large so that

$$1 - \mathbf{P}_{x_0,i}^N \{\beta_N > t\} = 1 - \mathbf{P}_{x_0,i} \{\beta_N > t\} < \frac{\varepsilon}{12 \|f\|}. \qquad (2.51)$$

Also by Lemma 2.23, the function $x \mapsto \mathbf{P}_{x,i}^N \{\beta_N > t\}$ is continuous. Hence there exists some $\delta_1 > 0$ such that whenever $|x - x_0| < \delta_1$ we have

$$\begin{aligned}
1 - \mathbf{P}_{x,i}^N \{\beta_N > t\} &\leq 1 - \mathbf{P}_{x_0,i}^N \{\beta_N > t\} \\
&\quad + \left| \mathbf{P}_{x_0,i}^N \{\beta_N > t\} - \mathbf{P}_{x,i}^N \{\beta_N > t\} \right| \qquad (2.52) \\
&< \frac{\varepsilon}{6 \|f\|}.
\end{aligned}$$

Note also that by virtue of Lemma 2.22, there exists some $\delta_2 > 0$ such that

$$\left| \mathbf{E}_{x,i}^N f(X(t), \alpha(t)) - \mathbf{E}_{x_0,i}^N f(X(t), \alpha(t)) \right| < \varepsilon/3, \quad \text{if } |x - x_0| < \delta_2. \quad (2.53)$$

Finally, it follows from (2.50)–(2.53) that whenever $|x - x_0| < \delta$, where $\delta = \min\{\delta_1, \delta_2\}$, we have

$$\begin{aligned}
&\left| \mathbf{E}_{x,i} f(X(t), \alpha(t)) - \mathbf{E}_{x_0,i} f(X(t), \alpha(t)) \right| \\
&\leq \left| \mathbf{E}_{x,i} f(X(t), \alpha(t)) - \mathbf{E}_{x,i}^N f(X(t), \alpha(t)) \right| \\
&\quad + \left| \mathbf{E}_{x_0,i} f(X(t), \alpha(t)) - \mathbf{E}_{x_0,i}^N f(X(t), \alpha(t)) \right| \\
&\quad + \left| \mathbf{E}_{x,i}^N f(X(t), \alpha(t)) - \mathbf{E}_{x_0,i}^N f(X(t), \alpha(t)) \right| \\
&< \varepsilon.
\end{aligned}$$

This concludes the proof of the theorem. $\square$

A slight modification of the standard argument in [38, Theorem 13.5] leads to the following theorem. We omit the detailed proof here.

Theorem 2.25. *Assume* (A2.2)–(A2.4). *Let* $U := D \times J \subset \mathbb{R}^r \times \mathcal{M}$, *where* $D \subset \mathbb{R}^r$ *is nonempty, open, and bounded with sufficiently smooth boundary* ∂D. *Then the process* $(X(\cdot \wedge \tau_U), \alpha(\cdot \wedge \tau_U))$ *possesses the Feller property, where* τ_U *is the first exit time from* U.

We end this section with a brief discussion of strong Feller processes and $\mathcal{L}$-harmonic functions. More properties of $\mathcal{L}$-harmonic functions are treated in Chapter 3. As in [52], for any $U = D \times J$, where $D \subset \mathbb{R}^r$ is a nonempty domain, and $J \subset \mathcal{M}$, a Borel measurable function $u : U \mapsto \mathbb{R}$ is said to be $\mathcal{L}$-*harmonic* in U if u is bounded in compact subsets of U and that for all

$(x, i) \in U$ and any $V = \widetilde{D} \times \widetilde{J}$ with $\widetilde{D} \subset\subset D$ being a neighborhood of x and $i \in \widetilde{J} \subset J$, we have

$$u(x, i) = \mathbf{E}_{x,i} u(X(\tau_V), \alpha(\tau_V)),$$

where τ_V denotes the first exit time of the process $(X(t), \alpha(t))$ from V, and $\widetilde{D} \subset\subset D$ means $\widetilde{D} \subset \overline{\widetilde{D}} \subset D$ and $\overline{\widetilde{D}}$ is compact, with $\overline{\widetilde{D}} = \widetilde{D} \cup \partial\widetilde{D}$ denoting the closure of D. It is now obvious that if the process $(X(t), \alpha(t))$ is strong Feller, then any $\mathcal{L}$-harmonic function is continuous. Therefore we have

Proposition 2.26. *Assume* (A2.2)–(A2.4). *Then any $\mathcal{L}$-harmonic function is continuous in its domain.*

2.7 Continuous and Smooth Dependence on the Initial Data x

When one deals with a continuous-time dynamic system modeled by an ordinary differential equation together with appropriate initial data, the well-posedness is crucial. The well-posedness appears in ordinary differential equations and partial differential equations, together with initial and/or boundary data. They are time-honored phenomena, which naturally carry over to stochastic differential equations as well as stochastic differential equations with random switching. A problem for the associated switching diffusion is well posed if there is a unique solution for the initial value problem and the solution continuously depends on the initial data.

Continuous dependence on the initial data can be obtained using the Feller property by choosing the function $u(\cdot)$ appropriately. In the subsequent development, we also need to use smooth dependence on the initial data, which is even more difficult to obtain. In this section, we devote our attention to this smoothness property. Again we need to use the notion of multi-index. Recall that a vector $\beta = (\beta_1, \ldots, \beta_r)$ with nonnegative integer components is referred to as a multi-index. Put

$$|\beta| = \beta_1 + \cdots + \beta_r,$$

and define D_x^β as

$$D_x^\beta = \frac{\partial^\beta}{\partial x^\beta} = \frac{\partial^{|\beta|}}{\partial x_1^{\beta_1} \cdots \partial x_r^{\beta_r}}.$$

The main result of this section is the following theorem.

Theorem 2.27. *Denote by $(X^{x,\alpha}(t), \alpha^{x,\alpha}(t))$ the solution to the system given by (2.2) and (2.3). Assume that the conditions of Theorem 2.1 are satisfied and that for each $i \in \mathcal{M}$, $b(\cdot, i)$ and $\sigma(\cdot, i)$ have continuous partial derivatives with respect to the variable x up to the second order and that*

$$\left| D_x^\beta b(x, i) \right| + \left| D_x^\beta \sigma(x, i) \right| \le K(1 + |x|^\gamma), \tag{2.54}$$

where K and γ are positive constants and β is a multi-index with $|\beta| \leq 2$. Then $X^{x,\alpha}(t)$ is twice continuously differentiable in mean square with respect to x.

For ease of presentation, we prove Theorem 2.27 for the case when $X(t)$ is one-dimensional. The multidimensional case can be handled similarly. To proceed, we need to introduce a few more notations. Let $\Delta \neq 0$ be small and denote $\widetilde{x} = x + \Delta$. As in Lemma 2.14, let $(X(t), \alpha(t))$ be the switching diffusion process satisfying (2.2) and (2.3) with initial condition (x, α) and $(\widetilde{X}(t), \widetilde{\alpha}(t))$ be the process starting from $(\widetilde{x}, \alpha)$ (i.e., $(X(0), \alpha(0)) = (x, \alpha)$) and $(\widetilde{X}(0), \widetilde{\alpha}(0)) = (\widetilde{x}, \alpha)$, resp.).

Fix any $T > 0$ and let $0 < t < T$. Put

$$Z^{\Delta}(t) = Z^{x,\Delta,\alpha}(t) := \frac{\widetilde{X}(t) - X(t)}{\Delta}. \tag{2.55}$$

Then we have

$$
\begin{aligned}
Z^{\Delta}(t) &= 1 + \frac{1}{\Delta} \int_0^t [b(\widetilde{X}(s), \widetilde{\alpha}(s)) - b(X(s), \alpha(s))]ds \\
&\quad + \frac{1}{\Delta} \int_0^t [\sigma(\widetilde{X}(s), \widetilde{\alpha}(s)) - b(X(s), \alpha(s))]dw(s), \\
&= 1 + \phi^{\Delta}(t) + \frac{1}{\Delta} \int_0^t [b(\widetilde{X}(s), \alpha(s)) - b(X(s), \alpha(s))]ds \\
&\quad + \frac{1}{\Delta} \int_0^t [\sigma(\widetilde{X}(s), \alpha(s)) - \sigma(X(s), \alpha(s))]dw(s),
\end{aligned}
\tag{2.56}
$$

where

$$
\begin{aligned}
\phi^{\Delta}(t) &= \frac{1}{\Delta} \int_0^t [b(\widetilde{X}(s), \widetilde{\alpha}(s)) - b(\widetilde{X}(s), \alpha(s))]ds \\
&\quad + \frac{1}{\Delta} \int_0^t [\sigma(\widetilde{X}(s), \widetilde{\alpha}(s)) - \sigma(\widetilde{X}(s), \alpha(s))]dw(s).
\end{aligned}
$$

To proceed, we first prove a lemma.

Lemma 2.28. *Under the conditions of Theorem 2.27,*

$$\lim_{\Delta \to 0} \mathbf{E} \sup_{0 \leq t \leq T} |\phi^{\Delta}(t)|^2 = 0.$$

Proof. By virtue of Hölder's inequality and Doob's martingale inequality (A.19),

$$
\begin{aligned}
\mathbf{E} \sup_{0 \leq t \leq T} |\phi^{\Delta}(t)|^2 &\leq \frac{2T}{\Delta^2} \mathbf{E} \int_0^T |b(\widetilde{X}(s), \widetilde{\alpha}(s)) - b(\widetilde{X}(s), \alpha(s))|^2 ds \\
&\quad + \frac{8}{\Delta^2} \mathbf{E} \left| \int_0^T [\sigma(\widetilde{X}(s), \widetilde{\alpha}(s)) - \sigma(\widetilde{X}(s), \alpha(s))]dw(s) \right|^2.
\end{aligned}
$$

We treat each of the terms above separately. Choose $\eta = \Delta^{\gamma_0}$ with $\gamma_0 > 2$ and partition the interval $[0,T]$ by η. We obtain

$$
\mathbf{E} \int_0^T |b(\widetilde{X}(s),\widetilde{\alpha}(s)) - b(\widetilde{X}(s),\alpha(s))|^2 ds
$$

$$
= \mathbf{E} \sum_{k=0}^{\lfloor T/\eta\rfloor-1} \int_{k\eta}^{k\eta+\eta} |b(\widetilde{X}(s),\widetilde{\alpha}(s)) - b(\widetilde{X}(s),\alpha(s))|^2 ds
$$

$$
\leq \sum_{k=0}^{\lfloor T/\eta\rfloor-1} \Big[K\mathbf{E} \int_{k\eta}^{k\eta+\eta} |b(\widetilde{X}(s),\widetilde{\alpha}(s)) - b(\widetilde{X}(\eta k),\widetilde{\alpha}(s))|^2 ds \qquad (2.57)
$$

$$
+ K\mathbf{E} \int_{k\eta}^{k\eta+\eta} |b(\widetilde{X}(\eta k),\widetilde{\alpha}(s)) - b(\widetilde{X}(\eta k),\alpha(s))|^2 ds
$$

$$
+ K\mathbf{E} \int_{k\eta}^{k\eta+\eta} |b(\widetilde{X}(\eta k),\alpha(s)) - b(\widetilde{X}(s),\alpha(s))|^2 ds \Big].
$$

Note that the constant K in (2.57) does not depend on $k = 0,1,\ldots,\lfloor T/\eta\rfloor$ or η. The exact value of K may be different in each occurrence. We use this convention throughout the rest of the section.

By Lipschitz continuity and the tightness type of estimate (2.26), we obtain

$$
\mathbf{E} \int_{k\eta}^{k\eta+\eta} |b(\widetilde{X}(s),\widetilde{\alpha}(s)) - b(\widetilde{X}(\eta k),\widetilde{\alpha}(s))|^2 ds
$$

$$
\leq K \int_{k\eta}^{k\eta+\eta} \mathbf{E}|\widetilde{X}(s) - \widetilde{X}(\eta k)|^2 ds \qquad (2.58)
$$

$$
\leq K \int_{k\eta}^{k\eta+\eta} (s - \eta k)ds \leq K\eta^2.
$$

Likewise, we can deal with the term on the last line of (2.57), and obtain

$$
\mathbf{E} \int_{k\eta}^{k\eta+\eta} |b(\widetilde{X}(\eta k),\alpha(\eta k)) - b(\widetilde{X}(s),\alpha(s))|^2 ds \leq K\eta^2. \qquad (2.59)
$$

To treat the term on the next to the last line of (2.57), note that for $k = 0,1,\ldots,\lfloor T/\eta\rfloor - 1$,

$$
\mathbf{E} \int_{k\eta}^{k\eta+\eta} |b(\widetilde{X}(\eta k),\widetilde{\alpha}(s)) - b(\widetilde{X}(\eta k),\alpha(s))|^2 ds
$$

$$
\leq K\mathbf{E} \int_{k\eta}^{k\eta+\eta} |b(\widetilde{X}(\eta k),\widetilde{\alpha}(s)) - b(\widetilde{X}(\eta k),\widetilde{\alpha}(\eta k))|^2 ds \qquad (2.60)
$$

$$
+ K\mathbf{E} \int_{k\eta}^{k\eta+\eta} |b(\widetilde{X}(\eta k),\widetilde{\alpha}(\eta k)) - b(\widetilde{X}(\eta k),\alpha(s))|^2 ds.
$$

For the term on the second line of (2.60) and $k = 0, 1, \ldots, \lfloor T/\eta \rfloor - 1$,

$$
\begin{aligned}
&\mathbf{E} \int_{k\eta}^{k\eta+\eta} |b(\widetilde{X}(\eta k), \widetilde{\alpha}(s)) - b(\widetilde{X}(\eta k), \widetilde{\alpha}(\eta k))|^2 ds \\
&= \mathbf{E} \int_{k\eta}^{k\eta+\eta} |b(\widetilde{X}(\eta k), \widetilde{\alpha}(s)) - b(\widetilde{X}(\eta k), \widetilde{\alpha}(\eta k))|^2 I_{\{\widetilde{\alpha}(s) \neq \widetilde{\alpha}(\eta k)\}} ds \\
&= \mathbf{E} \sum_{i \in \mathcal{M}} \sum_{j \neq i} \int_{k\eta}^{k\eta+\eta} |b(\widetilde{X}(\eta k), i) - b(\widetilde{X}(\eta k), j)|^2 I_{\{\widetilde{\alpha}(s)=j\}} I_{\{\widetilde{\alpha}(\eta k)=i\}} ds \\
&\leq K\mathbf{E} \sum_{i \in \mathcal{M}} \sum_{j \neq i} \int_{k\eta}^{k\eta+\eta} [1 + |\widetilde{X}(\eta k)|^2] I_{\{\widetilde{\alpha}(\eta k)=i\}} \\
&\quad \times \mathbf{E}[I_{\{\widetilde{\alpha}(s)=j\}} | \widetilde{X}(\eta k), \widetilde{\alpha}(\eta k) = i] ds \\
&\leq K\mathbf{E} \sum_{i \in \mathcal{M}} \int_{k\eta}^{k\eta+\eta} [1 + |\widetilde{X}(\eta k)|^2] I_{\{\widetilde{\alpha}(\eta k)=i\}} \\
&\quad \times \Big[\sum_{j \neq i} q_{ij}(\widetilde{X}(\eta k))(s - \eta k) + o(s - \eta k) \Big] ds \\
&\leq K \int_{k\eta}^{k\eta+\eta} O(\eta) ds \leq K\eta^2.
\end{aligned}
$$

In the above, we used Proposition 2.3 and the boundedness of $Q(x)$.
Next, we show that for $k = 1, \ldots, \lfloor T/\eta \rfloor - 1$,

$$
\mathbf{E} \int_{k\eta}^{k\eta+\eta} |b(\widetilde{X}(\eta k), \widetilde{\alpha}(\eta k)) - b(\widetilde{X}(\eta k), \alpha(s))|^2 ds \leq K\eta^2. \tag{2.61}
$$

To do so, we use the technique of basic coupling of Markov processes (see e.g., Chen [23, p. 11]). For $x, \widetilde{x} \in \mathbb{R}^r$, and $i, j \in \mathcal{M}$, consider the measure $\Lambda((x,j),(\widetilde{x},i)) = |x - \widetilde{x}| + d(j,i)$, where $d(j,i) = 0$ if $j = i$ and $d(j,i) = 1$ if $j \neq i$. That is, $\Lambda(\cdot, \cdot)$ is a measure obtained by piecing the usual Euclidean length of two vectors and the discrete measure together. Let $(\alpha(t), \widetilde{\alpha}(t))$ be a discrete random process with a finite state space $\mathcal{M} \times \mathcal{M}$ such that

$$
\begin{aligned}
&\mathbf{P}\Big[(\alpha(t+h), \widetilde{\alpha}(t+h)) = (j,i) \big| (\alpha(t), \widetilde{\alpha}(t)) = (k,l), (X(t), \widetilde{X}(t)) = (x, \widetilde{x})\Big] \\
&= \begin{cases} \widetilde{q}_{(k,l)(j,i)}(x, \widetilde{x})h + o(h), & \text{if } (k,l) \neq (j,i), \\ 1 + \widetilde{q}_{(k,l)(k,l)}(x, \widetilde{x})h + o(h), & \text{if } (k,l) = (j,i), \end{cases}
\end{aligned}
\tag{2.62}
$$

where $h \to 0$, and the matrix $(\widetilde{q}_{(k,l)(j,i)}(x, \widetilde{x}))$ is the basic coupling of matrices $Q(x) = (q_{kl}(x))$ and $Q(\widetilde{x}) = (q_{kl}(\widetilde{x}))$ satisfying

$$
\begin{aligned}
\widetilde{Q}(x,\widetilde{x})\widetilde{f}(k,l) &= \sum_{(j,i)\in \mathcal{M}\times\mathcal{M}} q_{(k,l)(j,i)}(x,\widetilde{x})(\widetilde{f}(j,i) - \widetilde{f}(k,l)) \\
&= \sum_{j}(q_{kj}(x) - q_{lj}(\widetilde{x}))^{+}(\widetilde{f}(j,l) - \widetilde{f}(k,l)) \\
&\quad + \sum_{j}(q_{lj}(\widetilde{x}) - q_{kj}(x))^{+}(\widetilde{f}(k,j) - \widetilde{f}(k,l)) \\
&\quad + \sum_{j}(q_{kj}(x) \wedge q_{lj}(\widetilde{x}))(\widetilde{f}(j,j) - \widetilde{f}(k,l)),
\end{aligned} \tag{2.63}
$$

for any function $\widetilde{f}(\cdot,\cdot)$ defined on $\mathcal{M}\times\mathcal{M}$. Note that for $s\in[\eta k,\eta k+\eta)$, $\widetilde{\alpha}(s)$ can be written as

$$
\widetilde{\alpha}(s) = \sum_{l\in\mathcal{M}} l I_{\{\widetilde{\alpha}(s)=l\}}.
$$

Owing to the coupling defined above and noting the transition probabilities (2.62), for $i_1, i, j, l \in \mathcal{M}$ with $j\neq i$ and $s\in[\eta k,\eta k+\eta)$, we have

$$
\begin{aligned}
&\mathbf{E}[I_{\{\alpha(s)=j\}}\big|\alpha(\eta k) = i_1, \widetilde{\alpha}(\eta k) = i, X(\eta k) = x, \widetilde{X}(\eta k) = \widetilde{x}] \\
&= \sum_{l\in\mathcal{M}}\mathbf{E}[I_{\{\alpha(s)=j\}}I_{\{\widetilde{\alpha}(s)=l\}}\big|\alpha(\eta k) = i_1, \widetilde{\alpha}(\eta k) = i, X(\eta k) = x, \widetilde{X}(\eta k) = \widetilde{x}] \\
&= \sum_{l\in\mathcal{M}}\widetilde{q}_{(i_1,i)(j,l)}(x,\widetilde{x})(s-\eta k) + o(s-\eta k) = O(\eta).
\end{aligned}
$$

$$\tag{2.64}$$

By virtue of (2.64), we obtain

$$
\begin{aligned}
&\mathbf{E}\int_{k\eta}^{k\eta+\eta}|b(\widetilde{X}(\eta k),\widetilde{\alpha}(\eta k)) - b(\widetilde{X}(\eta k),\alpha(s))|^2\,ds \\
&= \mathbf{E}\int_{k\eta}^{k\eta+\eta}|b(\widetilde{X}(\eta k),\alpha(s)) - b(\widetilde{X}(\eta k),\widetilde{\alpha}(\eta k))|^2 I_{\{\alpha(s)\neq\widetilde{\alpha}(\eta k)\}}\,ds \\
&= \mathbf{E}\sum_{i\in\mathcal{M}}\sum_{j\neq i}\int_{k\eta}^{k\eta+\eta}|b(\widetilde{X}(\eta k),i) - b(\widetilde{X}(\eta k),j))|^2 I_{\{\alpha(s)=j\}}I_{\{\widetilde{\alpha}(\eta k)=i\}}\,ds \\
&\leq K\mathbf{E}\sum_{i,i_1\in\mathcal{M}}\sum_{j\neq i}\int_{k\eta}^{k\eta+\eta}[1 + |\widetilde{X}(\eta k)|^2]I_{\{\widetilde{\alpha}(\eta k)=i,\alpha(\eta k)=i_1\}} \\
&\quad \times \mathbf{E}[\alpha(s) = j\big|\alpha(\eta k) = i_1, \widetilde{\alpha}(\eta k) = i, X(\eta k) = x, \widetilde{X}(\eta k) = \widetilde{x}]\,ds \\
&= O(\eta^2).
\end{aligned}
$$

Using the assumption $\widetilde{\alpha}(0) = \alpha(0) = \alpha$ and noting $\widetilde{X}(0) = \widetilde{x}$, we obtain

$$
\begin{aligned}
\mathbf{E} &\int_0^\eta |b(\widetilde{X}(0), \widetilde{\alpha}(0)) - b(\widetilde{X}(0), \alpha(s))|^2 ds \\
&= \mathbf{E} \int_0^\eta |b(\widetilde{x}, \alpha(0)) - b(\widetilde{x}, \alpha(s))|^2 ds \\
&= \mathbf{E} \int_0^\eta \sum_{j \neq \alpha} |b(\widetilde{x}, \alpha) - b(\widetilde{x}, j)|^2 I_{\{\alpha(s)=j\}} ds \\
&= \int_0^\eta \sum_{j \neq \alpha} |b(\widetilde{x}, \alpha) - b(\widetilde{x}, j)|^2 \Big[q_{\alpha j}(\widetilde{x})s + o(s) \Big] ds \leq K\eta^2.
\end{aligned}
\tag{2.65}
$$

Thus, it follows that for $k = 0, 1, \ldots, \lfloor T/\eta \rfloor - 1$,

$$
\mathbf{E} \int_{k\eta}^{k\eta+\eta} |b(\widetilde{X}(\eta k), \widetilde{\alpha}(s)) - b(\widetilde{X}(\eta k), \alpha(s))|^2 ds \leq K\eta^2.
\tag{2.66}
$$

Using the estimates (2.58), (2.59), and (2.66) in (2.57), we obtain

$$
\begin{aligned}
\mathbf{E} &\int_0^T |b(\widetilde{X}(s), \widetilde{\alpha}(s)) - b(\widetilde{X}(s), \alpha(s))|^2 ds \\
&\leq \sum_{k=0}^{\lfloor T/\eta \rfloor - 1} K\eta^2 \leq K\eta.
\end{aligned}
\tag{2.67}
$$

Likewise, we obtain

$$
\mathbf{E} \left| \int_0^T [\sigma(\widetilde{X}(s), \widetilde{\alpha}(s)) - \sigma(\widetilde{X}(s), \alpha(s))] dw(s) \right|^2 \leq K\eta.
\tag{2.68}
$$

Putting (2.67) and (2.68) into $\phi^\Delta(t)$, and noting $\gamma_0 > 2$, we obtain

$$
\mathbf{E} \sup_{0 \leq t \leq T} |\phi^\Delta(t)|^2 \leq K \frac{\eta}{\Delta^2} = K\Delta^{\gamma_0 - 2} \to 0 \quad \text{as} \quad \Delta \to 0.
\tag{2.69}
$$

The lemma is proved.   □

Remark 2.29. In deriving (2.65), the condition $\alpha(0) = \widetilde{\alpha}(0) = \alpha$ is used crucially. If the initial data are not the same, there will be a contribution of a nonzero term resulting in difficulties in obtaining the differentiability.

Proposition 2.30. *Using the conditions of Lemma 2.14 except $Q(x)$ being allowed to be x-dependent, the conclusion of Lemma 2.14 continues to hold.*

Proof. As before, let $(X(t), \alpha(t))$ denote the switching diffusion process satisfying (2.2) and (2.3) with initial condition (x, α) and $(\widetilde{X}(t), \widetilde{\alpha}(t))$ be the process starting from $(\widetilde{x}, \alpha)$ (i.e., $(X(0), \alpha(0)) = (x, \alpha)$) and $(\widetilde{X}(0), \widetilde{\alpha}(0)) =$

$(\widetilde{x}, \alpha)$ respectively). Let $T > 0$ be fixed and denote $\Delta = \widetilde{x} - x$. Then we have $\widetilde{X}(t) - X(t) = \Delta + A(t) + B(t)$, and hence

$$\sup_{t \in [0,T]} \left| \widetilde{X}(t) - X(t) \right|^2 \leq 3\Delta^2 + 3 \sup_{t \in [0,T]} |A(t)|^2 + 3 \sup_{t \in [0,T]} |B(t)|^2,$$

where

$$
\begin{aligned}
A(t) := & \int_0^t [b(\widetilde{X}(s), \widetilde{\alpha}(s)) - b(\widetilde{X}(s), \alpha(s))] ds \\
& + \int_0^t [\sigma(\widetilde{X}(s), \widetilde{\alpha}(s)) - b(\widetilde{X}(s), \alpha(s))] dw(s) = \Delta \phi^\Delta(t),
\end{aligned}
$$

and

$$
\begin{aligned}
B(t) := & \int_0^t [b(\widetilde{X}(s), \alpha(s)) - b(X(s), \alpha(s))] ds \\
& + \int_0^t [\sigma(\widetilde{X}(s), \alpha(s)) - b(X(s), \alpha(s))] dw(s).
\end{aligned}
$$

It follows from (2.69) that

$$\mathbf{E} \sup_{t \in [0,T]} |A(t)|^2 \leq K \Delta^2 \Delta^{\gamma_0 - 2} = K \Delta^{\gamma_0} = o(\Delta^2).$$

Meanwhile, we can apply the conclusion of Lemma 2.14 to obtain that

$$\mathbf{E} \sup_{t \in [0,T]} |B(t)|^2 \leq K \Delta^2.$$

Therefore, we have

$$\mathbf{E} \sup_{t \in [0,T]} \left| \widetilde{X}(t) - X(t) \right|^2 \leq K \Delta^2 + o(\Delta^2) \leq K |\widetilde{x} - x|^2.$$

This finishes the proof of the proposition.   □

A direct consequence of Proposition 2.30 is the mean square continuity of the solution of the switching diffusion with respect to x; that is, for any $T > 0$,

$$\lim_{y \to x} \mathbf{E} |X^{y,\alpha}(t) - X^{x,\alpha}(t)|^2 = 0, \quad \text{for each } \alpha \in \mathcal{M}, \text{ and } t \in [0, T].$$

That is, the continuous dependence on the initial data x is obtained. We state this fact below.

Corollary 2.31. *Assume the conditions of Theorem 2.27. Then $X^{x,\alpha}(t)$ is continuous in mean square with respect to x.*

Proof of Theorem 2.27. With Lemma 2.28 and Proposition 2.30 at our hands, we proceed to prove Theorem 2.27. Because $b(\cdot, j)$ is twice continuously differentiable with respect to x, we can write

$$\frac{1}{\Delta} \int_0^t [b(\widetilde{X}(s), \alpha(s)) - b(X(s), \alpha(s))]ds$$

$$= \frac{1}{\Delta} \int_0^t \int_0^1 \frac{d}{dv} b(X(s) + v(\widetilde{X}(s) - X(s)), \alpha(s))dvds$$

$$= \int_0^t \left[\int_0^1 b_x(X(s) + v(\widetilde{X}(s) - X(s)), \alpha(s))dv \right] Z^{\Delta}(s)ds,$$

where $Z^{\Delta}(t)$ is defined in (2.55) and $b_x(\cdot)$ denotes the partial derivative of $b(\cdot)$ with respect to x (i.e., $b_x = (\partial/\partial x)b$). It follows from Lemma 2.30 that for any $s \in [0, T]$,

$$\widetilde{X}(s) - X(s) \to 0$$

in probability as $\Delta \to 0$. This implies that

$$\int_0^1 b_x(X(s) + v(\widetilde{X}(s) - X(s)), \alpha(s))dv \to b_x(X(s), \alpha(s)) \qquad (2.70)$$

in probability as $\Delta \to 0$. Similarly, we have

$$\frac{1}{\Delta} \int_0^t [\sigma(\widetilde{X}(s), \alpha(s)) - \sigma(X(s), \alpha(s))]dw(s)$$

$$= \int_0^t \left[\int_0^1 \sigma_x(X(s) + v(\widetilde{X}(s) - X(s)), \alpha(s))dv \right] Z^{\Delta}(s)dw(s)$$

and

$$\int_0^1 \sigma_x(X(s) + v(\widetilde{X}(s) - X(s)), \alpha(s))dv \to \sigma_x(X(s), \alpha(s)) \qquad (2.71)$$

in probability as $\Delta \to 0$. Let $\zeta(t) := \zeta^{x,\alpha}(t)$ be the solution of

$$\zeta(t) = 1 + \int_0^t b_x(X(s), \alpha(s))\zeta(s)ds + \int_0^t \sigma_x(X(s), \alpha(s))\zeta(s)dw(s), \quad (2.72)$$

where b_x and σ_x denote the partial derivatives of b and σ with respect to x, respectively. Then (2.56), (2.69)–(2.71), and [47, Theorem 5.5.2] imply that

$$\mathbf{E}\left|Z^{\Delta}(t) - \zeta(t)\right|^2 \to 0 \quad \text{as} \quad \Delta \to 0 \qquad (2.73)$$

and $\zeta(t) = \zeta^{x,\alpha}(t)$ is mean square continuous with respect to x. Therefore, $(\partial/\partial x)X(t)$ exists in the mean square sense and $(\partial/\partial x)X(t) = \zeta(t)$.

Likewise, we can show that $(\partial^2/\partial x^2)X^{x,\alpha}(t)$ exists in the mean square sense and is mean square continuous with respect to x. The proof of the theorem is thus concluded.   □

Corollary 2.32. *Under the assumptions of Theorem 2.27, the mean square derivatives $(\partial/\partial x_j)X^{x,\alpha}(t)$ and $(\partial^2/\partial x_j\partial x_k)X^{x,\alpha}(t)$, $j,k = 1,\ldots,n$, are mean square continuous with respect to t.*

Proof. As in the proof of Theorem 2.27, we consider only the case when $X(t)$ is real valued. Also, we use the same notations here as those in the proof of Theorem 2.27. To see that $\zeta(t)$ is continuous in the mean square sense, we first observe that for any $t \in [0,T]$,

$$\mathbf{E}\,|\zeta(t)|^2 \leq 2\mathbf{E}\,|\zeta(t) - Z^{\Delta}(t)|^2 + 2\mathbf{E}\,|Z^{\Delta}(t)|^2 .$$

It follows from (2.69), the Lipschitz condition, and Lemma 2.30 that

$$
\begin{aligned}
\mathbf{E}\,|Z^{\Delta}(t)|^2 \leq{}& 3\mathbf{E}\,|\phi^{\Delta}(t)|^2 + 3\mathbf{E}\left|\frac{1}{\Delta}\int_0^t [b(\widetilde{X}(u),\alpha(u)) - b(X(u),\alpha(u))]du\right|^2 \\
&+ 3\mathbf{E}\left|\frac{1}{\Delta}\int_0^t [\sigma(\widetilde{X}(u),\alpha(u)) - \sigma(X(u),\alpha(u))]dw(u)\right|^2 \\
\leq{}& K + 3t\frac{1}{|\Delta|^2}\mathbf{E}\int_0^t \left|b(\widetilde{X}(u),\alpha(u)) - b(X(u),\alpha(u))\right|^2 du \\
&+ 3\mathbf{E}\frac{1}{|\Delta|^2}\int_0^t \left|\sigma(\widetilde{X}(u),\alpha(u)) - \sigma(X(u),\alpha(u))\right|^2 du \\
\leq{}& K + 3K(T+1)\frac{1}{|\Delta|^2}\mathbf{E}\int_0^t \left|\widetilde{X}(u) - X(u)\right|^2 du \\
\leq{}& C = C(x,T,K).
\end{aligned}
$$

Hence we have from (2.73) that

$$\sup_{t\in[0,T]} \mathbf{E}\,|\zeta(t)|^2 \leq C = C(x,T,K) < \infty. \tag{2.74}$$

Thus $\zeta(t)$ is mean square continuous if we can show that

$$\mathbf{E}\,|\zeta(t) - \zeta(s)|^2 \to 0 \quad \text{as} \quad |s-t| \to 0.$$

To this end, we note that for any $s,t \in [0,T]$,

$$
\begin{aligned}
\mathbf{E}\,|\zeta(t) - \zeta(s)|^2 \leq{}& 3\mathbf{E}\Big[\,|\zeta(t) - Z^{\Delta}(t)|^2 + |\zeta(s) - Z^{\Delta}(s)|^2 \\
&+ |Z^{\Delta}(t) - Z^{\Delta}(s)|^2\Big].
\end{aligned}
$$

In view of (2.73), we need only prove that

$$\mathbf{E}\,|Z^{\Delta}(t) - Z^{\Delta}(s)|^2 \to 0 \quad \text{as} \quad |s-t| \to 0.$$

Without loss of generality, assume that $s < t$. Then by (2.56), we have

$$
\begin{aligned}
&\mathbf{E} \left| Z^\Delta(t) - Z^\Delta(s) \right|^2 \\
&\leq 3\mathbf{E} \left| \phi^\Delta(t) - \phi^\Delta(s) \right|^2 \\
&\quad + 3\mathbf{E} \left| \frac{1}{\Delta} \int_s^t b(\widetilde{X}(u), \alpha(u)) - b(X(u), \alpha(u)) du \right|^2 \\
&\quad + 3\mathbf{E} \left| \frac{1}{\Delta} \int_s^t \sigma(\widetilde{X}(u), \alpha(u)) - \sigma(X(u), \alpha(u)) dw(u) \right|^2 .
\end{aligned}
\tag{2.75}
$$

It follows from the Cauchy-Schwartz inequality, the Lipschitz condition, and Lemma 2.30 that

$$
\begin{aligned}
&\mathbf{E} \left| \frac{1}{\Delta} \int_s^t b(\widetilde{X}(u), \alpha(u)) - b(X(u), \alpha(u)) du \right|^2 \\
&\leq (t-s) \frac{1}{|\Delta|^2} \mathbf{E} \int_s^t \left| b(\widetilde{X}(u), \alpha(u)) - b(X(u), \alpha(u)) \right|^2 du \\
&\leq (t-s) \frac{1}{|\Delta|^2} \mathbf{E} \int_s^t K \left| \widetilde{X}(u) - X(u) \right|^2 du \\
&\leq KC(t-s)^2,
\end{aligned}
\tag{2.76}
$$

where K is the Lipschitz constant and C is a constant independent of t, s, or Δ. Similarly, we can show that

$$
\begin{aligned}
&\mathbf{E} \left| \frac{1}{\Delta} \int_s^t \sigma(\widetilde{X}(u), \alpha(u)) - \sigma(X(u), \alpha(u)) dw(u) \right|^2 \\
&= \mathbf{E} \frac{1}{|\Delta|^2} \int_s^t \left| \sigma(\widetilde{X}(u), \alpha(u)) - \sigma(X(u), \alpha(u)) \right|^2 du \\
&\leq KC(t-s).
\end{aligned}
\tag{2.77}
$$

Next, using the same argument as that of Lemma 2.28, we can show that

$$
\mathbf{E} \left| \phi^\Delta(t) - \phi^\Delta(s) \right|^2 \leq K(t-s).
\tag{2.78}
$$

Thus it follows from (2.75)–(2.78) that

$$
\mathbf{E} \left| Z^\Delta(t) - Z^\Delta(s) \right|^2 = O(|t-s|) \to 0 \quad \text{as} \quad |t-s| \to 0
$$

and hence $\zeta(t)$ is mean square continuous with respect to t.

Likewise, we can show that $(\partial^2/\partial x^2) X^{x,\alpha}(t)$ is mean square continuous with respect to t. This concludes the proof. $\qquad\square$

2.8 A Remark Regarding Nonhomogeneous Markov Processes

Throughout the book, for notational simplicity, we have decided to mainly concern ourselves with time-homogeneous switching diffusion processes. As a consequence, the drift and diffusion coefficients are all independent of t.

The discussion throughout the book can be extended to nonhomogeneous Markov processes. The setup can be changed as follows. Suppose that $b(\cdot,\cdot,\cdot) : \mathbb{R} \times \mathbb{R}^r \times \mathcal{M} \mapsto \mathbb{R}^r$ and that $\sigma(\cdot,\cdot,\cdot) : \mathbb{R} \times \mathbb{R}^r \times \mathcal{M} \mapsto \mathbb{R}^d$. Let $w(\cdot)$ be an $\mathbb{R}^d$-valued standard Brownian motion defined in the filtered probability space $(\Omega, \mathcal{F}, \{\mathcal{F}_t\}, \mathbf{P})$, and $\alpha(\cdot)$ be a pure jump process whose generator is given by $Q(x)$ as before such that for $f(\cdot,\cdot,\cdot) : \mathbb{R} \times \mathbb{R}^r \times \mathcal{M}$,

$$Q(x)f(t,x,\cdot)(\imath) = \sum_{\jmath \in \mathcal{M}} q_{\imath\jmath}(x)(f(t,x,\jmath) - f(t,x,\imath)), \quad \text{for each } \imath \in \mathcal{M}.$$

$$(2.79)$$

In lieu of (2.2), we may consider the two-component process $(X(\cdot), \alpha(\cdot))$ satisfying

$$X(t) = X(0) + \int_0^t b(s, X(s), \alpha(s))ds + \int_0^t \sigma(s, X(s), \alpha(s))dw(s), \quad (2.80)$$

and as $\Delta \to 0$,

$$\mathbf{P}\{\alpha(t + \Delta) = \jmath | X(t) = x, \alpha(t) = \imath, X(s), \alpha(s), s \le t\}$$
$$= q_{\imath\jmath}(x)\Delta + o(\Delta), \quad \text{for } \imath \ne \jmath.$$

$$(2.81)$$

It follows that the associated operator can be defined as: For each $\iota \in \mathcal{M}$, $f(\cdot,\cdot,\iota) \in C^{1,2}$ where $C^{1,2}$ denotes the class of functions whose first-order derivative with respect to t and second-order partial derivatives with respect to x are continuous,

$$\mathcal{L}f(t,x,\iota) = \sum_{i=1}^r b_i(t,x,\iota)\frac{\partial f(t,x,\iota)}{\partial x_i}$$
$$+ \frac{1}{2}\sum_{i,j=1}^r a_{ij}(t,x,\iota)\frac{\partial^2 f(t,x,\iota)}{\partial x_i \partial x_j}$$
$$+ Q(x)f(t,x,\cdot)(\iota).$$

$$(2.82)$$

Now, the generalized Itô lemma is changed to

$$f(t, X(t), \alpha(t)) - f(0, X(0), \alpha(0))$$
$$= \int_0^t \left(\frac{\partial}{\partial s} + \mathcal{L}\right) f(s, X(s), \alpha(s))ds + M_1(t) + M_2(t),$$

$$(2.83)$$

where

$$M_1(t) = \int_0^t \langle \nabla f(s, X(s), \alpha(s)), \sigma(s, X(s), \alpha(s))dw(s)\rangle,$$
$$M_2(t) = \int_0^t \int_{\mathbb{R}} [f(s, X(s), \alpha(0) + h(X(s), \alpha(s), z))$$
$$- f(s, X(s), \alpha(s))]\mu(ds, dz).$$

In view of the generalized Itô formula, for any $f(\cdot, \cdot, \imath) \in C^{1,2}$ with $\imath \in \mathcal{M}$, and τ_1, τ_2 being bounded stopping times such that $0 \leq \tau_1 \leq \tau_2$ a.s., if $f(t, X(t), \alpha(t))$ and $\mathcal{L}f(t, X(t), \alpha(t))$ etc. are bounded on $t \in [\tau_1, \tau_2]$ with probability 1, then Dynkin's formula becomes

$$\mathbf{E}f(\tau_2, X(\tau_2), \alpha(\tau_2)) = \mathbf{E}f(\tau_1, X(\tau_1), \alpha(\tau_1)) + \mathbf{E} \int_{\tau_1}^{\tau_2} \mathcal{L}f(s, X(s), \alpha(s))ds.$$

(2.84)

Moreover, for each $f(\cdot, \cdot, \imath) \in C_b^{1,2}$ or $f(\cdot, \cdot, \imath) \in C_0^{1,2}$,

$$M_f(t) = f(t, X(t), \alpha(t)) - f(0, X(0), \alpha(0))$$
$$- \int_0^t \left(\frac{\partial}{\partial s} + \mathcal{L} \right) f(s, X(s), \alpha(s))ds$$

is a martingale. With the setup changed to nonhomogeneous Markov processes, most of the subsequent results will carry over. Nevertheless, modifications are necessary to take into account the added complexity due to nonhomogeneous processes. In order to present the main results and ideas without much notational complication, we confine ourselves to the homogeneous switching diffusions throughout the book.

2.9 Notes

The connection between generators of Markov processes and martingales is explained in Ethier and Kurtz [43]. An account of piecewise-deterministic processes is in Davis [30]. Results on basic probability theory may be found in Chow and Teicher [27]; the theory of stochastic processes can be found in Gihman and Skorohod [53], Khasminskii [83], and Liptser and Shiryayev [110], among others. More detailed discussions regarding martingales and diffusions are in Elliott [40]; an in-depth study of stochastic differential equations and diffusion processes is contained in Ikeda and Watanabe [72]. Concerning the existence of solutions to stochastic differential equations with switching, using Poisson random measures (see Skorohod [150] also Basak, Bisi, and Ghosh [6] and Mao and Yuan [120]), it can be shown that there is a unique solution for each initial condition by following the approach of Ikeda and Watanabe [72] with the appropriate use of the stopping times. However, the Picard iteration method does not work, which is further explained when we study the numerical solutions (see Chapter 5). In Section 2.7, we dealt with smoothness properties of solutions of stochastic differential equations with x-dependent switching. It is interesting to note that even the time-honored concept of well-posedness cannot be easily obtained. Once x-dependence is added, the difficulty rises considerably.

3

Recurrence

3.1 Introduction

This chapter is concerned with recurrence of switching diffusion processes. Because practical systems in applications are often in operation for a relatively long time, it is of foremost importance to understand the systems' asymptotic behavior. By asymptotic behavior, we mean the properties of the underlying processes in a neighborhood of "∞" and in a neighborhood of an equilibrium point. Properties concerning a neighborhood of ∞ are treated in this chapter, whereas stability of an equilibrium point is dealt with in Chapter 7.

Dealing with dynamic systems, one often wishes to examine if the underlying system is sensitive to perturbations. In accordance with LaSalle and Lefschetz [107], a deterministic system $\dot{x} = h(t, x)$ satisfying appropriate conditions, is Lagrange stable if the solutions are ultimately uniformly bounded. When diffusions are used, one may wish to add probability modifications such as "almost surely" or "in probability" to the aforementioned uniform boundedness. Nevertheless, for instance, if almost sure boundedness is used, it will exclude many systems due to the presence of Brownian motion. Thus as pointed out by Wonham [160], such boundedness is inappropriate; an alternative notion of stability in a certain weak sense should be used. In lieu of requiring the system to be bounded, one aims to find conditions under which the systems return to a prescribed compact region in finite time. This chapter focuses on weak-sense stability for switching diffusions. We define recurrence, positive recurrence, and null recurrence;

we also develop Liapunov-function-based criteria together with more easily verifiable conditions on the coefficients of the processes for positive recurrence as well as nonrecurrence, and null recurrence. Despite the growing interest in treating regime-switching systems, the results regarding such issues as recurrence and positive recurrence (or weak stochastic stability as coined by Wonham [160]) are still scarce. These are not simple extensions of their diffusion counterpart. Due to the coupling and interactions, elliptic systems instead of a single elliptic equation must be treated. Moreover, even the classical approaches such as Liapunov function methods and Dynkin's formula are still applicable for switching diffusions, the analysis is more delicate than the diffusion counterparts, and requires careful handling of discrete-event component $\alpha(\cdot)$.

The rest of the chapter is arranged as follows. In Section 3.2, in addition to introducing certain notations, we also provide definitions of recurrence, transience, positive recurrence, and null recurrence, as well as some preliminary results. Section 3.3 focuses on recurrence and transience. Section 3.4 proceeds with the study of positive and null recurrence. We present results of necessary and sufficient conditions for recurrence using Liapunov functions. We also consider the case under "linearization" for the continuous component. Section 3.5 is devoted to a number of examples as applications of the general results. Section 3.6, containing the proofs of several technical lemmas, is provided to facilitate reading. Discussions and further remarks are made in Section 3.7.

3.2   Formulation and Preliminaries

3.2.1   Switching Diffusion

Recall that $(\Omega, \mathcal{F}, \{\mathcal{F}_t\}_{t \geq 0}, \mathbf{P})$ is a complete probability space with a filtration $\{\mathcal{F}_t\}_{t \geq 0}$ satisfying the usual condition (i.e., it is right continuous with $\mathcal{F}_0$ containing all $\mathbf{P}$-null sets). Let $x \in \mathbb{R}^r$, $\mathcal{M} = \{1, \ldots, m_0\}$, and $Q(x) = (q_{ij}(x))$ an $m_0 \times m_0$ matrix depending on x satisfying that for any $x \in \mathbb{R}^r$, $q_{ij}(x) \geq 0$ for $i \neq j$ and $\sum_{j=1}^{m_0} q_{ij}(x) = 0$. For any twice continuously differentiable function $h(\cdot, i)$, $i \in \mathcal{M}$, define $\mathcal{L}$ by

$$
\begin{aligned}
\mathcal{L}h(x, i) &= \frac{1}{2} \sum_{j,k=1}^{r} a_{jk}(x, i) \frac{\partial^2 h(x, i)}{\partial x_j \partial x_k} + \sum_{j=1}^{r} b_j(x, i) \frac{\partial h(x, i)}{\partial x_j} + Q(x) h(x, \cdot)(i) \\
&= \frac{1}{2} \mathrm{tr}(a(x, i) \nabla^2 h(x, i)) + b'(x, i) \nabla h(x, i) + Q(x) h(x, \cdot)(i),
\end{aligned}
$$

(3.1)

where $\nabla h(\cdot, i)$ and $\nabla^2 h(\cdot, i)$ denote the gradient and Hessian of $h(\cdot, i)$, respectively, $b'(x, i) \nabla h(x, i)$ denotes the usual inner product on $\mathbb{R}^r$ with z'

being the transpose of z for $z \in \mathbb{R}^{\iota_1 \times \iota_2}$ with ι_1, $\iota_2 \geq 1$, and

$$
\begin{aligned}
Q(x)h(x,\cdot)(i) &= \sum_{j=1}^{m_0} q_{ij}(x)h(x,j) \\
&= \sum_{j \in \mathcal{M}} q_{ij}(x)(h(x,j) - h(x,i)), \quad i \in \mathcal{M}.
\end{aligned}
\tag{3.2}
$$

Consider a Markov process $Y(t) = (X(t), \alpha(t))$, whose associated operator is given by $\mathcal{L}$. Note that $Y(t)$ has two components, an r-dimensional continuous component $X(t)$ and a discrete component $\alpha(t)$ taking value in $\mathcal{M} = \{1, \ldots, m_0\}$.

Recall that the process $Y(t) = (X(t), \alpha(t))$ may be described by the following pair of equations:

$$
\begin{aligned}
dX(t) &= b(X(t), \alpha(t))dt + \sigma(X(t), \alpha(t))dw(t), \\
X(0) &= x, \quad \alpha(0) = \alpha,
\end{aligned}
\tag{3.3}
$$

and

$$
\begin{aligned}
\mathbf{P}\{\alpha(t + \Delta) &= j | \alpha(t) = i, X(s), \alpha(s), s \leq t\} \\
&= q_{ij}(X(t))\Delta + o(\Delta), \quad i \neq j,
\end{aligned}
\tag{3.4}
$$

where $w(t)$ is a d-dimensional standard Brownian motion, $b(\cdot, \cdot) : \mathbb{R}^r \times \mathcal{M} \mapsto \mathbb{R}^r$, and $\sigma(\cdot, \cdot) : \mathbb{R}^r \times \mathcal{M} \mapsto \mathbb{R}^{r \times d}$ satisfying $\sigma(x, i)\sigma'(x, i) = a(x, i)$. Note that (3.3) depicts the system dynamics and (3.4) delineates the probability structure of the jump process. Note that if $\alpha(\cdot)$ is a continuous-time Markov chain independent of the Brownian motion $w(\cdot)$ and $Q(x) = Q$ or $Q(x) = Q(t)$ (independent of x), then equation (3.3) together with the generator Q or $Q(t)$ is sufficient to characterize the underlying process. As long as there is an x-dependence, equation (3.4) is needed in delineating the dynamics of the switching diffusion.

In this chapter, our study is carried out with the use of the operator $\mathcal{L}$ given in (3.1). Throughout the chapter, we assume that both $b(\cdot, i)$ and $\sigma(\cdot, i)$ satisfy the usual local Lipschitz and linear growth conditions for each $i \in \mathcal{M}$ and that $Q(\cdot)$ is bounded and continuous. As described in Theorem 2.1, the system (3.3)–(3.4) has a unique strong solution. In what follows, denote the solution of (3.3)–(3.4) by $(X^{x,\alpha}(t), \alpha^{x,\alpha}(t))$ when we emphasize the dependence on initial data. To study recurrence and ergodicity of the process $Y(t) = (X(t), \alpha(t))$, we further assume that the following condition (A3.1) holds throughout the chapter. For convenience, we also put the boundedness and continuity of $Q(\cdot)$ in (A3.1).

(A3.1) The operator $\mathcal{L}$ satisfies the following conditions. For each $i \in \mathcal{M}$, $a(x, i) = (a_{jk}(x, i))$ is symmetric and satisfies

$$
\kappa_1 |\xi|^2 \leq \xi' a(x, i)\xi \leq \kappa_1^{-1}|\xi|^2, \quad \text{for all } \xi \in \mathbb{R}^r,
\tag{3.5}
$$

with some constant $\kappa_1 \in (0,1]$ for all $x \in \mathbb{R}^r$. $Q(\cdot) : \mathbb{R}^r \mapsto \mathbb{R}^{m_0 \times m_0}$ is a bounded and continuous function. Moreover, $Q(x)$ is irreducible for each $x \in \mathbb{R}^r$.

3.2.2 Definitions of Recurrence and Positive Recurrence

This subsection is devoted to the definitions of recurrence, positive recurrence, and null recurrence. First, we introduce the following notation and conventions. For any $D \subset \mathbb{R}^r$, $J \subset \mathcal{M}$, and $U = D \times J \subset \mathbb{R}^r \times \mathcal{M}$, denote

$$\begin{aligned} \tau_U &:= \inf\{t \geq 0 : (X(t), \alpha(t)) \notin U\}, \\ \sigma_U &:= \inf\{t \geq 0 : (X(t), \alpha(t)) \in U\}. \end{aligned} \tag{3.6}$$

In particular, if $U = D \times \mathcal{M}$ is a "cylinder," we set

$$\begin{aligned} \tau_D &:= \inf\{t \geq 0 : X(t) \notin D\}, \\ \sigma_D &:= \inf\{t \geq 0 : X(t) \in D\}. \end{aligned} \tag{3.7}$$

Definition 3.1. Recurrence and positive and null recurrence are defined as follows.

- *Recurrence and Transience.* For $U := D \times J$, where $J \subset \mathcal{M}$ and $D \subset \mathbb{R}^r$ is an open set with compact closure, let

$$\sigma_U^{x,\alpha} = \inf\{t : (X^{x,\alpha}(t), \alpha^{x,\alpha}(t)) \in U\}.$$

  A regular process $(X^{x,\alpha}(\cdot), \alpha^{x,\alpha}(\cdot))$ is *recurrent* with respect to U if $\mathbf{P}\{\sigma_U^{x,\alpha} < \infty\} = 1$ for any $(x, \alpha) \in D^c \times \mathcal{M}$, where D^c denotes the complement of D; otherwise, the process is *transient* with respect to U.

- *Positive Recurrence and Null Recurrence.* A recurrent process with finite mean recurrence time for some set $U = D \times J$, where $J \subset \mathcal{M}$ and $D \subset \mathbb{R}^r$ is a bounded open set with compact closure, is said to be *positive recurrent* with respect to U; otherwise, the process is *null recurrent* with respect to U.

3.2.3 Preparatory Results

We first prove the following theorem, which asserts that under assumption (A3.1), the process $Y(t) = (X(t), \alpha(t))$ will exit every bounded "cylinder" with finite mean exit time.

Theorem 3.2. *Let* $D \subset \mathbb{R}^r$ *be a nonempty open set with compact closure* $\overline{D}$. *Let* $\tau_D := \inf\{t \geq 0 : X(t) \notin D\}$. *Then*

$$\mathbf{E}_{x,i}\tau_D < \infty, \quad \text{for any} \quad (x,i) \in D \times \mathcal{M}. \tag{3.8}$$

Proof. First, note that from the uniform ellipticity condition (3.5), we have

$$\kappa_1 \leq a_{11}(x, i) \leq \kappa_1^{-1}, \quad \text{for any} \quad (x, i) \in D \times \mathcal{M}. \tag{3.9}$$

For each $i \in \mathcal{M}$, consider

$$W(x, i) = k - (x_1 + \beta)^c,$$

where the constants k, c (with $c \geq 2$), and β are to be specified, and $x_1 = e_1' x$ is the first component of x with $e_1 = (1, 0, \ldots, 0)'$ being the standard unit vector. Direct computation leads to

$$\mathcal{L}W(x, i) = -c(x_1 + \beta)^{c-2} \left[b_1(x, i)(x_1 + \beta) + \frac{c-1}{2} a_{11}(x, i) \right].$$

Set

$$c = \frac{2}{\kappa_1} \left(\sup_{(x,i) \in \overline{D} \times \mathcal{M}} |b_1(x, i)(x_1 + \beta)| + 1 \right) + 1.$$

Then we have from (3.9) that

$$\frac{c-1}{2} a_{11}(x, i) + b_1(x, i)(x_1 + \beta)$$
$$\geq \frac{c-1}{2} \kappa_1 - \sup_{(x,i) \in \overline{D} \times \mathcal{M}} |b_1(x, i)(x_1 + \beta)| \geq 1.$$

Meanwhile, since $x \in D \subset \overline{D}$ and $\overline{D}$ is compact, we can choose β such that $1 \leq x_1 + \beta \leq M$ for all $x \in D$, where M is some positive constant. Thus we have $(x_1 + \beta)^{c-2} \geq 1^{c-2} = 1$. Finally, we choose k large enough so that $W(x, i) = k - (x_1 + \beta)^c > 0$ for all $(x, i) \in D \times \mathcal{M}$. Therefore, $W(x, i), i \in \mathcal{M}$ are Liapunov functions satisfying

$$\mathcal{L}W(x, i) \leq -c, \quad \text{for all} \quad (x, i) \in D \times \mathcal{M}. \tag{3.10}$$

Now let $\tau_D(t) = t \wedge \tau_D := \min\{t, \tau_D\}$. Then we have from Dynkin's formula and (3.10) that

$$\mathbf{E}_{x,i} W(X(\tau_D(t)), \alpha(\tau_D(t))) - W(x, i)$$
$$= \mathbf{E}_{x,i} \int_0^{\tau_D(t)} \mathcal{L}W(X(u), \alpha(u)) du \leq -c \mathbf{E}_{x,i} \tau_D(t).$$

Because the function $W(\cdot, \cdot)$ is nonnegative, we have

$$\mathbf{E}_{x,i} \tau_D(t) \leq \frac{1}{c} W(x, i). \tag{3.11}$$

Because

$$\mathbf{E}_{x,i} \tau_D(t) = \mathbf{E}_{x,i} \tau_D I_{\{\tau_D \leq t\}} + \mathbf{E}_{x,i} t I_{\{\tau_D > t\}},$$

we have from (3.11) that

$$tP_{x,i}[\tau_D > t] \leq \frac{1}{c}W(x,i).$$

Letting $t \to \infty$, we obtain

$$\mathbf{P}_{x,i}[\tau_D = \infty] = 0 \quad \text{or} \quad \mathbf{P}_{x,i}[\tau_D < \infty] = 1.$$

This yields that $\tau_D(t) \to \tau_D$ almost surely (a.s.) $\mathbf{P}_{x,i}$ as $t \to \infty$. Now applying Fatou's lemma, as $t \to \infty$, we obtain

$$\mathbf{E}_{x,i}\tau_D \leq \frac{1}{c}W(x,i) < \infty,$$

as desired.   □

Remark 3.3. A closer examination of the proof shows that the conclusion of Theorem 3.2 remains valid if we replace the uniform ellipticity condition (3.5) by a weaker condition: There exist some $\iota = 1, 2, \ldots, r$ and positive constant κ such that

$$a_{\iota\iota}(x,i) \geq \kappa \quad \text{for any} \ (x,i) \in D \times \mathcal{M}. \tag{3.12}$$

Let us recall the definition of $\mathcal{L}$-harmonic functions. For any $U = D \times J$, where $D \subset \mathbb{R}^r$ is a nonempty domain, and $J \subset \mathcal{M}$, a Borel measurable function $u : U \mapsto \mathbb{R}$ is said to be $\mathcal{L}$-*harmonic* in U if u is bounded in compact subsets of U and that for all $(x,i) \in U$ and any $V = \widetilde{D} \times \widetilde{J}$ with $\widetilde{D} \subset\subset D$ being a neighborhood of x and $i \in \widetilde{J} \subset J$, we have

$$u(x,i) = \mathbf{E}_{x,i}u(X(\tau_V), \alpha(\tau_V)),$$

where τ_V denotes the first exit time of the process $(X(t), \alpha(t))$ from V, and $\widetilde{D} \subset\subset D$ means $\widetilde{D} \subset \overline{\widetilde{D}} \subset D$ and $\overline{\widetilde{D}}$ is compact, with $\overline{\widetilde{D}} = \widetilde{D} \cup \partial\widetilde{D}$ denoting the closure of D.

Lemma 3.4. *For any* $U = D \times J \subset \mathbb{R}^r \times \mathcal{M}$, *where* $D \subset \mathbb{R}^r$ *is a nonempty domain, the functions*

$$f(x,i) = \mathbf{P}_{x,i}\{\tau_U < \infty\} \quad and \quad g(x,i) = \mathbf{E}_{x,i}\phi(X(\tau_U), \alpha(\tau_U))$$

are $\mathcal{L}$-*harmonic in* U, *where* ϕ *is any bounded and Borel measurable function on* $\partial D \times \mathcal{M}$.

Proof. Fix any $(x,i) \in U$. Consider any $V = \widetilde{D} \times \widetilde{J} \subset U$ such that $x \in \widetilde{D} \subset\subset D$ and $i \in \widetilde{J} \subset J$. Then it follows from the strong Markov property that

$$f(x,i) = \mathbf{E}_{x,i}[I_{\{\tau_U < \infty\}}] = \mathbf{E}_{x,i}\left[\mathbf{E}_{x,i}[I_{\{\tau_U < \infty\}}]\big|\mathcal{F}_{\tau_V}\right]$$
$$= \mathbf{E}_{x,i}\left[\mathbf{E}_{X(\tau_V),\alpha(\tau_V)}[I_{\{\tau_U < \infty\}}]\right] = \mathbf{E}_{x,i}f(X(\tau_V), \alpha(\tau_V)).$$

This shows that f is $\mathcal{L}$-harmonic in U. A very similar argument shows that g is also $\mathcal{L}$-harmonic in U. □

Following the well-known arguments in [38, Vol. II, Chapter 13], we obtain the following two lemmas. (Note that Lemma 3.5 was also proved in [52, Lemma 4.3], and in [24], when the operator $\mathcal{L}$ is in divergence form.)

Lemma 3.5. *Assume* (A3.1). *Let* $U = D \times \mathcal{M} \subset \mathbb{R}^r \times \mathcal{M}$ *and* $f : U \mapsto \mathbb{R}$, *where* $D \subset \mathbb{R}^r$ *is a nonempty domain. Then*

$$\mathcal{L}f(x,i) = 0 \quad \text{for any} \ (x,i) \in U \tag{3.13}$$

if and only if f *is* $\mathcal{L}$-*harmonic in* U. *Moreover, assume that* ∂D *is sufficiently smooth,* $\overline{D}$ *is compact, and* $\varphi(\cdot,i)$ *is an arbitrary continuous function on* ∂D *for any* $i \in \mathcal{M}$. *Then*

$$u(x,i) := \mathbf{E}_{x,i}\varphi(X(\tau_U), \alpha(\tau_U)) \tag{3.14}$$

is the unique solution of the differential equation (3.13) *with boundary condition*

$$\lim_{x \to x_0, x \in D} u(x,i) = \varphi(x_0, i) \quad \text{for any} \ (x_0, i) \in \partial D \times \mathcal{M}. \tag{3.15}$$

Proof. We prove the lemma in several steps.

Step 1. Assume (3.13). Let $(x,i) \in V \subset U$, V, and τ_V be as in the proof of Lemma 3.4 and $t > 0$. Then Dynkin's formula and (3.13) lead to

$$\mathbf{E}_{x,i}f(X(\tau_V \wedge t), \alpha(\tau_V \wedge t)) = f(x,i) + \mathbf{E}_{x,i} \int_0^{\tau_V \wedge t} \mathcal{L}f(X(s), \alpha(s))ds$$
$$= f(x,i).$$

Note that Theorem 3.2 implies that $\mathbf{P}_{x,i}\{\tau_V < \infty\} = 1$. Letting $t \to \infty$, we obtain by virtue of the bounded convergence theorem that

$$f(x,i) = \mathbf{E}_{x,i}f(X(\tau_V), \alpha(\tau_V)).$$

This shows that f is $\mathcal{L}$-harmonic in U.

Step 2. Assume that f is $\mathcal{L}$-harmonic in U. Note that by virtue of Proposition 2.26, f is continuous in U. Consider $V = \widetilde{D} \times \mathcal{M} \subset U$ with $\widetilde{D} \subset\subset D$ and $\partial\widetilde{D}$ sufficiently smooth. Then by virtue of [41], the boundary value problem

$$\begin{cases} \mathcal{L}\widetilde{f}(x,i) = 0, & (x,i) \in V, \\ \widetilde{f}(x,i) = f(x,i), & (x,i) \in \partial\widetilde{D} \times \mathcal{M}, \end{cases} \tag{3.16}$$

has a unique classical solution. We show that $\widetilde{f}$ agrees with f in V. In fact, for any $(x,i) \in V$, the same argument as in Step 1 shows that

$$\widetilde{f}(x,i) = \mathbf{E}_{x,i}\widetilde{f}(X(\tau_V), \alpha(\tau_V)).$$

But the boundary condition in (3.16) implies that

$$\mathbf{E}_{x,i}\widetilde{f}(X(\tau_V),\alpha(\tau_V)) = \mathbf{E}_{x,i}f(X(\tau_V),\alpha(\tau_V)).$$

Therefore the assumption that f is $\mathcal{L}$-harmonic in U further leads to

$$\widetilde{f}(x,i) = \mathbf{E}_{x,i}f(X(\tau_V),\alpha(\tau_V)) = f(x,i).$$

This shows that $f(\cdot,i) \in C^2(D)$ for each $i \in \mathcal{M}$ and that f satisfies the differential equation (3.13).

Step 3. Now assume ∂D is sufficiently smooth, $\overline{D}$ is compact, and $\varphi(\cdot,i)$ is an arbitrary continuous function on ∂D for any $i \in \mathcal{M}$. We show that the function u defined in (3.14) is the unique solution of (3.13) with boundary condition (3.15). Indeed, it follows from Lemma 3.4 that u is $\mathcal{L}$-harmonic in U. Then we have from Step 2 that u satisfies the differential equation (3.13). Finally, the boundary condition (3.15) is satisfied by the assumptions that ∂D is sufficiently smooth, $\overline{D}$ is compact, and that ϕ is continuous. This completes the proof of the lemma.   □

Using a similar argument, we can prove the following lemma.

Lemma 3.6. *Let $U = D \times \mathcal{M} \subset \mathbb{R}^r \times \mathcal{M}$, where $D \subset \mathbb{R}^r$ is a nonempty open set with compact closure. Suppose that $g(\cdot,i) \in C_b(\overline{D})$ and $f(\cdot,\cdot) : \overline{D} \times \mathcal{M} \mapsto \mathbb{R}$. Then f solves the boundary value problem*

$$\begin{cases} \mathcal{L}f(x,i) = -g(x,i), & (x,i) \in D \times \mathcal{M} \\ f(x,i) = 0, & (x,i) \in \partial D \times \mathcal{M} \end{cases}$$

if and only if

$$f(x,i) = \mathbf{E}_{x,i} \int_0^{\tau_U} g(X(t),\alpha(t))dt, \quad for \ all \ (x,i) \in D \times \mathcal{M}.$$

Using Lemmas 3.5 and 3.6, we proceed to prove that if the process $Y(t) = (X(t),\alpha(t))$ is recurrent (resp., positive recurrent) with respect to some "cylinder" $D \times \mathcal{M} \subset \mathbb{R}^r \times \mathcal{M}$, then it is recurrent (resp., positive recurrent) with respect to any "cylinder" $E \times \mathcal{M} \subset \mathbb{R}^r \times \mathcal{M}$, where D is any nonempty domain in $\mathbb{R}^r$ with compact closure. These results are proved in the following two lemmas. To preserve the flow of presentation, the proofs are postponed to Section 3.6.

Lemma 3.7. *Let $D \subset \mathbb{R}^r$ be a nonempty open set with compact closure. Suppose that*

$$\mathbf{P}_{x,i}\{\sigma_D < \infty\} = 1 \quad for \ any \ (x,i) \in D^c \times \mathcal{M}. \tag{3.17}$$

Then for any nonempty open set $E \subset \mathbb{R}^r$, we have

$$\mathbf{P}_{x,i}\{\sigma_E < \infty\} = 1 \quad for \ any \ (x,i) \in E^c \times \mathcal{M}.$$

Lemma 3.8. *Let $D \subset \mathbb{R}^r$ be a nonempty open set with compact closure. Suppose that*

$$\mathbf{E}_{x,i}\sigma_D < \infty \text{ for any } (x,i) \in D^c \times \mathcal{M}. \tag{3.18}$$

Then for any nonempty open set $E \subset \mathbb{R}^r$, we have

$$\mathbf{E}_{x,i}\sigma_E < \infty \quad \text{for any } (x,i) \in E^c \times \mathcal{M}.$$

The following lemma shows that if the process $Y(t) = (X(t), \alpha(t))$ reaches the "cylinder" $D \times \mathcal{M}$ in finite time a.s. $\mathbf{P}_{x,i}$, then it will visit the set $D \times \{\ell\}$ in finite time a.s. $\mathbf{P}_{x,i}$ for any $\ell \in \mathcal{M}$. Its proof together with the proof of Lemma 3.10 is placed in Section 3.6 as well.

Lemma 3.9. *Let $D \subset \mathbb{R}^r$ be a nonempty open set with compact closure satisfying*

$$\mathbf{P}_{y,j}\{\sigma_D < \infty\} = 1 \quad \text{for any } (y,j) \in D^c \times \mathcal{M}. \tag{3.19}$$

Then for any $(x,i) \in \mathbb{R}^r \times \mathcal{M}$,

$$\mathbf{P}_{x,i}\{\sigma_{D \times \{\ell\}} < \infty\} = 1 \quad \text{for any } \ell \in \mathcal{M}. \tag{3.20}$$

With Lemma 3.9, we can now prove that if the process $Y(t) = (X(t), \alpha(t))$ is positive recurrent with respect to some "cylinder" $D \times \mathcal{M}$, then it is positive recurrent with respect to the set $D \times \{\ell\} \subset \mathbb{R}^r \times \mathcal{M}$.

Lemma 3.10. *Let $D \subset \mathbb{R}^r$ be a nonempty open set with compact closure satisfying*

$$\mathbf{E}_{y,j}\sigma_D < \infty \quad \text{for any } (y,j) \in D^c \times \mathcal{M}. \tag{3.21}$$

Then for any $(x,i) \in \mathbb{R}^r \times \mathcal{M}$,

$$\mathbf{E}_{x,i}\sigma_{D \times \{\ell\}} < \infty \quad \text{for any } \ell \in \mathcal{M}. \tag{3.22}$$

Remark 3.11. By virtue of Lemmas 3.7–3.10, under assumption (A3.1), the process $Y(t) = (X(t), \alpha(t))$ is recurrent (resp., positive recurrent) with respect to some "cylinder" $D \times \mathcal{M}$ if and only if it is recurrent (resp., positive recurrent) with respect to the product set $D \times \{\ell\} \subset \mathbb{R}^r \times \mathcal{M}$ for any $\ell \in \mathcal{M}$. Also we have proved that the properties of recurrence and positive recurrence are independent of the choice of the set D. We summarize these in the following theorem.

Theorem 3.12. *Suppose that (A3.1) holds. Then the following assertions hold:*

- *The process $(X(t), \alpha(t))$ is recurrent (resp., positive recurrent) with respect to $D \times \mathcal{M}$ if and only if it is recurrent (resp., positive recurrent) with respect to $D \times \{\ell\}$, where $D \subset \mathbb{R}^r$ is a nonempty open set with compact closure and $\ell \in \mathcal{M}$.*

- *If the process $(X(t), \alpha(t))$ is recurrent (resp., positive recurrent) with respect to some $U = D \times \mathcal{M}$, where $D \subset \mathbb{R}^r$ is a nonempty open set with compact closure, then it is recurrent (resp., positive recurrent) with respect to any $\widetilde{U} = \widetilde{D} \times \mathcal{M}$, where $\widetilde{D} \subset \mathbb{R}^r$ is any nonempty open set.*

Remark 3.13. In view of Theorem 3.12, we make the following remarks.

- Recurrence is a property independent of the region chosen; henceforth, a g process $(X(t), \alpha(t))$ with the associated generator $\mathcal{L}$ satisfying (A3.1) is said to be *recurrent*, if it is recurrent with respect to some $U = D \times \{\ell\}$, where $D \subset \mathbb{R}^r$ is a nonempty bounded open set and $\ell \in \mathcal{M}$; otherwise it is said to be *transient*.

- Henceforth, we call a recurrent process $(X(t), \alpha(t))$ *positive recurrent* if it is positive recurrent with respect to some bounded domain $U = D \times \{\ell\} \subset \mathbb{R}^r \times \mathcal{M}$; otherwise, we have a *null recurrent* process.

3.3   Recurrence and Transience

3.3.1   Recurrence

To study the recurrence of the process $(X(t), \alpha(t))$, we first present the following criterion based upon the existence of certain Liapunov functions.

Theorem 3.14. *Assume that there exists a nonempty bounded open set $D \subset \mathbb{R}^r$ such that there exists $V(\cdot, \cdot) : D^c \times \mathcal{M} \mapsto \mathbb{R}^+$ satisfying*

$$V_n := \inf_{|x| \geq n, i \in \mathcal{M}} V(x, i) \to \infty, \quad as \ n \to \infty,$$

$$\mathcal{L}V(x, i) \leq 0, \quad for \ all \ (x, i) \in D^c \times \mathcal{M}. \tag{3.23}$$

Then the process $(X(t), \alpha(t))$ is recurrent.

Proof. Fix any $(x, \alpha) \in D^c \times \mathcal{M}$. Define

$$\sigma_D = \sigma_D^{x,\alpha} := \inf\{t \geq 0 : X^{x,\alpha}(t) \in D\}$$

and

$$\sigma_D^{(n)}(t) := \sigma_D \wedge t \wedge \beta_n,$$

where $\beta_n = \inf\{t : |X^{(n)}(t)| \geq n\}$ is as in (2.14). Then it follows from Dynkin's formula that

$$\mathbf{E}V(X(\sigma_D^{(n)}(t)), \alpha(\sigma_D^{(n)}(t))) - V(x, \alpha)$$

$$= \mathbf{E} \int_0^{\sigma_D^{(n)}(t)} \mathcal{L}V(X(u), \alpha(u)) du \leq 0.$$

Consequently,

$$\mathbf{E}V(X(\sigma_D^{(n)}(t)), \alpha(\sigma_D^{(n)}(t))) \leq V(x, \alpha).$$

Note that as $t \to \infty$, $\sigma_D \wedge t \wedge \beta_n \to \sigma_D \wedge \beta_n$ a.s. Hence Fatou's lemma implies that

$$\mathbf{E}V(X(\sigma_D \wedge \beta_n), \alpha(\sigma_D \wedge \beta_n)) \leq V(x, \alpha).$$

Then we have

$$V(x, \alpha) \geq \mathbf{E}V(X(\sigma_D \wedge \beta_n), \alpha(\sigma_D \wedge \beta_n))$$
$$\geq \mathbf{E}V(X(\beta_n), \alpha(\beta_n))I_{\{\beta_n < \sigma_D\}} \geq V_n \mathbf{P}\{\beta_n < \sigma_D\}.$$

Hence it follows from (3.23) that as $n \to \infty$,

$$\mathbf{P}\{\beta_n < \sigma_D\} \leq \frac{V(x, \alpha)}{V_n} \to 0.$$

Note that

$$\mathbf{P}\{\sigma_D = \infty\} \leq \mathbf{P}\{\beta_n < \sigma_D\}.$$

Thus we have $\mathbf{P}\{\sigma_D < \infty\} = 1$, as desired.   □

The above result is based on a Liapunov function argument. However, constructing Liapunov functions is generally difficult. It would be nice if we could place certain conditions on the coefficients of processes. The following theorem is an attempt in this direction.

Theorem 3.15. *Either one of the following conditions implies that the process $(X(t), \alpha(t))$ is recurrent.*

(i) *There exist constants $\gamma > 0$ and $c_i \in \mathbb{R}$ with $i \in \mathcal{M}$ such that for $(x, i) \in \{x \in \mathbb{R}^r : |x| \geq 1\} \times \mathcal{M}$,*

$$\frac{x'b(x, i)}{|x|^2} + \frac{\text{tr}(a(x, i))}{2|x|^2} + (\gamma - 2)\frac{x'a(x, i)x}{2|x|^4} - \frac{1}{k - \gamma c_i}\sum_{j=1}^{m_0} q_{ij}(x)c_j \leq 0,$$

(3.24)

where k is a positive constant sufficiently large so that $k - \gamma c_i > 0$ for each $i \in \mathcal{M}$.

(ii) *There exist a positive constant γ and symmetric and positive definite matrices P_i for $i \in \mathcal{M}$ such that for $(x, i) \in \{x \in \mathbb{R}^r : |x| \geq 1\} \times \mathcal{M}$,*

$$\frac{x'P_ib(x, i)}{x'P_ix} + \frac{\text{tr}(\sigma'(x, i)P_i\sigma(x, i))}{2x'P_ix}$$
$$+ (\gamma - 2)\frac{|\sigma(x, i)'P_ix|^2}{|x'P_ix|^2} + \sum_{j=1}^{m_0} q_{ij}(x)\frac{|x'P_jx|^{\gamma/2}}{|x'P_ix|^{\gamma/2}} \leq 0.$$

(3.25)

Proof. (i) For each $i \in \mathcal{M}$, define a Liapunov function as

$$V(x,i) = (k - \gamma c_i)|x|^{\gamma}.$$

Then direct computation shows that for $x \neq 0$, we have

$$V(x,i) = (k - \gamma c_i)\gamma|x|^{\gamma-2}x,$$
$$\nabla^2 V(x,i) = (k - \gamma c_i)\gamma \left[|x|^{\gamma-2}I + (\gamma - 2)|x|^{\gamma-4}xx' \right].$$

Hence it follows that

$$\mathcal{L}V(x,i) = (k - \gamma c_i)\gamma|x|^{\gamma} \left\{ \frac{x'b(x,i)}{|x|^2} + \frac{\mathrm{tr}(a(x,i))}{|x|^2} - (\gamma - 2)\frac{x'a(x,i)x}{|x|^4} \right.$$
$$\left. - \frac{1}{k - \gamma c_i} \sum_{j=1}^{m_0} q_{ij}(x)c_j \right\}.$$

Therefore Theorem 3.14 implies the desired conclusion.

Assertion (ii) can be established using a similar argument as in (i) by considering

$$W(x,i) = (x'P_i x)^{\gamma/2} \quad \text{for } (x,i) \in \{x \in \mathbb{R}^r : |x| \geq 1\} \times \mathcal{M}$$

and verifying that

$$\nabla W(x,i) = \gamma(x'P_i x)^{(\gamma-2)/2}P_i x,$$
$$\nabla^2 W(x,i) = \gamma(x'P_i x)^{(\gamma-2)/2}P_i + \gamma(\gamma-2)(x'P_i x)^{(\gamma-4)/2}P_i xx'P_i.$$

The details are omitted for brevity. $\qquad \square$

Lemma 3.16. *If there exists some $(x_0, \ell) \in \mathbb{R}^r \times \mathcal{M}$ such that for any $\varepsilon > 0$,*

$$\mathbf{P}_{x_0,\ell} \left\{ (X(t_n), \alpha(t_n)) \in B(x_0, \varepsilon) \times \{\ell\}, \text{ for a sequence } t_n \uparrow \infty \right\} = 1,$$
$$(3.26)$$

then for any $U := D \times \{j\} \subset \mathbb{R}^r \times \mathcal{M}$, where $D \subset \mathbb{R}^r$ is a nonempty bounded domain and $j \in \mathcal{M}$, we have

$$\mathbf{P}_{x_0,\ell} \{ \sigma_U < \infty \} = 1.$$

In particular, if (3.26) is true for any $(x,i) \in \mathbb{R}^r \times \mathcal{M}$, then the process $(X(t), \alpha(t))$ is recurrent.

Proof. It is enough to consider two cases, namely, when $x_0 \notin D$ and when $x_0 \in D$ with $j \neq \ell$. For the first case, the proof in Lemma 3.7 can be adopted to show that $\mathbf{P}_{x_0,\ell} \{ \sigma_D < \infty \} = 1$. Then by virtue of Lemma 3.9, it follows that $\mathbf{P}_{x_0,\ell} \{ \sigma_{D \times \{j\}} < \infty \} = 1$. The second case follows from a slight modification of the argument in the proof of Lemma 3.9. $\qquad \square$

Lemma 3.17. *If the process* $(X(t), \alpha(t))$ *is recurrent, then for any* $(x, i) \in \mathbb{R}^r \times \mathcal{M}$ *and* $\varepsilon > 0$, *we have*

$$\mathbf{P}_{x,i}\left\{(X(t_n), \alpha(t_n)) \in B(x, \varepsilon) \times \{i\}, \text{ for a sequence } t_n \uparrow \infty\right\} = 1. \quad (3.27)$$

Proof. Denote $B = B(x, \varepsilon)$, $B_1 = B(x, \varepsilon/2)$, and $B_2 = B(x, 2\varepsilon)$. Define a sequence of stopping times by

$$\eta_1 := \inf\{t \geq 0 : X(t) \notin B_2\};$$

and for $n = 1, 2, \ldots$,

$$\eta_{2n} := \inf\{t \geq \eta_{2n-1} : (X(t), \alpha(t)) \in B_1 \times \{i\}\},$$
$$\eta_{2n+1} := \inf\{t \geq \eta_{2n} : X(t) \notin B_2\}.$$

Note that the process $(X(t), \alpha(t))$ is recurrent, in particular, with respect to $B_1 \times \{i\}$. This, together with Theorem 3.2, implies that $\eta_n < \infty$ a.s. $\mathbf{P}_{x,i}$. Thus (3.27) follows. □

Combining Lemmas 3.16 and 3.17, we obtain the following theorem.

Theorem 3.18. *The process* $(X(t), \alpha(t))$ *is recurrent if and only if every point* $(x, i) \in \mathbb{R}^r \times \mathcal{M}$ *is recurrent in the sense that for any* $\varepsilon > 0$,

$$\mathbf{P}_{x,i}\left\{(X(t_n), \alpha(t_n)) \in B(x, \varepsilon) \times \{i\}, \text{ for a sequence } t_n \uparrow \infty\right\} = 1.$$

Theorem 3.18 enables us to provide another criterion for recurrence in terms of mean sojourn time. This is motivated by the results in [109]. To this end, for any $U := D \times J \subset \mathbb{R}^r \times \mathcal{M}$ and any $\lambda \geq 0$, define

$$R_\lambda(x, i, U) := \mathbf{E}_{x,i}\left[\int_0^\infty e^{-\lambda t} I_U(X(t), \alpha(t)) dt\right]. \quad (3.28)$$

In particular, $R_0(x, i, U)$ denotes the mean sojourn time of the process $(X^{x,i}(t), \alpha^{x,i}(t))$ in the domain U. We state a proposition below, whose proof, being similar to a result in [109], is relegated to Section 3.6.

Proposition 3.19. *Assume* (A3.1). *If for some point* $(x_0, \ell) \in \mathbb{R}^r \times \mathcal{M}$ *and any* $\rho > 0$,

$$R_0(x_0, \ell, B(x_0, \rho) \times \{\ell\}) = \infty, \quad (3.29)$$

then (x_0, ℓ) *is a recurrent point; that is, for every* $\rho > 0$,

$$\mathbf{P}_{x_0,\ell}\left\{(X(t_n), \alpha(t_n)) \in B(x_0, \rho) \times \{\ell\}, \text{ for a sequence } t_n \uparrow \infty\right\} = 1. \quad (3.30)$$

In particular, if (3.29) *is true for any* $(x, i) \in \mathbb{R}^r \times \mathcal{M}$, *then the process* $(X(t), \alpha(t))$ *is recurrent.*

3.3.2 Transience

We first argue that if the process $(X(t), \alpha(t))$ is transient, then the norm of the continuous component $|X(t)| \to \infty$ a.s. as $t \to \infty$, and vice versa. Then we provide two criteria for transience in terms of mean sojourn time and Liapunov functions, respectively. If the coefficients $b(x, i)$ and $\sigma(x, i)$ for $i \in \mathcal{M}$ are linearizable in x, we also obtain easily verifiable conditions for transience.

Theorem 3.20. *The process $(X(t), \alpha(t))$ is transient if and only if*

$$\lim_{t \to \infty} |X(t)| = \infty \quad a.s. \quad \mathbf{P}_{x,\alpha} \quad for \ any \ (x, \alpha) \in \mathbb{R}^r \times \mathcal{M}.$$

The proof of Theorem 3.20 follows the argument of Bhattacharya [15]. We next obtain a criterion for transience under certain conditions. To preserve the flow of the presentation, the proofs of both Theorem 3.20 and Proposition 3.21 are placed in Section 3.6.

Proposition 3.21. *Assume that the following conditions hold.*

- *For $i = 1, 2, \ldots, m_0$, the coefficients $b(\cdot, i)$, $\sigma(\cdot, i)$, and $Q(\cdot)$ are Hölder continuous with exponent $0 < \gamma \leq 1$.*

- *$Q(x)$ is irreducible for each $x \in \mathbb{R}^r$.*

- *For each $i \in \mathcal{M}$, $a(x, i) = \sigma(x, i)\sigma'(x, i)$ is symmetric and satisfies*

$$\langle a(x, i)\xi, \xi \rangle \geq \kappa |\xi|^2, \quad for \ all \ \ \xi \in \mathbb{R}^r, \tag{3.31}$$

with some positive constant $\kappa \in \mathbb{R}$ for all $x \in \mathbb{R}^r$.

If for some $U := D \times J \subset \mathbb{R}^r \times \mathcal{M}$ containing the point (x_0, ℓ), where D is a nonempty open and bounded set, $R_0(x_0, \ell, U) < \infty$, then the process $(X(t), \alpha(t))$ is transient.

Remark 3.22. It follows from Propositions 3.19 and 3.21 that the process $(X(t), \alpha(t))$ is recurrent if and only if for every $(x, i) \in \mathbb{R}^r \times \mathcal{M}$ and every $\rho > 0$, we have

$$R_0(x, i, B(x, \rho) \times \{i\}) = \infty.$$

Next we obtain a sufficient condition for transience in terms of the existence of a Liapunov function.

Theorem 3.23. *Assume there exists a nonempty bounded domain $D \subset \mathbb{R}^r$ and a function $V(\cdot, \cdot) : D^c \times \mathcal{M} \mapsto \mathbb{R}$ satisfying*

$$\begin{aligned}
&\sup_{(x,i) \in \partial D \times \mathcal{M}} V(x, i) \leq 0, \\
&\mathcal{L}V(x, i) \geq 0 \ \ for \ any \ \ (x, i) \in D^c \times \mathcal{M}, \\
&\sup_{(x,i) \in D^c \times \mathcal{M}} V(x, i) \leq M < \infty, \\
&V(y, \ell) > 0 \ \ for \ some \ \ (y, \ell) \in D^c \times \mathcal{M}.
\end{aligned} \tag{3.32}$$

Then the process $(X(t), \alpha(t))$ is either transient or not regular.

Proof. Assuming the process $(X(t), \alpha(t))$ is regular, we need to show that it is transient. Fix $(y, \ell) \in D^c \times \mathcal{M}$ with $V(y, \ell) > 0$. Define the stopping times $\sigma_D = \sigma_D^{y,\ell}$, $\beta_n = \beta_n^{y,\ell}$, and $\beta_n \wedge \sigma_D \wedge t$ as in the proof of Theorem 3.14, with $n > (n_0 \vee |y|)$, where n_0 is an integer such that $D \subset \{x : |x| < n_0\}$ and $\beta_n^{y,\ell} = \inf\{t : |X^{y,\ell}(t)| = n\}$. By virtue of Dynkin's formula and (3.32),

$$\mathbf{E} V(X(\beta_n \wedge \sigma_D \wedge t), \alpha(\beta_n \wedge \sigma_D \wedge t)) - V(y, \ell)$$
$$= \mathbf{E} \int_0^{\beta_n \wedge \sigma_D \wedge t} \mathcal{L} V(X(u), \alpha(u)) du \geq 0.$$

Therefore, we have from (3.32) that

$$V(y, \ell) \leq \mathbf{E} V(X(\beta_n \wedge \sigma_D \wedge t), \alpha(\beta_n \wedge \sigma_D \wedge t))$$
$$= \mathbf{E} V(X(\sigma_D), \alpha(\sigma_D)) I_{\{\sigma_D \leq \beta_n \wedge t\}}$$
$$+ \mathbf{E} V(X(\beta_n \wedge t), \alpha(\beta_n \wedge t)) I_{\{\sigma_D > \beta_n \wedge t\}}$$
$$\leq M \mathbf{P} \{\sigma_D > \beta_n \wedge t\}.$$

Let $A_n := \{\omega \in \Omega : \sigma_D(\omega) > \beta_n(\omega) \wedge t\}$. Then

$$\mathbf{P}(A_n) \geq \frac{V(y, \ell)}{M}, \quad \text{for any } n \geq n_0.$$

Note that $\beta_n \leq \beta_{n+1}$ implies $A_n \supset A_{n+1}$. This, together with the regularity yields that

$$\bigcap_{n=n_0}^{\infty} A_n = \lim_{n \to \infty} A_n = \{\sigma_D > t\}.$$

Therefore,

$$\mathbf{P} \{\sigma_D > t\} \geq \frac{V(y, \ell)}{M}.$$

Finally, by letting $t \to \infty$, we obtain that

$$\mathbf{P} \{\sigma_D = \infty\} > 0.$$

Thus the process $(X(t), \alpha(t))$ is transient. This completes the proof. $\square$

To proceed, we focus on linearizable (in the continuous component) systems. In addition to condition (A3.1), we also assume that the following condition holds.

(A3.2) For each $i \in \mathcal{M}$, there exist $b(i)$, $\sigma_j(i) \in \mathbb{R}^{r \times r}, j = 1, 2, \ldots, d$, and $\widehat{Q} = (\widehat{q}_{ij})$, a generator of a continuous-time Markov chain

$\widehat{\alpha}(t)$ such that as $|x| \to \infty$,

$$
\begin{aligned}
\frac{b(x,i)}{|x|} &= b(i)\frac{x}{|x|} + o(1),\\
\frac{\sigma(x,i)}{|x|} &= (\sigma_1(i)x, \sigma_2(i)x, \ldots, \sigma_d(i)x)\frac{1}{|x|} + o(1),\\
Q(x) &= \widehat{Q} + o(1),
\end{aligned}
\tag{3.33}
$$

where $o(1) \to 0$ as $|x| \to \infty$. Moreover, $\widehat{\alpha}(t)$ is irreducible with stationary distribution denoted by $\pi = (\pi_1, \pi_2, \ldots, \pi_{m_0}) \in \mathbb{R}^{1 \times m_0}$.

As an application of Theorem 3.23, we have the following easily verifiable condition for transience under conditions (A3.1) and (A3.2).

Theorem 3.24. *Assume* (A3.1) *and* (A3.2). *If for each* $i \in \mathcal{M}$,

$$
\lambda_{\min}\left(b(i) + b'(i) + \sum_{j=1}^{d}\sigma_j(i)\sigma_j'(i)\right) - \frac{1}{2}\sum_{j=1}^{d}\left[\rho(\sigma_j(i)+\sigma_j'(i))\right]^2 > 0, \tag{3.34}
$$

then the process $(X(t),\alpha(t))$ *is transient.*

Proof. Let $D = \{x \in \mathbb{R}^r : |x| < k\}$, where k is a sufficiently large positive number. Consider $W(x,i) = k^\beta - |x|^\beta$, $(x,i) \in D^c \times \mathcal{M}$, where $\beta < 0$ is a sufficiently small constant. Then we have $W(x,i) = 0$ for all $(x,i) \in \partial D \times \mathcal{M}$ and $k^\beta \geq W(x,i) > 0$ for all $(x,i) \in D^c \times \mathcal{M}$. Thus conditions (3.32) are verified. Detailed computations reveal that for $x \neq 0$, we have

$$
\begin{aligned}
\nabla W(x,i) &= -\beta|x|^{\beta-2}x,\\
\nabla^2 W(x,i) &= -\beta\left[|x|^{\beta-2}I + (\beta-2)|x|^{\beta-4}xx'\right].
\end{aligned}
$$

Hence it follows from (3.33) that for all $(x,i) \in D^c \times \mathcal{M}$, we have

$$
\begin{aligned}
\mathcal{L}W(x,i) = -\beta|x|^\beta\Bigg\{&\frac{x'b(i)x}{|x|^2} + \frac{1}{2}\sum_{j=1}^{d}\left(\frac{x'\sigma_j'(i)\sigma_j(i)x}{|x|^2}\right.\\
&\left. + (\beta-2)\frac{(x'\sigma_j'(i)x)^2}{|x|^4}\right) + o(1)\Bigg\}.
\end{aligned}
\tag{3.35}
$$

Note that

$$
\frac{x'b(i)x}{|x|^2} + \frac{1}{2}\sum_{j=1}^{d}\frac{x'\sigma_j'(i)\sigma_j(i)x}{|x|^2} = \frac{x'\left(b'(i)+b(i)+\sum_{j=1}^{d}\sigma_j'(i)\sigma_j(i)\right)x}{2|x|^2}
$$

$$
\geq \frac{1}{2}\lambda_{\min}\left(b(i)+b'(i)+\sum_{j=1}^{d}\sigma_j'(i)\sigma_j(i)\right).
\tag{3.36}
$$

Meanwhile, for any symmetric matrix A with real eigenvalues $\lambda_1 \geq \lambda_2 \geq \cdots \geq \lambda_n$, using the transformation $x = Uy$ with U being a real orthogonal matrix satisfying $U'AU = \text{diag}(\lambda_1, \lambda_2, \ldots, \lambda_n)$ (see [59, Theorem 8.1.1]), we have

$$|x'Ax| = |\lambda_1 y_1^2 + \lambda_2 y_2^2 + \cdots + \lambda_n y_n^2| \leq \rho_A |y|^2 = \rho_A |x|^2. \quad (3.37)$$

Therefore it follows from (3.35)–(3.37) that for all $(x, i) \in D^c \times \mathcal{M}$,

$$\mathcal{L}W(x, i) \geq -\frac{1}{2}\beta |x|^\beta \left\{ \lambda_{\min}\left(b(i) + b'(i) + \sum_{j=1}^d \sigma_j'(i)\sigma_j(i) \right) \right.$$
$$\left. -\frac{1}{2}\sum_{j=1}^d \left[\rho(\sigma_j'(i) + \sigma_j(i)) \right]^2 + o(1) + O(\beta) \right\}$$
$$\geq 0,$$

where in the last step, we used the fact that $\beta < 0$ and condition (3.34). Hence the second equation in (3.32) of Theorem 3.23 is valid. Thus Theorem 3.23 implies that the process $(X(t), \alpha(t))$ is not recurrent.   □

3.4   Positive and Null Recurrence

This section takes up the positive recurrence issue. It entails the use of appropriate Liapunov functions. Recall that the process $Y(t) = (X(t), \alpha(t))$ is recurrent (resp., positive recurrent) with respect to some "cylinder" $D \times \mathcal{M}$ if and only if it is recurrent (resp., positive recurrent) with respect to $D \times \{\ell\}$, where $D \subset \mathbb{R}^r$ is a nonempty open set with compact closure and $\ell \in \mathcal{M}$. Thus the properties of recurrence or positive recurrence do not depend on the choice of the open set $D \subset \mathbb{R}^r$ or $\ell \in \mathcal{M}$. The result in Example 3.35 is quite interesting, which shows that the combination of a transient diffusion and a positive recurrent diffusion can be a positive recurrent switching diffusion.

3.4.1   General Criteria for Positive Recurrence

Theorem 3.25. *A necessary and sufficient condition for positive recurrence with respect to a domain $U = D \times \{\ell\} \subset \mathbb{R}^r \times \mathcal{M}$ is: For each $i \in \mathcal{M}$, there exists a nonnegative function $V(\cdot, i) : D^c \mapsto \mathbb{R}$ such that $V(\cdot, i)$ is twice continuously differentiable and that*

$$\mathcal{L}V(x, i) = -1, \quad (x, i) \in D^c \times \mathcal{M}. \quad (3.38)$$

Let $u(x, i) = \mathbf{E}_{x,i}\sigma_D$. Then $u(x, i)$ is the smallest positive solution of

$$\begin{cases} \mathcal{L}u(x, i) = -1, & (x, i) \in D^c \times \mathcal{M}, \\ u(x, i) = 0, & (x, i) \in \partial D \times \mathcal{M}, \end{cases} \quad (3.39)$$

where ∂D denotes the boundary of D.

Proof. The proof is divided in three steps.

Step 1: Show that the process $Y(t) = (X(t), \alpha(t))$ is positive recurrent if there exists a nonnegative function $V(\cdot, \cdot)$ satisfying the conditions of the theorem. Choose n_0 to be a positive integer sufficiently large so that $D \subset \{|x| < n_0\}$. Fix any $(x, i) \in D^c \times \mathcal{M}$. For any $t > 0$ and $n \in \mathbb{N}$ with $n > n_0$, we define

$$\sigma_D^{(n)}(t) = \sigma_D \wedge t \wedge \beta_n,$$

where $\beta_n = \inf\{t : |X^{(n)}(t)| \geq n\}$ is defined as in (2.14) and σ_D is the first entrance time $X(t)$ to D. That is, $\sigma_D = \inf\{t : X(t) \in D\}$. Now Dynkin's formula and equation (3.38) imply that

$$\mathbf{E}_{x,i} V(X(\sigma_D^{(n)}(t)), \alpha(\sigma_D^{(n)}(t))) - V(x, i)$$
$$= \mathbf{E}_{x,i} \int_0^{\sigma_D^{(n)}(t)} \mathcal{L}V(X(s), \alpha(s)) ds = -\mathbf{E}_{x,i} \sigma_D^{(n)}(t).$$

Note that the function V is nonnegative, hence we have $\mathbf{E}_{x,i} \sigma_D^{(n)}(t) \leq V(x, i)$. Meanwhile, because the process $Y(t) = (X(t), \alpha(t))$ is regular, it follows that $\sigma_D^{(n)}(t) \to \sigma_D(t)$ a.s. as $n \to \infty$, where $\sigma_D(t) = \sigma_D \wedge t$. By virtue of Fatou's lemma, we obtain

$$\mathbf{E}_{x,i} \sigma_D(t) \leq V(x, i). \tag{3.40}$$

Now the argument after equation (3.11) in the proof of Theorem 3.2 yields that $\mathbf{E}_{x,i} \sigma_D \leq V(x, i) < \infty$. Then Lemma 3.10 implies that $\mathbf{E}_{x,i} \sigma_U = \mathbf{E}_{x,i} \sigma_{D \times \{\ell\}} < \infty$. Since $(x, i) \in D^c \times \mathcal{M}$ is arbitrary, we conclude that $Y(t)$ is positive recurrent with respect to U.

Step 2: Show that $u(x, i) := \mathbf{E}_{x,i} \sigma_D$ is the smallest positive solution of (3.39). To this end, let n_0 be defined as before, that is, a positive integer sufficiently large so that $D \subset \{|x| < n_0\}$. For $n \geq n_0$, set $\sigma_D^{(n)} = \sigma_D \wedge \beta_n$. Clearly, we have $\sigma_D^{(n)} \leq \sigma_D^{(n+1)}$ for all $n \geq n_0$. Then the regularity of the process $Y(t)$ implies that $\sigma_D^{(n)} \nearrow \sigma_D$ a.s. as $n \to \infty$. Hence the monotone convergence theorem implies that as $n \to \infty$,

$$\mathbf{E}_{x,i} \sigma_D^{(n)} \nearrow \mathbf{E}_{x,i} \sigma_D. \tag{3.41}$$

Note that $\mathbf{E}_{x,i} \sigma_D < \infty$ from Step 1. Meanwhile, Lemma 3.6 implies that the function $u_n(x, i) = \mathbf{E}_{x,i} \sigma_D^{(n)}$ solves the boundary value problem

$$\mathcal{L}u_n(x, i) = -1,$$
$$u_n(x, i)|_{x \in \partial D} = 0, \tag{3.42}$$
$$u_n(x, i)|_{|x|=n} = 0, \quad i \in \mathcal{M}.$$

Thus the function $v_n(x, i) := u_{n+1}(x, i) - u_n(x, i)$ is $\mathcal{L}$-harmonic in the domain $(D^c \cap \{|x| < n\}) \times \mathcal{M}$. Since $\sigma_D^{(n)} \leq \sigma_D^{(n+1)}$, it follows that $\mathbf{E}_{x,i}\sigma_D^{(n)} \leq \mathbf{E}_{x,i}\sigma_D^{(n+1)}$ and hence $v_n(x, i) \geq 0$. Now (3.41) implies that

$$u(x, i) = u_{n_0}(x, i) + \sum_{k=n_0}^{\infty} v_k(x, i). \tag{3.43}$$

Using Harnack's inequality for $\mathcal{L}$-elliptic systems of equations (see [3, 25], and also [158] for general references on elliptic systems), it can be shown by a slight modification of the well-known arguments (see, e.g., [56, pp. 21–22]) that the sum of a convergent series of positive $\mathcal{L}$-harmonic functions is also an $\mathcal{L}$-harmonic function. Hence we conclude that $u(x, i)$ is twice continuously differentiable and satisfies equation (3.39). To verify that $u(x, i)$ is the smallest positive solution of (3.39), let $w(x, i)$ be any positive solution of (3.39). Note that $u_n(x, i) = \mathbf{E}_{x,i}\sigma_D^{(n)}$ satisfies the boundary conditions

$$u_n(x, i)|_{x \in \partial D} = 0,$$

$$u_n(x, i)|_{|x|=n} = 0, \quad i \in \mathcal{M}.$$

Then the functions $u_n(x, i) - w(x, i)$ for $i \in \mathcal{M}$ are $\mathcal{L}$-harmonic and satisfy $u_n(x, i) - w(x, i) = 0$ for $(x, i) \in \partial D \times \mathcal{M}$ and $u_n(x, i) - w(x, i) < 0$ for $(x, i) \in \{|x| = n\} \times \mathcal{M}$. Hence it follows from the maximum principle for $\mathcal{L}$-elliptic system of equations [138, p. 192] that $u_n(x, i) \leq w(x, i)$ in $(D^c \cap \{|x| < n\}) \times \mathcal{M}$ for all $n \geq n_0$. Letting $n \to \infty$, we obtain $u(x, i) \leq w(x, i)$, as desired.

Step 3: Show that there exists a nonnegative function V satisfying the conditions of the theorem if the process $Y(t) = (X(t), \alpha(t))$ is positive recurrent with respect to the domain $U = D \times \{\ell\}$. Then $\mathbf{E}_{x,i}\sigma_D < \infty$ for all $(x, i) \in D^c \times \mathcal{M}$ and consequently equation (3.43) and Harnack's inequality for $\mathcal{L}$-elliptic system of equations [3, 25] imply that the bounded monotone increasing sequence $u_n(x, i)$ converges uniformly on every compact subset of $D^c \times \mathcal{M}$. Moreover, its limit $u(x, i)$ satisfies the equation $\mathcal{L}u(x, i) = -1$ for each $i \in \mathcal{M}$. Therefore the function $V(x, i) := u(x, i)$ satisfies equation (3.38). This completes the proof of the theorem. $\square$

Theorem 3.26. *A necessary and sufficient condition for positive recurrence with respect to a domain* $U = D \times \{\ell\} \subset \mathbb{R}^r \times \mathcal{M}$ *is: For each* $i \in \mathcal{M}$, *there exists a nonnegative function* $V(\cdot, i) : D^c \mapsto \mathbb{R}$ *such that* $V(\cdot, i)$ *is twice continuously differentiable and that for some* $\gamma > 0$,

$$\mathcal{L}V(x, i) \leq -\gamma, \quad (x, i) \in D^c \times \mathcal{M}. \tag{3.44}$$

Proof. Necessity: This part follows immediately from the necessity of Theorem 3.25 with $\gamma = -1$.

Sufficiency: Suppose that there exists a nonnegative function V satisfying the conditions of the theorem. Define the stopping time $\sigma_D^{(n)}(t) = \sigma_D \wedge t \wedge \beta_n$ as in the proof of Theorem 3.25. Now Dynkin's formula and equation (3.44) imply that for any $(x, i) \in D^c \times \mathcal{M}$,

$$
\mathbf{E}_{x,i} V(X(\sigma_D^{(n)}(t)), \alpha(\sigma_D^{(n)}(t))) - V(x, i)
$$
$$
= \mathbf{E}_{x,i} \int_0^{\sigma_D^{(n)}(t)} \mathcal{L}V(X(s), \alpha(s)) ds \leq -\gamma \mathbf{E}_{x,i} \sigma_D^{(n)}(t).
$$

Hence we have by the nonnegativity of the function V that $\mathbf{E}_{x,i} \sigma_D^{(n)}(t) \leq V(x, i)/\gamma$. Meanwhile, the regularity of the process $Y(t) = (X(t), \alpha(t))$ implies that $\sigma_D^{(n)}(t) \to \sigma_D(t)$ a.s. as $n \to \infty$, where $\sigma_D(t) = \sigma_D \wedge t$. Therefore Fatou's lemma leads to $\mathbf{E}_{x,i} \sigma_D(t) \leq V(x, i)/\gamma$. Moreover, from the proof of Theorem 3.25, $\sigma_D(t) \to \sigma_D$ a.s. as $t \to \infty$. Thus we obtain

$$
\mathbf{E}_{x,i} \sigma_D \leq \frac{1}{\gamma} V(x, i)
$$

by applying Fatou's lemma again. Then Lemma 3.10 implies that $\mathbf{E}_{x,i} \sigma_U = \mathbf{E}_{x,i} \sigma_{D \times \{\ell\}} < \infty$. Because $(x, i) \in D^c \times \mathcal{M}$ is arbitrary, we conclude that $Y(t)$ is positive recurrent with respect to U. This completes the proof of the theorem. $\qquad \square$

Example 3.27. Let us continue our discussion from Example 1.1. Consider (1.1) and assume that for each $\alpha \in \mathcal{M} = \{1, 2, \dots, m\}$ and $i, j = 1, 2, \dots, n$ with $j \neq i$, $a_{ii}(\alpha) > 0$ and $a_{ij}(\alpha) \geq 0$. Then for each $i = 1, \dots, n$ and $\alpha \in \mathcal{M}$, we have

$$
-a_{ii}(\alpha)x_i^2 + \left(r_i(\alpha) + \sum_{j=1}^n a_{ji}(\alpha) \right) x_i - b_i(\alpha)
$$
$$
\leq \frac{\left(r_i(\alpha) + \sum_{j=1}^n a_{ji}(\alpha) \right)^2}{4a_{ii}(\alpha)} - b_i(\alpha) := \widehat{K}_i(\alpha).
$$

Denote

$$
K_i(\alpha) := \widehat{K}_i(\alpha) \vee 0, \quad \text{for each } i = 1, \dots, n \text{ and } \alpha = 1, \dots, m.
$$

We can further find a number $\rho_i(\alpha) > 0$ sufficiently large so that

$$
-a_{ii}(\alpha)x_i^2 + \left(r_i(\alpha) + \sum_{j=1}^n a_{ji}(\alpha) \right) x_i - b_i(\alpha)
$$
$$
\leq -\sum_{j=1}^n K_j(\alpha) - 1, \quad \text{for any } x_i > \rho_i(\alpha),
$$

Denote
$$\rho := \max\left\{\rho_i(\alpha), i = 1, \ldots, n, \alpha = 1, \ldots, m\right\}. \tag{3.45}$$

Then the solution $x(t)$ to (1.1) is positive recurrent with respect to the domain
$$E_\rho := \left\{x \in \mathbb{R}^n_+ : 0 < x_i < \rho, i = 1, 2, \ldots, n\right\}.$$

We refer the reader to [189] for a verbatim proof.

3.4.2 Path Excursions

Applications of the positive recurrence criteria enable us to establish path excursions of the underlying processes. Suppose that $Y(t) = (X(t), \alpha(t))$ is positive recurrent, and that the Liapunov functions $V(x, i)$ (with $i \in \mathcal{M}$) are given in Theorem 3.26, so is the set D. Let D_0 be a bounded open set with compact closure satisfying $D \subset D_0$, and τ_1 be the first exit time of $(X(t), \alpha(t))$ from $D_0 \times \mathcal{M}$; that is, $\tau_1 = \min\{t > 0 : X(t) \notin D_0\}$. Define $\sigma_1 = \min\{t > \tau_1 : X(t) \in D_0\}$. We can obtain

$$\mathbf{P}\left(\sup_{\tau_1 \leq t \leq \sigma_1} V(X(t), \alpha(t)) \geq \gamma\right) \leq \frac{\mathbf{E}V(X(\tau_1), \alpha(\tau_1))}{\gamma}, \quad \text{for } \gamma > 0,$$

$$\mathbf{E}(\sigma_1 - \tau_1) \leq \frac{\mathbf{E}V(X(\tau_1), \alpha(\tau_1))}{\gamma},$$

$$\tag{3.46}$$

where γ is as given in Theorem 3.26. The idea can be continued in the following way for $k \geq 1$. Define

$$\tau_{k+1} = \min\{t > \sigma_k : X(t) \notin D_0\}$$

and

$$\sigma_{k+1} = \min\{t \geq \tau_{k+1} : X(t) \in D_0\}.$$

Then we can obtain estimates of $\mathbf{E}(\sigma_{k+1} - \tau_{k+1})$, which gives us the expected difference of $(k+1)$st return time and exit time.

3.4.3 Positive Recurrence under Linearization

This subsection is devoted to positive recurrence of regime-switching diffusions under linearization with respect to the continuous component. By linearizable systems, we mean such systems that are linearizable with respect to the continuous component. These systems are important inasmuch as linearization is widely used in many applications because it is much easier to deal with linear systems.

With Theorem 3.28 at hand, we proceed to study positive recurrence of linearizable (in the continuous component) systems as described in condition (A3.2). We are ready to present an easily verifiable condition for positive recurrence.

Theorem 3.28. *Assume* (A3.1) *and* (A3.2). *If*

$$\sum_{i=1}^{m_0} \pi_i \lambda_{\max}\left(b(i) + b'(i) + \sum_{j=1}^{d} \sigma_j(i)\sigma_j'(i) \right) < 0, \qquad (3.47)$$

then the process $(X(t), \alpha(t))$ *is positive recurrent.*

Proof. For notational simplicity, define the column vector

$$\mu = (\mu_1, \mu_2, \ldots, \mu_{m_0})' \in \mathbb{R}^{m_0}$$

with

$$\mu_i = \frac{1}{2}\lambda_{\max}\left(b(i) + b'(i) + \sum_{j=1}^{d} \sigma_j(i)\sigma_j'(i) \right).$$

Let

$$\beta := -\pi\mu = -\frac{1}{2}\sum_{i=1}^{m_0} \pi_i \left(b(i) + b'(i) + \sum_{j=1}^{d} \sigma_j(i)\sigma_j'(i) \right).$$

Note that $\beta > 0$ by (3.47). Because

$$\pi(\mu + \beta\mathbb{1}) = \pi\mu + \beta \cdot \pi\mathbb{1} = -\beta + \beta = 0,$$

condition (A3.2) and Lemma A.12 yield that the equation

$$\widehat{Q}c = \mu + \beta\mathbb{1}$$

has a solution $c = (c_1, c_2, \ldots, c_{m_0})' \in \mathbb{R}^{m_0}$. Thus we have

$$\mu_i - \sum_{j=1}^{m_0} \widehat{q}_{ij} c_j = -\beta, \quad i \in \mathcal{M}. \qquad (3.48)$$

For each $i \in \mathcal{M}$, consider the Liapunov function

$$V(x, i) = (1 - \gamma c_i)|x|^{\gamma},$$

where $0 < \gamma < 1$ is sufficiently small so that $1 - \gamma c_i > 0$ for each $i \in \mathcal{M}$. It is readily seen that for each $i \in \mathcal{M}$, $V(\cdot, i)$ is continuous, nonnegative, and has continuous second partial derivatives with respect to x in any deleted neighborhood of 0. Detailed calculations reveal that for $x \neq 0$, we have

$$\nabla V(x, i) = (1 - \gamma c_i)\gamma|x|^{\gamma-2}x,$$
$$\nabla^2 V(x, i) = (1 - \gamma c_i)\gamma \left[|x|^{\gamma-2}I + (\gamma - 2)|x|^{\gamma-4}xx' \right].$$

Meanwhile, it follows from (3.33) that

$$\frac{a(x, i)}{|x|^2} = \frac{\sigma(x, i)\sigma'(x, i)}{|x|^2} = \sum_{j=1}^{d} \sigma_j(i)xx'\sigma_j'(i)\frac{1}{|x|^2} + o(1),$$

where $o(1) \to 0$ as $|x| \to \infty$. Therefore, we have that

$$
\begin{aligned}
\mathcal{L}V(x,i) = \frac{\gamma}{2}(1 - \gamma c_i) \sum_{j=1}^{d} &\Big(x'\sigma_j'(i)|x|^{\gamma-2} I\sigma_j(i)x \\
&+ x'\sigma_j'(i)(\gamma - 2)|x|^{\gamma-4} xx'\sigma_j(i)x \Big) \\
&+ (1 - \gamma c_i)\gamma|x|^{\gamma-2} x'b(i)x + o(|x|^\gamma) \\
&- \sum_{j \neq i} q_{ij}(x)|x|^\gamma \gamma(c_j - c_i) \\
= \gamma(1 - \gamma c_i)|x|^\gamma &\left\{ \frac{1}{2} \sum_{j=1}^{d} \left(\frac{x'\sigma_j'(i)\sigma_j(i)x}{|x|^2} + (\gamma - 2)\frac{(x'\sigma_j'(i)x)^2}{|x|^4} \right) \right. \\
&\left. + \frac{x'b(i)x}{|x|^2} - \sum_{j \neq i} q_{ij}(x) \frac{c_j - c_i}{1 - \gamma c_i} + o(1) \right\},
\end{aligned}
$$

(3.49)

with $o(1) \to 0$ as $|x| \to \infty$. Note that

$$
\begin{aligned}
&\frac{x'b(i)x}{|x|^2} + \frac{1}{2} \sum_{j=1}^{d} \frac{x'\sigma_j'(i)\sigma_j(i)x}{|x|^2} \\
&\quad \leq \frac{1}{2}\lambda_{\max}\left(b(i) + b'(i) + \sum_{j=1}^{d} \sigma_j'(i)\sigma_j(i) \right) = \mu_i.
\end{aligned}
$$

(3.50)

Next, using condition (A3.2),

$$
\begin{aligned}
\sum_{j \neq i} q_{ij}(x) \frac{c_j - c_i}{1 - \gamma c_i} \\
= \sum_{j=1}^{m_0} q_{ij}(x)c_j + \sum_{j \neq i} q_{ij}(x) \frac{c_i(c_j - c_i)}{1 - \gamma c_i}\gamma \\
= \sum_{j=1}^{m_0} \widehat{q}_{ij} c_j + O(\gamma) + o(1),
\end{aligned}
$$

(3.51)

where $O(\gamma)/\gamma$ is bounded and $o(1) \to 0$ as $|x| \to \infty$ and $\gamma \to 0$. Hence it follows from (3.49)–(3.51) that when $|x| > R$ with R sufficiently large and $0 < \gamma < 1$ sufficiently small, we have

$$
\mathcal{L}V(x,i) \leq \gamma(1 - \gamma c_i)|x|^\gamma \left\{ \mu_i - \sum_{j=1}^{m_0} \widehat{q}_{ij} c_j + o(1) + O(\gamma) \right\}.
$$

Furthermore, by virtue of (3.48), we have

$$
\mathcal{L}V(x,i) \leq \gamma(1 - \gamma c_i)|x|^\gamma \left(-\beta + o(1) + O(\gamma) \right) \leq -K < 0,
$$

for any $(x, i) \in \mathbb{R}^r \times \mathcal{M}$ with $|x| > r$, where K is a positive constant. Therefore we conclude from Theorem 3.28 that the process $(X(t), \alpha(t))$ is positive recurrent.   □

The above result further specializes to the following corollary.

Corollary 3.29. *Suppose that the continuous component $X(t)$ is one-dimensional and that as $|x| \to \infty$,*

$$\frac{b(x, i)}{|x|} = b_i \frac{x}{|x|} + o(1),$$
$$\frac{\sigma(x, i)}{|x|} = \sigma_i \frac{x}{|x|} + o(1), \tag{3.52}$$

for some constants b_i, σ_i, $i \in \mathcal{M}$. If

$$\pi b - \frac{1}{2}\pi \sigma^2 := \sum_{i=1}^{m_0} \pi_i \left(b_i - \frac{\sigma_i^2}{2} \right) < 0, \tag{3.53}$$

where $\pi = (\pi_1, \dots, \pi_{m_0}) \in \mathbb{R}^{1 \times m_0}$ is as in condition (A3.2), then the process $(X(t), \alpha(t))$ is positive recurrent.

Next we develop a sufficient condition for nonpositive recurrence.

Theorem 3.30. *Suppose that there exists a nonempty bounded domain D such that there exist functions $V(x, i)$ and $W(x, i)$ defined on $D^c \times \mathcal{M}$ satisfying*

(a) *$V(x, i) \geq 0$ for all $(x, i) \in D^c \times \mathcal{M}$, and for some positive constant k,*

$$0 \leq \mathcal{L}V(x, i) \leq k \quad \text{for all} \quad (x, i) \in D^c \times \mathcal{M};$$

(b) *$W(x, i) \leq 0$ for all $(x, i) \in \partial D \times \mathcal{M}$, and*

$$\mathcal{L}W(x, i) \geq 0 \quad \text{for all} \quad (x, i) \in D^c \times \mathcal{M};$$

(c) *for an increasing sequence of bounded domains $E_n \supset D$ with boundaries Γ_n,*

$$\frac{\inf_{(x,i) \in \Gamma_n \times \mathcal{M}} V(x, i)}{\sup_{(x,i) \in \Gamma_n \times \mathcal{M}} W(x, i)} = R_n \to \infty \quad \text{as} \quad n \to \infty. \tag{3.54}$$

If there exists some $(x, \alpha) \in D^c \times \mathcal{M}$ satisfying $W(x, \alpha) > 0$, then the process $(X(t), \alpha(t))$ is not positive recurrent. That is, $\mathbf{E}\sigma_D = \infty$ for all $(x, \alpha) \in D^c \times \mathcal{M}$ such that $W(x, \alpha) > 0$, where

$$\sigma_D = \sigma_D^{x,\alpha} = \inf \{ t \geq 0 : X^{x,\alpha}(t) \in D \}.$$

Proof. Consider the function $V - R_nW$ in $(D^c \cap E_n) \times \mathcal{M}$. Then from conditions (a), (b), and (c), we have

$$\mathcal{L}(V - R_nW)(x, i) \le k,$$

$$(V - R_nW)(x, i)|_{x \in \partial D} \ge 0,$$

$$(V - R_nW)(x, i)|_{x \in \Gamma_n} \ge 0.$$

Fix any $(x, \alpha) \in D^c \times \mathcal{M}$ satisfying $W(x, \alpha) > 0$. Then it follows that $\mathbf{P}\{\sigma_D \wedge \beta_n < \infty\} = 1$ for any $n > (n_0 \vee |y|)$, where n_0 is a sufficiently large integer such that $D \subset \{x \in \mathbb{R}^r : |x| < n_0\}$ and β_n is as in (2.14). Hence Dynkin's formula implies that

$$\mathbf{E}(V - R_nW)(X(\sigma_D \wedge \beta_n), \alpha(\sigma_D \wedge \beta_n)) - (V - R_nW)(x, \alpha)$$

$$= \mathbf{E}\int_0^{\sigma_D \wedge \beta_n} \mathcal{L}(V - R_nW)(X(u), \alpha(u))du$$

$$\le k\mathbf{E}[\sigma_D \wedge \beta_n].$$

That is,

$$\mathbf{E}\sigma_D \wedge \beta_n \ge \frac{1}{k}[R_nW(x, \alpha) - V(x, \alpha)].$$

Therefore, we have from (3.54) that

$$\mathbf{E}\sigma_D \wedge \beta_n \to \infty, \quad \text{as} \quad n \to \infty.$$

Meanwhile, note that $\sigma_D \ge \sigma_D \wedge \beta_n$. We have

$$\mathbf{E}\sigma_D = \infty.$$

Therefore the process $(X(t), \alpha(t))$ is not positive recurrent by virtue of Theorem 3.12. $\qquad\square$

3.4.4 Null Recurrence

Null recurrence is more complex. Even for diffusions alone, the available results are scarce. For simplicity here, we assume that the continuous component of the system is a diffusion without drift term.

Theorem 3.31. *Consider a real-valued regime-switching diffusion process*

$$\begin{cases} dX(t) = \sigma(X(t), \alpha(t))dw(t) \\ \mathbf{P}\{\alpha(t + \Delta) = j | \alpha(t) = i, X(s), \alpha(s), s \le t\} = q_{ij}(X(t))\Delta + o(\Delta). \end{cases}$$

(3.55)

If for some constants $0 \le \beta \le 1$, $k_1 > 0$, and $k_2 > 0$,

$$\sigma^2(x, i) \le k_1|x|^{1-\beta}, \quad \text{for all} \quad (x, i) \in \{x \in \mathbb{R} : |x| \ge k_2\} \times \mathcal{M}, \quad (3.56)$$

then the process $(X(t), \alpha(t))$ defined by (3.55) is null recurrent.

Proof. In fact, by Theorem 3.12, it suffices to prove that the process is null recurrent with respect to $D \times \{\ell\}$, where $D = (-k_2, k_2)$ and $\ell \in \mathcal{M}$. For each $i \in \mathcal{M}$, the Liapunov function

$$V(x, i) = \begin{cases} |x|^{1+\beta}, & \text{if } \beta > 0, \\ |x|(\ln|x| - 1) + K, & \text{if } \beta = 0, \end{cases}$$

where K is a sufficiently large positive constant, and

$$W(x, i) = |x| - k_2 \text{ for } (x, i) \in D^c \times \mathcal{M}.$$

Then detailed computations show that the nonnegative function W satisfies (3.23) in Theorem 3.14. This implies that the process is recurrent with respect to $D \times \{\ell\}$.

Similarly, by virtue of (3.56), detailed computations show that conditions (a)–(c) of Theorem 3.30 are satisfied with the functions V and W as chosen above. It thus follows from Theorem 3.30 that the process $(X(t), \alpha(t))$ is not positive recurrent with respect to $D \times \{\ell\}$. The details are omitted. Therefore the process defined by (3.55) is null recurrent. $\qquad \square$

Remark 3.32. If for some positive constants ϱ, k_1, and k_2, we have

$$\sigma^2(x, i) \geq k_1 |x|^{1+\varrho}, \text{ for all } (x, i) \in \{x \in \mathbb{R} : |x| \geq k_2\} \times \mathcal{M}, \quad (3.57)$$

then the process $(X(t), \alpha(t))$ defined by (3.55) is positive recurrent.

To verify this, we choose $1 > \varsigma > 0$ such that $\varsigma + \varrho > 1$ and define $V(x, i) = |x|^\varsigma$ for $(x, i) \in \{x \in \mathbb{R} : |x| \geq k_2\} \times \mathcal{M}$. Then we can show

$$\mathcal{L}V(x, i) = \frac{1}{2}\varsigma(\varsigma - 1)|x|^{\varsigma - 2}\sigma^2(x, i) \leq \frac{1}{2}k_1\varsigma(\varsigma - 1)k_2^{\varsigma + \varrho - 1} < 0,$$

where $(x, i) \in \{x \in \mathbb{R} : |x| \geq k_2\} \times \mathcal{M}$. Therefore, it follows from Theorem 3.28 that the process $(X(t), \alpha(t))$ is positive recurrent.

3.5   Examples

In this section, we provide several examples to illustrate the results obtained thus far.

Example 3.33. Suppose that for each $x \in \mathbb{R}^r$ and each $i \in \mathcal{M}$, there exist positive constants c and γ such that for all x with $|x| \geq c$,

$$b'(x, i)\frac{x}{|x|} < -\gamma; \quad (3.58)$$

where $|x|$ denotes the norm of x. That is, the drifts are pointed inward. Then the process $Y(t) = (X(t), \alpha(t))$ is positive recurrent.

First note that (3.5) implies that for all x with $|x| \geq r/(\gamma\kappa_1)$, we have

$$\operatorname{tr}(a(x,i)) = \sum_{j=1}^{r} e_j' a(x,i) e_j \leq \sum_{j=1}^{r} \kappa_1^{-1} \leq \gamma |x|.$$

By Theorem 3.12, it is enough to prove that the process $Y(t) = (X(t), \alpha(t))$ is positive recurrent with respect to the domain $U := \{|x| < \varrho\} \times \{\ell\}$ for some $\ell \in \mathcal{M}$, where $\varrho := \max\{c, r/(\gamma\kappa_1)\}$. To this end, consider the function

$$V(x,i) = \frac{1}{2} x' x, \quad \text{for each } i \in \mathcal{M} \text{ and for all } |x| \geq \varrho.$$

For each $i \in \mathcal{M}$, $\nabla V(\cdot, i) = x$ and $\nabla^2 V(\cdot, i) = I$, where I is the $r \times r$ identity matrix. Thus by the definition of $\mathcal{L}$, we have for all $(x,i) \in \{|x| \geq \varrho\} \times \mathcal{M}$ that

$$\begin{aligned}
\mathcal{L}V(x,i) &= \frac{1}{2}\operatorname{tr}(a(x,i)) + b'(x,i)\frac{x}{|x|}|x| \\
&< \frac{1}{2}\gamma|x| - \gamma|x| \\
&= -\frac{1}{2}\gamma|x| \leq -\frac{1}{2}\kappa\varrho.
\end{aligned}$$

Then the conclusion follows from Theorem 3.26 immediately.

Remark 3.34. Suppose that the diffusion component $X(t)$ of the process $Y(t) = (X(t), \alpha(t))$ is one-dimensional and that there exist constants $c_0 > 0$ and $c_1 > 0$ such that for each $i \in \mathcal{M}$,

$$b(x,i) \begin{cases} < -c_1, & \text{for } x > c_0, \\ > c_1, & \text{for } x < -c_0. \end{cases} \tag{3.59}$$

Then the process $Y(t) = (X(t), \alpha(t))$ is positive recurrent. In fact, the conclusion follows immediately if we observe that (3.59) satisfies (3.58). Alternatively, we can verify this directly by defining the Liapunov function $V(x,i) = |x|$ for each $i \in \mathcal{M}$.

Example 3.35. To illustrate the utility of Theorem 3.26, consider a real-valued process

$$dX(t) = b(X(t), \alpha(t))dt + \sigma(X(t), \alpha(t))dw(t), \tag{3.60}$$

where $\alpha(t)$ is a two-state random jump process, with x-dependent generator

$$Q(x) = \begin{pmatrix} -\dfrac{1}{3} - \dfrac{1}{4}\cos x & \dfrac{1}{3} + \dfrac{1}{4}\cos x \\ \dfrac{7}{3} + \dfrac{1}{2}\sin x & -\dfrac{7}{3} - \dfrac{1}{2}\sin x \end{pmatrix},$$

and

$$b(x,1) = -x, \quad \sigma(x,1) = 1, \quad b(x,2) = x, \quad \sigma(x,2) = 1.$$

Thus (3.60) can be regarded as the result of the following two diffusions:

$$dX(t) = -X(t)dt + dw(t), \qquad (3.61)$$

and

$$dX(t) = X(t)dt + dw(t), \qquad (3.62)$$

switching back and forth from one to the other according to the movement of $\alpha(t)$.

Note that (3.61) is positive recurrent whereas (3.62) is a transient diffusion process. But, the switching diffusion (3.60) is positive recurrent. We verify these as follows. Consider the Liapunov function $V(x,1) = |x|$. Let $\mathcal{L}_1$ be the operator associated with (3.61). Then we have for all $|x| \geq 1$, $\mathcal{L}_1 V(x,1) = -x \operatorname{sign}(x) = -|x| \leq -1 < 0$. It follows from [83, Theorem 3.7.3] that (3.61) is positive recurrent. Recall that the real-valued diffusion process

$$dX(t) = b(X(t))dt + \sigma(X(t))dw(t)$$

with $\sigma(x) \neq 0$ for all $x \in \mathbb{R}$, is recurrent if and only if

$$\int_0^x \exp\left\{ -2 \int_0^u \frac{b(z)}{\sigma^2(z)} dz \right\} du \to \pm\infty \qquad (3.63)$$

as $x \to \pm\infty$; see [83, p. 105]. Direct computation shows that (3.62) fails to satisfy this condition and hence is transient.

Next, we use Theorem 3.26 to demonstrate that the switching diffusion (3.60) is positive recurrent for appropriate Q. Consider Liapunov functions

$$V(x,1) = |x|, \quad V(x,2) = \frac{7}{3}|x|.$$

We have

$$\mathcal{L}V(x,1) = -x \cdot \operatorname{sign} x + \left(\frac{1}{3} + \frac{1}{4} \cos x \right) \left(\frac{7}{3} - 1 \right) |x| \leq -\frac{2}{9}|x| \leq -\frac{2}{9},$$

$$\mathcal{L}V(x,2) = x \cdot \frac{7}{3} \operatorname{sign} x + \left(\frac{7}{3} + \frac{1}{2} \sin x \right) \left(1 - \frac{7}{3} \right) |x| \leq -\frac{1}{9}|x| \leq -\frac{1}{9},$$

for all $|x| \geq 1$. Thus the switching diffusion (3.60) is positive recurrent by Theorem 3.26.

Example 3.36. To illustrate the result of Corollary 3.29, we consider a real-valued process given by

$$\begin{cases} dX(t) = b(X(t), \alpha(t))dt + \sigma(X(t), \alpha(t))dw(t) \\ \mathbf{P}\{\alpha(t+\Delta) = j | \alpha(t) = i, X(s), \alpha(s), s \leq t\} = q_{ij}(X(t))\Delta + o(\Delta), \end{cases}$$

$$(3.64)$$

with the following specifications. The jump component $\alpha(t)$ has three states and is generated by

$$Q(x) = \begin{pmatrix} -\dfrac{2+|x|}{1+x^2} - \dfrac{1+3x^2}{2+x^2} & \dfrac{2+|x|}{1+x^2} & \dfrac{1+3x^2}{2+x^2} \\[3mm] 1 - \dfrac{\sin x}{1+x^2} & \dfrac{\sin x}{1+x^2} - \dfrac{\cos x}{2+|x|} - 3 & 2 + \dfrac{\cos x}{2+|x|} \\[3mm] \dfrac{1+\cos^2 x}{2+x^2} & 2 & -\dfrac{\cos^2 x + 1}{2+x^2} - 2 \end{pmatrix},$$

and the drift and diffusion coefficients are given by

$$b(x,1) = 3x - 1, \; b(x,2) = \frac{3}{2}x + 1, \; b(x,3) = -x + \frac{x}{1+|x|},$$

$$\sigma(x,1) = \sqrt{3 + x^2}, \; \sigma(x,2) = \sqrt{2 - \sin x + 2x^2}, \; \sigma(x,3) = 3 + \sqrt{4 + x^2}.$$

Hence associated with (3.64), there are three diffusions

$$dX(t) = (3X(t) - 1)dt + \sqrt{3 + X^2(t)}\,dw(t), \tag{3.65}$$

$$dX(t) = \left(\frac{3}{2}X(t) + 1\right)dt + \sqrt{2 - \sin(X(t)) + 2X^2(t)}\,dw(t), \tag{3.66}$$

$$dX(t) = \left(\frac{X(t)}{1+|X(t)|} - X(t)\right)dt + \left(3 + \sqrt{4 + X^2(t)}\right)dw(t), \tag{3.67}$$

switching back and forth from one to another according to the movement of the jump component $\alpha(t)$.

Note that (3.67) is positive recurrent, whereas (3.65) and (3.66) are transient diffusions. But due to the stabilization effect of $\alpha(t)$, the switching diffusion (3.64) is positive recurrent. We verify these as follows. Consider the Liapunov function $V(x) = |x|^\gamma$ with $0 < \gamma < 1$ sufficiently small; let $\mathcal{L}_1$ be the operator associated with the third equation in (3.65). Detailed computation shows that for all $|x| \geq 1$, we have $\mathcal{L}_1 V(x) \leq -\frac{1}{2}\gamma < 0$. Thus it follows from [83, Theorem 3.7.3] that (3.67) is positive recurrent. Detailed computations show that (3.65) and (3.66) fail to satisfy (3.63) and hence these diffusions are transient.

Next we use Corollary 3.29 to demonstrate that the switching diffusion (3.64) is positive recurrent. In fact, it is readily seen that as $x \to \infty$, the constants $b(i)$ and $\sigma^2(i)$, $i = 1, 2, 3$, as in (3.52) are given by

$$b(1) = 3, \quad b(2) = \frac{3}{2}, \quad b(3) = -1,$$
$$\sigma^2(1) = 1, \quad \sigma^2(2) = 2, \quad \sigma^2(3) = 1. \tag{3.68}$$

In addition, as $|x| \to \infty$, $Q(x)$ tends to

$$\widehat{Q} = \begin{pmatrix} -3 & 0 & 3 \\ 1 & -3 & 2 \\ 0 & 2 & -2 \end{pmatrix}.$$

By solving the system of equations

$$\begin{cases} \pi \widehat{Q} = 0, \\ \pi \mathbb{1} = 1, \end{cases}$$

we obtain the stationary distribution π associated with $\widetilde{Q}$

$$\pi = \left(\frac{2}{17}, \frac{6}{17}, \frac{9}{17} \right). \tag{3.69}$$

Thus by virtue of (3.68) and (3.69), observe that

$$\sum_{i=1}^{3} \pi_i \left(b(i) - \frac{\sigma^2(i)}{2} \right) = -\frac{23}{34} < 0.$$

Thus Corollary 3.29 implies that (3.64) is positive recurrent; see the sample path demonstrated in Figure 3.1.

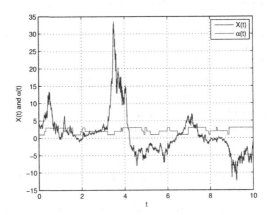

FIGURE 3.1. Sample path of switching diffusion (3.64) with initial condition $(x, \alpha) = (3, 1)$.

Example 3.37. Consider a two-dimensional (in the continuous component) regime-switching diffusion

$$\begin{cases} dX(t) = b(X(t), \alpha(t))dt + \sigma(X(t), \alpha(t))dw(t) \\ \mathbf{P}\{\alpha(t + \Delta) = j | \alpha(t) = i, X(s), \alpha(s), s \le t\} = q_{ij}(X(t))\Delta + o(\Delta) \end{cases} \tag{3.70}$$

with $(X(t), \alpha(t)) \in \mathbb{R}^2 \times \{1, 2\}$. The discrete component $\alpha(t)$ is generated by

$$Q(x_1, x_2) = \begin{pmatrix} -3 + \dfrac{2(1 + \cos(x_1^2))}{3 + x_1^2 + x_2^2} & 3 - \dfrac{2(1 + \cos(x_1^2))}{3 + x_1^2 + x_2^2} \\ 1 + \dfrac{1}{\sqrt{1 + x_1^2 + x_2^2}} & -1 - \dfrac{1}{\sqrt{1 + x_1^2 + x_2^2}} \end{pmatrix},$$

and

$$b(x_1, x_2, 1) = \begin{pmatrix} -x_1 + 2x_2 \\ 2x_2 \end{pmatrix},$$

$$\sigma(x_1, x_2, 1) = \begin{pmatrix} 3 + \sin x_1 + \sqrt{2 + x_2^2} & 0 \\ 0 & 2 - \cos x_2 + \sqrt{1 + x_1^2} \end{pmatrix},$$

$$b(x_1, x_2, 2) = \begin{pmatrix} -3x_1 - x_2 \\ x_1 - 2x_2 \end{pmatrix},$$

$$\sigma(x_1, x_2, 2) = \begin{pmatrix} 1 & 1 \\ 0 & 10 + \sqrt{3 + x_1^2} \end{pmatrix}.$$

Associated with the regime-switching diffusion (3.70), there are two diffusions

$$dX(t) = b(X(t), 1)dt + \sigma(X(t), 1)dw(t), \tag{3.71}$$

and

$$dX(t) = b(X(t), 2)dt + \sigma(X(t), 2)dw(t), \tag{3.72}$$

switching back and forth from one to another according to the movement of the jump component $\alpha(t)$, where $w(t) = (w_1(t), w_2(t))'$ is a two-dimensional standard Brownian motion. By selecting appropriate Liapunov functions as in [83, Section 3.8], or using the criteria in [15], we can verify that (3.71) is transient and (3.72) is positive recurrent.

Next we use Theorem 3.28 to show that the switching diffusion (3.70) is positive recurrent owing to the presence of the stabilizing effect of the discrete component $\alpha(t)$. Note that the matrices $b(i)$ and $\sigma_j(i)$, $i, j = 1, 2$, and $\widehat{Q}$ as in condition (A3.2) are

$$b(1) = \begin{pmatrix} -1 & 2 \\ 0 & 2 \end{pmatrix}, \quad b(2) = \begin{pmatrix} -3 & -1 \\ 1 & -2 \end{pmatrix}, \quad \sigma_1(2) = 0,$$

$$\sigma_1(1) = \sigma_2(2) = \sigma_2'(1) = \begin{pmatrix} 0 & 1 \\ 0 & 0 \end{pmatrix}, \quad \text{and } \widehat{Q} = \begin{pmatrix} -3 & 3 \\ 1 & -1 \end{pmatrix}.$$

Thus the stationary distribution associated with $\widehat{Q}$ is $\pi = (0.25, 0.75)$, and

$$\lambda_{\max}(b(1) + b'(1) + \sigma_1(1)\sigma_1'(1) + \sigma_2(1)\sigma_2'(1)) = 5.6056,$$
$$\lambda_{\max}(b(2) + b'(2) + \sigma_1(2)\sigma_1'(2) + \sigma_2(2)\sigma_2'(2)) = -4.$$

This yields that

$$\sum_{i=1}^{2} \pi_i \lambda_{\max}(b(i) + b'(i) + \sigma_1(i)\sigma_1'(i) + \sigma_2(i)\sigma_2'(i)) = -1.5986 < 0.$$

Therefore, we conclude from Theorem 3.28 that the switching diffusion (3.70) is positive recurrent. For comparison, we begin by considering a sample path of the switching diffusion. Next treating t as a parameter and eliminating it from the two components x_1 and x_2, we plot a curve of x_2 versus x_1 in the "phase space." Borrowing the terminology from ordinary differential equations, and abusing the terminology slightly, we still call such plots phase portraits henceforth. The phase portrait Figure 3.2 confirms our findings. For comparison and better visualization, we also present the componentwise sample paths in Figure 3.3 (a) and (b).

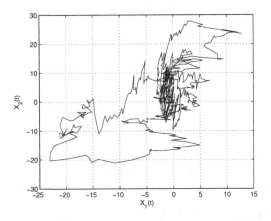

FIGURE 3.2. Phase portrait of switching diffusion (3.70) with initial condition (x, α), where $x = [2.5, 2.5]'$ and $\alpha = 1$.

3.6 Proofs of Several Results

Proof of Lemma 3.7. It suffices to prove the lemma when $E \cup \partial E \subset D$ and ∂E is sufficiently smooth. Fix any $(x, i) \in E^c \times \mathcal{M}$. Let $G \subset \mathbb{R}^r$ be an open

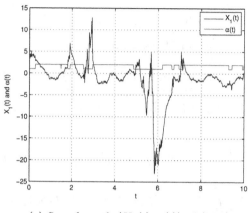

(a) Sample path $(X_1(t), \alpha(t))$ of (3.70).

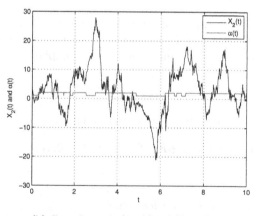

(b) Sample path $(X_2(t), \alpha(t))$ of (3.70).

FIGURE 3.3. Componentwise sample path of switching diffusion (3.70) with initial condition (x, α), where $x = [2.5, 2.5]'$ and $\alpha = 1$.

and bounded set with sufficiently smooth boundary such that $D \cup \partial D \subset G$. Without loss of generality, we may further assume that $(x, i) \in G \times \mathcal{M}$. Define a sequence of stopping times by

$$\varsigma_1 := \inf\{t \geq 0 : X(t) \in \partial G\}, \tag{3.73}$$

and for $n = 1, 2, \ldots$,

$$\begin{aligned}
\varsigma_{2n} &:= \inf\{t \geq \varsigma_{2n-1} : X(t) \in \partial D\}, \\
\varsigma_{2n+1} &:= \inf\{t \geq \varsigma_{2n} : X(t) \in \partial G\}.
\end{aligned} \tag{3.74}$$

It follows from (3.17) and Theorem 3.2 that $\varsigma_n < \infty$ a.s. $\mathbf{P}_{x,i}$ for $n = 1, 2, \ldots$ Let $H := G - \overline{E}$ and define $u(x, i) := \mathbf{P}_{x,i}\{X(\tau_H) \in \partial E\}$. Note that $u(x, j)|_{x \in \partial E} = 1$ and $u(x, j)|_{x \in \partial G} = 0$ for all $j \in \mathcal{M}$. Therefore, it follows that

$$\begin{aligned}
u(x, i) &= \sum_{j=1}^{m_0} \int_{\partial E} \mathbf{P}_{x,i}\{(X(\tau_H), \alpha(\tau_H)) \in (dy \times \{j\})\} u(y, j) \\
&\quad + \sum_{j=1}^{m_0} \int_{\partial G} \mathbf{P}_{x,i}\{(X(\tau_H), \alpha(\tau_H)) \in (dy \times \{j\})\} u(y, j) \\
&= \mathbf{E}_{x,i} u(X(\tau_H), \alpha(\tau_H)).
\end{aligned}$$

Thus $u(x, i) \geq 0$ is $\mathcal{L}$-harmonic in $H \times \mathcal{M}$ by Lemma 3.5. Moreover u is not identically zero since $u(x, i) = 1$ for $(x, i) \in \partial E \times \mathcal{M}$. Therefore the maximum principle for $\mathcal{L}$-harmonic functions [52] implies that

$$\inf_{(x,i) \in K \times \mathcal{M}} u(x, i) \geq \delta_1 > 0, \tag{3.75}$$

where K is some compact subset of H containing x and ∂D. Define

$$A_0 := \{X(t) \in \partial E, \text{ for some } t \in [0, \varsigma_1)\}, \tag{3.76}$$

and for $n = 1, 2, \ldots$,

$$A_n := \{X(t) \in \partial E, \text{ for some } t \in [\varsigma_{2n}, \varsigma_{2n+1})\}. \tag{3.77}$$

Note that the event A_0^c implies that $X(\tau_H) = X(\varsigma_1) \in \partial G$. Hence we have from (3.75) that

$$\mathbf{P}_{x,i}(A_0^c) \leq \mathbf{P}_{x,i}(X(\tau_H) \in \partial G) = 1 - u(x, i) \leq 1 - \delta_1.$$

Then it follows from the strong Markov property and (3.75) that

$$\mathbf{P}_{x,i}\left\{ \bigcap_{k=0}^{n} A_k^c \right\} \leq (1 - \delta_1)^{n+1}. \tag{3.78}$$

Thus, we have

$$\mathbf{P}_{x,i}\{\sigma_E = \infty\} = \mathbf{P}_{x,i}\{X(t) \notin \partial E, \text{ for any } t \geq 0\}$$

$$\leq \lim_{n \to \infty} \mathbf{P}_{x,i}\left\{\bigcap_{k=0}^{n} A_k^c\right\}$$

$$\leq \lim_{n \to \infty} (1 - \delta_1)^{n+1} = 0.$$

It follows that $\mathbf{P}_{x,i}\{\sigma_E < \infty\} = 1$ as desired. □

Proof of Lemma 3.8. As in Lemma 3.7, it is enough to prove the lemma when $E \cup \partial E \subset D$ and ∂E is sufficiently smooth. Fix any $(x,i) \in E^c \times \mathcal{M}$. Let $G \subset \mathbb{R}^r$ be an open and bounded set with sufficiently smooth boundary such that $D \cup \partial D \subset G$. As in the proof of Lemma 3.7, we may further assume that $(x,i) \in G \times \mathcal{M}$. Define stopping times $\varsigma_1, \varsigma_2, \dots$ and events $A_0, A_1, A_2, \dots$ as in (3.73), (3.74), (3.76), and (3.77) in the proof of Lemma 3.7. It follows from (3.18) and Lemma 3.7 that $\mathbf{P}_{x,i}\{\sigma_E < \infty\} = 1$. Note that if $\varsigma_{2n} < \sigma_E < \varsigma_{2n+1}$, then the event $\bigcap_{k=0}^{n-1} A_k^c$ happens a.s. Hence, it follows from (3.78) that

$$\mathbf{P}_{x,i}\{\varsigma_{2n} < \sigma_E < \varsigma_{2n+1}\} \leq \mathbf{P}_{x,i}\left\{\bigcap_{k=0}^{n-1} A_k^c\right\} \leq (1 - \delta_1)^n.$$

Therefore, we have

$$\mathbf{E}_{x,i}\tau_{E^c} = \mathbf{E}_{x,i}\sigma_E I_{\{0 < \sigma_E < \varsigma_1\}} + \sum_{n=1}^{\infty} \mathbf{E}_{x,i}\sigma_E I_{\{\varsigma_{2n} < \sigma_E < \varsigma_{2n+1}\}}$$

$$\leq \mathbf{P}_{x,i}[0 < \sigma_E < \varsigma_1]\mathbf{E}_{x,i}\varsigma_1 + \sum_{n=1}^{\infty} \mathbf{P}_{x,i}[\varsigma_{2n} < \sigma_E < \varsigma_{2n+1}]\mathbf{E}_{x,i}\varsigma_{2n+1}$$

$$\leq \sum_{n=0}^{\infty} (1 - \delta_1)^n \mathbf{E}_{x,i}\varsigma_{2n+1},$$

where I_A is the indicator of the set A. In what follows, denote by M_i ($i = 1, 2, 3$) positive real numbers. Because $(x,i) \in G \times \mathcal{M}$, it follows from Theorem 3.2 that $\mathbf{E}_{x,i}\varsigma_1 = \mathbf{E}_{x,i}\tau_G \leq M_1 < \infty$. Consequently, using σ_D and τ_G defined in (3.7) (where τ_G is defined with G replacing D),

$$\mathbf{E}_{x,i}\varsigma_3 = \mathbf{E}_{x,i}\varsigma_1 + \mathbf{E}_{x,i}\mathbf{E}_{X(\varsigma_1),\alpha(\varsigma_1)}(\varsigma_3 - \varsigma_1)$$

$$\leq M_1 + \sup_{(y,j)\in\partial G\times\mathcal{M}} \mathbf{E}_{y,j}\sigma_D + \sup_{(z,k)\in\partial D\times\mathcal{M}} \mathbf{E}_{z,k}\tau_G$$

$$\leq M_1 + M_2 + M_3 \leq 2M,$$

where $M = \max\{M_1, M_2 + M_3\} < \infty$. Note that in the above deductions, we used equation (3.18) and Theorem 3.2. Likewise, in general, we have

$\mathbf{E}_{x,i}\varsigma_{2n+1} \leq (n+1)M$ for any $n = 1, 2, \ldots$ Therefore, it follows that

$$\mathbf{E}_{x,i}\sigma_E \leq \sum_{n=0}^{\infty}(1 - \delta_1)^n(n+1)M < \infty.$$

This completes the proof of the lemma.   □

Proof of Lemma 3.9. Fix any $\ell \in \mathcal{M}$. It suffices to prove (3.20) when $(x, i) \in D \times (\mathcal{M} - \{\ell\})$ because the process $Y(t) = (X(t), \alpha(t))$, starting from $(y, j) \in D^c \times \mathcal{M}$, will reach $D \times \mathcal{M}$ in finite time a.s. $\mathbf{P}_{y,j}$ by (3.19). Choose $\varepsilon > 0$ sufficiently small such that $B \subset \overline{B} \subset B_1 \subset \overline{B_1} \subset D$, where

$$B = B(x, \varepsilon) = \{y \in \mathbb{R}^r : |y - x| < \varepsilon\}, \quad \text{and} \quad B_1 = B(x, 2\varepsilon). \qquad (3.79)$$

Redefine

$$\varsigma_1 := \inf\{t \geq 0 : X(t) \in \partial B\}, \qquad (3.80)$$

and for $n = 1, 2, \ldots,$

$$\begin{aligned}\varsigma_{2n} &:= \inf\{t \geq \varsigma_{2n-1} : X(t) \in \partial B_1\}, \\ \varsigma_{2n+1} &:= \inf\{t \geq \varsigma_{2n} : X(t) \in \partial B\}.\end{aligned} \qquad (3.81)$$

Note that equation (3.19), Theorem 3.2, and Lemma 3.7 imply that $\varsigma_n < \infty$ a.s. $\mathbf{P}_{x,i}$. Set

$$u(x, i) := \mathbf{P}_{x,i}\left\{\sigma_{\overline{B} \times \{\ell\}} < \tau_{B_1}\right\}.$$

As in the proof of Lemma 3.7, we can verify that $u(x, i)$ is $\mathcal{L}$-harmonic in $B_1 \times \mathcal{M}$. Moreover, u is not identically zero, because $u(x, \ell)|_{x \in \partial B} = 1$. Therefore, the maximum principle [52] implies that

$$\inf_{(x,i) \in \overline{B} \times \mathcal{M}} u(x, i) \geq \delta_2 > 0. \qquad (3.82)$$

Redefine

$$A_0 := \{\alpha(t) = \ell, \text{ for some } t \in [0, \varsigma_2)\}, \qquad (3.83)$$

and for $n = 1, 2, \ldots,$

$$A_n := \{\alpha(t) = \ell, \text{ for some } t \in [\varsigma_{2n+1}, \varsigma_{2n+2})\}. \qquad (3.84)$$

Using almost the same argument as in the proof of Lemma 3.7, we obtain that

$$\mathbf{P}_{x,i}(A_0^c) \leq 1 - \delta_2, \quad \text{and} \quad \mathbf{P}_{x,i}\left\{\bigcap_{k=0}^{n} A_k^c\right\} \leq (1 - \delta_2)^{n+1}. \qquad (3.85)$$

Thus, we have

$$\mathbf{P}_{x,i}\left\{(X(t),\alpha(t)) \notin D \times \{\ell\},\ \text{for any}\ t \geq 0\right\}$$
$$\leq \mathbf{P}_{x,i}\left\{(X(t),\alpha(t)) \notin \overline{B_1} \times \{\ell\},\ \text{for any}\ t \geq 0\right\}$$
$$\leq \lim_{n\to\infty} \mathbf{P}_{x,i}\left\{\bigcap_{k=0}^{n} A_k^c\right\}$$
$$\leq \lim_{n\to\infty}(1-\delta_2)^{n+1} = 0.$$

As a result,

$$\mathbf{P}_{x,i}\{\sigma_{D\times\{\ell\}} = \infty\} = \mathbf{P}_{x,i}\left\{(X(t),\alpha(t)) \notin D \times \{\ell\},\ \text{for any}\ t \geq 0\right\} = 0,$$

or $\mathbf{P}_{x,i}\{\sigma_{D\times\{\ell\}} < \infty\} = 1$. This completes the proof of the lemma. □

Proof of Lemma 3.10. Fix any $\ell \in \mathcal{M}$. As in Lemma 3.9, it is enough to prove (3.22) when $(x,i) \in D \times (\mathcal{M} - \{\ell\})$. Let the balls B and B_1, stopping times $\varsigma_1, \varsigma_2, \ldots$, and events $A_0, A_1, \ldots$ be as in (3.79)–(3.81), (3.83), and (3.84) in the proof of Lemma 3.9. It follows from equation (3.21) and Lemma 3.9 that $\mathbf{P}_{x,i}\{\sigma_{D\times\{\ell\}} < \infty\} = 1$. Observe that if $\varsigma_{2n} \leq \sigma_{D\times\{\ell\}} < \varsigma_{2n+2}$, then the event $\bigcap_{k=0}^{n-1} A_k^c$ happens a.s. Hence we have from (3.85) that

$$\mathbf{P}_{x,i}\{\varsigma_{2n} \leq \sigma_{D\times\{\ell\}} < \varsigma_{2n+2}\} \leq \mathbf{P}_{x,i}\left\{\bigcap_{k=0}^{n-1} A_k^c\right\} \leq (1-\delta_2)^n.$$

It follows that

$$\mathbf{E}_{x,i}\sigma_{D\times\{\ell\}} = \mathbf{E}_{x,i}\sigma_{D\times\{\ell\}}I_{\{0 \leq \sigma_{D\times\{\ell\}} < \varsigma_2\}}$$
$$+ \sum_{n=1}^{\infty} \mathbf{E}_{x,i}\sigma_{D\times\{\ell\}}I_{\{\varsigma_{2n} \leq \sigma_{D\times\{\ell\}} < \varsigma_{2n+2}\}}$$
$$\leq \mathbf{P}_{x,i}[0 \leq \sigma_{D\times\{\ell\}} < \varsigma_2]\mathbf{E}_{x,i}\varsigma_2$$
$$+ \sum_{n=1}^{\infty} \mathbf{P}_{x,i}[\varsigma_{2n} \leq \sigma_{D\times\{\ell\}} < \varsigma_{2n+2}]\mathbf{E}_{x,i}\varsigma_{2n+2}$$
$$\leq \sum_{n=0}^{\infty}(1-\delta_2)^n \mathbf{E}_{x,i}\varsigma_{2n+2}.$$

Following almost the same argument as that for the proof of Lemma 3.8, we can show that $\mathbf{E}_{x,i}\varsigma_{2n} \leq nM$ for some positive constant M. Consequently,

$$\mathbf{E}_{x,i}\sigma_{D\times\{\ell\}} \leq \sum_{n=0}^{\infty}(1-\delta_2)^n(n+1)M < \infty.$$

The proof of the lemma is thus completed. □

Proof of Proposition 3.19. In view of Theorem 3.18, it is enough to prove the first assertion only. Denote $B = B(x_0, \rho)$ and $U = B \times \{\ell\}$. Define for any $T > 0$ and any $0 < \varepsilon < \rho$ a sequence of stopping times by $\varsigma_0 := 0$, and for $n \geq 1$,

$$\varsigma_n := \inf \{t > \varsigma_{n-1} + T : (X(t), \alpha(t)) \in B(x_0, \varepsilon) \times \{\ell\}\}.$$

(We use the convention that $\inf \{\emptyset\} = \infty$.) Then by virtue of the strong Markov property, we can show

$$\sup_{x \in B(x_0, \varepsilon)} \mathbf{P}_{x,\ell} \{\varsigma_n < \infty\} \leq \Big(\sup_{x \in B(x_0, \varepsilon)} \mathbf{P}_{x,\ell} \{\varsigma_1 < \infty\} \Big)^n. \tag{3.86}$$

Hence we have

$$\begin{aligned} R_0(x_0, \ell, U) &= \sum_{n=0}^{\infty} \mathbf{E}_{x_0, \ell} \left[I_{\{\varsigma_n < \infty\}} \int_{\varsigma_n}^{\varsigma_{n+1}} I_U(X(t), \alpha(t)) dt \right] \\ &= \sum_{n=0}^{\infty} \mathbf{E}_{x_0, \ell} \left[I_{\{\varsigma_n < \infty\}} \mathbf{E}_{X(\varsigma_n), \alpha(\varsigma_n)} \left(\int_0^{\varsigma_1} I_U(X(t), \alpha(t)) dt \right) \right]. \end{aligned}$$

Note that

$$\int_0^{\varsigma_1} I_U(X(t), \alpha(t)) dt = \int_0^T I_U(X(t), \alpha(t)) dt + \int_T^{\varsigma_1} I_U(X(t), \alpha(t)) dt \leq T.$$

It follows from (3.86) that

$$\begin{aligned} R_0(x_0, \ell, U) &\leq T \sum_{n=0}^{\infty} \mathbf{P}_{x_0, \ell} \{\varsigma_n < \infty\} \\ &\leq T \sum_{n=0}^{\infty} \Big(\sup_{x \in B(x_0, \ell)} \mathbf{P}_{x,\ell} \{\varsigma_1 < \infty\} \Big)^n. \end{aligned}$$

But $R_0(x_0, \ell, U) = \infty$, thus

$$\sup_{x \in B(x_0, \varepsilon)} \mathbf{P}_{x,\ell} \left\{ \sigma_{B(x_0,\varepsilon) \times \{\ell\}}^T < \infty \right\} = \sup_{x \in B(x_0, \ell)} \mathbf{P}_{x,\ell} \{\varsigma_1 < \infty\} = 1,$$

where $\sigma_{B(x_0,\varepsilon) \times \{\ell\}}^T := \varsigma_1 = \inf \{t > T : (X(t), \alpha(t)) \in B(x_0, \varepsilon) \times \{\ell\}\}$. Because $B(x_0, \varepsilon) \subset B$,

$$\mathbf{P}_{x,\ell} \left\{ \sigma_{B(x_0,\varepsilon) \times \{\ell\}}^T < \infty \right\} \leq \mathbf{P}_{x,\ell} \left\{ \sigma_{B \times \{\ell\}}^T < \infty \right\},$$

and $\sup_{x \in B(x_0, \varepsilon)} \mathbf{P}_{x,\ell} \left\{ \sigma_{B \times \{\ell\}}^T < \infty \right\} = 1$. Finally, because $(X(t), \alpha(t))$ is strong Feller, we obtain by letting $\varepsilon \to 0$ that

$$\mathbf{P}_{x_0, \ell} \left\{ \sigma_{B \times \{\ell\}}^T < \infty \right\} = 1$$

for any $T > 0$. Thus (3.30) follows. This completes the proof of the proposition. $\qquad\square$

Proof of Theorem 3.20. Sufficiency. This is clear by a contradiction argument.

Necessity. Assume the process $(X(t), \alpha(t))$ is transient. Fix any $(x, \alpha) \in \mathbb{R}^r \times \mathcal{M}$. Let $\rho \in \mathbb{R}$ be sufficiently large such that $\rho > 1 \vee |x|$. The process $(X(t), \alpha(t))$ is transient, therefore it is transient with respect to the "cylinder" $B(0, \rho) \times \mathcal{M}$. Thus there exists some $(y_0, j_0) \in (\mathbb{R}^r - B(0, \rho)) \times \mathcal{M}$ such that

$$\mathbf{P}_{y_0, j_0} \left\{ \sigma_{B(0, \rho)} < \infty \right\} < 1. \tag{3.87}$$

Assume $|y_0| = r_0 > \rho$. Then by virtue of Lemma 3.4, the function $(y, j) \mapsto \mathbf{P}_{y, j} \left\{ \sigma_{B(0, \rho)} < \infty \right\}$ is $\mathcal{L}$-harmonic. Hence it follows from the maximum principle for $\mathcal{L}$-harmonic functions [52] and (3.87) that

$$\sup_{|y| = r_0, j \in \mathcal{M}} \mathbf{P}_{y, j} \left\{ \sigma_{B(0, \rho)} < \infty \right\} = \delta < 1. \tag{3.88}$$

Then using the standard argument (see [15, Theorem 3.2]), we can show that

$$\mathbf{P}_{x, \alpha} \left\{ \liminf_{t \to \infty} |X(t)| > \rho - 1 \right\} = 1.$$

Because this is true for all $\rho > 0$,

$$\mathbf{P}_{x, \alpha} \left\{ |X(t)| \to \infty \text{ as } t \to \infty \right\} = 1,$$

as desired. $\qquad\square$

Proof of Proposition 3.21. By virtue of Theorem 3.18, it is enough to show that the point (x_0, ℓ) is transient in the sense that there exists some $\varepsilon_0 > 0$ and a finite time $T_0 > 0$ such that

$$\mathbf{P}_{x_0, \ell} \left\{ (X(t), \alpha(t)) \notin B(x_0, \varepsilon_0) \times \{\ell\}, \text{ for all } t \geq T_0 \right\} = 1. \tag{3.89}$$

Clearly $R_0(x_0, \ell, U) = \mathbf{E}_{x_0, \ell} \int_0^\infty I_U(X(t), \alpha(t)) dt > 0$. Thus there exists some $t > 0$ such that

$$\mathbf{E}_{x_0, \ell} \int_0^t I_U(X(t), \alpha(t)) dt > 0.$$

Because the process $(X(t), \alpha(t))$ is strong Feller by virtue of Theorem 2.24, we conclude that there exists a neighborhood $E \subset D$ of x_0 such that

$$\inf_{y \in E} \mathbf{E}_{y, \ell} \int_0^t I_U(X(t), \alpha(t)) dt > 0.$$

Hence we have

$$\delta := \inf_{x \in E} R_0(x, \ell, U) > 0. \tag{3.90}$$

For each $T > 0$, define

$$\sigma^T_{E \times \{\ell\}} = \inf \{t > T : (X(t), \alpha(t)) \in E \times \{\ell\}\}. \tag{3.91}$$

Then it follows from the strong Markov property that

$$R_0(x_0, \ell, U) \geq \mathbf{E}_{x_0, \ell} \int_T^\infty I_U(X(t), \alpha(t)) dt$$

$$\geq \mathbf{E}_{x_0, \ell} \left[I_{\left\{\sigma^T_{E \times \{\ell\}} < \infty\right\}} \int_{\sigma^T_{E \times \{\ell\}}}^\infty I_U(X(t), \alpha(t)) dt \right]$$

$$= \mathbf{E}_{x_0, \ell} \left[I_{\left\{\sigma^T_{E \times \{\ell\}} < \infty\right\}} R_0(X(\sigma^T_{E \times \{\ell\}}), \alpha(\sigma^T_{E \times \{\ell\}}), U) \right]$$

$$\geq \inf_{x \in E} R_0(x, \ell, U) \mathbf{P}_{x_0, \ell} \left\{\sigma^T_{E \times \{\ell\}} < \infty\right\}.$$

Hence we have from (3.90) and (3.91) that

$$\mathbf{P}_{x_0, \ell} \left\{\sigma^T_{E \times \{\ell\}} < \infty\right\} \leq \frac{1}{\delta} \mathbf{E}_{x_0, \ell} \int_T^\infty I_U(X(t), \alpha(t)) dt.$$

For any $\varepsilon > 0$, in view of the assumption $R_0(x_0, \ell, U) < \infty$, we can choose some $\widetilde{T}_0 > 0$ such that

$$\mathbf{E}_{x_0, \ell} \int_{\widetilde{T}_0}^\infty I_U(X(t), \alpha(t)) dt < \delta \varepsilon$$

and hence

$$\mathbf{P}_{x_0, \ell} \left\{\sigma^{\widetilde{T}_0}_{E \times \{\ell\}} < \infty\right\} < \varepsilon.$$

That is,

$$\lim_{T \to \infty} \mathbf{P}_{x_0, \ell} \left\{\sigma^T_{E \times \{\ell\}} < \infty\right\}$$
$$= \mathbf{P}_{x_0, \ell} \left\{\sigma^T_{E \times \{\ell\}} < \infty, \text{ for every } T > 0\right\} = 0.$$

Therefore there exists some $T_0 > 0$ such that

$$\mathbf{P}_{x_0, \ell} \left\{\sigma^{T_0}_{E \times \{\ell\}} = \infty\right\} > 0. \tag{3.92}$$

Hence (3.89) follows.   □

3.7 Notes

Under general conditions, necessary and sufficient conditions for recurrence, nonrecurrence, and positive recurrence have been studied in this chapter. We refer the reader to Chapter 2; see also Skorohod [150] for related stochastic differential equations involving Poisson measures describing the evolution of the switching processes. In our formulation, the finite-state process depicts a random environment that has right-continuous sample paths and that cannot be described by a diffusion. Consequently, both continuous dynamics (diffusions) and discrete events (jumps) coexist yielding hybrid dynamic systems, which provide a more realistic formulation for many applications. The discrete events are frequently used to provide more realistic models and to capture random evolutions. For instance, the switching may be used to describe stochastic volatility resulting from market modes and interest rates, as well as other economic factors in modeling financial markets, to enhance the versatility in risk management practice, to better understand ruin probability in insurance, and to carry out dividend optimization tasks.

Regime-switching diffusions have received much attention lately. For instance, optimal controls of switching diffusions were studied in [11] using a martingale problem formulation; jump-linear systems were treated in [78]; stability of semi-linear stochastic differential equations with Markovian switching was considered in [6]; ergodic control problems of switching diffusions were studied in [52]; stability of stochastic differential equations with Markovian switching was dealt with in [116, 136, 183]; asymptotic expansions for solutions of integro-differential equations for transition densities of singularly perturbed switching-diffusion processes were developed in [74]; switching diffusions were used for stock liquidation models in [184]. For some recent applications of hybrid systems in communication networks, air traffic management, control problems, and so on, we refer the reader to [67, 68, 123, 137, 155] and references therein.

In [6, 116, 183, 184], $Q(x) = Q$, a constant matrix. In such cases, $\alpha(\cdot)$ is a continuous-time Markov chain. Moreover, it is assumed that the Markov chain $\alpha(\cdot)$ is independent of Brownian motion. In our formulation, x-dependent $Q(x)$ is considered, and as a result, the transition rates of the discrete event $\alpha(\cdot)$ depend on the continuous dynamic $X(\cdot)$, as depicted in (3.4). Although the pair $(X(\cdot), \alpha(\cdot))$ is a Markov process, for x-dependent $Q(x)$, only for each fixed x, the discrete-event process $\alpha(\cdot)$ is a Markov chain. Such formulation enables us to describe complex systems and their inherent uncertainty and randomness in the environment. However, it adds much difficulty in analysis. Our formulation is motivated by the fact that in many applications, the discrete event and continuous dynamic are intertwined. It would be useful to relax the independence assumption of the discrete-event process and Brownian motion.

As seen in this chapter, the study of switching diffusions is connected with systems of partial differential equations. The works [1, 39, 46, 56, 60, 105, 108, 158] and references therein provide a systematic treatment of partial differential equations and systems of partial differential equations. These tools are handy to use.

4
Ergodicity

4.1 Introduction

Continuing with the study of basic properties of switching-diffusion processes, this chapter is concerned with ergodicity. Many applications in control and optimization require minimizing an expected cost of certain objective functions. Treating average cost per unit time problems, we often wish to "replace" the time-dependent instantaneous measure by a steady-state (or ergodic) measure. Thus we face the following questions: Do the systems possess an ergodic property? Under what conditions do the systems have the desired ergodicity? Significant effort has been devoted to approximating such expected values by replacing the instantaneous measures with stationary measures when the time horizon is long enough. To justify such a replacement, ergodicity is needed. For diffusion processes, we refer the readers to, for example, [10, 103] among others for the study of ergodic control problems. In what follows, we study ergodicity and reveal the main features of the ergodic measures. We carry out our study on ergodicity by constructing cycles and using induced discrete-time Markov chains.

We consider the two-component process $Y(t) = (X(t), \alpha(t))$ as in Chapter 3. Let $w(t)$ be a d-dimensional standard Brownian motion, $b(\cdot, \cdot) : \mathbb{R}^r \times \mathcal{M} \mapsto \mathbb{R}^r$, and $\sigma(\cdot, \cdot) : \mathbb{R}^r \times \mathcal{M} \mapsto \mathbb{R}^{r \times d}$ satisfying $\sigma(x, i)\sigma'(x, i) = a(x, i)$. For $t \geq 0$, let $X(t) \in \mathbb{R}^r$ and $\alpha(t) \in \mathcal{M}$ such that

$$dX(t) = b(X(t), \alpha(t))dt + \sigma(X(t), \alpha(t))dw(t),$$
$$X(0) = x, \quad \alpha(0) = \alpha, \tag{4.1}$$

G.G. Yin and C. Zhu, *Hybrid Switching Diffusions: Properties and Applications*, Stochastic Modelling and Applied Probability 63, DOI 10.1007/978-1-4419-1105-6_4, © Springer Science + Business Media, LLC 2010

and

$$\mathbf{P}\{\alpha(t + \Delta) = j | \alpha(t) = i, X(s), \alpha(s), s \leq t\}$$
$$= q_{ij}(X(t))\Delta + o(\Delta), \quad i \neq j. \tag{4.2}$$

The rest of the chapter is arranged as follows. Section 4.2 begins with the discussion of ergodicity. The analysis is carried out by using cycles and induced Markov chains in discrete time. Then the desired result is obtained together with the representation of the stationary density. Section 4.3 takes up the issue of making a switching-diffusion process ergodic by means of feedback controls. Section 4.4 discusses some ramifications, and Section 4.5 obtains asymptotic normality when the continuous component belongs to a compact set. Section 4.6 presents some further remarks.

4.2   Ergodicity

In this section, we study the ergodic properties of the process $Y(t) = (X(t), \alpha(t))$ under the assumption that the process is positive recurrent with respect to some bounded domain $U = E \times \{\ell\}$, where $E \subset \mathbb{R}^r$ and $\ell \in \mathcal{M}$ are fixed throughout this section. We also assume that the boundary ∂E of E is sufficiently smooth. Let the operator $\mathcal{L}$ satisfy (A3.1). Then it follows from Theorem 3.12 that the process is positive recurrent with respect to any nonempty open set.

Let $D \subset \mathbb{R}^r$ be a bounded open set with sufficiently smooth boundary ∂D such that $E \cup \partial E \subset D$. Let $\varsigma_0 = 0$ and define the stopping times $\varsigma_1, \varsigma_2, \ldots$ inductively as follows: ς_{2n+1} is the first time after ς_{2n} at which the process $Y(t) = (X(t), \alpha(t))$ reaches the set $\partial E \times \{\ell\}$ and ς_{2n+2} is the first time after ς_{2n+1} at which the path reaches the set $\partial D \times \{\ell\}$. Now we can divide an arbitrary sample path of the process $Y(t) = (X(t), \alpha(t))$ into cycles:

$$[\varsigma_0, \varsigma_2), [\varsigma_2, \varsigma_4), \ldots, [\varsigma_{2n}, \varsigma_{2n+2}), \ldots \tag{4.3}$$

Figure 4.1 presents a demonstration of such cycles when the discrete component $\alpha(\cdot)$ has three states.

The process $Y(t) = (X(t), \alpha(t))$ is positive recurrent with respect to $E \times \{\ell\}$ and hence positive recurrent with respect to $D \times \{\ell\}$ by Theorem 3.12. It follows that all the stopping times $\varsigma_0 < \varsigma_1 < \varsigma_2 < \varsigma_3 < \varsigma_4 < \cdots$ are finite almost surely (a.s.). Because the process $Y(t) = (X(t), \alpha(t))$ is positive recurrent, we may assume without loss of generality that $Y(0) = (X(0), \alpha(0)) = (x, \ell) \in \partial D \times \{\ell\}$. It follows from the strong Markov property of the process $Y(t) = (X(t), \alpha(t))$ that the sequence $\{Y_n\}$ is a Markov chain on $\partial D \times \{\ell\}$, where $Y_n = Y(\varsigma_{2n}) = (X_n, \ell)$, $n = 0, 1, \ldots$ Let $\widetilde{P}(x, A)$ denote the one-step transition probabilities of this Markov chain; that is,

$$\widetilde{P}(x, A) = \mathbf{P}\left(Y_1 \in (A \times \{\ell\}) \mid Y_0 = (x, \ell)\right)$$

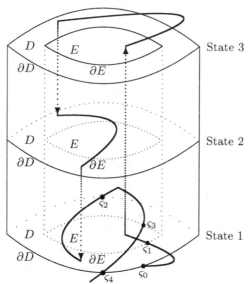

FIGURE 4.1. A sample path of the process $Y(t) = (X(t), \alpha(t))$ when $m_0 = 3$.

for any $x \in \partial D$ and $A \in \mathcal{B}(\partial D)$, where $\mathcal{B}(\partial D)$ denotes the collection of Borel measurable sets on ∂D. Note that the process $Y(t) = (X(t), \alpha(t))$, starting from (x, ℓ), may jump many times before it reaches the set (A, ℓ); see [150] for more details. Denote by $\widetilde{P}^{(n)}(x, A)$ the n-step transition probability of the Markov chain for any $n \geq 1$. For any Borel measurable function $f : \mathbb{R}^r \mapsto \mathbb{R}$, set

$$\mathbf{E}_x f(X_1) := \mathbf{E}_{x,\ell} f(X_1) = \int_{\partial D} f(y) \widetilde{P}(x, dy). \tag{4.4}$$

Throughout this section, we write $\mathbf{E}_x$ in lieu of $\mathbf{E}_{x,\ell}$ for simplicity. We show that the process $Y(t) = (X(t), \alpha(t))$ possesses a unique stationary distribution. To this end, we need the following lemma.

Lemma 4.1. *The Markov chain $Y_i = (X_i, \ell)$ has a unique stationary distribution $m(\cdot)$ such that the n-step transition matrix $\widetilde{P}^{(n)}(x, A)$ satisfies*

$$\left| \widetilde{P}^{(n)}(x, A) - m(A) \right| < \lambda^n, \quad \text{for any } A \in \mathcal{B}(\partial D), \tag{4.5}$$

for some constant $0 < \lambda < 1$.

Proof. Note that

$$\widetilde{P}(x, A) = \mathbf{P}\{Y_1 \in (A \times \{\ell\}) | Y_0 = (x, \ell)\}$$
$$= \int_{\partial D} \mathbf{P}_{x,\ell}\{(X(\varsigma_1), \alpha(\varsigma_1)) \in (dy \times \{\ell\})\}$$
$$\times \mathbf{P}_{y,\ell}\{(X(\varsigma_2), \alpha(\varsigma_2)) \in (A \times \{\ell\})\}.$$

Using the harmonic measure defined and Lemmas 2.2 and 2.3 in Chen and Zhao [25], relating the kernel and surface area (similar to the solution for the diffusion process without switching in the form of double-layer potential given in the first displayed equation in [83, p. 97]) and the harmonic measure, we can finish the proof of this lemma analogously to that of [83, Lemma 4.4.1]. The details are omitted.   □

Remark 4.2. Note that

$$
\begin{aligned}
&(X^{s,X(s),\alpha(s)}(t), \alpha^{s,X(s),\alpha(s)}(t)) \\
&= (X^{0,X(0),\alpha(0)}(t+s), \alpha^{0,X(0),\alpha(0)}(t+s)),
\end{aligned}
\tag{4.6}
$$

where $(X^{0,X(0),\alpha(0)}(u), \alpha^{0,X(0),\alpha(0)}(u))$ denotes the sample path of the process $(X(\cdot), \alpha(\cdot))$ with initial point $(X(0), \alpha(0))$ at time $t = 0$, and a similar definition for $(X^{s,X(s),\alpha(s)}(t), \alpha^{s,X(s),\alpha(s)}(t))$. When no confusion arises, we simply write

$$
(X(u), \alpha(u)) = (X^{0,X(0),\alpha(0)}(u), \alpha^{0,X(0),\alpha(0)}(u)).
$$

Let τ be an $\mathcal{F}_t$ stopping time with $\mathbf{E}_{x,i}\tau < \infty$ and let $f : \mathbb{R}^r \times \mathcal{M} \mapsto \mathbb{R}$ be a Borel measurable function. Then

$$
\begin{aligned}
&\mathbf{E}_{x,i} \int_0^\tau f(X(s+t), \alpha(s+t))ds \\
&= \mathbf{E}_{x,i} \int_0^\tau \mathbf{E}_{X(s),\alpha(s)} f(X(s+t), \alpha(s+t))ds.
\end{aligned}
\tag{4.7}
$$

Now we can construct the stationary distribution of the process $Y(t) = (X(t), \alpha(t))$ explicitly.

Theorem 4.3. *The positive recurrent process* $Y(t) = (X(t), \alpha(t))$ *has a unique stationary distribution* $\nu(\cdot, \cdot) = (\nu(\cdot, i) : i \in \mathcal{M})$.

Proof. Recall that the cycles were defined in (4.3). Let $A \in \mathcal{B}(\mathbb{R}^r)$ and $i \in \mathcal{M}$. Denote by $\tau^{A \times \{i\}}$ the time spent by the path of $Y(t) = (X(t), \alpha(t))$ in the set $(A \times \{i\})$ during the first cycle. Set

$$
\widehat{\nu}(A, i) := \int_{\partial D} m(dx)\mathbf{E}_x \tau^{A \times \{i\}},
\tag{4.8}
$$

where $m(\cdot)$ is the stationary distribution of $Y_i = (X_i, \ell)$, whose existence is guaranteed by Lemma 4.1. It is easy to verify that $\widehat{\nu}(\cdot, \cdot)$ is a positive measure defined on $\mathcal{B}(\mathbb{R}^r) \times \mathcal{M}$. Thus for any bounded Borel measurable function $g(\cdot) : \mathbb{R}^r \mapsto \mathbb{R}$, it follows from (4.4) and Fubini's theorem that

$$
\int_{\partial D} \mathbf{E}_x g(X_1) m(dx) = \int_{\partial D} m(dx) \int_{\partial D} g(y)\widetilde{P}(x, dy) = \int_{\partial D} g(y)m(dy).
\tag{4.9}
$$

Now we claim that for any bounded and continuous function $f(\cdot,\cdot)$,

$$\sum_{j=1}^{m_0} \int_{\mathbb{R}^r} f(y,j)\widehat{\nu}(dy,j) = \int_{\partial D} m(dx)\mathbf{E}_x \int_0^{\varsigma_2} f(X(t),\alpha(t))dt \qquad (4.10)$$

holds. In fact, if f is an indicator function, that is, $f(y,j)I_{A\times\{i\}}(y,j)$ for some $A \in \mathcal{B}(\mathbb{R}^r)$ and $i \in \mathcal{M}$, then from (4.8),

$$\sum_{j=1}^{m_0} \int_{\mathbb{R}^r} I_{A\times\{i\}}(y,j)\widehat{\nu}(dy,j)$$
$$= \widehat{\nu}(A,i) = \int_{\partial D} m(dx)\mathbf{E}_x \tau^{A\times\{i\}}$$
$$= \int_{\partial D} m(dx)\mathbf{E}_x \int_0^{\varsigma_2} I_{A\times\{i\}}(X(t),\alpha(t))dt.$$

Similarly, (4.10) holds for f being a simple function of the form

$$f(y,j) = \sum_{p=1}^{n} c_p I_{U_p}(y,j),$$

where $U_p \subset \mathbb{R}^r \times \mathcal{M}$. Finally, if f is a bounded and continuous function, equation (4.10) follows by approximating f by simple functions. It follows from equations (4.10), (4.6), and (4.7) that

$$\sum_{i=1}^{m_0} \int_{\mathbb{R}^r} \mathbf{E}_{x,i} f(X(t),\alpha(t))\widehat{\nu}(dx,i)$$
$$= \int_{\partial D} m(dx)\mathbf{E}_x \int_0^{\varsigma_2} \mathbf{E}_{X(s),\alpha(s)} f(X(t+s),\alpha(t+s))ds$$
$$= \int_{\partial D} m(dx)\mathbf{E}_x \int_0^{\varsigma_2} f(X(t+s),\alpha(t+s))ds$$
$$= \int_{\partial D} m(dx)\mathbf{E}_x \int_0^{\varsigma_2} f(X(u),\alpha(u))du$$
$$+ \int_{\partial D} m(dx)\mathbf{E}_x \int_{\varsigma_2}^{t+\varsigma_2} f(X(u),\alpha(u))du$$
$$- \int_{\partial D} m(dx)\mathbf{E}_x \int_0^{t} f(X(u),\alpha(u))du.$$

Applying (4.9) with

$$g(x) = \mathbf{E}_x \int_{\varsigma_2}^{\varsigma_2+t} f(X(u),\alpha(u))du,$$

we obtain

$$
\int_{\partial D} m(dx)\mathbf{E}_x \int_{\varsigma_2}^{\varsigma_2+t} f(X(u), \alpha(u))du
$$
$$
= \int_{\partial D} m(dx)\mathbf{E}_x \mathbf{E}_{X_1, \ell} \int_{\varsigma_2}^{\varsigma_2+t} f(X(u+\varsigma_2), \alpha(u+\varsigma_2))du
$$
$$
= \int_{\partial D} m(dx)\mathbf{E}_x \int_0^t f(X(u), \alpha(u))du.
$$

Note that in the above deduction, we used equation (4.6) again. Therefore, the above two equations and (4.10) yield that

$$
\sum_{i=1}^{m_0} \int_{\mathbb{R}^r} \mathbf{E}_{x,i} f(X(t), \alpha(t))\widehat{\nu}(dx, i) = \sum_{i=1}^{m_0} \int_{\mathbb{R}^r} f(x, i)\widehat{\nu}(dx, i).
$$

Thus, the normalized measure

$$
\nu(A, i) = \frac{\widehat{\nu}(A, i)}{\sum_{j=1}^{m_0} \widehat{\nu}(\mathbb{R}^r, j)}, \quad i \in \mathcal{M}, \tag{4.11}
$$

defines the desired stationary distribution. The theorem thus follows.   □

Theorem 4.4. *Denote by* $\mu(\cdot, \cdot)$ *the stationary density associated with the stationary distribution* $\nu(\cdot, \cdot)$ *constructed in Theorem 4.3 and let* $f(\cdot, \cdot)$: $\mathbb{R}^r \times \mathcal{M} \mapsto \mathbb{R}$ *be a Borel measurable function such that*

$$
\sum_{i=1}^{m_0} \int_{\mathbb{R}^r} |f(x, i)|\mu(x, i)dx < \infty. \tag{4.12}
$$

Then

$$
\mathbf{P}_{x,i} \left(\lim_{T \to \infty} \frac{1}{T} \int_0^T f(X(t), \alpha(t))dt = \overline{f} \right) = 1, \tag{4.13}
$$

for any $(x, i) \in \mathbb{R}^r \times \mathcal{M}$, *where*

$$
\overline{f} = \sum_{i=1}^{m_0} \int_{\mathbb{R}^r} f(x, i)\mu(x, i)dx. \tag{4.14}
$$

Proof. We first prove (4.13) if the initial distribution is the stationary distribution of the Markov chain $Y_i = (X_i, \alpha_i)$; that is,

$$
\mathbf{P}\{(X(0), \alpha(0)) \in (A \times \{\ell\})\} = m(A) \tag{4.15}
$$

for any $A \in \mathcal{B}(\partial D)$. Consider the sequence of random variables

$$
\eta_n = \int_{\varsigma_{2n}}^{\varsigma_{2n+2}} f(X(t), \alpha(t))dt. \tag{4.16}
$$

Then it follows from equation (4.15) that $\{\eta_n\}$ is a strictly stationary sequence. Also from equations (4.8) and (4.10), we have

$$\mathbf{E}\eta_n = \sum_{i=1}^{m_0} \int_{\mathbb{R}^r} f(x, i)\widehat{\nu}(dx, i), \tag{4.17}$$

for all $n = 0, 1, 2, \ldots$ Meanwhile, equation (4.5) implies that the sequence η_n is metrically transitive. Let $\upsilon(T)$ denote the number of cycles completed up to time T. That is,

$$\upsilon(T) := \max\left\{ n \in \mathbb{N} : \sum_{k=1}^{n}(\varsigma_{2k} - \varsigma_{2k-2}) \leq T \right\}.$$

Then we can decompose $\int_0^T f(X(t), \alpha(t))dt$ into

$$\int_0^T f(X(t), \alpha(t))dt = \sum_{n=0}^{\upsilon(T)} \eta_n + \int_{\varsigma_{2\upsilon(T)}}^T f(X(t), \alpha(t))dt, \tag{4.18}$$

with η_n given in equation (4.16). We may assume without loss of generality that $f(x, i) \geq 0$ (for the general case, we can write $f(x, i)$ as a difference of two nonnegative functions). Then it follows from equation (4.18) that

$$\sum_{n=0}^{\upsilon(T)} \eta_n \leq \int_0^T f(X(t), \alpha(t))dt \leq \sum_{n=0}^{\upsilon(T)+1} \eta_n.$$

Inasmuch as the sequence $\{\eta_n\}$ is stationary and metrically transitive, the law of large numbers for such sequences implies that

$$\mathbf{P}\left\{ \frac{1}{n}\sum_{k=0}^{n} \eta_k \to \sum_{i=1}^{m_0} \int_{\mathbb{R}^r} f(x, i)\widehat{\nu}(dx, i), \quad \text{as } n \to \infty \right\} = 1. \tag{4.19}$$

In particular, if $f(x, i) \equiv 1$, then the above equation reduces to

$$\mathbf{P}\left\{ \frac{\varsigma_{2n+2}}{n} \to \sum_{i=1}^{m_0} \widehat{\nu}(\mathbb{R}^r, i), \quad \text{as } n \to \infty \right\} = 1. \tag{4.20}$$

Note that the positive recurrence of the process $Y(t) = (X(t), \alpha(t))$ implies that $\upsilon(T) \to \infty$ as $T \to \infty$. Clearly, $\upsilon(T)/(\upsilon(T) + 1) \to 1$ almost surely as $T \to \infty$. Thus, it follows from (4.20) that as $T \to \infty$,

$$\frac{\varsigma_{2\upsilon(T)}}{\varsigma_{2\upsilon(T)+2}} = \frac{\dfrac{\varsigma_{2\upsilon(T)}}{\upsilon(T)}}{\dfrac{\varsigma_{2\upsilon(T)+2}}{\upsilon(T)+1}}\frac{\upsilon(T)}{\upsilon(T)+1} \to 1 \quad \text{a.s.} \tag{4.21}$$

Meanwhile, since $\varsigma_{2\upsilon(T)} \le T \le \varsigma_{2\upsilon(T)+2}$, we have

$$\frac{\varsigma_{2\upsilon(T)}}{\varsigma_{2\upsilon(T)+2}} \le \frac{\varsigma_{2\upsilon(T)}}{T} \le \frac{\varsigma_{2\upsilon(T)}}{\varsigma_{2\upsilon(T)}} = 1.$$

Therefore, we have from (4.21) that

$$\frac{\varsigma_{2\upsilon(T)}}{T} \to 1 \quad \text{a.s. as } T \to \infty. \tag{4.22}$$

Moreover, (4.20) implies that

$$\frac{\upsilon(T)}{\varsigma_{2\upsilon(T)}} \to \frac{1}{\sum_{i=1}^{m_0} \widehat{\nu}(\mathbb{R}^r, i)} \quad \text{a.s. as } T \to \infty. \tag{4.23}$$

Now using equations (4.19), (4.22), and (4.23), we obtain

$$\mathbf{P}\bigg\{ \frac{1}{T}\int_0^T f(X(t), \alpha(t))dt = \frac{\int_0^T f(X(t), \alpha(t))dt}{\upsilon(T)} \frac{\upsilon(T)}{\varsigma_{2\upsilon(T)}} \frac{\varsigma_{2\upsilon(T)}}{T}$$

$$\xrightarrow{T\to\infty} \sum_{i=1}^{m_0} \int_{\mathbb{R}^r} f(x, i)\nu(dx, i) \bigg\} = 1.$$

Finally, we note that

$$\int_{\mathbb{R}^r} f(x, i)\nu(dx, i) = \int_{\mathbb{R}^r} f(x, i)\mu(x, i)dx$$

by the definition of $\mu(\cdot, \cdot)$. Thus, equation (4.13) holds. This proves (4.13) if the initial distribution is (4.15).

Let $(x, i) \in \mathbb{R}^r \times \mathcal{M}$. Because the process $Y(t) = (X(t), \alpha(t))$ is positive recurrent with respect to the domain $D \times \{\ell\}$, we have

$$\mathbf{P}_{x,i}\bigg[\lim_{T\to\infty} \frac{1}{T}\int_0^T f(X(t), \alpha(t))dt = \overline{f} \bigg]$$

$$= \mathbf{P}_{x,i}\bigg[\lim_{T\to\infty} \frac{1}{T}\int_{\varsigma_2}^T f(X(t), \alpha(t))dt = \overline{f} \bigg].$$

We can further write the latter as

$$\mathbf{P}_{x,i}\bigg[\lim_{T\to\infty} \frac{1}{T}\int_{\varsigma_2}^T f(X(t), \alpha(t))dt = \overline{f} \bigg]$$

$$= \int_{\partial D} \mathbf{P}_{x,i}[(X(\varsigma_2), \alpha(\varsigma_2)) \in (dy, \ell)]$$

$$\times \mathbf{P}_{y,\ell}\bigg[\lim_{T\to\infty} \frac{1}{T}\int_0^T f(X(t), \alpha(t))dt = \overline{f} \bigg]$$

$$= \mathbf{P}_{y,\ell}\bigg[\lim_{T\to\infty} \frac{1}{T}\int_0^T f(X(t), \alpha(t))dt = \overline{f} \bigg],$$

where the last line above follows from the use of the invariant distribution. The following illustrates that starting from an arbitrary point (x, i) and arbitrary initial distribution is asymptotically equivalent to starting with the initial distribution being the stationary distribution. Therefore, (4.13) holds for all $(x, i) \in \mathbb{R}^r \times \mathcal{M}$. This completes the proof of the theorem. $\square$

As a consequence of Theorem 4.4, we obtain the following corollary.

Corollary 4.5. *Let the assumptions of Theorem 4.4 be satisfied and let $u(t, x, i)$ be the solution of the Cauchy problem*

$$\begin{cases} \dfrac{\partial u(t, x, i)}{\partial t} = \mathcal{L}u(t, x, i), & t > 0, (x, i) \in \mathbb{R}^r \times \mathcal{M}, \\ u(0, x, i) = f(x, i), & (x, i) \in \mathbb{R}^r \times \mathcal{M}, \end{cases} \quad (4.24)$$

where $f(\cdot, i) \in C_b(\mathbb{R}^r)$ for each $i \in \mathcal{M}$. Then as $T \to \infty$,

$$\frac{1}{T} \int_0^T u(t, x, i)dt \to \sum_{i=1}^{m_0} \int_{\mathbb{R}^r} f(x, i)\mu(x, i)dx. \quad (4.25)$$

Proof. By virtue of Lemma 2.21, $u(t, x, i) = \mathbf{E}_{x,i} f(X(t), \alpha(t))$. Thus we have

$$\frac{1}{T} \int_0^T u(t, x, i)dt = \mathbf{E}_{x,i} \left(\frac{1}{T} \int_0^T f(X(t), \alpha(t))dt \right). \quad (4.26)$$

Meanwhile, (4.13) implies that

$$\frac{1}{T} \int_0^T f(X(t), \alpha(t))dt \overset{T \to \infty}{\longrightarrow} \sum_{i=1}^{m_0} \int_{\mathbb{R}^r} f(x, i)\mu(x, i)dx \text{ a.s.}$$

with respect to the probability $\mathbf{P}_{x,i}$. Then equation (4.25) follows from the dominated convergence theorem. $\square$

4.3 Feedback Controls for Weak Stabilization

Many applications in control and optimization require minimizing an expected cost of a certain objective function. For example, one often wishes to minimize an average cost per unit time cost function. The computation could be difficult, complicated, and time consuming. Significant effort has been devoted to approximating such expected values by replacing the instantaneous measure with stationary measures when the time horizon is long enough. To justify such a replacement, ergodicity is needed. As we have proved in the previous section, a positive recurrent regime-switching diffusion possesses an ergodic measure. One question of both theoretical and practical interest is: If a switching diffusion is not ergodic, can we design suitable controls so that the controlled regime-switching diffusion becomes positive recurrent and hence ergodic. Because positive recurrence was named as weak stability by Wonham [160], the problem of interest can be stated as: Can we find suitable controls to stabilize (in the weak sense) the regime-switching diffusion system.

In this section, our goal is to design suitable controls so that the resulting regime-switching diffusion is positive recurrent and hence ergodic. Consider the following regime switching diffusion (4.1) with the transition of the jump process specified by (4.2) and initial condition

$$X(0) = x, \quad \alpha(0) = \alpha.$$

Consider also its controlled system

$$dX(t) = b(X(t), \alpha(t))dt + B(\alpha(t))u(X(t), \alpha(t))dt + \sigma(X(t), \alpha(t))dw(t),$$
$$(4.27)$$

where $B(i) \in \mathbb{R}^{r \times r}, i \in \mathcal{M}$ are constant matrices, and $u(\cdot, \cdot) : \mathbb{R}^r \times \mathcal{M} \mapsto \mathbb{R}^r$ denotes the feedback control to be identified.

It is well known that the regime-switching diffusion (4.1) can be regarded as the following m_0 single diffusions

$$dX(t) = b(X(t), i)dt + \sigma(X(t), i)dw(t), \quad i \in \mathcal{M}, \qquad (4.28)$$

coupled by the discrete-event component $\alpha(t)$ according to the transition laws specified in (4.2). Often the system is observable only when it operates in some modes but not all. Accordingly, it is natural to decompose the discrete state space $\mathcal{M}$ into two disjoint subsets $\mathcal{M}_1$ and $\mathcal{M}_2$, namely, $\mathcal{M} = \mathcal{M}_1 \cup \mathcal{M}_2$, where for each mode $i \in \mathcal{M}_2$, the process (4.28) cannot be stabilized by feedback control, but it can be stabilized for each $i \in \mathcal{M}_1$. Thus we consider feedback control of the following form

$$u(X(t), \alpha(t)) = -L(\alpha(t))X(t),$$

where for each $i \in \mathcal{M}$, $L(i) \in \mathbb{R}^{r \times r}$ is a constant matrix. Moreover, if $i \in \mathcal{M}_2$, $L(i) = 0$. Thus (4.27) can be rewritten as

$$dX(t) = [b(X(t), \alpha(t)) - B(\alpha(t))L(\alpha(t))X(t)] \, dt + \sigma(X(t), \alpha(t))dw(t).$$
$$(4.29)$$

Assume conditions (A3.1) and (A3.2). In other words, we assume that the nonlinear system is locally linearizable in a neighborhood of ∞. Then we have the following theorem by applying Theorem 3.28 to (4.29).

Theorem 4.6. *If for each $i \in \mathcal{M}_1$, there exists a constant matrix $L(i) \in$*

$\mathbb{R}^{r \times r}$ *such that*

$$\sum_{i \in \mathcal{M}_1} \pi_i \lambda_{\max}\left(b(i) - B(i)L(i) + b'(i) - L'(i)B'(i) + \sum_{j=1}^{d} \sigma_j(i)\sigma_j'(i) \right)$$

$$+ \sum_{i \in \mathcal{M}_2} \pi_i \lambda_{\max}\left(b(i) + b'(i) + \sum_{j=1}^{d} \sigma_j(i)\sigma_j'(i) \right) < 0,$$

$$(4.30)$$

then the resulting controlled regime-switching system (4.27) *is weakly stabilizable. That is, the controlled regime-switching diffusion is positive recurrent.*

Theorem 4.6 ensures that under simple conditions, there are many choices for the matrices $L(i)$ with $i \in \mathcal{M}_1$ in order to make the regime-switching diffusion (2.2) be positive recurrent. Take, for example,

$$L(i) = \theta_i I, \quad i \in \mathcal{M}_1,$$

where I is the $r \times r$ identity matrix and θ_i are nonnegative constants to be determined. Hence it follows that for $i \in \mathcal{M}_1$, we have

$$\lambda_{\max}\left(b(i) - B(i)L(i) + b'(i) - L'(i)B'(i) + \sum_{j=1}^{d} \sigma_j(i)\sigma_j'(i) \right)$$

$$\leq \lambda_{\max}\left(b(i) + b'(i) + \sum_{j=1}^{d} \sigma_j(i)\sigma_j'(i) \right) + \lambda_{\max}(-B(i)L(i) - L'(i)B'(i))$$

$$= \lambda_{\max}\left(b(i) + b'(i) + \sum_{j=1}^{d} \sigma_j(i)\sigma_j'(i) \right) - \theta_i \lambda_{\min}(B(i) + B'(i)).$$

$$(4.31)$$

Assume that for some $\iota \in \mathcal{M}_1$ such that the symmetric matrix $B(\iota) + B'(\iota)$ is positive definite. Hence $\lambda_{\min}(B(\iota) + B'(\iota)) > 0$. Then if $\theta_\iota > 0$ is sufficiently large, and $\theta_i = 0$ for $i \neq \iota$, we have from (4.31) that the left-hand side of (4.30) is less than 0. Thus the resulting controlled regime-switching diffusion (4.27) is positive recurrent. We summarize the discussion in the following theorem.

Theorem 4.7. *If for some $\iota \in \mathcal{M}_1$, the symmetric matrix $B(\iota) + B'(\iota)$ is positive definite, then there exists a feedback control $u(\cdot, \cdot)$ such that the controlled regime-switching diffusion* (4.27) *is positive recurrent.*

Example 4.8. In this example, we apply Theorems 4.6 and 4.7 to stabilize (in the weak sense) a switching diffusion. Consider a two-dimensional (in the continuous component) regime-switching diffusion

$$\begin{cases} dX(t) = b(X(t), \alpha(t))dt + \sigma(X(t), \alpha(t))dw(t) \\ \mathbf{P}\{\alpha(t + \Delta) = j | \alpha(t) = i, X(s), \alpha(s), s \leq t\} = q_{ij}(X(t))\Delta + o(\Delta) \end{cases}$$

$$(4.32)$$

and its controlled system

$$\begin{cases} dX(t) = b(X(t), \alpha(t)) + B(\alpha(t))u(X(t), \alpha(t))dt + \sigma(X(t), \alpha(t))dw(t) \\ \mathbf{P}\{\alpha(t + \Delta) = j | \alpha(t) = i, X(s), \alpha(s), s \le t\} = q_{ij}(X(t))\Delta + o(\Delta), \end{cases}$$
$$(4.33)$$

respectively. Suppose that

$$b(x_1, x_2, 1) = \begin{pmatrix} x_1 + x_2 \\ 2x_2 \end{pmatrix},$$

$$\sigma(x_1, x_2, 1) = \begin{pmatrix} \sqrt{3 + 2x_2^2} & 5 \\ 0 & \sqrt{4 + x_2^2} - \sin(x_1 x_2) \end{pmatrix},$$

$$b(x_1, x_2, 2) = \begin{pmatrix} 2x_1 + x_2 \\ -x_1 + 3x_2 \end{pmatrix},$$

$$\sigma(x_1, x_2, 2) = \begin{pmatrix} -2 + \sqrt{1 + x_2^2} & 0 \\ -3 & \sqrt{2 + x_1^2} \end{pmatrix},$$

and

$$B(1) = \begin{pmatrix} -2 & 3 \\ 1 & -1 \end{pmatrix}, \quad B(2) = \begin{pmatrix} 2 & -3 \\ 1 & 3 \end{pmatrix}.$$

The generator of $\alpha(\cdot)$ is given by

$$Q(x) = \begin{pmatrix} -2 - \dfrac{\sin x_1 \cos(x_1^2) + \sin(x_2^2)}{1 + \sqrt{x_1^2 + x_2^2}} & 2 + \dfrac{\sin x_1 \cos(x_1^2) + \sin(x_2^2)}{1 + \sqrt{x_1^2 + x_2^2}} \\ 1 - \dfrac{2x_2 - \cos x_1}{3 + x_1^2 + x_2^2} & -1 + \dfrac{2x_2 - \cos x_1}{3 + x_1^2 + x_2^2} \end{pmatrix}.$$

Thus associated with the regime-switching diffusion (4.32), there are two diffusions

$$dX(t) = b(X(t), 1)dt + \sigma(X(t), 1)dw(t), \tag{4.34}$$

and

$$dX(t) = b(X(t), 2)dt + \sigma(X(t), 2)dw(t) \tag{4.35}$$

switching back and forth from one to another according to the movement of the discrete component $\alpha(t)$, where $w(t) = (w_1(t), w_2(t))'$ is a two-dimensional standard Brownian motion. Assume that the system is observable when the discrete component $\alpha(\cdot)$ is in state 2. Detailed calculations using the methods in [83, Section 3.8] or [15] allow us to verify that both (4.34) and (4.35) are transient diffusions; see also the phase portraits in Figure 4.2 (a) and (b).

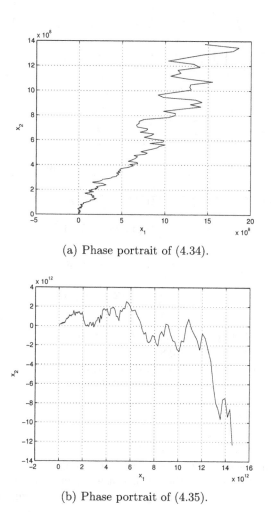

(a) Phase portrait of (4.34).

(b) Phase portrait of (4.35).

FIGURE 4.2. Phase portraits of transient diffusions in (4.34) and (4.35) with initial condition x, where $x = [-5, 3]'$.

Next we use Theorem 3.24 to verify that the switching diffusion (4.32) is transient. To this end, we compute the matrices $b(i)$, $\sigma_j(i)$ for $i, j = 1, 2$ and $\widehat{Q}$ as in condition (A3.2):

$$b(1) = \begin{pmatrix} 1 & 1 \\ 0 & 2 \end{pmatrix}, \quad \sigma_1(1) = \begin{pmatrix} 0 & \sqrt{2} \\ 0 & 0 \end{pmatrix}, \quad \sigma_2(1) = \begin{pmatrix} 0 & 0 \\ 0 & 1 \end{pmatrix},$$

$$b(2) = \begin{pmatrix} 2 & 1 \\ -1 & 3 \end{pmatrix}, \quad \sigma_1(2) = \begin{pmatrix} 0 & 1 \\ 0 & 0 \end{pmatrix}, \quad \sigma_2(2) = \begin{pmatrix} 0 & 0 \\ 1 & 0 \end{pmatrix},$$

and

$$\widehat{Q} = \begin{pmatrix} -2 & 2 \\ 1 & -1 \end{pmatrix}.$$

Thus the stationary distribution is $\pi = (1/3, \ 2/3)$ and

$$\lambda_{\min}(b(1) + b'(1) + \sigma_1(1)\sigma_1'(1) + \sigma_2(1)\sigma_2'(1))$$
$$-\frac{1}{2} \sum_{j=1}^{2} \left[\rho(\sigma_j(1) + \sigma_j'(1)) \right]^2 = 0.3820,$$

$$\lambda_{\min}(b(2) + b'(2) + \sigma_1(2)\sigma_1'(2) + \sigma_2(2)\sigma_2'(2))$$
$$-\frac{1}{2} \sum_{j=1}^{2} \left[\rho(\sigma_j(2) + \sigma_j'(2)) \right]^2 = 4.$$

Hence Theorem 3.24 implies that the switching diffusion (4.32) is transient. Figure 4.3 (a) confirms our analysis.

By our assumption, the switching diffusion is observable when the discrete component $\alpha(\cdot)$ is in state 2. Note that

$$B(2) + B'(2) = \begin{pmatrix} 4 & -2 \\ -2 & 6 \end{pmatrix}$$

is symmetric and positive definite. Therefore it follows from Theorem 4.7 that (4.33) is stabilizable in the weak sense. In fact, as we discussed in Section 4.3, if we take $\theta_1 = 0$ and $\theta_2 = -4$, then direct computation leads to

$$\pi_2 \lambda_{\max}(b(2) + b'(2) - 4B(2) - 4B'(2) + \sigma_1(2)\sigma_1'(2) + \sigma_2(2)\sigma_2'(2))$$
$$+ \pi_1 \lambda_{\max}(b(1) + b'(1) + \sigma_1(1)\sigma_1'(1) + \sigma_2(1)\sigma_2'(1)) = -1.7647 < 0.$$

Therefore the controlled switching diffusion (4.33) is positive recurrent under the feedback control law

$$u(x, 1) = 0 \quad \text{and} \quad u(x, 2) = -4x. \tag{4.36}$$

Figure 4.3(b) demonstrates the phase portrait of (4.33) under the feedback control (4.36). We also demonstrate the component-wise sample path of (4.33) under the feedback control (4.36) in Figure 4.4(a) and (b).

4.4 Ramifications

Remark on a Tightness Result

Under positive recurrence, we may obtain tightness (or boundedness in the sense of in probability) of the underlying process. Suppose that $(X(t), \alpha(t))$ is positive recurrent. We can use the result in Chapter 3 about path excursion (3.46) to prove that for any compact set $\overline{D}$ (the closure of the open set D), the set

$$\bigcup_{x \in \overline{D}} \{(X(t), \alpha(t)) : t \geq 0, \ X(0) = x, \ \alpha(0) = \alpha\}$$

is tight (or bounded in probability). The idea is along the line of the diffusion counterpart; see [102].

Suppose that for each $i \in \mathcal{M}$, there is a Liapunov function $V(\cdot, i)$ such that $\min_x V(x, i) = 0$, $V(x, i) \to \infty$ as $|x| \to \infty$. Let $a_1 > a_0 > 0$ and $X(0) = x$ and $\alpha(0) = \alpha$. Using the argument in Chapter 3, because recurrence is independent of the chosen set, we can work with a fixed $\ell \in \mathcal{M}$, and consider the sets

$$B_{a_0} = \{x \in \mathbb{R}^r : V(x, \ell) \leq a_0\},$$
$$B_{a_1} = \{x \in \mathbb{R}^r : V(x, \ell) \leq a_1\}.$$

Then B_{a_0} will be visited by the switching diffusion infinitely often a.s. Because the switching is positive recurrent, Theorem 3.26 implies that there is a $\kappa > 0$ such that $\mathcal{L}V(x, i) \leq -\kappa$ for all $(x, i) \in B_{a_0}^c$.

Define a sequence of stopping times recursively as

$$\tau_1 = \min\{t : X(t) \in B_{a_0}\}$$
$$\tau_{2n} = \min\{t > \tau_{2n-1} : X(t) \in \partial B_{a_1}\}$$
$$\tau_{2n+1} = \min\{t > \tau_{2n} : X(t) \in \partial B_{a_0}\}.$$

Using the argument as in (3.46), we can obtain

$$\mathbf{E}_{\tau_{2n}}(\tau_{2n+1} - \tau_{2n})I_{\{\tau_{2n} < \infty\}} \leq \frac{a_1}{\kappa} I_{\{\tau_{2n}\}},$$
$$I_{\{\tau_n < \infty\}}\mathbf{P}_{\tau_n}\left(\sup_{\tau_n \leq t \leq \tau_{n+1}} V(X(t)) \geq a\right) \to 0 \ \text{ as } \ a \to \infty,$$

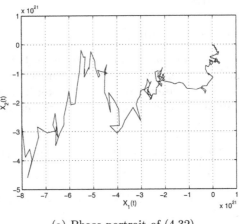

(a) Phase portrait of (4.32).

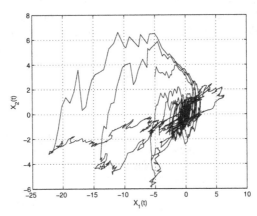

(b) Phase portrait of (4.33).

FIGURE 4.3. Phase portraits of switching diffusion (4.32) and its controlled system (4.33) under feedback control law (4.36) with initial condition (x, α), where $x = [-5, 3]'$ and $\alpha = 2$.

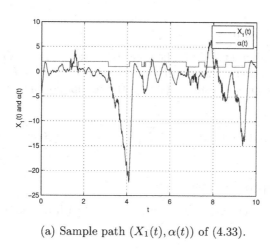

(a) Sample path $(X_1(t), \alpha(t))$ of (4.33).

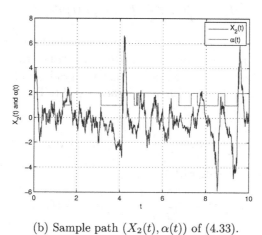

(b) Sample path $(X_2(t), \alpha(t))$ of (4.33).

FIGURE 4.4. Componentwise sample path of (4.33) under feedback control law (4.36) with initial condition (x, α), where $x = [-5, 3]'$ and $\alpha = 2$.

uniformly in $n \geq 1$. Then there exist $\Delta_i > 0$ for $i = 0, 1$ satisfying $I_{\{\tau_{2n-1} < \infty\}} \mathbf{P}_{\tau_{2n-1}}(\tau_{2n} - \tau_{2n-1} \geq \Delta_0) \geq \Delta_1 I_{\{\tau_{2n-1} < \infty\}}$. Working with the estimates up to now and using the argument as in [102, pp. 147–148], we can show that for any $\Delta > 0$ and compact set B, there is an a_Δ such that $\mathbf{P}_x(V(X(t), i) \geq a_\Delta) \leq \Delta$ for any $x \in B$ and $t \geq 0$. The condition $V(x, i) \to \infty$ as $|x| \to \infty$ then implies the desired tightness claim.

Occupation Measures

To illustrate another utility of Theorem 4.4, take $f(x, i) = I_{\{B \times J\}}(x, i)$, the indicator function of the set $B \times J$, where $B \subset \mathbb{R}^r$ and $J \subset \mathcal{M}$. Then Theorem 4.4 becomes a result regarding the occupation measure. In fact, we have

$$\frac{1}{T} \int_0^T I_{\{B \times J\}}(X(t), \alpha(t)) dt \to \sum_{i \in J} \int_B \mu(x, i) dx \quad \text{a.s. as } T \to \infty.$$

Stochastic Approximation

Consider a parameter optimization problem. We wish to find θ_*, a vector-valued parameter so that the cost function

$$J(\theta) = \lim_{T \to \infty} \mathbf{E} \frac{1}{T} \int_0^T \widehat{J}(\theta, Y(t)) dt$$

is minimized, where $Y(t)$ is a positive recurrent switching diffusion as considered in this chapter and for each θ, $\widehat{J}(\theta, \cdot, \cdot)$ satisfies the conditions of Theorem 4.4. For simplicity, we assume that the gradient of $\widehat{J}(\cdot, x, i)$ with respect to θ is available for each x and each $i \in \mathcal{M}$. Then we consider a constant stepsize recursive algorithm

$$\theta_{n+1} = \theta_n - \varepsilon \frac{1}{T} \int_{nT}^{nT+T} \nabla \widehat{J}(\theta_n, Y(t)) dt,$$

or a decreasing stepsize algorithm

$$\theta_{n+1} = \theta_n - \varepsilon_n \frac{1}{T} \int_{nT}^{nT+T} \nabla \widehat{J}(\theta_n, Y(t)) dt,$$

where $\varepsilon > 0$, and $\varepsilon_n \to 0$ as $n \to \infty$ and $\sum_n \varepsilon_n = \infty$. Modifications and variants are possible. For example, additional measurement noise may be included, and the gradient of $\widehat{J}(\cdot)$ may be replaced by its gradient estimates. The motivation for such algorithms stems from optimization of average cost per unit time problems arising from parameter estimations in switching systems of SDEs, manufacturing systems, and queueing networks; see related work in [104, Chapter 9] and [174]. The ergodicity of the switching diffusion is crucial in the study of the asymptotic behavior of the algorithms.

4.5   Asymptotic Distribution

Based on ergodicity of hybrid diffusions, this section is concerned with a centered and scaled sequence. It begins with a lemma on the convergence of the transition density to the invariant density, followed by the verification of a couple of lemmas. Then a result concerning the correlation function is presented. Finally, an asymptotic distribution result is established. Here we work with a compact state space for the x component. To be more specific, we assume that the state space is a torus $\mathbb{S}$ in $\mathbb{R}^r$. More general compact manifolds can also be considered with essentially the same argument.

The following lemma can be obtained in the same spirit as that of [74]. We omit the detailed argument and refer the reader to the reference.

Lemma 4.9. *Denote by $p(y, j; t, x, i)$ the transition density of the process $(X(t), \alpha(t))$ with*

$$\mathbf{P}_{y,j}(X(t) \in S, \alpha(t) = i) = \mathbf{P}(X(t) \in S, \alpha(t) = i | X(0) = y, \alpha(0) = j))$$
$$= \int_S p(y, j; t, x, i) dx,$$

where $S \in \mathcal{B}(\mathbb{S})$. Then $p(y, j; t, x, i)$ converges exponentially fast to $\mu(\cdot, \cdot)$. That is, for some $C > 0$ and $\kappa_3 > 0$,

$$|p(y, j; t, x, i) - \mu(x, i)| \leq C \exp(-\kappa_3 t) \quad for \ any \ (y, j), \ (x, i) \in \mathbb{S} \times \mathcal{M},$$
$$(4.37)$$

where $\mu(x, i)$ is the invariant density as in [74], which is the solution of the system of Kolmogorov–Fokker–Planck equations

$$\mathcal{L}^* g(x, i) = 0, \quad for \ each \ i \in \mathcal{M},$$
$$\sum_{i=1}^{m_0} \int_{\mathbb{S}} g(x, i) dx = 1,$$
$$(4.38)$$

where $\mathcal{L}^*$ is the adjoint operator of $\mathcal{L}$ defined in (3.1).

Lemma 4.10. *Let the assumptions of Lemma 4.9 be satisfied. Suppose that $f(\cdot, \cdot) : \mathbb{S} \times \mathcal{M} \mapsto \mathbb{R}$ is a Borel measurable function. Then for any $0 \leq s \leq t$, we have*

$$\left| \mathbf{E}_{X(s), \alpha(s)} f(X(t), \alpha(t)) - \mathbf{E} f(X(t), \alpha(t)) \right| \leq C \exp(-\kappa_3(t - s)), \quad (4.39)$$

for some positive constant C and κ_3 given by (4.37), where $\mathbf{E} f(X(t), \alpha(t))$ is the expectation with respect to the stationary measure

$$\mathbf{E} f(X(t), \alpha(t)) = \sum_{i=1}^{m_0} \int_{\mathbb{S}} f(x, i) \mu(x, i) dx \overset{\text{def}}{=} f_{av}. \quad (4.40)$$

Proof. Note that

$$f(X(t), \alpha(t)) = \sum_{i=1}^{m_0} f(X(t), i) I_{\{\alpha(t)=i\}}.$$

Inasmuch as we work with $\mathbb{S}$, a torus in $\mathbb{R}^r$,

$$\sum_{i=1}^{m_0} \int_{\mathbb{S}} |f(x, i)| dx < \infty.$$

Moreover,

$$\mathbf{E}_{X(s),\alpha(s)} f(X(t), \alpha(t))$$
$$= \sum_{i=1}^{m_0} \mathbf{E}_{X(s),\alpha(s)} f(X(t), i) I_{\{\alpha(t)=i\}}$$
$$= \sum_{i=1}^{m_0} \int_{\mathbb{S}} f(x, i) \mathbf{P}_{X(s),\alpha(s)} (X(t) \in dx, \alpha(t) = i)$$
$$= \sum_{i=1}^{m_0} \int_{\mathbb{S}} f(x, i) p(X(s), \alpha(s); t - s, x, i) dx.$$

Thus, in view of (4.40), we have from (4.37) that

$$\left| \mathbf{E}_{X(s),\alpha(s)} f(X(t), \alpha(t)) - \mathbf{E} f(X(t), \alpha(t)) \right|$$
$$= \left| \sum_{i=1}^{m_0} \int_{\mathbb{S}} f(x, i)[p(X(s), \alpha(s), t - s, x, i) - \mu(x, i)] dx \right|$$
$$\leq C \exp(-\kappa_3 (t - s)).$$

Hence the lemma follows.   □
    Using Lemma 4.10, we obtain another lemma.

Lemma 4.11. *Assume the conditions of Lemma 4.10. Then the following assertion holds:*

$$|\mathbf{E}[f(X(t), \alpha(t)) - f_{av}][f(X(s), \alpha(s)) - f_{av}]|$$
$$\leq C \exp(-\kappa_3 |t - s|) \tag{4.41}$$

for some positive constant C.

Proof. Assume without loss of generality that $s \leq t$. Note that from (4.40),

$\mathbf{E}f(X(u), \alpha(u)) - f_{av} = 0$ for any $u \geq 0$. Hence we have

$$
\begin{aligned}
&|\mathbf{E}[f(X(t), \alpha(t)) - f_{av}][f(X(s), \alpha(s)) - f_{av}]| \\
&= \big| \mathbf{E}[f(X(t), \alpha(t)) - f_{av}][f(X(s), \alpha(s)) - f_{av}] \\
&\quad - \mathbf{E}[f(X(t), \alpha(t)) - f_{av}]\mathbf{E}[f(X(s), \alpha(s)) - f_{av}]\big| \\
&= \big| \mathbf{E}\{[f(X(s), \alpha(s)) - f_{av}](\mathbf{E}_{X(s),\alpha(s)}[f(X(t), \alpha(t)) - f_{av}] \\
&\quad - \mathbf{E}[f(X(t), \alpha(t)) - f_{av}])\}\big| \\
&\leq C\mathbf{E}^{1/2}|f(X(s), \alpha(s)) - f_{av}|^2 \\
&\quad \times \mathbf{E}^{1/2}\big|\mathbf{E}_{X(s),\alpha(s)}[f(X(t), \alpha(t)) - f_{av}] - \mathbf{E}[f(X(t), \alpha(t)) - f_{av}]\big|^2 \\
&\leq C\exp(-\kappa_3(t - s))
\end{aligned}
$$

by virtue of Lemma 4.10. In the above, from the next to the last line to the last line, we used C as a generic positive constant whose values may be different for different appearances. The proof of the lemma is concluded. □

In what follows, denote

$$
\rho(t - s) = \mathbf{E}[f(X(t), \alpha(t)) - f_{av}][f(X(s), \alpha(s)) - f_{av}], \quad 0 \leq s \leq t \leq T.
\tag{4.42}
$$

This is the covariance function. The process is time homogeneous, thus it is a function of the time difference only.

Lemma 4.12. *Assume the conditions of Lemma 4.10 hold. Then*

$$
\int_0^\infty |\rho(t)|dt < \infty.
\tag{4.43}
$$

Proof. In view of (4.42), we have from Lemma 4.11 that

$$
\begin{aligned}
|\rho(t)| &= |\rho(s + t - s)| \\
&= |\mathbf{E}[f(X(s + t), \alpha(s + t)) - f_{av}][f(X(s), \alpha(s)) - f_{av}]| \\
&\leq C\exp(-\kappa_3 t).
\end{aligned}
$$

Then equation (4.43) follows immediately.   □

Theorem 4.13. *Let the conditions of Lemma 4.10 be satisfied. Then*

$$
\frac{1}{\sqrt{T}} \int_0^T [f(X(t), \alpha(t)) - f_{av}]dt \quad \text{converges in distribution to } N(0, \sigma_{av}^2),
\tag{4.44}
$$

a normal random variable with mean 0 and variance

$$
\sigma_{av}^2 = 2 \int_0^\infty \rho(t)dt.
$$

Proof. Define

$$\zeta(T) = \int_0^T [f(X(t), \alpha(t)) - f_{av}]dt, \quad \text{and} \quad \xi(T) = \frac{1}{\sqrt{T}}\zeta(T). \qquad (4.45)$$

Note that

$$\mathbf{E}\xi(T) = \mathbf{E}\frac{1}{\sqrt{T}}\int_0^T [f(X(t), \alpha(t)) - f_{av}]dt$$

$$= \frac{1}{\sqrt{T}}\int_0^T \mathbf{E}[f(X(t), \alpha(t)) - f_{av}]dt = 0.$$

To calculate the asymptotic variance, note that

$$\mathbf{E}\left[\frac{1}{\sqrt{T}}\int_0^T [f(X(t), \alpha(t)) - f_{av}]dt\right]^2$$

$$= \mathbf{E}\frac{1}{T}\int_0^T \int_0^T [f(X(t), \alpha(t)) - f_{av}][f(X(s), \alpha(s)) - f_{av}]dsdt$$

$$= 2\mathbf{E}\frac{1}{T}\int_0^T \int_0^t [f(X(t), \alpha(t)) - f_{av}][f(X(s), \alpha(s)) - f_{av}]dsdt. \qquad (4.46)$$

Equation (4.42) and changing variables lead to

$$\mathbf{E}\frac{1}{T}\int_0^T \int_0^t [f(X(t), \alpha(t)) - f_{av}][f(X(s), \alpha(s)) - f_{av}]dsdt$$

$$= \frac{1}{T}\int_0^T \int_0^t \rho(t-s)dsdt$$

$$= \frac{1}{T}\int_0^T \int_u^T \rho(u)dtdu \qquad (4.47)$$

$$= \int_0^T \left(1 - \frac{u}{T}\right)\rho(u)du.$$

Choose some $0 < \Delta < 1$. Then by virtue of Lemma 4.12, it follows that as $T \to \infty$,

$$\int_0^T \frac{u}{T}\rho(u)du = \int_0^{T^\Delta} \frac{u}{T}\rho(u)du + \int_{T^\Delta}^T \frac{u}{T}\rho(u)du$$

$$\leq \frac{T^\Delta}{T}\int_0^{T^\Delta} \rho(u)du + \frac{T}{T}\int_{T^\Delta}^T \rho(u)du$$

$$\to 0.$$

Combining the above, we arrive at

$$\mathbf{E}\frac{1}{T}\int_0^T \int_0^t [f(X(t), \alpha(t)) - f_{av}][f(X(s), \alpha(s)) - f_{av}]dsdt$$

$$\to \int_0^\infty \rho(t)dt \quad \text{as} \quad T \to \infty.$$

Therefore, we have

$$\mathbf{E}[\xi(T)]^2 \to \sigma_{av}^2 = 2 \int_0^\infty \rho(t)dt \quad \text{as} \quad T \to \infty. \qquad (4.48)$$

The desired result thus follows.   □

4.6   Notes

Many applications of diffusion processes without switching require using invariant distributions of the underlying processes. One of them is the formulation of two-time-scale diffusions. Dealing with diffusions having fast and slow components, under the framework of diffusion approximation, it has been shown that in the limit, the slow component is averaged out with respect to the invariant measure of the fast component; see Khasminskii [82] for a weak convergence limit result, and Papanicolaou, Stroock, and Varadhan [131] for using a martingale problem formulation. In any event, a crucial step in these references is the use of the invariance measure. In this chapter, in contrast to the diffusion counterpart, we treat regime-switching diffusions, where in addition to the continuous component, there are discrete events. Our interest has been devoted to obtaining ergodicity.

For regime-switching diffusions, asymptotic stability for the density of the two-state random process $(X(t), \alpha(t))$ was established in [136]; asymptotic stability in distribution (or the convergence to the stationary measures of the switching diffusions) for the process $(X(t), \alpha(t))$ was obtained in [6, 183]. Here, we addressed ergodicity for $(X(t), \alpha(t))$ under different conditions from those in [6, 136, 183].

Taking into consideration of the many applications in which discrete events and continuous dynamics are intertwined and the discrete-event process depends on the continuous state, we allow the discrete component $\alpha(\cdot)$ to have x-dependent generator $Q(x)$. Another highlight is that we obtain the explicit representation of the invariant measure of the process $(X(t), \alpha(t))$ by considering certain cylinder sets and by defining cycles appropriately. As a byproduct, we demonstrate a strong law of large numbers type theorem for positive recurrent regime-switching diffusions. It reveals that positive recurrence and ergodicity of switching diffusions are equivalent.

In this chapter, we first developed ergodicity for positive recurrent regime-switching diffusions. Focusing on a compact space, we then obtained asymptotic distributions as a consequence of the ergodicity of the process. The asymptotic distributions are important in treating limit ergodic control problems as well as in applications of Markov chain Monte Carlo. A crucial step in the proof is the verification of the centered and scaled process being a mixing process with exponential mixing rate.

A number of important problems remain open. Obtaining large deviation-type bounds is a worthwhile undertaking, which will have an important impact on studying the associated control and optimization problems.

Part II

Numerical Solutions and Approximation

5

Numerical Approximation

5.1 Introduction

As is the case for deterministic dynamic systems or stochastic differential equations, closed-form solutions for switching diffusions are often difficult to obtain, and numerical approximation is frequently a viable or possibly the only alternative. Being extremely important, numerical methods have drawn much attention. To date, a number of works (e.g., [120, 121, 122, 171]) have focused on numerical approximations where the switching process is independent of the continuous component and is modeled by a continuous-time Markov chain. In addition to the numerical methods, approximation to invariant measures and non-Lipschitz data were dealt with. Nevertheless, it is necessary to be able to handle the coupling and dependence of the continuous states and discrete events. This chapter is devoted to numerical approximation methods for switching diffusions whose switching component is x-dependent. Section 5.2 presents the setup of the problem. Section 5.3 suggests numerical algorithms. Section 5.4 establishes the convergence of the numerical algorithms. Section 5.5 proceeds with a couple of examples. Section 5.6 gives a few remarks concerning the rates of convergence of the algorithms and the study on decreasing stepsize algorithms. Finally Section 5.7 concludes the chapter.

G.G. Yin and C. Zhu, *Hybrid Switching Diffusions: Properties and Applications*,
Stochastic Modelling and Applied Probability 63, DOI 10.1007/978-1-4419-1105-6_5,
© Springer Science + Business Media, LLC 2010

5.2 Formulation

Let $\mathcal{M} = \{1, \ldots, m_0\}$ and consider the hybrid diffusion system

$$
\begin{aligned}
dX(t) &= b(X(t), \alpha(t))dt + \sigma(X(t), \alpha(t))dw(t), \\
X(0) &= X_0, \quad \alpha(0) = \alpha_0,
\end{aligned}
\tag{5.1}
$$

and

$$
\begin{aligned}
\mathbf{P}(\alpha(t + \Delta) &= j | \alpha(t) = i, X(s), \alpha(s), s \le t) \\
&= q_{ij}(X(t))\Delta + o(\Delta), \quad i \ne j,
\end{aligned}
\tag{5.2}
$$

where $w(\cdot)$ is an r-dimensional standard Brownian motion, $X(t) \in \mathbb{R}^r$, $b(\cdot, \cdot) : \mathbb{R}^r \times \mathcal{M} \mapsto \mathbb{R}^r$, $\sigma(\cdot, \cdot) : \mathbb{R}^r \times \mathcal{M} \mapsto \mathbb{R}^{r \times r}$ are appropriate functions satisfying certain regularity conditions, and $Q(x) = (q_{ij}(x)) \in \mathbb{R}^{m_0 \times m_0}$ satisfies that for each x, $q_{ij}(x) \ge 0$ for $i \ne j$, $\sum_{j=1}^{m_0} q_{ij}(x) = 0$ for each $i \in \mathcal{M}$. There is an associated operator for the switching diffusion process defined as follows. For each $i \in \mathcal{M}$ and suitable smooth function $h(\cdot, i)$, define an operator

$$
\begin{aligned}
\mathcal{L}h(x, i) = {}& \nabla h'(x, i)b(x, i) + \frac{1}{2}\text{tr}[\nabla^2 h(x, i)\sigma(x, i)\sigma'(x, i)] \\
& + \sum_{j=1}^{m_0} q_{ij}(x)h(x, j).
\end{aligned}
\tag{5.3}
$$

In this chapter, our aim is to construct numerical approximation schemes for solving (5.1). Note that in our setup, $Q(x)$, the generator of the switching process $\alpha(t)$, taking values in $\mathcal{M} = \{1, \ldots, m_0\}$, is state dependent. It is the x-dependence of $Q(x)$ that makes the analysis much more difficult. One of the main difficulties is that due to the continuous-state dependence, $\alpha(t)$ and $X(t)$ are dependent; $\alpha(t)$ is a Markov chain only for a fixed x but is otherwise non-Markovian. The essence of our approach is to treat the pair of processes $(X(t), \alpha(t))$ jointly; the two-component process turns out to be Markovian. Nevertheless, much care needs to be exercised in handling the mixture distributions. To proceed, we use the following conditions.

(A5.1) The function $Q(\cdot) : \mathbb{R}^r \mapsto \mathbb{R}^{m_0 \times m_0}$ is bounded and continuous.

(A5.2) The functions $b(\cdot, \cdot)$ and $\sigma(\cdot, \cdot)$ satisfy
   (a) $|b(x, \alpha)| \le K(1 + |x|)$, $|\sigma(x, \alpha)| \le K(1 + |x|)$, and
   (b) $|b(x, \alpha) - b(z, \alpha)| \le K_0|x - z|$ and $|\sigma(x, \alpha) - \sigma(z, \alpha)| \le K_0|x - z|$ for some $K > 0$ and $K_0 > 0$ and for all $x, z \in \mathbb{R}^r$ and $\alpha \in \mathcal{M}$.

We first construct Euler's scheme with a constant stepsize for approximating solutions of switching diffusions. It should be mentioned in particular, our analysis differs from the usual approach. To obtain convergence of the algorithms, we first obtain weak convergence of the algorithm by means of martingale problem formulation. This is particularly suited for x-dependent switching processes because the convergence result would be much more difficult to obtain otherwise. We then obtain the convergence in the sense of L^2. As a demonstration, we provide numerical experiments to delineate sample path properties of the approximating solutions. Further discussion of uniform convergence in the sense of L^2 and the associated rates of convergence are derived. In addition, we present a decreasing step-size algorithm. Also provided in this chapter is a brief discussion of rates of convergence.

5.3   Numerical Algorithms

To approximate (5.1), choosing a sequence of independent and identically distributed random variables $\{\xi_n\}$ with mean 0 and finite variance, we propose the following algorithm

$$X_{n+1} = X_n + \varepsilon b(X_n, \alpha_n) + \sqrt{\varepsilon}\sigma(X_n, \alpha_n)\xi_n. \tag{5.4}$$

We proceed to describe the terms involved above. We would like to have α_n be a discrete-time stochastic process that approximates $\alpha(t)$ in an appropriate sense. It is natural when $X_{n-1} = x$, that α_n has a transition probability matrix $\exp(Q(x)\varepsilon)$. It is easily seen that the transition matrix may be approximated further by $I + \varepsilon Q(x) + O(\varepsilon^2)$ by virtue of the boundedness and the continuity of $Q(\cdot)$. Based on this observation, in what follows, we discard the $O(\varepsilon^2)$ term and simply use $I + \varepsilon Q(x)$ for the transition matrix for α_n when $X_{n-1} = x$. To approximate the Brownian motion, we use $\{\xi_n\}$, a sequence of independent and identically distributed random variables with 0 mean and finite variance. We put what has been said above into the following assumption.

(A5.3) In (5.4), for each n, when $X_{n-1} = x$, α_n has the transition matrix $I + \varepsilon Q(x)$, and $\{\xi_n\}$ is a sequence of independent and identically distributed variables such that ξ_n is independent of the σ-algebra $\mathcal{G}_n$ generated by $\{X_k, \alpha_k : k \leq n\}$, that $\mathbf{E}\xi_n = 0$, $\mathbf{E}|\xi_n|^p < \infty, p \geq 2$, and that $\mathbf{E}\xi_n\xi_n' = I$.

Remark 5.1. One of the features of (5.4) is that it is easily implementable. In lieu of discretizing a Brownian motion, we generate a sequence of independent and identically distributed normal random variables to approximate the Brownian motion. This facilitates the computational task. In addition, instead of using transition matrix $\exp(\varepsilon Q(x))$ for a fixed x, we use another fold of approximation $I + \varepsilon Q(x)$ based on a truncated Taylor series. All of these stem from consideration of numerical computation and Monte Carlo implementation. For simplicity, we have chosen ξ_n to be Gaussian. In fact, any sequence with mean 0 and finite second moment can be used. Moreover, correlated sequence $\{\xi_n\}$ can be used as well. However, from a Monte Carlo perspective, there seems to be no strong reason that we need to use correlated sequences.

5.4 Convergence of the Algorithm

5.4.1 Moment Estimates

Throughout this chapter, we assume the stepsize $\varepsilon < 1$. Note that by T/ε, we mean the integer part of T/ε, that is, $\lfloor T/\varepsilon \rfloor$. However, for simplicity, we do not use the floor function notation most of the time and retain this notation only if it is necessary. We first obtain an estimate on the pth moment of $\{X_n\}$. This is stated as follows.

Lemma 5.2. *Under* (A5.1)–(A5.3), *for any fixed* $p \geq 2$ *and* $T > 0$,

$$\sup_{0 \leq n \leq T/\varepsilon} \mathbf{E}|X_n|^p \leq (|X_0|^p + KT) \exp(KT) < \infty. \tag{5.5}$$

Proof. Define $U(x) = |x|^p$ and use $\mathbf{E}_n$ to denote the conditional expectation with respect to the σ-algebra $\mathcal{G}_n$, where $\mathcal{G}_n$ was given in (A5.3). Note that

$$\mathbf{E}_n \sigma(X_n, \alpha_n)\xi_n = \sigma(X_n, \alpha_n)\mathbf{E}_n \xi_n = 0 \quad \text{and}$$
$$\mathbf{E}_n |\sigma(X_n, \alpha_n)|^2 |\xi_n|^2 = |\sigma(X_n, \alpha_n)|^2 \mathbf{E}_n |\xi_n|^2 \leq K|\sigma(X_n, \alpha_n)|^2,$$

where K is a generic positive constant. Thus

$$\begin{aligned}
\mathbf{E}_n U(X_{n+1}) - U(X_n) &= \mathbf{E}_n \nabla U'(X_n)[X_{n+1} - X_n] \\
&\quad + \mathbf{E}_n (X_{n+1} - X_n)' \nabla^2 U(X_n^+)(X_{n+1} - X_n) \\
&\leq \varepsilon \nabla U(X_n) b(X_n, \alpha_n) + K\varepsilon |X_n|^{p-2}(1 + |X_n|^2) \\
&\leq K\varepsilon(1 + |X_n|^p),
\end{aligned}$$
$$\tag{5.6}$$

where ∇U and $\nabla^2 U$ denotes the gradient and Hessian of U with respect to to x, and X_n^+ denotes a vector on the line segment joining X_n and X_{n+1}. Note that in the last line of (5.6), we have used the linear growth in x for both $b(\cdot, \cdot)$ and $\sigma(\cdot, \cdot)$. Because $U(X_n) = |X_n|^p$, we obtain

$$\mathbf{E}_n |X_{n+1}|^p \leq |X_n|^p + K\varepsilon + K\varepsilon |X_n|^p.$$

Taking the expectation on both sides and iterating on the resulting recursion, we obtain

$$\mathbf{E}|X_{n+1}|^p \le |X_0|^p + K\varepsilon n + K\varepsilon \sum_{k=0}^{n} \mathbf{E}|X_k|^p.$$

An application of Gronwall's inequality yields that

$$\mathbf{E}|X_{n+1}|^p \le (|X_0|^p + KT)\exp(KT)$$

as desired.   □

Remark 5.3. In view of the estimate above, $\{X_n : 0 \le n \le T/\varepsilon\}$ is tight in $\mathbb{R}^r$ by means of the well-known Tchebyshev inequality. That is, for each $\eta > 0$, there is a K_η satisfying $K_\eta > \sqrt{(1/\eta)}$ such that

$$\mathbf{P}(|X_n| > K_\eta) \le \frac{\sup_{0 \le n \le T/\varepsilon} \mathbf{E}|X_n|^2}{K_\eta^2} \le K\eta.$$

This indicates that the sequence of iterates is "mass preserving" or no probability is lost. To proceed, take continuous-time interpolations defined by

$$X^\varepsilon(t) = X_n, \ \alpha^\varepsilon(t) = \alpha_n, \ \text{ for } \ t \in [n\varepsilon, n\varepsilon + \varepsilon). \tag{5.7}$$

We show that $X^\varepsilon(\cdot)$ and $\alpha^\varepsilon(\cdot)$ are tight in suitable function spaces.

Lemma 5.4. *Assume* (A5.1)–(A5.3). *Define*

$$\chi_n = (I_{\{\alpha_n = 1\}}, \dots, I_{\{\alpha_n = m\}}) \in \mathbb{R}^{1 \times m} \ \text{ and}$$
$$\chi^\varepsilon(t) = \chi_n, \ \text{ for } \ t \in [\varepsilon n, \varepsilon n + \varepsilon).$$

Then for any $t, s > 0$,

$$\mathbf{E}[\chi^\varepsilon(t+s) - \chi^\varepsilon(t)|\mathcal{F}_t^\varepsilon] = O(s), \tag{5.8}$$

where $\mathcal{F}_t^\varepsilon$ *denotes the* σ-*algebra generated by* $\{X^\varepsilon(u), \alpha^\varepsilon(u) : u \le t\}$.

Proof. First note that by the boundedness and the continuity of $Q(\cdot)$, for each $i \in \mathcal{M}$,

$$\sum_{j=1}^{m_0} \mathbf{E}[I_{\{\alpha_{k+1}=j\}} - I_{\{\alpha_k=i\}}|\mathcal{G}_k]$$
$$= \sum_{j \ne i, j \in \mathcal{M}} \mathbf{E}[I_{\{\alpha_{k+1}=j\}}|\mathcal{G}_k]$$
$$= \sum_{j \ne i, j \in \mathcal{M}} \varepsilon q_{ij}(X_k) I_{\{\alpha_k=i\}} = O(\varepsilon) I_{\{\alpha_k=i\}}.$$

Note that we have used the ijth entry of $I + \varepsilon Q(X_k)$,

$$(I + \varepsilon Q(X_k))_{ij} = \delta_{ij} + \varepsilon q_{ij}(X_k),$$

where $\mathcal{G}_n$ is given in (A5.3), and

$$\delta_{ij} = \begin{cases} 1, & \text{if } i = j, \\ 0, & \text{otherwise.} \end{cases}$$

It then follows that there is a random function $\widetilde{g}(\cdot)$ such that

$$\begin{aligned}
\mathbf{E}\Big[\sum_{k=t/\varepsilon}^{(t+s)/\varepsilon - 1} [\chi_{k+1} - \chi_k] \Big| \mathcal{F}_t^\varepsilon \Big] \\
= \mathbf{E}\Big[\sum_{k=t/\varepsilon}^{(t+s)/\varepsilon - 1} \mathbf{E}[\chi_{k+1} - \chi_k | \mathcal{G}_k] \Big| \mathcal{F}_t^\varepsilon \Big] \\
= \widetilde{g}(t + s - t) \\
= \widetilde{g}(s),
\end{aligned}$$
(5.9)

and that

$$\mathbf{E}\widetilde{g}(s) = O(s).$$

In the above, we have used the convention that t/ε and $(t + s)/\varepsilon$ denote the integer parts of t/ε and $(t + s)/\varepsilon$, respectively. Inasmuch as

$$\mathbf{E}\Big[\chi^\varepsilon(t + s) - \chi^\varepsilon(t) - \sum_{k=t/\varepsilon}^{(t+s)/\varepsilon - 1} [\chi_{k+1} - \chi_k] \Big| \mathcal{F}_t^\varepsilon \Big] = 0,$$

it follows from (5.9) that

$$\mathbf{E}[\chi^\varepsilon(t + s) | \mathcal{F}_t^\varepsilon] = \chi^\varepsilon(t) + \widetilde{g}(s).$$

The desired result then follows.   □

Lemma 5.5. *Under the conditions of Lemma 5.4, $\{\alpha^\varepsilon(\cdot)\}$ is tight.*

Proof. By virtue of Lemma 5.4,

$$\begin{aligned}
\mathbf{E}[|\chi^\varepsilon(t + s) - \chi^\varepsilon(t)|^2 | \mathcal{F}_t^\varepsilon] \\
= \mathbf{E}[\chi^\varepsilon(t + s)\chi^{\varepsilon,\prime}(t + s) - 2\chi^\varepsilon(t + s)\chi^{\varepsilon,\prime}(t) + \chi^\varepsilon(t)\chi^{\varepsilon,\prime}(t) | \mathcal{F}_t^\varepsilon] \\
= \sum_{i=1}^{m_0} \mathbf{E}[I_{\{\alpha_{(t+s)/\varepsilon}=i\}} - 2I_{\{\alpha_{(t+s)/\varepsilon}=i\}} I_{\{\alpha_{t/\varepsilon}=i\}} + I_{\{\alpha_{t/\varepsilon}=i\}} | \mathcal{F}_t^\varepsilon].
\end{aligned}$$
(5.10)

The estimates in Lemma 5.4 then imply that

$$\lim_{s \to 0} \limsup_{\varepsilon \to 0} \mathbf{E}[\mathbf{E}|\chi^{\varepsilon}(t + s) - \chi^{\varepsilon}(t)|^2 \big| \mathcal{F}_t^{\varepsilon}] = 0.$$

The tightness criterion in [102, p. 47] yields that $\{\chi^{\varepsilon}(\cdot)\}$ is tight. Consequently, $\{\alpha^{\varepsilon}(\cdot)\}$ is tight.   □

Lemma 5.6. *Assume that the conditions of Lemma 5.5 are satisfied. Then $\{X^{\varepsilon}(\cdot)\}$ is tight in $D^r([0,\infty) : \mathbb{R}^r)$, the space of functions that are right continuous and have left limits, endowed with the Skorohod topology.*

Proof. For any $\eta > 0$, $t \geq 0$, $0 \leq s \leq \eta$, we have

$$\mathbf{E}|X^{\varepsilon}(t + s) - X^{\varepsilon}(t)|^2$$

$$= \mathbf{E}\left| \varepsilon \sum_{k=t/\varepsilon}^{(t+s)/\varepsilon-1} b(X_k, \alpha_k) + \sqrt{\varepsilon} \sum_{k=t/\varepsilon}^{(t+s)/\varepsilon-1} \sigma(X_k, \alpha_k)\xi_k \right|^2$$

$$\leq K\varepsilon^2 \sum_{k=t/\varepsilon}^{(t+s)/\varepsilon-1} (1 + \mathbf{E}|X_k|^2) + K\varepsilon \sum_{k=t/\varepsilon}^{(t+s)/\varepsilon-1} \mathbf{E}|\sigma(X_k, \alpha_k)|^2 \mathbf{E}|\xi_k|^2$$

$$\leq K\varepsilon^2 \sum_{k=t/\varepsilon}^{(t+s)/\varepsilon-1} \left(1 + \sup_{t/\varepsilon \leq k \leq (t+s)/\varepsilon-1} \mathbf{E}|X_k|^2\right)$$

$$+ K\varepsilon \sum_{k=t/\varepsilon}^{(t+s)/\varepsilon-1} \left(1 + \sup_{t/\varepsilon \leq k \leq (t+s)/\varepsilon-1} \mathbf{E}|X_k|^2\right)$$

$$\leq O\left(\frac{t+s}{\varepsilon} - \frac{t}{\varepsilon}\right) O(\varepsilon) = O(s).$$

(5.11)

In the above, we have used Lemma 5.2 to ensure that

$$\sup_{t/\varepsilon \leq k \leq (t+s)/\varepsilon-1} \mathbf{E}|X_k|^2 < \infty.$$

Therefore, (5.11) leads to

$$\lim_{\eta \to 0} \limsup_{\varepsilon \to 0} \mathbf{E}|X^{\varepsilon}(t + s) - X^{\varepsilon}(t)|^2 = 0.$$

The tightness of $\{X^{\varepsilon}(\cdot)\}$ then follows from [102, p. 47].   □

Combining Lemmas 5.4-5.6, we obtain the following tightness result.

Lemma 5.7. *Under assumptions (A5.1)–(A5.3), $\{X^{\varepsilon}(\cdot), \alpha^{\varepsilon}(\cdot)\}$ is tight in $D([0,\infty) : \mathbb{R}^r \times \mathcal{M})$.*

5.4.2 Weak Convergence

The main result of this section is the following theorem.

Theorem 5.8. *Under (A5.1)–(A5.3), $(X^\varepsilon(\cdot), \alpha^\varepsilon(\cdot))$ converges weakly to $(X(\cdot), \alpha(\cdot))$, which is a process with generator given by (5.3).*

Proof. Because $(X^\varepsilon(\cdot), \alpha^\varepsilon(\cdot))$ is tight, by Prohorov's theorem (see Theorem A.20 of this book and also [43, 104]), we may select a convergent subsequence. For simplicity, we still denote the subsequence by $(X^\varepsilon(\cdot), \alpha^\varepsilon(\cdot))$ with the limit denoted by $(\widetilde{X}(\cdot), \widetilde{\alpha}(\cdot))$. We proceed to characterize the limit.

By Skorohod representation (see Theorem A.21 and also [43, 104]), without loss of generality and without changing notation, we may assume that $(X^\varepsilon(\cdot), \alpha^\varepsilon(\cdot))$ converges to $(\widetilde{X}(\cdot), \widetilde{\alpha}(\cdot))$ w.p.1, and the convergence is uniform on each bounded interval. We proceed to characterize the limit process.

Step 1: We first work with the marginal of the switching component, and characterize the limit of $\alpha^\varepsilon(\cdot)$. The weak convergence of $\alpha^\varepsilon(\cdot)$ to $\widetilde{\alpha}(\cdot)$ yields that $\chi^\varepsilon(\cdot)$ converges to $\chi(\cdot)$ weakly. For each $t > 0$ and $s > 0$, each positive integer κ, each $0 \le t_\iota \le t$ with $\iota \le \kappa$, and each bounded and continuous function $\rho_\iota(\cdot, i)$ for each $i \in \mathcal{M}$,

$$\mathbf{E} \prod_{\iota=1}^{\kappa} \rho_\iota(X^\varepsilon(t_\iota), \alpha^\varepsilon(t_\iota)) \left[\chi^\varepsilon(t+s) - \chi^\varepsilon(t) - \sum_{k=t/\varepsilon}^{(t+s)/\varepsilon - 1} (\chi_{k+1} - \chi_k) \right] = 0. \quad (5.12)$$

The weak convergence of $\chi^\varepsilon(\cdot)$ to $\chi(\cdot)$ and the Skorohod representation imply that

$$\lim_{\varepsilon \to 0} \mathbf{E} \prod_{\iota=1}^{\kappa} \rho_\iota(X^\varepsilon(t_\iota), \alpha^\varepsilon(t_\iota))[\chi^\varepsilon(t+s) - \chi^\varepsilon(t)]$$
$$= \mathbf{E} \prod_{\iota=1}^{\kappa} \rho_\iota(\widetilde{X}(t_\iota), \widetilde{\alpha}(t_\iota))[\chi(t+s) - \chi(t)].$$

Pick out a sequence $\{n_\varepsilon\}$ of nonnegative real numbers such that

$$n_\varepsilon \to \infty \quad \text{as} \quad \varepsilon \to 0 \quad \text{but} \quad \delta_\varepsilon = \varepsilon n_\varepsilon \to 0.$$

Use Ξ_l^ε, the set of indices defined by

$$\Xi_l^\varepsilon = \{k : ln_\varepsilon \le k \le ln_\varepsilon + n_\varepsilon - 1\}, \quad (5.13)$$

as a base for partition. Then the continuity together with the boundedness of $Q(\cdot)$ implies that

$$
\lim_{\varepsilon \to 0} \mathbf{E} \prod_{\iota=1}^{\kappa} \rho_\iota(X^\varepsilon(t_\iota), \alpha^\varepsilon(t_\iota)) \Big[\sum_{k=t/\varepsilon}^{(t+s)/\varepsilon-1} (\chi_{k+1} - \chi_k) \Big]
$$

$$
= \lim_{\varepsilon \to 0} \mathbf{E} \prod_{\iota=1}^{\kappa} \rho_\iota(X^\varepsilon(t_\iota), \alpha^\varepsilon(t_\iota)) \Big[\sum_{k=t/\varepsilon}^{(t+s)/\varepsilon-1} (\mathbf{E}(\chi_{k+1}|\mathcal{G}_k) - \chi_k) \Big]
$$

$$
= \lim_{\varepsilon \to 0} \mathbf{E} \prod_{\iota=1}^{\kappa} \rho_\iota(X^\varepsilon(t_\iota), \alpha^\varepsilon(t_\iota)) \Big[\sum_{ln_\varepsilon=t/\varepsilon}^{(t+s)/\varepsilon-1} \sum_{k=ln_\varepsilon}^{ln_\varepsilon+n_\varepsilon-1} \chi_k(I + \varepsilon Q(X_k) - I) \Big]
$$

$$
= \lim_{\varepsilon \to 0} \mathbf{E} \prod_{\iota=1}^{\kappa} \rho_\iota(X^\varepsilon(t_\iota), \alpha^\varepsilon(t_\iota)) \Big[\sum_{ln_\varepsilon=t/\varepsilon}^{(t+s)/\varepsilon-1} \delta_\varepsilon \frac{1}{n_\varepsilon} \sum_{k=ln_\varepsilon}^{ln_\varepsilon+n_\varepsilon-1} \chi_k Q(X_{ln_\varepsilon}) \Big].
$$

$$(5.14)$$

Note that

$$
\lim_{\varepsilon \to 0} \mathbf{E} \prod_{\iota=1}^{\kappa} \rho_\iota(X^\varepsilon(t_\iota), \alpha^\varepsilon(t_\iota)) \Big[\sum_{ln_\varepsilon=t/\varepsilon}^{(t+s)/\varepsilon-1} \delta_\varepsilon \frac{1}{n_\varepsilon} \sum_{k=ln_\varepsilon}^{ln_\varepsilon+n_\varepsilon-1} [\chi_k - \chi_{ln_\varepsilon}]Q(X_{ln_\varepsilon}) \Big]
$$

$$
= \lim_{\varepsilon \to 0} \mathbf{E} \prod_{\iota=1}^{\kappa} \rho_\iota(X^\varepsilon(t_\iota), \alpha^\varepsilon(t_\iota))
$$
$$
\times \Big[\sum_{ln_\varepsilon=t/\varepsilon}^{(t+s)/\varepsilon-1} \delta_\varepsilon \frac{1}{n_\varepsilon} \sum_{k=ln_\varepsilon}^{ln_\varepsilon+n_\varepsilon-1} \mathbf{E}[\chi_k - \chi_{ln_\varepsilon}|\mathcal{G}_{ln_\varepsilon}]Q(X_{ln_\varepsilon}) \Big]
$$

$$
= \lim_{\varepsilon \to 0} \mathbf{E} \prod_{\iota=1}^{\kappa} \rho_\iota(X^\varepsilon(t_\iota), \alpha^\varepsilon(t_\iota))
$$
$$
\times \Big[\sum_{ln_\varepsilon=t/\varepsilon}^{(t+s)/\varepsilon-1} \delta_\varepsilon \frac{1}{n_\varepsilon} \sum_{k=ln_\varepsilon}^{ln_\varepsilon+n_\varepsilon-1} \mathbf{E}\chi_{ln_\varepsilon}[(I + \varepsilon Q(X_{ln_\varepsilon}))^{k-ln_\varepsilon} - I]Q(X_{ln_\varepsilon}) \Big]
$$

$$
= 0.
$$

$$(5.15)$$

Therefore,

$$
\lim_{\varepsilon \to 0} \mathbf{E} \prod_{\iota=1}^{\kappa} \rho_\iota(X^\varepsilon(t_\iota), \alpha^\varepsilon(t_\iota)) \Big[\sum_{ln_\varepsilon=t/\varepsilon}^{(t+s)/\varepsilon-1} \delta_\varepsilon \frac{1}{n_\varepsilon} \sum_{k=ln_\varepsilon}^{ln_\varepsilon+n_\varepsilon-1} \chi_k Q(X_{ln_\varepsilon}) \Big]
$$

$$
= \lim_{\varepsilon \to 0} \mathbf{E} \prod_{\iota=1}^{\kappa} \rho_\iota(X^\varepsilon(t_\iota), \alpha^\varepsilon(t_\iota)) \Big[\sum_{ln_\varepsilon=t/\varepsilon}^{(t+s)/\varepsilon-1} \delta_\varepsilon \chi_{ln_\varepsilon} Q(X_{ln_\varepsilon}) \Big]
$$

$$
= \mathbf{E} \prod_{\iota=1}^{\kappa} \rho_\iota(\widetilde{X}(t_\iota), \widetilde{\alpha}(t_\iota)) \int_t^{t+s} \chi(u) Q(\widetilde{X}(u)) du.
$$

$$(5.16)$$

Moreover, the limit does not depend on the chosen subsequence. Thus,

$$\mathbf{E} \prod_{\iota=1}^{\kappa} \rho_\iota(\widetilde{X}(t_\iota), \widetilde{\alpha}(t_\iota)) \left[\chi(t+s) - \chi(t) - \int_t^{t+s} \chi(u) Q(\widetilde{X}(u)) du \right] = 0. \quad (5.17)$$

Therefore, the limit process $\widetilde{\alpha}(\cdot)$ has a generator $Q(\widetilde{X}(\cdot))$.

Step 2: For t, s, κ, t_ι as chosen before, for each bounded and continuous function $\rho_\iota(\cdot, i)$, and for each twice continuously differentiable function with compact support $h(\cdot, i)$ with $i \in \mathcal{M}$, we show that

$$\begin{aligned}
\mathbf{E} \prod_{\iota=1}^{\kappa} &\rho_\iota(\widetilde{X}(t_\iota), \widetilde{\alpha}(t_\iota))[h(\widetilde{X}(t+s), \widetilde{\alpha}(t+s)) \\
&- h(\widetilde{X}(t), \widetilde{\alpha}(t)) - \int_t^{t+s} \mathcal{L}h(\widetilde{X}(u), \widetilde{\alpha}(u)) du] = 0.
\end{aligned} \quad (5.18)$$

This yields that

$$h(\widetilde{X}(t), \widetilde{\alpha}(t)) - \int_0^t \mathcal{L}h(\widetilde{X}(u), \widetilde{\alpha}(u)) du,$$

is a continuous-time martingale, which in turn implies that $(\widetilde{X}(\cdot), \widetilde{\alpha}(\cdot))$ is a solution of the martingale problem with operator $\mathcal{L}$ defined in (5.3).

To establish the desired result, we work with the sequence $(X^\varepsilon(\cdot), \alpha^\varepsilon(\cdot))$. Again, we use the sequence $\{n_\varepsilon\}$ as in Step 1. By virtue of the weak convergence and the Skorohod representation, it is readily seen that

$$\begin{aligned}
\mathbf{E} \prod_{\iota=1}^{\kappa} &\rho_\iota(X^\varepsilon(t_\iota), \alpha^\varepsilon(t_\iota))[h(X^\varepsilon(t+s), \alpha^\varepsilon(t+s)) - h(X^\varepsilon(t), \alpha^\varepsilon(t))] \\
&\to \mathbf{E} \prod_{\iota=1}^{\kappa} \rho_\iota(\widetilde{X}(t_\iota), \widetilde{\alpha}(t_\iota))[h(\widetilde{X}(t+s), \widetilde{\alpha}(t+s)) - h(\widetilde{X}(t), \widetilde{\alpha}(t))]
\end{aligned}$$

$$(5.19)$$

as $\varepsilon \to 0$. On the other hand, direct calculation shows that

$$\begin{aligned}
\mathbf{E} \prod_{\iota=1}^{\kappa} &\rho_\iota(X^\varepsilon(t_\iota), \alpha^\varepsilon(t_\iota))[h(X^\varepsilon(t+s), \alpha^\varepsilon(t+s)) - h(X^\varepsilon(t), \alpha^\varepsilon(t))] \\
&= \mathbf{E} \prod_{\iota=1}^{\kappa} \rho_\iota(X^\varepsilon(t_\iota), \alpha^\varepsilon(t_\iota)) \\
&\quad \times \Bigg\{ \sum_{ln_\varepsilon=t/\varepsilon}^{(t+s)/\varepsilon-1} \Big[[h(X_{ln_\varepsilon+n_\varepsilon}, \alpha_{ln_\varepsilon+n_\varepsilon}) - h(X_{ln_\varepsilon+n_\varepsilon}, \alpha_{ln_\varepsilon})] \\
&\quad + [h(X_{ln_\varepsilon+n_\varepsilon}, \alpha_{ln_\varepsilon}) - h(X_{ln_\varepsilon}, \alpha_{ln_\varepsilon})] \Big] \Bigg\}.
\end{aligned}$$

$$(5.20)$$

Step 3: Still use the notation Ξ_ℓ^ε defined in (5.13). For the terms on the last line of (5.20), we have

$$
\lim_{\varepsilon \to 0} \mathbf{E} \prod_{\iota=1}^{\kappa} \rho_\iota(X^\varepsilon(t_\iota), \alpha^\varepsilon(t_\iota))
$$

$$
\times \sum_{ln_\varepsilon = t/\varepsilon}^{(t+s)/\varepsilon - 1} [h(X_{ln_\varepsilon + n_\varepsilon}, \alpha_{ln_\varepsilon}) - h(X_{ln_\varepsilon}, \alpha_{ln_\varepsilon})]
$$

$$
= \lim_{\varepsilon \to 0} \mathbf{E} \prod_{\iota=1}^{\kappa} \rho_\iota(X^\varepsilon(t_\iota), \alpha^\varepsilon(t_\iota)) \tag{5.21}
$$

$$
\times \left\{ \sum_{ln_\varepsilon = t/\varepsilon}^{(t+s)/\varepsilon - 1} \left[\varepsilon \nabla h'(X_{ln_\varepsilon}, \alpha_{ln_\varepsilon}) \sum_{k=ln_\varepsilon}^{ln_\varepsilon + n_\varepsilon - 1} b(X_k, \alpha_{ln_\varepsilon}) \right.\right.
$$

$$
\left.\left. + \frac{\varepsilon}{2} \sum_{k=ln_\varepsilon}^{ln_\varepsilon + n_\varepsilon - 1} \mathrm{tr}[\nabla^2 h(X_{ln_\varepsilon}, \alpha_{ln_\varepsilon}) \sigma(X_k, \alpha_{ln_\varepsilon}) \sigma'(X_k, \alpha_{ln_\varepsilon})] \right] \right\}.
$$

By the continuity of $b(\cdot, i)$ for each $i \in \mathcal{M}$ and the choice of n_ε,

$$
\lim_{\varepsilon \to 0} \mathbf{E} \prod_{\iota=1}^{\kappa} \rho_\iota(X^\varepsilon(t_\iota), \alpha^\varepsilon(t_\iota))
$$

$$
\times \left\{ \sum_{ln_\varepsilon = t/\varepsilon}^{(t+s)/\varepsilon - 1} \delta_\varepsilon \nabla h'(X_{ln_\varepsilon}, \alpha_{ln_\varepsilon}) \right.
$$

$$
\left. \times \frac{1}{n_\varepsilon} \sum_{k=ln_\varepsilon}^{ln_\varepsilon + n_\varepsilon - 1} [b(X_k, \alpha_{ln_\varepsilon}) - b(X_{ln_\varepsilon}, \alpha_{ln_\varepsilon})] \right\}
$$

$$
= 0.
$$

Thus, in evaluating the limit, for $ln_\varepsilon \le k \le ln_\varepsilon + n_\varepsilon - 1$, $b(X_k, \alpha_{ln_\varepsilon})$ can be replaced by $b(X_{ln_\varepsilon}, \alpha_{ln_\varepsilon})$.

The choice of n_ε implies that $\varepsilon l n_\varepsilon \to u$ as $\varepsilon \to 0$ yielding $\varepsilon k \to u$ for all $ln_\varepsilon \le k \le ln_\varepsilon + n_\varepsilon$. Consequently, by weak convergence and the Skorohod representation, we obtain

$$
L_1 \stackrel{\text{def}}{=} \lim_{\varepsilon \to 0} \mathbf{E} \prod_{\iota=1}^{\kappa} \rho_\iota(X^\varepsilon(t_\iota), \alpha^\varepsilon(t_\iota)) \left\{ \sum_{ln_\varepsilon = t/\varepsilon}^{(t+s)/\varepsilon - 1} \varepsilon \nabla h'(X_{ln_\varepsilon}, \alpha_{ln_\varepsilon}) \right.
$$

$$
\left. \times \sum_{k=ln_\varepsilon}^{ln_\varepsilon + n_\varepsilon - 1} b(X_k, \alpha_{ln_\varepsilon}) \right\}
$$

$$
\tag{5.22}
$$

$$
= \lim_{\varepsilon \to 0} \mathbf{E} \prod_{\iota=1}^{\kappa} \rho_\iota(X^\varepsilon(t_\iota), \alpha^\varepsilon(t_\iota))
$$

$$
\times \left\{ \sum_{ln_\varepsilon = t/\varepsilon}^{(t+s)/\varepsilon - 1} \delta_\varepsilon \nabla h'(X_{ln_\varepsilon}, \alpha_{ln_\varepsilon}) \frac{1}{n_\varepsilon} \sum_{k=ln_\varepsilon}^{ln_\varepsilon + n_\varepsilon - 1} b(X_{ln_\varepsilon}, \alpha_{ln_\varepsilon}) \right\}.
$$

Thus,

$$
\begin{aligned}
L_1 &= \lim_{\varepsilon \to 0} \mathbf{E} \prod_{\iota=1}^{\kappa} \rho_\iota(X^\varepsilon(t_\iota), \alpha^\varepsilon(t_\iota)) \\
&\quad \times \Big\{ \sum_{ln_\varepsilon = t/\varepsilon}^{(t+s)/\varepsilon - 1} \delta_\varepsilon \nabla h'(X_{ln_\varepsilon}, \alpha_{ln_\varepsilon}) b(X^\varepsilon(l\delta_\varepsilon), \alpha^\varepsilon(l\delta_\varepsilon)) \Big\} \\
&= \mathbf{E} \prod_{\iota=1}^{\kappa} \rho_\iota(\widetilde{X}(t_\iota), \widetilde{\alpha}(t_\iota)) \Big\{ \int_t^{t+s} \nabla h'(\widetilde{X}(u), \widetilde{\alpha}(u)) b(\widetilde{X}(u), \widetilde{\alpha}(u)) du \Big\}.
\end{aligned}
\tag{5.23}
$$

In the above, treating such terms as $b(X^\varepsilon(l\delta_\varepsilon), \alpha^\varepsilon(l\delta_\varepsilon))$, we can approximate $X^\varepsilon(\cdot)$ by a process taking finitely many values using a standard approximation argument (see, e.g., [104, p. 169] for more details).

Similar to (5.23), we also obtain

$$
\begin{aligned}
L_2 &\overset{\text{def}}{=} \lim_{\varepsilon \to 0} \mathbf{E} \prod_{\iota=1}^{\kappa} \rho_\iota(X^\varepsilon(t_\iota), \alpha^\varepsilon(t_\iota)) \\
&\quad \times \Big\{ \frac{\varepsilon}{2} \sum_{ln_\varepsilon = t/\varepsilon}^{(t+s)/\varepsilon - 1} \sum_{k=ln_\varepsilon}^{ln_\varepsilon + n_\varepsilon - 1} \mathrm{tr}[\nabla^2 h(X_{ln_\varepsilon}, \alpha_{ln_\varepsilon}) \sigma(X_k, \alpha_{ln_\varepsilon}) \sigma'(X_k, \alpha_{ln_\varepsilon})] \Big\} \\
&= \mathbf{E} \prod_{\iota=1}^{\kappa} \rho_\iota(\widetilde{X}(t_\iota), \widetilde{\alpha}(t_\iota)) \\
&\quad \times \Big\{ \int_t^{t+s} \frac{1}{2} \mathrm{tr}[\nabla^2 h(\widetilde{X}(u), \widetilde{\alpha}(u)) \sigma(\widetilde{X}(u), \widetilde{\alpha}(u)) \sigma'(\widetilde{X}(u), \widetilde{\alpha}(u))] du \Big\}.
\end{aligned}
\tag{5.24}
$$

Step 4: We next examine the terms on the next to the last line of (5.20). First, again using the continuity, weak convergence, and the Skorohod representation, it can be shown that

$$
\begin{aligned}
&\lim_{\varepsilon \to 0} \mathbf{E} \prod_{\iota=1}^{\kappa} \rho_\iota(X^\varepsilon(t_\iota), \alpha^\varepsilon(t_\iota)) \Big[\sum_{ln_\varepsilon = t/\varepsilon}^{(t+s)/\varepsilon - 1} [h(X_{ln_\varepsilon + n_\varepsilon}, \alpha_{ln_\varepsilon + n_\varepsilon}) \\
&\quad - h(X_{ln_\varepsilon + n_\varepsilon}, \alpha_{ln_\varepsilon})] \Big] \\
&= \lim_{\varepsilon \to 0} \mathbf{E} \prod_{\iota=1}^{\kappa} \rho_\iota(X^\varepsilon(t_\iota), \alpha^\varepsilon(t_\iota)) \\
&\quad \times \Big[\sum_{ln_\varepsilon = t/\varepsilon}^{(t+s)/\varepsilon - 1} [h(X_{ln_\varepsilon}, \alpha_{ln_\varepsilon + n_\varepsilon}) - h(X_{ln_\varepsilon}, \alpha_{ln_\varepsilon})] \Big].
\end{aligned}
\tag{5.25}
$$

That is, as far as asymptotic analysis is concerned, owing to the choice of $\{n_\varepsilon\}$ and the continuity of $h(\cdot, i)$, the term

$$
h(X_{ln_\varepsilon + n_\varepsilon}, \alpha_{ln_\varepsilon + n_\varepsilon}) - h(X_{ln_\varepsilon + n_\varepsilon}, \alpha_{ln_\varepsilon})
$$

in the next to the last line of (5.20) can be replaced by

$$h(X_{ln_\varepsilon}, \alpha_{ln_\varepsilon + n_\varepsilon}) - h(X_{ln_\varepsilon}, \alpha_{ln_\varepsilon})$$

with an error tending to 0 in probability as $\varepsilon \to 0$ uniformly in t. It follows that

$$\lim_{\varepsilon \to 0} \mathbf{E} \prod_{\iota=1}^{\kappa} \rho_\iota(X^\varepsilon(t_\iota), \alpha^\varepsilon(t_\iota)) \left\{ \sum_{ln_\varepsilon=t/\varepsilon}^{(t+s)/\varepsilon-1} [h(X_{ln_\varepsilon}, \alpha_{ln_\varepsilon+n_\varepsilon}) - h(X_{ln_\varepsilon}, \alpha_{ln_\varepsilon})] \right\}$$

$$= \lim_{\varepsilon \to 0} \mathbf{E} \prod_{\iota=1}^{\kappa} \rho_\iota(X^\varepsilon(t_\iota), \alpha^\varepsilon(t_\iota))$$

$$\times \left\{ \sum_{ln_\varepsilon=t/\varepsilon}^{(t+s)/\varepsilon-1} \sum_{k=ln_\varepsilon}^{ln_\varepsilon+n_\varepsilon-1} [h(X_{ln_\varepsilon}, \alpha_{k+1}) - h(X_{ln_\varepsilon}, \alpha_k)] \right\}$$

$$= \lim_{\varepsilon \to 0} \mathbf{E} \prod_{\iota=1}^{\kappa} \rho_\iota(X^\varepsilon(t_\iota), \alpha^\varepsilon(t_\iota))$$

$$\times \left\{ \sum_{ln_\varepsilon=t/\varepsilon}^{(t+s)/\varepsilon-1} \sum_{k=ln_\varepsilon}^{ln_\varepsilon+n_\varepsilon-1} \sum_{i=1}^{m_0} \sum_{i_1=1}^{m_0} \mathbf{E}\Big[[h(X_{ln_\varepsilon}, i) I_{\{\alpha_{k+1}=i\}} \right.$$

$$\left. - h(X_{ln_\varepsilon}, i_1) I_{\{\alpha_k=i_1\}}] \Big| \mathcal{G}_k \Big] \right\}.$$

(5.26)

Note that for $k \geq ln_\varepsilon$,

$$\mathbf{E}\Big[[h(X_{ln_\varepsilon}, i) I_{\{\alpha_{k+1}=i\}} - h(X_{ln_\varepsilon}, i_1) I_{\{\alpha_k=i_1\}}] \Big| \mathcal{G}_k \Big]$$

$$= [h(X_{ln_\varepsilon}, i) \mathbf{P}(\alpha_{k+1} = i | \mathcal{G}_k, \alpha_k = i_1) - h(X_{ln_\varepsilon}, i_1)] I_{\{\alpha_k=i_1\}}$$

$$= [h(X_{ln_\varepsilon}, i)(\delta_{i_1 i} + \varepsilon q_{i_1 i}(X_k)) - h(X_{ln_\varepsilon}, i_1)] I_{\{\alpha_k=i_1\}}$$

$$= \varepsilon h(X_{ln_\varepsilon}, i) q_{i_1 i}(X_k) I_{\{\alpha_k=i_1\}}.$$

(5.27)

Using (5.27) in (5.26) and noting the continuity and boundedness of $Q(\cdot)$, we can replace $q_{i_1 i}(X_k)$ by $q_{i_1 i}(X_{ln_\varepsilon})$ yielding the same limit. Then as in (5.15) and (5.16), replace $I_{\{\alpha_k=i_1\}}$ by $I_{\{\alpha^\varepsilon(\varepsilon ln_\varepsilon)=i_1\}}$, again yielding the same limit. Thus, we have

$$\lim_{\varepsilon \to 0} \mathbf{E} \prod_{\iota=1}^{\kappa} \rho_\iota(X^\varepsilon(t_\iota), \alpha^\varepsilon(t_\iota)) \left\{ \sum_{ln_\varepsilon=t/\varepsilon}^{(t+s)/\varepsilon-1} [h(X_{ln_\varepsilon}, \alpha_{ln_\varepsilon+n_\varepsilon}) - h(X_{ln_\varepsilon}, \alpha_{ln_\varepsilon})] \right\}$$

$$= \mathbf{E} \prod_{\iota=1}^{\kappa} \rho_\iota(\tilde{X}(t_\iota), \tilde{\alpha}(t_\iota)) \left\{ \int_t^{t+s} Q(\tilde{X}(u)) h(\tilde{X}(u), \cdot)(\tilde{\alpha}(u)) du \right\},$$

(5.28)

where for each $i_1 \in \mathcal{M}$,

$$
\begin{aligned}
Q(x)h(x,\cdot)(i_1) &= \sum_{i=1}^{m_0} q_{i_1 i}(x)h(x,i) \\
&= \sum_{i \neq i_1} q_{i_1 i}(x)(h(x,i) - h(x,i_1)).
\end{aligned}
$$

Step 5: Combining Steps 1–4, we arrive at that $(\widetilde{X}(\cdot), \widetilde{\alpha}(\cdot))$, the weak limit of $(X^{\varepsilon}(\cdot), \alpha^{\varepsilon}(\cdot))$ is a solution of the martingale problem with operator $\mathcal{L}$ defined in (5.3). Using characteristic functions, we can show as in [176, Lemma 7.18], $(X(\cdot), \alpha(\cdot))$, the solution of the martingale problem with operator $\mathcal{L}$, is unique in the sense of in distribution. Thus $(X^{\varepsilon}(\cdot), \alpha^{\varepsilon}(\cdot))$ converges to $(X(\cdot), \alpha(\cdot))$ as desired, which concludes the proof of the theorem.
□

In addition, we can obtain the following convergence result as a corollary.

Corollary 5.9. *Under the conditions of Theorem 5.8, the sequence of processes $(X^{\varepsilon}(\cdot), \alpha^{\varepsilon}(\cdot))$ converges to $(X(\cdot), \alpha(\cdot))$ in the sense*

$$
\sup_{0 \leq t \leq T} \mathbf{E}|X^{\varepsilon}(t) - X(t)|^2 \to 0 \quad as \quad \varepsilon \to 0. \tag{5.29}
$$

Remark 5.10. With a little more effort, we can obtain strong convergence (in the usual sense used in the numerical solutions of stochastic differential equations). The steps involved can be outlined as follows. (a) We consider two sequences of approximations $(X^{\varepsilon}(\cdot), \alpha^{\varepsilon}(\cdot))$ and $(X^{\eta}(\cdot), \alpha^{\eta}(\cdot))$ with the same initial data. (b) Define for sufficiently small $\varepsilon > 0$ and $\eta > 0$,

$$
\begin{aligned}
\widetilde{X}^{\varepsilon}(t) &= X_0 + \int_0^t b(X^{\varepsilon}(s), \alpha^{\varepsilon}(s))ds + \int_0^t \sigma(X^{\varepsilon}(s), \alpha^{\varepsilon}(s))dw(s), \\
\widetilde{X}^{\eta}(t) &= X_0 + \int_0^t b(X^{\eta}(s), \alpha^{\eta}(s))ds + \int_0^t \sigma(X^{\eta}(s), \alpha^{\eta}(s))dw(s).
\end{aligned}
$$

That is, they are two approximations of the solution with the use of different stepsizes. Then we can show

$$
\mathbf{E} \sup_{t \in [0,T]} |\widetilde{X}^{\varepsilon}(t) - \widetilde{X}^{\eta}(t)|^2 \to 0 \quad as \quad \varepsilon \to 0 \ \ and \ \ \eta \to 0.
$$

The main ingredient is the application of Doob's martingale inequality. (c) Let $\{\varepsilon_n\}$ be a sequence of positive real numbers satisfying $\varepsilon_n \to 0$. We show that $\{X^{\varepsilon_n}(t) : t \in [0,T]\}$ is an L^2 Cauchy sequence of random elements. We then conclude $\mathbf{E} \sup_{t \in [0,T]} |X^{\varepsilon}(t) - X(t)|^2 \to 0$ as $\varepsilon \to 0$. (d) Moreover, we can use the above results to give an alternative proof of the existence and uniqueness of the solution of (5.1) together with (5.2) (or (2.2) together with (2.3)); see Remark 2.2 for different approaches for proving the existence and uniqueness of the solution.

5.5   Examples

Here we present two examples for demonstration. It would be ideal if we could compare the numerical solutions using our algorithms with the analytic solutions. Unfortunately, due to the complexity of the x-dependent switching process, closed-form solutions are not available. We are thus contended with the numerical solutions. In both examples, we use the state-dependent generator $Q(x)$ given by

$$Q(x) = \begin{pmatrix} -5\cos^2 x & 5\cos^2 x \\ 10\cos^2 x & -10\cos^2 x \end{pmatrix}.$$

Example 5.11. Consider a jump linear system. Suppose that

$$\sigma(x,1) = 2x \quad \text{and} \quad \sigma(x,2) = x.$$

Let

$$b(x,i) = A(i)x \quad \text{with} \quad A(1) = -3.3 \quad \text{and} \quad A(2) = -2.7.$$

We use the constant stepsize algorithm (5.4). Specify the initial conditions as $X_0 = 5$ and $\alpha_0 = 1$, and use the constant stepsize 0.001. A sample path of the computed iterations is depicted in Figure 5.1.

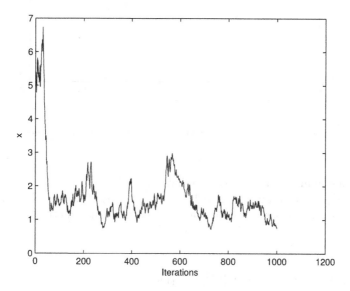

FIGURE 5.1. A sample path of a numerical approximation to a jump-linear system.

Example 5.12. This example is concerned with switching diffusion with nonlinear drifts with respect to the continuous-state variable. We use the following specifications,

$$\sigma(x,1) = 0.5x, \quad \sigma(x,2) = 0.2x.$$

For each $i \in \{1,2\}$, consider the nonlinear functions

$$b(x,1) = -(2 + \sin x) \quad \text{and} \quad b(x,2) = -(1 + \sin x \cos x).$$

Using the same initial data and stepsize as in Example 5.11, the calculation is carried out. The computational result is displayed in Figure 5.2.

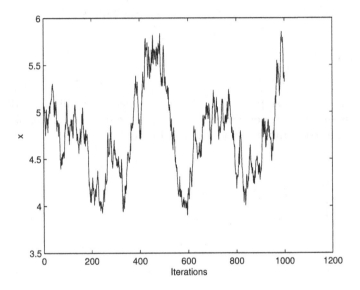

FIGURE 5.2. A sample path of switching diffusion with nonlinear drifts.

For these numerical examples, we have tried different stepsizes. They all produced similar sample path behavior as displayed above. For numerical experiment purposes, we have also tested different functions $b(\cdot,\cdot)$ and $\sigma(\cdot,\cdot)$.

5.6   Discussions and Remarks

This section provides some discussions on issues related to numerical approximation. First, rates of convergence are discussed, and then decreasing stepsize algorithms are studied.

5.6.1   Remarks on Rates of Convergence

Because we are dealing with a numerical algorithm, it is desirable that we have some estimation error bounds on the rates of convergence. This section takes up this issue. In Kloeden and Platen [94, p. 323], the rate of convergence was defined as follows. For a finite time $T > 0$, if there exists a positive constant K that does not depend on ε such that $\mathbf{E}|X^\varepsilon(T) - X(T)| \leq K\varepsilon^\gamma$ for some $\gamma > 0$ then the approximation X^ε is said to converge strongly to X with order γ. Here we adopt the more recent approach in Mao and Yuan [120], and concentrate on uniform convergence in the sense of error bounds of the form $\mathbf{E}\sup_{0 \leq t \leq T}|X^\varepsilon(t) - X(t)|^2$.

We assume the conditions of Theorem 5.8 are satisfied. Also we make use of Remark 5.10. In what follows, to simplify the discussion, we take $\sqrt{\varepsilon}\xi_n = w(\varepsilon(n + 1)) - w(\varepsilon n)$. (Independent and identically distributed "white" noise sequence $\{\xi_n\}$ can be used, which makes the notation more complex.) It is straightforward that the piecewise constant interpolation of (5.4) leads to

$$X^\varepsilon(t) = X_0 + \int_0^t b(X^\varepsilon(s), \alpha^\varepsilon(s))ds + \int_0^t \sigma(X^\varepsilon(s), \alpha^\varepsilon(s))dw(s). \quad (5.30)$$

The representation (5.30) enables us to compare the solution (5.1) with that of the discrete iterations.

Comparing the interpolation of the iterates and the solution of (5.1), we obtain

$$\begin{aligned}
&\mathbf{E}\sup_{0 \leq t \leq T}|X^\varepsilon(t) - X(t)|^2 \\
&\leq 2\mathbf{E}\sup_{0 \leq t \leq T}\left|\int_0^t [b(X(s), \alpha(s)) - b(X^\varepsilon(s), \alpha^\varepsilon(s))]ds\right|^2 \\
&\quad + 2\mathbf{E}\sup_{0 \leq t \leq T}\left|\int_0^t [\sigma(X(s), \alpha(s)) - \sigma(X^\varepsilon(s), \alpha^\varepsilon(s))]dw\right|^2 \quad (5.31) \\
&\leq 2T\mathbf{E}\int_0^T |b(X(s), \alpha(s)) - b(X^\varepsilon(s), \alpha^\varepsilon(s))|^2 ds \\
&\quad + 8\mathbf{E}\int_0^T |\sigma(X(s), \alpha(s)) - \sigma(X^\varepsilon(s), \alpha^\varepsilon(s))|^2 ds.
\end{aligned}$$

Note that in (5.31), the first inequality is obtained from the familiar inequality $(a + b)^2 \leq 2(a^2 + b^2)$ for two real numbers a and b. The first term on the right-side of the second inequality follows from Hölder's inequality, and the second term is a consequence of the well-known Doob martingale inequality (see (A.19) in the appendix). To proceed, we treat the drift and diffusion terms separately. Note that

$$
\begin{aligned}
\mathbf{E} \int_0^T &|b(X(s), \alpha(s)) - b(X^\varepsilon(s), \alpha^\varepsilon(s))|^2 ds \\
&\leq \mathbf{E} \int_0^T |b(X(s), \alpha(s)) - b(X^\varepsilon(s), \alpha(s))|^2 ds \\
&\quad + \mathbf{E} \int_0^T |b(X^\varepsilon(s), \alpha(s)) - b(X^\varepsilon(s), \alpha^\varepsilon(s))|^2 ds \\
&\leq K \int_0^T \mathbf{E}|X^\varepsilon(s) - X(s)|^2 ds \\
&\quad + \mathbf{E} \int_0^T [1 + |X^\varepsilon(s)|^2] I_{\{\alpha(s) \neq \alpha^\varepsilon(s)\}} ds.
\end{aligned}
\tag{5.32}
$$

The first inequality in (5.32) follows from the familiar triangle inequality, and the second inequality is a consequence of the Lipschitz continuity, the Cauchy inequality, and the linear growth condition. We now concentrate on the last term in (5.32). Using discrete iteration, we have

$$
\begin{aligned}
\mathbf{E} \int_0^T &[1 + |X^\varepsilon(s)|^2] I_{\{\alpha(s) \neq \alpha^\varepsilon(s)\}} ds \\
&= \sum_{k=0}^{\lfloor t/\varepsilon \rfloor - 1} \mathbf{E} \int_{\varepsilon k}^{\varepsilon k + \varepsilon} [1 + |X^\varepsilon(s)|^2] I_{\{\alpha(s) \neq \alpha^\varepsilon(s)\}} ds.
\end{aligned}
$$

Using nested conditioning, we further obtain

$$
\begin{aligned}
\mathbf{E} \int_{\varepsilon k}^{\varepsilon k + \varepsilon} &[1 + |X^\varepsilon(s)|^2] I_{\{\alpha(s) \neq \alpha^\varepsilon(s)\}} ds \\
&= \mathbf{E} \int_{\varepsilon k}^{\varepsilon k + \varepsilon} [1 + |X_k|^2] \mathbf{E}[I_{\{\alpha(s) \neq \alpha^\varepsilon(s)\}} | \mathcal{F}_{\varepsilon k}] ds \\
&= \mathbf{E}\Big([1 + |X_k|^2] \int_{\varepsilon k}^{\varepsilon k + \varepsilon} \sum_{i \in \mathcal{M}} I_{\{\alpha_k^\varepsilon = i\}} \sum_{j \neq i} [q_{ij}(X(\varepsilon k))(s - \varepsilon k) \\
&\quad + o(s - \varepsilon k)] ds \Big) \\
&\leq K\varepsilon \int_{\varepsilon k}^{\varepsilon k + \varepsilon} ds \leq K\varepsilon^2.
\end{aligned}
$$

Thus the moment estimate of $\mathbf{E}|X(t)|^2$ yields that

$$
\begin{aligned}
\mathbf{E} \int_0^T &|b(X(s), \alpha(s)) - b(X^\varepsilon(s), \alpha^\varepsilon(s))|^2 ds \\
&\leq K\varepsilon + K \int_0^T \mathbf{E}|X^\varepsilon(s) - X(s)|^2 ds.
\end{aligned}
\tag{5.33}
$$

Likewise, for the term involving diffusion, we also obtain

$$
\mathbf{E} \int_0^T |\sigma(X(s), \alpha(s)) - \sigma(X^\varepsilon(s), \alpha^\varepsilon(s))|^2 dw
$$
$$
\leq K\varepsilon + K \int_0^T \mathbf{E}|X^\varepsilon(s) - X(s)|^2 ds. \tag{5.34}
$$

Using (5.31)–(5.34), we obtain

$$
\mathbf{E} \sup_{0 \leq t \leq T} |X^\varepsilon(t) - X(t)|^2
$$
$$
\leq K\varepsilon + \int_0^T \mathbf{E} \sup_{0 \leq s \leq T} |X^\varepsilon(s) - X(s)|^2 ds. \tag{5.35}
$$

An application of Gronwall's inequality leads to

$$
\mathbf{E} \sup_{0 \leq t \leq T} |X^\varepsilon(t) - X(t)|^2 \leq K\varepsilon.
$$

Thus, we conclude that the discrete iterates converge strongly in the L^2 sense with an error bound of the order $O(\varepsilon)$. We state it as a result below.

Theorem 5.13. *Assume* (A5.1)–(A5.3). *Then the sequence* $(X^\varepsilon(\cdot), \alpha^\varepsilon(\cdot))$ *converges to* $(X(\cdot), \alpha(\cdot))$ *in the sense*

$$
\mathbf{E} \sup_{0 \leq t \leq T} |X^\varepsilon(t) - X(t)|^2 \to 0 \quad as \quad \varepsilon \to 0. \tag{5.36}
$$

Moreover, we have the following rate of convergence estimate

$$
\mathbf{E} \sup_{0 \leq t \leq T} |X^\varepsilon(t) - X(t)|^2 = O(\varepsilon). \tag{5.37}
$$

5.6.2   Remarks on Decreasing Stepsize Algorithms

So far the development is based on using constant stepsize algorithms. To approximate (5.1), we could also use a decreasing stepsize algorithm of the form

$$
X_{n+1} = X_n + \varepsilon_n b(X_n, \alpha_n) + \sqrt{\varepsilon_n} \sigma(X_n, \alpha_n) \xi_n. \tag{5.38}
$$

Compared with (5.4), for $X_{n-1} = x$, α_n is a finite-state process with transition matrix $I + \varepsilon_n Q(x)$, Instead of (A5.3), we assume the following condition.

(A5.4) In (5.38), $\{\varepsilon_n\}$ is a sequence of decreasing stepsizes satisfying $\varepsilon_n \geq 0$, $\varepsilon_n \to 0$ as $n \to \infty$ and $\sum_n \varepsilon_n = \infty$. The $\{\xi_n\}$ is a sequence of independent and identically distributed normal random variables such that ξ_n is independent of the σ-algebra $\mathcal{G}_n$ generated by $\{X_k, \alpha_k : k \leq n\}$, and that $\mathbf{E}\xi_n = 0$, $\mathbf{E}|\xi_n|^2 < \infty$, and $\mathbf{E}\xi_n \xi_n' = I$. Moreover, for $X_{n-1} = x$, α_n is a finite-state process with transition matrix $I + \varepsilon_n Q(x)$.

Define

$$t_n = \sum_{k=0}^{n-1} \varepsilon_k, \quad m(t) = \max\{n : t_n \le t\},$$

and continuous-time interpolations

$$X^0(t) = X_n, \quad \alpha^0(t) = \alpha_n, \quad \text{for } t \in [t_n, t_{n+1}).$$
$$X^n(t) = X^0(t + t_n), \quad \alpha^n(t) = \alpha^0(t + t_n).$$

Using essentially the same approach as in the development of Theorem 5.8 together with the ideas from stochastic approximation [104], we obtain the following result.

Theorem 5.14. *Under* (A5.1), (A5.2), *and* (A5.4), $(X^n(\cdot), \alpha^n(\cdot))$ *converges to* $(X(\cdot), \alpha(\cdot))$ *weakly, which is a solution of the martingale problem with operator* $\mathcal{L}$ *defined in* (5.3). *Moreover,*

$$\mathbf{E} \sup_{0 \le t \le T} |X^n(t) - X(t)|^2 \to 0 \quad as \quad n \to \infty. \tag{5.39}$$

Furthermore, the rate of convergence is given by

$$\mathbf{E} \sup_{0 \le t \le T} |X^n(t) - X(t)|^2 = O(\varepsilon_n) \quad as \quad n \to \infty. \tag{5.40}$$

5.7 Notes

Numerical methods for stochastic differential equations have been studied extensively, for example, in [94, 126] among others. A comprehensive study of the early results is contained in Kloeden and Platen [94]. Accelerated rates of convergence are given in Milstein [126]. As a natural extension, further results are considered in Milstein and Tretyakov [127], among others. Numerical solutions for stochastic differential equations modulated by Markovian switching have also been well studied; recent progress for switching diffusions are contained in [120] and references therein.

   Although numerical methods for Markov modulated switching diffusions have been considered by many researchers, less is known for processes with continuous-state-dependent switching. In fact, the study of switching diffusions with state-dependent switching is still in its infancy. There are certain difficulties. For example, the usual Picard iterations cannot be used, which uses the Lipschitz conditions crucially. When Markovian regime-switching processes are treated, with the given generator of the switching process, we can pre-generate the switching process throughout the iterations. However, in the state-dependent switching case, since the generation of the x-dependent switching processes is different in every step, we can no longer pre-generate the switching process without interacting with the continuous-state process and use the Lipschitz condition directly.

Part of the results of this chapter are based on Yin, Mao, Yuan, and Cao [172]. The approach uses local analysis and weak convergence methods. It relies on the solutions of associated martingale problems. The approach is different from the usual techniques developed in the literature for numerical solutions of stochastic differential equations to date. The idea and the techniques used are interesting in their own right. Rates of convergence are then ascertained together with the development of the unusual strong convergence and decreasing stepsize algorithms.

6

Numerical Approximation to Invariant Measures

6.1 Introduction

Continuing with the development in Chapter 5, this chapter is devoted to additional properties of numerical approximation algorithms for switching diffusions, where continuous dynamics are intertwined with discrete events. In this chapter, we establish that if the invariant measure exists, under suitable conditions, the sequence of iterates obtained using Euler–Maruyama approximation converges to the invariant measure.

Here for simplicity, the discrete events are formulated as a finite-state continuous-time Markov chain that can accommodate a set of possible regimes, across which the dynamic behavior of the systems may be markedly different. For simplicity, we have chosen to present the result for $Q(x) = Q$. One of the motivations for the use of a constant matrix Q is: We may view it as an approximation to x-dependent $Q(x)$ in the sense of $Q(x) = Q + o(1)$ as $|x| \to \infty$. This is based on the results in Chapters 3 and 4. Because positive recurrence implies ergodicity, we could concentrate on a neighborhood of ∞. Then effectively, Q, the constant matrix is the one having the most contributions to the asymptotic properties in which we are interested. Thus, we can "replace" $Q(x)$ by Q in the first approximation. At any given instance, in lieu of a fixed regime, the system parameters can take one of several possible regimes (configurations). As the Markov chain sojourns in a given state for a random duration, the system dynamics are governed by a diffusion process in accordance with the associated stochastic differential equation. Subsequently, the Markov chain jumps into another state, and

G.G. Yin and C. Zhu, *Hybrid Switching Diffusions: Properties and Applications*,
Stochastic Modelling and Applied Probability 63, DOI 10.1007/978-1-4419-1105-6_6,
© Springer Science + Business Media, LLC 2010

the dynamic system switches to another diffusion process associated with the new state, and so on. Instead of staying in a fixed configuration following one diffusion process, the system jumps back and forth among a set of possible configurations, resulting in a hybrid system of diffusions. In this chapter, we consider switching diffusions given by

$$dX(t) = b(X(t), \alpha(t))dt + \sigma(X(t), \alpha(t))dw(t), \qquad (6.1)$$

where $\alpha(t)$ is a continuous-time Markov chain that is generated by Q and that has state space $\mathcal{M} = \{1, \ldots, m_0\}$. Associated with (6.1), there is an operator defined by

$$
\begin{aligned}
\mathcal{L}f(x, \imath) = &\sum_{i=1}^{r} b_i(x, \imath) \frac{\partial f(x, \imath)}{\partial x_i} \\
&+ \frac{1}{2} \sum_{i,j=1}^{r} a_{ij}(x, \imath) \frac{\partial^2 f(x, \imath)}{\partial x_i \partial x_j} + Qf(x, \cdot)(\imath), \quad \imath \in \mathcal{M},
\end{aligned} \qquad (6.2)
$$

where

$$Qf(x, \cdot)(\imath) = \sum_{\jmath=1}^{m_0} q_{\imath\jmath} f(x, \jmath), \quad \imath \in \mathcal{M},$$

and

$$A(x, \imath) = (a_{ij}(x, \imath)) = \sigma(x, \imath)\sigma'(x, \imath).$$

As alluded to in the previous chapters, switching diffusions have provided many opportunities in terms of flexibility. The formulation allows the mathematical models to have multiple discrete configurations thereby making them more versatile. However, solving systems of diffusions with switching is still a challenging task, which often requires using numerical methods and/or approximation techniques. The Euler–Maruyama scheme is one of such approaches. In Section 6.2, we derive the tightness of the approximating sequence. To proceed, an important problem of both theoretical and practical concerns is: Whether the sequence of approximation converges to the invariant measure of the underlying system, provided that it exists. To answer this question, we derive the convergence to the invariant measures of the numerical approximation in Section 6.3. To obtain the results requires the convergence of the algorithm under the weak convergence framework. Rather than working with sample paths or numerically solving systems of Kolmogorov–Fokker–Planck equations, we focus on the corresponding measures and use a purely probabilistic argument. Why is such a consideration important from a practical point of view? Suppose that one considers an ergodic control problem of a hybrid-diffusion system with regime switching. Then it is desirable to "replace" the actual time-dependent measure by an invariant measure. The control problems often have to be solved using numerical approximations. Because solving the corresponding system of Kolmogorov–Fokker–Planck equations is computationally intensive, it is crucially important to be able to approximate the invariant measures numerically. For previous work on approximating invariant measures of diffusion processes without regime switching, we refer the reader to [102] and the many references cited there. Section 6.4 provides the proof of convergence for a decreasing stepsize algorithm and Section 6.5 concludes the chapter.

6.2 Tightness of Approximation Sequences

In this chapter, we concern ourselves with the following algorithms with a sequence of decreasing stepsizes $\{\varepsilon_n\}$,

$$X_{n+1} = X_n + \varepsilon_n b(X_n, \alpha_n) + \sqrt{\varepsilon_n}\sigma(X_n, \alpha_n)\xi_n, \qquad (6.3)$$

as well as algorithms with a constant stepsize ε,

$$X_{n+1} = X_n + \varepsilon b(X_n, \alpha_n) + \sqrt{\varepsilon}\sigma(X_n, \alpha_n)\xi_n. \qquad (6.4)$$

Various quantities are as given in the last chapter. Associated with (6.1), there is a martingale problem formulation. A process $(X(t), \alpha(t))$ is said to be a solution of the martingale problem with operator $\mathcal{L}$ defined in (6.2) if,

$$h(X(t), \alpha(t)) - \int_0^t \mathcal{L}h(X(s), \alpha(s))ds \qquad (6.5)$$

is a martingale for any real-valued function $h(\cdot)$ defined on $\mathbb{R}^r \times \mathcal{M}$ such that for each $i \in \mathcal{M}$, $h(\cdot, i) \in C_0^2$ (a collection of functions that are twice continuously differentiable w.r.t. the first variable with compact support).

To proceed, we state the following conditions first.

(A6.1) There is a unique solution of (6.1) for each initial condition.

(A6.2) For each $\alpha \in \mathcal{M}$, there is a Liapunov function $V(\cdot, \alpha)$ such that (a) $V(\cdot, \alpha)$ is twice continuously differentiable and $\nabla^2 V(\cdot, \alpha)$ is bounded uniformly; (b) $V(x, \alpha) \geq 0$; $|V(x, \alpha)| \to \infty$ as $|x| \to \infty$; (c) $\mathcal{L}V(x, \alpha) \leq -\lambda V(x, \alpha)$ for some $\lambda > 0$; (d) for each $x \in \mathbb{R}^r$, the following growth conditions hold:

$$|\nabla V'(x, \alpha)b(x, \alpha)| \leq K(1 + V(x, \alpha)),$$
$$|b(x, \alpha)|^2 \leq K(1 + V(x, \alpha)), \qquad (6.6)$$
$$|\sigma(x, \alpha)|^2 \leq K(1 + V(x, \alpha)).$$

(A6.3) The $\{\varepsilon_n\}$ is a sequence of decreasing stepsizes satisfying $\varepsilon_n \geq 0$, $\varepsilon_n \to 0$ as $n \to \infty$, and $\sum_n \varepsilon_n = \infty$. The $\{\xi_n\}$ is a sequence of independent and identically distributed random variables such that ξ_n is independent of the σ-algebra $\mathcal{G}_n$ generated by $\{X_k, \alpha_k : k \leq n\}$, and that $\mathbf{E}\xi_n = 0$, $\mathbf{E}|\xi_n|^2 < \infty$, and $\mathbf{E}\xi_n \xi_n' = I$.

Remark 6.1. Note that sufficient conditions for the existence and uniqueness of the solution of the switching diffusion were provided in Chapter 2. Here we simply assume them for convenience.

Condition (A6.2) requires the existence of Liapunov functions $V(\cdot, \alpha)$. Condition (A6.2)(d) is a growth condition on the functions $b(x, \alpha)$ and $\sigma(x, \alpha)$. If $b(\cdot, \alpha)$ and $\sigma(\cdot, \alpha)$ grow at most linearly, and the Liapunov function is quadratic, this condition is verified. Condition (c) requires the diffusion with regime switching (6.1) to be stable in the sense of Liapunov. Conditions listed in (A6.2) cover a large class of functions; see the related comments in [102, 104].

Remark 6.2. A quick glance at the algorithm reveals that (6.3) has a certain resemblance to a stochastic approximation algorithm, which has been the subject of extensive research for over five decades since the pioneering work of Robbins and Monro [139]. The most recent account on the subject and a state-of-the-art treatment can be found in Kushner and Yin [104] and references therein. In what follows, we use certain ideas from stochastic approximation methods to establish the limit of the discretization through suitable interpolations. Weak convergence methods are used to study the convergence of the algorithm and the associated invariant measure.

Remark 6.3. As alluded to in Chapter 5, it can be established that the sequence $\{X_n\}$ is tight by using the moment estimates together with Tchebeshev's inequality. Here we use an alternative approach based on Liapunov function methods. Note that this approach can be modified to a perturbed Liapunov function approach, which can be used to handle correlated random processes under suitable conditions. To illustrate the use of the Liapunov function method for the tightness, we present the result below together with a proof.

Theorem 6.4. *Assume* (A6.2) *and* (A6.3). *Then*

(i) *the iterates generated by* (6.3) *using decreasing stepsizes satisfy*

$$EV(X_n, \alpha_n) = O(1);$$

(ii) *the iterates generated by* (6.4) *using a constant stepsize satisfy*

$$\mathbf{E}V(X_n, \alpha_n) = O(1)$$

for n sufficiently large.

Proof. The proof uses Liapunov functions. We concern ourselves with the proof for algorithm (6.3) only. The results for (6.4) can be obtained similarly.

Henceforth, denote by $\widetilde{\mathcal{F}}_n$ and $\widetilde{\mathcal{F}}_n^\alpha$ the σ-algebras generated by $\{\xi_k, \alpha_k : k < n\}$ and $\{\alpha_n, \xi_k, \alpha_k : k < n\}$, and denote by $\mathbf{E}_n$ and $\mathbf{E}_n^\alpha$ the corresponding conditional expectations w.r.t. to $\widetilde{\mathcal{F}}_n$ and $\widetilde{\mathcal{F}}_n^\alpha$, respectively. Similarly, denote $\mathcal{F}_t = \sigma\{w(u), \alpha(u) : u \le t\}$ and $\mathbf{E}_t$ the conditional expectation w.r.t. $\mathcal{F}_t$, where $w(t)$ is an r-dimensional standard Brownian motion (having independent increments). Note that $\{\xi_n\}$ is a sequence of independent and identically distributed random vectors with 0 mean and covariance I. We have

$$\mathbf{E}_n^\alpha V(X_{n+1}, \alpha_{n+1}) - V(X_n, \alpha_n)$$
$$= \mathbf{E}_n^\alpha[V(X_{n+1}, \alpha_{n+1}) - V(X_{n+1}, \alpha_n)] \tag{6.7}$$
$$+ \mathbf{E}_n^\alpha V(X_{n+1}, \alpha_n) - V(X_n, \alpha_n).$$

We proceed to estimate each of the terms after the equality sign above.

Using the smoothness of $V(\cdot, \alpha)$, the independence of ξ_n with α_n, and the independent increment property, estimates (with details omitted) lead to

$$\mathbf{E}_n^\alpha V(X_{n+1}, \alpha_n) - V(X_n, \alpha_n)$$
$$= \varepsilon_n \nabla V'(X_n, \alpha_n) b(X_n, \alpha_n)$$
$$+ \frac{\varepsilon_n}{2} \text{tr}[\sigma'(X_n, \alpha_n) \nabla^2 V(X_n, \alpha_n) \sigma(X_n, \alpha_n)] \tag{6.8}$$
$$+ O(\varepsilon_n^2)(1 + V(X_n, \alpha_n)).$$

As for the term on the second line of (6.7), using a truncated Taylor expansion, we obtain

$$\mathbf{E}_n^\alpha[V(X_{n+1}, \alpha_{n+1}) - V(X_{n+1}, \alpha_n)]$$
$$= \mathbf{E}_n^\alpha[V(X_n, \alpha_{n+1}) - V(X_n, \alpha_n)]$$
$$+ \mathbf{E}_n^\alpha \int_0^1 \nabla V'(X_n + s\Delta X_n, \alpha_n)(X_{n+1} - X_n) ds \tag{6.9}$$
$$= \varepsilon_n Q V(X_n, \cdot)(\alpha_n) + O(\varepsilon_n)(1 + V(X_n, \alpha_n)).$$

By virtue of (A6.2),

$$\nabla V'(X_n, \alpha_n) b(X_n, \alpha_n) + QV(X_n, \cdot)(\alpha_n)$$
$$+ \frac{1}{2} \text{tr}[\sigma'(X_n, \alpha_n) \nabla^2 V(X_n, \alpha_n) \sigma(X_n, \alpha_n)]$$
$$= \mathcal{L}V(X_n, \alpha_n) \le -\lambda V(X_n, \alpha_n).$$

Thus (6.7)–(6.9) yield

$$\mathbf{E}_n^\alpha V(X_{n+1}, \alpha_{n+1}) - V(X_n, \alpha_n)$$
$$\leq -\lambda \varepsilon_n V(X_n, \alpha_n) + O(\varepsilon_n) V(X_n, \alpha_n) + O(\varepsilon_n).$$

Taking the expectation and iterating on the resulting inequality lead to

$$EV(X_{n+1}, \alpha_{n+1})$$
$$\leq A_{n,-1} \mathbf{E} V(X_0, \alpha_0) + K \sum_{k=0}^{n} \varepsilon_k A_{nk} \mathbf{E} V(X_k, \alpha_k) + K \sum_{k=0}^{n} \varepsilon_k A_{nk},$$

$$(6.10)$$

where

$$A_{nk} = \begin{cases} \prod_{j=k+1}^{n}(1 - \lambda \varepsilon_j), & \text{if } n > k, \\ 1, & \text{otherwise.} \end{cases}$$

Note that

$$\sum_{k=0}^{n} \varepsilon_k A_{nk} = O(1).$$

An application of Gronwall's inequality implies

$$\mathbf{E} V(X_n, \alpha_n) \leq K \exp\left(\sum_{k=0}^{n} \varepsilon_k A_{nk} \right) = O(1).$$

Thus the theorem is proved.   □

Using the same kind of arguments as those of Chapter 5, we can obtain the convergence of the algorithm. Define

$$t_n = \sum_{l=0}^{n-1} \varepsilon_l.$$

Let $m(t)$ be the unique value of n such that $t_n \leq t < t_{n+1}$ for $t \geq 0$. Define $\alpha_n = \alpha(t_n)$. Let $X^0(t)$ and $\alpha^0(t)$ be the piecewise constant interpolations of X_n and α_n on $[t_n, t_{n+1})$, and $X^n(t)$ and $\alpha^n(t)$ be their shifts. That is,

$$X^0(t) = X_n, \quad \alpha^0(t) = \alpha_n \text{ if } 0 < t \in [t_n, t_{n+1}),$$
$$X^n(t) = X^0(t + t_n), \quad \alpha^n(t) = \alpha^0(t + t_n).$$

$$(6.11)$$

This shift is used to bring the asymptotic properties of the underlying sequence to the foreground. Note that $X^n(\cdot) \in D([0, \infty); \mathbb{R}^r)$, the space of $\mathbb{R}^r$-valued functions that are right continuous and have left limits endowed with the Skorohod topology [43, 102, 104]. In what follows, we show that $X^n(\cdot)$ converges weakly to $X(\cdot)$, the solution of (6.1). In fact, we work with the pair $(X^n(\cdot), \alpha^n(\cdot))$. By virtue of Theorem 5.14, for the constant stepsize algorithm, $\{X^\varepsilon(\cdot), \alpha^\varepsilon(\cdot)\}$ converges weakly to $(X(\cdot), \alpha(\cdot))$, which is a process with generator given by (6.2). To proceed, we establish a convergence result for the decreasing stepsize algorithm.

Theorem 6.5. *Consider algorithm* (6.4). *Assume* (A6.1)–(A6.3) *with* α_n *being a Markov chain with one-step transition probability matrix* $I + \varepsilon_n Q$. *Then the interpolated process* $(X^n(\cdot), \alpha^n(\cdot))$ *converges weakly to* $(X(\cdot), \alpha(\cdot))$, *which is a solution of* (6.1).

We note that because Q is independent of x, the limit process $\alpha(\cdot)$ is a Markov chain generated by Q. The proof of this theorem is still of independent interest although a constant stepsize algorithm has been proved in Chapter 5 for x-dependent $Q(x)$. However, the main result of this chapter is on convergence to the invariant measures, thus we postpone the proof until Section 6.4.

6.3 Convergence to Invariant Measures

Having the convergence of $\{X^n(\cdot), \alpha^n(\cdot)\}$ and hence the convergence of the algorithm, in this section, we examine the large-time behavior of the algorithm and address the issue: When will the sequence of measures of $(X^n(s_n + \cdot), \alpha^n(s_n + \cdot))$ converge to the invariant measure of the switching diffusion (6.1) if it exists, when $s_n \to \infty$ as $n \to \infty$. We first recall a couple of definitions.

For our two component process $(X(t), \alpha(t))$, let $C_b(\mathbb{R}^r)$ be the space of real-valued bounded and continuous functions defined on $\mathbb{R}^r$. A set $\mathcal{CD} \subset C_b(\mathbb{R}^r)$ is said to be convergence determining if

$$\sum_{i=1}^{m_0} \int f(x,i)\nu_n(dx,i) \to \sum_{i=1}^{m_0} \int f(x,i)\nu(dx,i)$$

for each $\iota \in \mathcal{M}$ and $f(\cdot, \iota) \in \mathcal{CD}$ implies that ν_n converges to ν weakly.

Remark 6.6. Different from the case of diffusion processes, in addition to the continuous component, we also have a discrete component. As a convention, with a slight abusing of notation, we use, for example, $\nu(t, dx \times \alpha)$ and $\nu(dx \times \alpha)$ instead of $\nu(t, dx \times \{\alpha\})$ and $\nu(dx \times \{\alpha\})$ throughout.

Recall that a measure $\nu(t, \cdot)$ of $(X(t), \alpha(t))$ is weakly stable or stable in distribution (see Kushner [102, p. 154]) if, for each $\delta > 0$ and arbitrary integer n_0, there exist an $\eta > 0$ and n_0^η such that for each $\alpha \in \mathcal{M}$ and any $\varphi(\cdot, \alpha) \in C_0(\mathbb{R}^r)$ (continuous functions with compact support), and for all $\nu(0, \cdot, \cdot)$

$$\left| \sum_{\alpha \in \mathcal{M}} \int \varphi_j(x,\alpha)\nu(0, dx \times \alpha) - \sum_{\alpha \in \mathcal{M}} \int \varphi_j(x,\alpha)\nu(dx \times \alpha) \right| < \eta, \ j \leq n_0^\eta,$$

implies that for all $t \geq 0$,

$$\left| \sum_{\alpha \in \mathcal{M}} \int \varphi_j(x,\alpha)\nu(t, dx \times \alpha) - \sum_{\alpha \in \mathcal{M}} \int \varphi_j(x,\alpha)\nu(dx \times \alpha) \right| < \delta, \ j \leq n_0.$$

Lemma 6.7. *The space of functions $C_b(\mathbb{R}^r)$ is convergence determining. The space of functions $C_0(\widehat{K})$, where $\widehat{K}$ is any compact subset of $\mathbb{R}^r$, is convergence determining.*

Proof. See Ethier and Kurtz [43, p. 112]. □

For each $\alpha \in \mathcal{M}$, suppose that $\{\varphi_k(\cdot, \alpha)\}$ is a sequence of uniformly continuous functions with compact support defined on $\widehat{K}$ for any compact $\widehat{K}$. Then $\{\varphi_k(\cdot, \alpha)\}$ is convergence determining according to Lemma 6.7. Moreover, the sequence ν_n converges weakly to ν as $n \to \infty$ is equivalent to

$$\sum_{\alpha \in \mathcal{M}} \int \varphi_k(x, \alpha) \nu_n(dx \times \alpha) \to \sum_{\alpha \in \mathcal{M}} \int \varphi_k(x, \alpha) \nu(dx \times \alpha) \text{ for each } k.$$

With the preparation above, we proceed with investigation of convergence to the invariant measure for numerical approximation. We begin by assuming that the solution of (6.1) has a unique invariant measure. This together with a couple of other conditions is given below; see Remark 6.8 for comments on these conditions.

(A6.4) The process $(X(\cdot), \alpha(\cdot))$ has a unique invariant measure $\nu(\cdot)$. Denote by $\nu(x, \alpha; t, \cdot)$ the measure of $(X(t), \alpha(t))$ with initial condition $(X(0), \alpha(0)) = (x, \alpha)$. As $t \to \infty$, $\nu(x, \alpha; t, \cdot)$ converges weakly to $\nu(\cdot)$ for each (x, α). For any compact $\widehat{K} \subset \mathbb{R}^r$ and for any $\varphi \in C(\mathbb{R}^r \times \mathcal{M})$,

$$\mathbf{E}_{x,\alpha}\varphi(X(t), \alpha(t)) \to \mathbf{E}_\nu\varphi$$
$$= \sum_{\alpha \in \mathcal{M}} \int \varphi(x, \alpha)\nu(dx \times \alpha). \tag{6.12}$$

uniformly on $\widehat{K} \times \mathcal{M}$, where $\mathbf{E}_\nu$ denotes the expectation with respect to the invariant measure ν.

(A6.5) For each $\alpha \in \mathcal{M}$, $X^{x,\alpha}(\cdot)$ is a Feller process with continuous coefficients on $[0, \infty)$ for each initial condition $X(0) = x$.

Remark 6.8. In accordance with the results in Chapter 4, the existence and uniqueness of an invariant measure of a switching diffusion is guaranteed by the positive recurrence of the process. Hence, sufficient conditions for positive recurrence of switching diffusions presented in Chapter 3 are the conditions to ensure the existence and uniqueness of the invariant measure. In addition, suppose that the continuous component lives in a compact set, that for each $\alpha \in \mathcal{M}$, $b(\cdot, \alpha)$ and $\sigma(\cdot, \alpha)$ satisfying the usual regularity condition and $\sigma(x, \alpha)\sigma'(x, \alpha)$ is positive definite, and that the generator Q is irreducible, with $\mathcal{L}^*$ denoting the adjoint operator of $\mathcal{L}$ given in (6.2). Then the system of equations

$$\mathcal{L}^*\mu(x,i) = 0, \quad i \in \mathcal{M},$$
$$\sum_{j=1}^{m_0} \int \mu(x,j)dx = 1 \tag{6.13}$$

has a unique solution, known as the invariant density; where $\mathcal{L}^*$ is the adjoint operator of $\mathcal{L}$; see Il'in, Khasminskii, and Yin [74] and references therein. Furthermore, in this case, the convergence to the invariant density takes places at an exponential rate.

In what follows, to highlight the dynamics starting at $(X(0), \alpha(0))$, by abusing notation slightly, we often write $\mathbf{E}_\nu \varphi$ on the right-hand side of (6.12) as $\mathbf{E}_\nu \varphi(X(0), \alpha(0))$. That is, for $\varphi(\cdot, \iota) \in C(\mathbb{R}^r)$ for each $\iota \in \mathcal{M}$,

$$\mathbf{E}_\nu \varphi(X(0), \alpha(0)) = \sum_{i \in \mathcal{M}} \int \varphi(y,i)\nu(dy,i).$$

It should be clear from the context.

Note that the Feller property assumed in (A6.5) has been established in Chapter 2 for switching diffusions with x-dependent $Q(x)$ under suitable conditions. It was noted that for Markovian switching diffusions, the Feller property can be obtained with a much simpler proof. For convenience, we put it here as a condition. As a direct consequence of these conditions, the following lemma holds.

Lemma 6.9. *Let K_c be a set of $\mathbb{R}^r$-valued random variables, which is tight and $(X(0), \alpha(0)) = (x, \alpha) \in K_c \times \mathcal{M}$. Under (A6.4) and (A6.5), for each $\alpha \in \mathcal{M}$ and any positive integer n_0, $0 = \delta_1 < \delta_2 < \cdots < \delta_{n_0}$, and each ι and any $\varphi(\cdot, \iota) \in C_b(\mathbb{R}^r)$,*

$$\mathbf{E}_{X(0),\alpha(0)}\varphi(X(t+\delta_j), \alpha(t+\delta_j), j \le n_0)$$
$$\to \mathbf{E}_\nu \varphi(X(\delta_j), \alpha(\delta_j), j \le n_0), \tag{6.14}$$

uniformly for $(X(0), \alpha(0)) \in K_c \times \mathcal{M}$ as $t \to \infty$.

Remark 6.10. Lemma 6.9 is a consequence of assumptions (A6.4) and (A6.5), and is handy to use in the subsequent development. When $n_0 = 1$,

$$\mathbf{E}_{X(0),\alpha(0)}\varphi(X(t), \alpha(t))$$

is bounded and continuous for each $t > 0$ because $\varphi(\cdot, \iota) \in C_b(\mathbb{R}^r)$ for each $\iota \in \mathcal{M}$. Suppose that $X(0) \in K_c$ and the distribution with initial condition $(X(0), \alpha(0))$ is denoted by $\nu(0, \cdot)$. Then the tightness of $X(0)$

and conditions (A6.4) and (A6.5) yield that

$$
\begin{aligned}
&\mathbf{E}_{X(0),\alpha(0)}\varphi(X(t),\alpha(t)) \\
&= \sum_{i=1}^{m_0} \int \nu(0, dx \times i)\mathbf{E}_{x,i}\varphi(X(t),\alpha(t)) \\
&\to \sum_{i=1}^{m_0} \int \nu(dx \times i)\mathbf{E}_{\nu}\varphi(X(0),\alpha(0)) \\
&= \mathbf{E}_{\nu}\varphi(X(0),\alpha(0)),
\end{aligned}
\tag{6.15}
$$

inasmuch as

$$
\sum_{i=1}^{m_0} \int \nu(dx \times i) = 1.
$$

Moreover, the convergence is uniform in $(X(0),\alpha(0))$ by the condition of uniform convergence on any compact x-set as given in (A6.4). Similar to the approach in Kushner [102, p. 155] (see also [101]), for general n_0, Lemma 6.9 can be proved by induction. The details are omitted.

Theorem 6.11. *Assume (A6.1)–(A6.5).*

(i) *For arbitrary positive integer n_0, $\varphi(\cdot,\iota) \in C(\mathbb{R}^r)$ for each $\iota \in \mathcal{M}$, and for any $\delta > 0$, there exist $t_0 < \infty$ and positive integer N_0 such that for all $t \geq t_0$ and $n \geq N_0$,*

$$
\begin{aligned}
&\Big|\mathbf{E}\varphi(X^n(t+\delta_j),\alpha^n(t+\delta_j),\ j \leq n_0) \\
&\quad - \mathbf{E}_{\nu}\varphi(X(\delta_j),\alpha(\delta_j),\ j \leq n_0)\Big| < \delta.
\end{aligned}
\tag{6.16}
$$

(ii) *Furthermore, for any sequence $s_n \to \infty$,*

$$
((X^n(s_n+\delta_1),\alpha^n(s_n+\delta_1)),\ldots,(X^n(s_n+\delta_{n_0}),\alpha^n(s_n+\delta_{n_0})))
$$

converges weakly to the stationary distribution of

$$
((X(\delta_1),\alpha(\delta_1)),\ldots,(X(\delta_{n_0}),\alpha(\delta_{n_0}))).
$$

Remark 6.12. Note that t_0 and N_0 above depend on δ and on $\varphi(\cdot)$.

Proof of Theorem 6.11. Suppose that (6.16) were not true. There would exist a subsequence $\{n_k\}$ and a sequence $s_{n_k} \to \infty$ such that

$$
\begin{aligned}
&\Big|\mathbf{E}\varphi(X^{n_k}(s_{n_k}+\delta_j),\alpha^{n_k}(s_{n_k}+\delta_j),\ j \leq n_0) \\
&\quad - \mathbf{E}_{\nu}\varphi(X(\delta_j),\alpha(\delta_j),\ j \leq n_0)\Big| \geq \delta > 0.
\end{aligned}
\tag{6.17}
$$

For a fixed $T > 0$, choose a further subsequence $\{k_\ell\}$ of $\{n_k\}$, and the corresponding sequence $(X^{k_\ell}(\cdot), \alpha^{k_\ell}(\cdot))$ such that $(X^{k_\ell}(s_{k_\ell} - T), \alpha^{k_\ell}(s_{k_\ell} - T))$ converges weakly to a random variable $(X(0), \alpha(0))$. Theorem 6.5 implies that $(X^{k_\ell}(s_{k_\ell} - T + \cdot), \alpha^{k_\ell}(s_{k_\ell} - T + \cdot))$ converges weakly to $(X(\cdot), \alpha(\cdot))$ with initial condition $(X(0), \alpha(0))$. Moreover,

$$\mathbf{E}\varphi(X^{k_\ell}(s_{k_\ell} - T + T + \delta_j), \alpha^{k_\ell}(s_{k_\ell} - T + T + \delta_j),\ j \le n_0)$$
$$\to \mathbf{E}\Big(\mathbf{E}_{X(0), \alpha(0)}\varphi(X(T + \delta_j), \alpha(T + \delta_j),\ j \le n_0)\Big)$$

as $k_\ell \to \infty$. Owing to (A6.5), the collection of all possible $X(0)$ over all $T > 0$ and weakly convergent subsequence is tight. Noting $\alpha(0) \in \mathcal{M}$, which is a finite integer, by Lemma 6.9, there exists $T_0 > 0$ such that for all $T \ge T_0$,

$$\Big|\mathbf{E}\Big(\mathbf{E}_{X(0), \alpha(0)}\varphi(X(T + \delta_j), \alpha(T + \delta_j),\ j \le n_0)\Big)$$
$$- \mathbf{E}_\nu\varphi(X(\delta_j), \alpha(\delta_j),\ j \le n_0)\Big| < \delta/2,$$

which contradicts (6.17).

Using Lemma 6.9 again, part (i) of the theorem implies that $(X^n(s_n + \cdot), \alpha^n(s_n + \cdot))$ converges weakly to the random variable with the invariant distribution $\nu(\cdot)$ as $s_n \to \infty$. Thus part (ii) of the assertion also follows. $\square$

For a constant stepsize algorithm (5.4), we can examine the associated invariant measures similar to the development for the decreasing stepsize algorithms. The following result can be derived. We state the result and omit the proof.

Theorem 6.13. *Consider algorithm* (6.3). *Assume the conditions of Theorem 6.11 are fulfilled. For arbitrary positive integer* n_0, $\varphi(\cdot, \iota) \in C_b(\mathbb{R}^r)$ *for each* $\iota \in \mathcal{M}$, *and for any* $\delta > 0$, *there exist* $t_0 < \infty$ *and positive integer* $\varepsilon_0 > 0$ *such that for all* $t \ge t_0$ *and* $\varepsilon \le \varepsilon_0$,

$$\Big|\mathbf{E}\varphi(X^\varepsilon(t + \delta_j), \alpha^\varepsilon(t + \delta_j),\ j \le n_0) - \mathbf{E}_\nu\varphi(X(\delta_j), \alpha(\delta_j),\ j \le n_0)\Big| < \delta.$$

Moreover, for any sequence $s_\varepsilon \to \infty$ *as* $\varepsilon \to 0$,

$$((X^\varepsilon(s_\varepsilon + \delta_1), \alpha^\varepsilon(s_\varepsilon + \delta_1)), \ldots, (X^\varepsilon(s_\varepsilon + \delta_{n_0}), \alpha^\varepsilon(s_\varepsilon + \delta_{n_0})))$$

converges weakly to the stationary distribution of

$$((X(\delta_1), \alpha(\delta_1)), \ldots, (X(\delta_{n_0}), \alpha(\delta_{n_0}))).$$

6.4   Proof: Convergence of Algorithm

Proof of Theorem 6.5. We can show that Lemma 5.4 continues to hold and $\{\alpha^n(\cdot)\}$ is tight with the constant stepsize replaced by the decreasing stepsizes.

We first show that the sequence of interest is tight in $D([0, \infty) : \mathbb{R}^r \times \mathcal{M})$, and then characterize the limit by means of martingale problem formulation. In the process of verifying the tightness, we need to show for each $0 < T < \infty$,

$$\lim_{K_0 \to \infty} \limsup_{n \to \infty} P(\sup_{t \leq T} |X^n(t)| \geq K_0) = 0. \tag{6.18}$$

Equation (6.18) is usually difficult to verify. Thus, we use a technical device, known as an N-truncation [104, p. 248], to overcome the difficulty.

We illustrate the use of the N-truncation device. The main idea is that for each $N < \infty$, we work with the sequence $X^{n,N}(\cdot)$ that is equal to $X^n(\cdot)$ up until the first exit from the N-sphere $S_N = \{x : |x| \leq N\}$ and is zero outside the $(N+1)$-sphere S_{N+1}. We then proceed to prove the truncated sequence is tight and obtain its limit. Finally, letting $N \to \infty$, a piecing together argument together with the uniqueness of the martingale problem enables us to complete the proof. To proceed, the proof is divided into a number of steps.

In lieu of (6.3), consider

$$X_{n+1}^N = X_n^N + \varepsilon_n b^N(X_n^N, \alpha_n) + \sqrt{\varepsilon_n} \sigma^N(X_n^N, \alpha_n)\xi_n, \tag{6.19}$$

where

$$b^N(x, \alpha) = b(x, \alpha)q^N(x), \quad \sigma^N(x, \alpha) = \sigma(x, \alpha)q^N(x), \tag{6.20}$$

and $q^N(x)$ is a smooth function satisfying $q^N(x) = 1$ when $x \in S_N$ and $q^N(x) = 0$ when $x \in \mathbb{R}^r - S_{N+1}$. Next, define $X^{0,N}(t) = X_n^N$ on $[t_n, t_{n+1})$ and $X^{n,N}(t) = X^{0,N}(t + t_n)$. Then $X^{n,N}(\cdot)$ is an N-truncation of $X^n(\cdot)$. The next lemma shows that $\{X^{n,N}(\cdot)\}$ is tight by the tightness criterion.

For any $0 < T < \infty$, any $\delta > 0$, $|t| \leq T$, and $0 < s \leq \delta$, we have

$$\mathbf{E}_{t_{m(t_n+t)}} \left| X^{n,N}(t+s) - X^{n,N}(t) \right|^2$$

$$\leq 2\mathbf{E}_{t_{m(t_n+t)}} \left| \sum_{k=m(t+t_n)}^{m(t+s+t_n)-1} b^N(X_k^N, \alpha_k)\varepsilon_k \right|^2 \tag{6.21}$$

$$+ 2\mathbf{E}_{t_{m(t_n+t)}} \left| \sum_{k=m(t+t_n)}^{m(t+s+t_n)-1} \sqrt{\varepsilon_k} \sigma^N(X_k^N, \alpha_k)\xi_k \right|^2,$$

where $\mathbf{E}_{t_{m(t_n+t)}}$ denotes the conditional expectation on the σ-algebra generated by $\{(X_j, \alpha_j) : j \leq m(t_n + t)\}$. The continuity of $b^N(\cdot, i)$ and $\sigma^N(\cdot, i)$ (for each $i \in \mathcal{M}$), the smoothness of $q^N(\cdot)$, and the boundedness of X_n^N

yield the boundedness of $b^N(\cdot, i)$ and $\sigma^N(\cdot, i)$. Thus,

$$
\mathbf{E}_{t_{m(t_n+t)}} \left| \sum_{k=m(t+t_n)}^{m(t+s+t_n)-1} b^N(X_k^N, \alpha_k)\varepsilon_k \right|^2
$$

$$
\leq K \sum_{l=m(t+t_n)}^{m(t+s+t_n)-1} \varepsilon_l \sum_{k=m(t+t_n)}^{m(t+s+t_n)-1} \varepsilon_k \leq O(s^2). \tag{6.22}
$$

In the above and hereafter, we use K to denote a generic positive constant; its values may vary for different appearances. It follows from (6.22),

$$
\lim_{\delta \to 0} \limsup_{n \to \infty} \mathbf{E} \left| \sum_{k=m(t+t_n)}^{m(t+s+t_n)-1} b^N(X_k^N, \alpha_k)\varepsilon_k \right|^2 = 0. \tag{6.23}
$$

By virtue of (A6.3), $\{\xi_n\}$ is an independent sequence with zero mean and covariance I. Without loss of generality, assume $l \leq k$. Then

$$
\mathbf{E}_l^\alpha \xi_k' \xi_l = \mathbf{E}[\xi_k' \xi_l | \alpha_l, \xi_j, \alpha_j; j < l]
$$

$$
= \begin{cases} 0, & \text{if } l \neq k, \\ 1, & \text{if } l = k. \end{cases}
$$

By the independence of X_n^N and ξ_n, and the independence of α_n and ξ_n,

$$
\mathbf{E}_{t_{m(t_n+t)}} \left| \sum_{k=m(t+t_n)}^{m(t+s+t_n)-1} \sqrt{\varepsilon_k} \sigma^N(X_k^N, \alpha_k)\xi_k \right|^2
$$

$$
\leq K \sum_{k=m(t+t_n)}^{m(t+s+t_n)-1} \varepsilon_k \leq O(s). \tag{6.24}
$$

Combining (6.22) and (6.24) and recalling that $0 \leq s < \delta$,

$$
\mathbf{E}_{t_{m(t_n+t)}} |X^{n,N}(t+s) - X^{n,N}(t)|^2 \leq \mathbf{E}_{t_{m(t_n+t)}} \gamma^{n,N}(s), \tag{6.25}
$$

where $\gamma^{n,N}(s)$ is a random variable satisfying

$$
\lim_{\delta \to 0} \limsup_{n \to \infty} \mathbf{E}\gamma^{n,N}(s) = 0.
$$

The criterion in [102, Theorem 3, p. 47] then implies the tightness of $\{X^{n,N}(\cdot)\}$. As a consequence of the tightness of $\{X^{n,N}(\cdot)\}$ and $\{\alpha^n(\cdot)\}$, the sequence of the interpolated pair of processes is tight.

Next, we use the martingale averaging techniques employed in Chapter 5 to show that $(X^{n,N}(\cdot), \alpha^n(\cdot))$ converges weakly to $(X^N(\cdot), \alpha(\cdot))$. This result is stated as a lemma below.

Lemma 6.14 *The pair $(X^N(\cdot), \alpha(\cdot))$ is the solution of the martingale problem with operator $\mathcal{L}^N$ obtained from $\mathcal{L}$ by replacing $b(\cdot)$ and $\sigma(\cdot)$ with $b^N(\cdot)$ and $\sigma^N(\cdot)$ (defined in (6.20)), respectively.*

Proof of Lemma 6.14. We need to verify that (6.5) holds. Without loss of generality, we work with $t \geq 0$. It suffices to show that for each $i \in \mathcal{M}$, any real-valued function $h(\cdot, i) \in C_0^2$, any $T < \infty$, $0 \leq t \leq T$, $s > 0$, arbitrary positive integer n_0, bounded and continuous functions $\varphi_j(\cdot, i)$ (with $j \leq n_0$), and any s_j satisfying $0 < s_j \leq t \leq t + s$,

$$
\begin{aligned}
\mathbf{E} \prod_{j=1}^{n_0} \varphi_j(X^N(s_j), \alpha(s_j)) \big[h(X^N(t+s), \alpha(t+s)) \\
- h(X^N(t), \alpha(t)) - \int_t^{t+s} \mathcal{L}^N h(X^N(u), \alpha(u)) du \big] = 0.
\end{aligned}
\tag{6.26}
$$

To obtain (6.26), let us begin with the process $(X^{n,N}(\cdot), \alpha^n(\cdot))$. By virtue of the weak convergence of $(X^{n,N}(\cdot), \alpha^n(\cdot))$ to $(X^N(\cdot), \alpha(\cdot))$ and the Skorohod representation, as $n \to \infty$,

$$
\begin{aligned}
\mathbf{E} \prod_{j=1}^{n_0} \varphi_j(X^{n,N}(s_j), \alpha^n(s_j)) \big(h(X^{n,N}(t+s), \alpha^n(t+s)) \\
- h(X^{n,N}(t), \alpha^n(t)) \big) \\
\to \mathbf{E} \prod_{j=1}^{n_0} \varphi_j(x^N(s_j), \alpha(s_j)) \big(h(X^N(t+s), \alpha(t+s)) - h(X^N(t), \alpha(t)) \big).
\end{aligned}
\tag{6.27}
$$

Choose δ_n and m_l such that $\delta_n \to 0$ as $n \to \infty$ and

$$
\frac{1}{\delta_n} \sum_{j=1}^{m_{l+1}-1} \varepsilon_j \to 1 \quad \text{as } n \to \infty.
$$

Use the notation Ξ given by

$$
\Xi = \{ l : m(t + t_n) \leq m_l \leq m_{l+1} - 1 \leq m(t + s + t_n) - 1 \}.
\tag{6.28}
$$

Then

$$
\begin{aligned}
h(X^{n,N}(t+s), \alpha^n(t+s)) - h(X^{n,N}(t), \alpha^n(t)) \\
= \sum_{l \in \Xi} [h(X_{m_{l+1}}^N, \alpha_{m_{l+1}}) - h(X_{m_{l+1}}^N, \alpha_{m_l})] \\
+ \sum_{l \in \Xi} [h(X_{m_{l+1}}^N, \alpha_{m_l}) - h(X_{m_l}^N, \alpha_{m_l})].
\end{aligned}
\tag{6.29}
$$

For the last term in (6.29),

$$\sum_{l\in\Xi}[h(X^N_{m_{l+1}},\alpha_{m_l}) - h(X^N_{m_l},\alpha_{m_l})]$$

$$= \sum_{l\in\Xi}\nabla h'(X^N_{m_l},\alpha_{m_l})[X^N_{m_{l+1}} - X^N_{m_l}] \qquad (6.30)$$

$$+\frac{1}{2}\sum_{l\in\Xi}[X^N_{m_{l+1}} - X^N_{m_l}]'\nabla^2 h(X^N_{m_l},\alpha_{m_l})[X^N_{m_{l+1}} - X^N_{m_l}] + \tilde{e}_n,$$

where $\nabla^2 h$ denotes the Hessian of $h(\cdot,\alpha)$ and $\tilde{e}_n$ represents the error incurred from the truncated Taylor expansion. By the continuity of $\nabla^2 h(\cdot,\alpha)$, the boundedness of $\{X^N_n\}$, it is readily seen that

$$\lim_{n\to\infty}\mathbf{E}\prod_{j=1}^{n_0}\varphi_j(X^{n,N}(s_j),\alpha^n(s_j))(|\tilde{e}_n|) = 0. \qquad (6.31)$$

Using (6.3) in (6.30),

$$\sum_{l\in\Xi}\nabla h'(X^N_{m_l},\alpha_{m_l})[X^N_{m_{l+1}} - X^N_{m_l}]$$

$$= \sum_{l\in\Xi}\nabla h'(X^N_{m_l},\alpha_{m_l})\sum_{k=m_l}^{m_{l+1}-1}b^N(X^N_k,\alpha_k)\varepsilon_k$$

$$+ \sum_{l\in\Xi}\nabla h'(X^N_{m_l},\alpha_{m_l})\sum_{k=m_l}^{m_{l+1}-1}\sqrt{\varepsilon_k}\sigma^N(X^N_k,\alpha_k)\xi_k.$$

The independence of ξ_k and α_k and the measurability of $(X^{n,N}(s_j),\alpha^n(s_j))$ for $j\le n_0$ with respect to $\mathcal{F}_{m(t_n+t)}$ imply

$$\mathbf{E}\prod_{j=1}^{n_0}\varphi_j(X^{n,N}(s_j),\alpha^n(s_j))\Big[\sum_{l\in\Xi}\nabla h'(X^N_{m_l},\alpha_{m_l})$$

$$\times \sum_{k=m_l}^{m_{l+1}-1}\sqrt{\varepsilon_k}\sigma^N(X^N_k,\alpha_k)\xi_k\Big]$$

$$= \mathbf{E}\prod_{j=1}^{n_0}\varphi_j(X^{n,N}(s_j),\alpha^n(s_j))\Big[\mathbf{E}_{m(t_n+t)}\sum_{l\in\Xi}\nabla h'(X^N_{m_l},\alpha_{m_l})$$

$$\times \sum_{k=m_l}^{m_{l+1}-1}\sqrt{\varepsilon_k}\sigma^N(X^N_k,\alpha_k)\mathbf{E}^\alpha_k\xi_k\Big] = 0.$$

Note that

$$
\lim_{n\to\infty} \mathbf{E} \prod_{j=1}^{n_0} \varphi_j(X^{n,N}(s_j), \alpha^n(s_j)) \Big[\sum_{l\in\Xi} \nabla h'(X^N_{m_l}, \alpha_{m_l})
$$
$$
\times \sum_{k=m_l}^{m_{l+1}-1} b^N(X^N_k, \alpha_k)\varepsilon_k \Big]
$$
$$
= \lim_{n\to\infty} \mathbf{E} \prod_{j=1}^{n_0} \varphi_j(x^{n,N}(s_j), \alpha^n(s_j)) \Big[\sum_{l\in\Xi} \nabla h'(X^N_{m_l}, \alpha_{m_l})\delta_n
$$
$$
\times \frac{1}{\delta_n} \sum_{k=m_l}^{m_{l+1}-1} b^N(X^N_{m_l}, \alpha_k)\varepsilon_k \Big]. \tag{6.32}
$$

Owing to the interpolation, for $k \in [m_l, m_{l+1} - 1]$, write α_k as $\alpha^n(u)$ with $u \in [t_{m_l}, t_{m_{l+1}-1})$. Then

$$
\frac{1}{\delta_n} \sum_{k=m_l}^{m_{l+1}-1} b^N(X^N_{m_l}, \alpha_k)\varepsilon_k
$$
$$
= \sum_{i=1}^{m_0} \frac{1}{\delta_n} \sum_{k=m_l}^{m_{l+1}-1} b^N(X^N_{m_l}, i) I_{\{\alpha_k=i\}} \varepsilon_k
$$
$$
= \sum_{i=1}^{m_0} \frac{1}{\delta_n} \sum_{k=m_l}^{m_{l+1}-1} b^N(X^N_{m_l}, i) I_{\{\alpha^n(u)=i\}} \varepsilon_k.
$$

When $t_{m_l} \to u$, $t_{m_{l+1}} \to u$ as well. In view of the weak convergence of $(X^{n,N}(\cdot), \alpha^n(\cdot))$ to $(X^N(\cdot), \alpha(\cdot))$ and the Skorohod representation, together with the continuity of $b^N(\cdot, \alpha)$ for each $\alpha \in \mathcal{M}$ then yield

$$
\mathbf{E} \prod_{j=1}^{n_0} \varphi_j(X^{n,N}(s_j), \alpha^n(s_j)) \Big[\sum_{l\in\Xi} \nabla h'(X^N_{m_l}, \alpha_{m_l}) \sum_{k=m_l}^{m_{l+1}-1} b^N(X^N_k, \alpha_k)\varepsilon_k \Big]
$$
$$
= \sum_{i=1}^{m_0} \mathbf{E} \prod_{j=1}^{n_0} \varphi_j(X^{n,N}(s_j), \alpha^n(s_j))
$$
$$
\times \Big[\sum_{l\in\Xi} \nabla h'(X^{n,N}(t_{m_l}), i) b^N(X^{n,N}(t_{m_l}), i) I_{\{\alpha^n(u)=i\}} \delta_n \Big]
$$
$$
\to \sum_{i=1}^{m_0} \mathbf{E} \prod_{j=1}^{n_0} \varphi_j(X^N(s_j), \alpha(s_j)) \Big[\int_t^{t+s} \nabla h'(X^N(u), \alpha(u)) b^N(X^N(u), i)
$$
$$
\times I_{\{\alpha(u)=i\}} du \Big].
$$

Thus

$$\mathbf{E}\prod_{j=1}^{n_0}\varphi_j(X^{n,N}(s_j),\alpha^n(s_j))[X_{m_{l+1}}^{n,N}-X_{m_l}^{n,N}]$$

$$\rightarrow \mathbf{E}\prod_{j=1}^{n_0}\varphi_j(X^N(s_j),\alpha(s_j))\Big[\int_t^{t+s}\nabla h'(X^N(u),\alpha(u))b^N(X^N(u),\alpha(u))du\Big].$$

(6.33)

Next, consider the term involving $\nabla^2 h(\cdot,\alpha)$ in (6.30). We have

$$\sum_{l\in\Xi}[X_{m_{l+1}}^N-X_{m_l}^N]'\nabla^2 h(X_{m_l}^N,\alpha_{m_l})[X_{m_{l+1}}^N-X_{m_l}^N]$$

$$=\sum_{l\in\Xi}\sum_{k_1=m_l}^{m_{l+1}-1}\sum_{k=m_l}^{m_{l+1}-1}\Big[\varepsilon_{k_1}\varepsilon_k b^{N,\prime}(X_{k_1}^N,\alpha_{k_1})$$

$$\times\nabla^2 h(X_{m_l}^N,\alpha_{m_l})b^N(X_k^N,\alpha_k)$$

(6.34)

$$+\varepsilon_{k_1}\sqrt{\varepsilon_k}b^{N,\prime}(X_{k_1}^N,\alpha_{k_1})\nabla^2 h(X_{m_l}^N,\alpha_{m_l})\sigma^{N,\prime}(X_k^N,\alpha_k)\xi_k$$

$$+\varepsilon_k\sqrt{\varepsilon_{k_1}}\xi_{k_1}'\sigma^{N,\prime}(X_{k_1}^N,\alpha_{k_1})\nabla^2 h(X_{m_l}^N,\alpha_{m_l})b^N(X_k^N,\alpha_k)$$

$$+\mathrm{tr}[\nabla^2 h(X_{m_l}^N,\alpha_{m_l})\sqrt{\varepsilon_k\varepsilon_{k_1}}\sigma^N(X_k^N,\alpha_k)\xi_k\xi_{k_1}'\sigma^{N,\prime}(X_{k_1}^N,\alpha_{k_1})]\Big].$$

By the boundedness of X_n^N, $b^N(\cdot)$, $\nabla^2 h(\cdot)$, and $\varphi_j(\cdot)$,

$$\mathbf{E}\Big|\prod_{j=1}^{n_0}\varphi_j(X^{n,N}(s_j),\alpha^n(s_j))\sum_{l\in\Xi}\sum_{k_1=m_l}^{m_{l+1}-1}\sum_{k=m_l}^{m_{l+1}-1}\varepsilon_{k_1}\varepsilon_k$$

$$\times b^{N,\prime}(X_{k_1}^N,\alpha_{k_1})\nabla^2 h(X_{m_l}^N,\alpha_{m_l})b^N(X_k^N,\alpha_k)\Big|$$

$$\leq K\sum_{l\in\Xi}\sum_{k_1=m_l}^{m_{l+1}-1}\varepsilon_{k_1}\sum_{k=m_l}^{m_{l+1}-1}\varepsilon_k$$

$$\leq K\sum_{l\in\Xi}\delta_n^2\frac{1}{\delta_n}\sum_{k_1=m_l}^{m_{l+1}-1}\varepsilon_{k_1}\frac{1}{\delta_n}\sum_{k=m_l}^{m_{l+1}-1}\varepsilon_k$$

$$\leq K\delta_n\rightarrow 0\quad\text{as }n\rightarrow\infty,$$

since

$$\sum_{l\in\Xi}\delta_n=\delta_n\Big(\frac{t+s}{\delta_n}-\frac{t}{\delta_n}\Big)=O(1)\quad\text{and}\quad\frac{1}{\delta_n}\sum_{k=m_l}^{m_{l+1}-1}\varepsilon_k=O(1).$$

Likewise, we also have

$$
\begin{aligned}
\mathbf{E}\Bigg| \sum_{l\in\Xi} & \sum_{k_1=m_l}^{m_{l+1}-1} \sum_{k=m_l}^{m_{l+1}-1} \varepsilon_{k_1}\sqrt{\varepsilon_k}\, b^{N,\prime}(X_{k_1}^N,\alpha_{k_1})\nabla^2 h(X_{m_l}^N,\alpha_{m_l}) \\
& \times \sigma^N(X_k^N,\alpha_k)\xi_k \Bigg| \\
\leq K \sum_{l\in\Xi} & \sum_{k_1=m_l}^{m_{l+1}-1} \varepsilon_{k_1}\mathbf{E}^{1/2}\Bigg|\sum_{k=m_l}^{m_{l+1}-1}\sqrt{\varepsilon_k}\,\sigma^N(X_k^N,\alpha_k)\xi_k\Bigg|^2 \\
\leq K \sum_{l\in\Xi} & \delta_n^{3/2}\frac{1}{\delta_n}\sum_{k_1=m_l}^{m_{l+1}-1}\varepsilon_{k_1} \\
\leq K \sum_{l\in\Xi} & \delta_n^{3/2}\to 0 \quad\text{as}\quad n\to\infty,
\end{aligned}
\tag{6.35}
$$

and

$$
\begin{aligned}
\mathbf{E}\prod_{j=1}^{n_0}\varphi_j(X^{n,N}(s_j),\alpha^n(s_j)) &\sum_{k_1=m_l}^{m_{l+1}-1}\sum_{k=m_l}^{m_{l+1}-1}\varepsilon_k\xi_{k_1}'\sigma^{N,\prime}(X_{k_1}^N,\alpha_{k_1}) \\
&\times\nabla^2 h(X_{m_l}^N,\alpha_{m_l})b^N(X_k^N,\alpha_k)\to 0 \quad\text{as}\quad n\to\infty.
\end{aligned}
\tag{6.36}
$$

The independence of $\{\xi_n\}$ and $\{\alpha_n\}$ and $\mathbf{E}_k^\alpha\xi_k\xi_k'=I$ yield that as $n\to\infty$,

$$
\begin{aligned}
\mathbf{E}\prod_{j=1}^{n_0}\varphi_j(X^{n,N}(s_j),\alpha^n(s_j)) &\sum_{k_1=m_l}^{m_{l+1}-1}\sum_{k=m_l}^{m_{l+1}-1}\sqrt{\varepsilon_k\varepsilon_{k_1}}\,\mathrm{tr}[\nabla^2 h(X_{m_l}^N,\alpha_{m_l}) \\
&\times\sigma^N(X_k^N,\alpha_k)\xi_k\xi_{k_1}'\sigma^{N,\prime}(X_{k_1}^N,\alpha_{k_1})] \\
=\mathbf{E}\prod_{j=1}^{n_0}\varphi_j(X^{n,N}(s_j),\alpha^n(s_j)) &\sum_{k=m_l}^{m_{l+1}-1}\varepsilon_k\mathrm{tr}[\nabla^2 h(X_{m_l}^N,\alpha_{m_l})\sigma^N(X_k^N,\alpha_k) \\
&\times\varepsilon_k[\mathbf{E}_k^\alpha\xi_k\xi_k']\sigma^{N,\prime}(X_k^N,\alpha_k)] \\
\to\mathbf{E}\prod_{j=1}^{n_0}\varphi_j(X^N(s_j),\alpha(s_j))\Big[&\int_t^{t+s}\mathrm{tr}[\nabla^2 h(X^N(u),\alpha(u))\sigma^N(X^N(u),\alpha(u)) \\
&\times\sigma^{N,\prime}(X^N(u),\alpha(u))]du\Big].
\end{aligned}
\tag{6.37}
$$

Combining (6.34)–(6.37), we arrive at that as $n \to \infty$,

$$\mathbf{E} \prod_{j=1}^{n_0} \varphi_j(X^{n,N}(s_j), \alpha^n(s_j)) \sum_{l \in \Xi} [X_{m_{l+1}}^N - X_{m_l}^N]' \nabla^2 h(X_{m_l}^N, \alpha_{m_l})$$

$$\times [X_{m_{l+1}}^N - X_{m_l}^N]$$

$$\to \mathbf{E} \prod_{j=1}^{n_0} \varphi_j(X^N(s_j), \alpha(s_j))$$

$$\times \left[\int_t^{t+s} \mathrm{tr}[\nabla^2 h(X^N(u), \alpha(u)) \sigma^N(X^N(u), \alpha(u)) \sigma^{N,\prime}(x^N(u), \alpha(u))] du \right].$$

$$(6.38)$$

By the smoothness of $h(\cdot, \alpha)$ for each $\alpha \in \mathcal{M}$, we can replace $X_{m_{l+1}}^N$ in the second line of (6.29) by $X_{m_l}^N$. In fact,

$$\sum_{l \in \Xi} [h(X_{m_{l+1}}^N, \alpha_{m_{l+1}}) - h(X_{m_{l+1}}^N, \alpha_{m_l})]$$

$$= \sum_{l \in \Xi} [h(X_{m_l}^N, \alpha_{m_{l+1}}) - h(X_{m_l}^N, \alpha_{m_l})] + o(1),$$

where $o(1) \to 0$ in probability as $n \to \infty$ uniformly in t. It follows that

$$\mathbf{E} \prod_{j=1}^{n_0} \varphi_j(X^{n,N}(s_j), \alpha^n(s_j)) \sum_{l \in \Xi} [h(X_{m_l}^N, \alpha_{m_{l+1}}) - h(X_{m_l}^N, \alpha_{m_l})]$$

$$= \mathbf{E} \prod_{j=1}^{n_0} \varphi_j(X^N(s_j), \alpha^n(s_j)) \sum_{l \in \Xi} \sum_{k=m_l}^{m_{l+1}-1} \mathbf{E}_k^\alpha [h(X_{m_l}^N, \alpha_{k+1}) - h(X_{m_l}^N, \alpha_k)],$$

$$(6.39)$$

with

$$\sum_{l \in \Xi} \sum_{k=m_l}^{m_{l+1}-1} \mathbf{E}_k^\alpha [h(X_{m_l}^N, \alpha_{k+1}) - h(X_{m_l}^N, \alpha_k)]$$

$$= \sum_{l \in \Xi} \sum_{k=m_l}^{m_{l+1}-1} \sum_{j_0=1}^{m_0} \sum_{i_0=1}^{m_0} [h(X_{m_l}^N, j_0) \mathbf{P}(\alpha_{k+1} = j_0 | \alpha_k = i_0)$$

$$- h(X_{m_l}^N, i_0) I_{\{\alpha_k = i_0\}}]$$

$$= \sum_{l \in \Xi} \sum_{k=m_l}^{m_{l+1}-1} \chi_k (P^{k,k+1} - I) H(X_{m_l}^N)$$

$$= \sum_{l \in \Xi} \sum_{k=m_l}^{m_{l+1}-1} \varepsilon_k \chi_k \frac{[\exp(\varepsilon_k Q) - I]}{\varepsilon_k} H(X_{m_l}^N)$$

$$= \sum_{l \in \Xi} \delta_n \frac{1}{\delta_n} \sum_{k=m_l}^{m_{l+1}-1} \varepsilon_k \chi_k \frac{[\exp(\varepsilon_k Q) - I]}{\varepsilon_k} H(X_{m_l}^N)$$

$$\to \int_t^{t+s} Q h(X^N(u), \cdot)(\alpha(u)) du \quad \text{as } n \to \infty,$$

where

$$H(x) = (h(x,1), \ldots, h(x,m_0))' \in \mathbb{R}^{m_0 \times 1}. \qquad (6.40)$$

Thus Lemma 6.14 is proved.   □

Completion of the proof of the theorem. We show the convergence of the untruncated process. We have demonstrated that the truncated process $\{(X^{n,N}(\cdot), \alpha^n(\cdot))\}$ converges to $(X^N(\cdot), \alpha(\cdot))$. Here we show that the untruncated sequence $\{(X^n(\cdot), \alpha^n(\cdot))\}$ also converges. The basic premise is the uniqueness of the martingale problem. By letting $N \to \infty$, we obtain the desired result. The argument is similar to that of [104, pp. 249–250]. We present only the basic idea here.

Let the measures induced by $(X(\cdot), \alpha(\cdot))$ and $(X^N(\cdot), \alpha(\cdot))$ be $P(\cdot)$ and $P^N(\cdot)$, respectively. The martingale problem with operator $\mathcal{L}$ has a unique solution (in the sense in distribution) for each initial condition, therefore $P(\cdot)$ is unique. For any $0 < T < \infty$ and $|t| \leq T$, $P(\cdot)$ and $P^N(\cdot)$ are the same on all Borel subsets of the set of paths in $D((-\infty, \infty); \mathbb{R}^r \times \mathcal{M})$ with values in $S_N \times \mathcal{M}$. By using

$$P(\sup_{|t| \leq T} |X(t)| \leq N) \to 1 \quad \text{as} \quad N \to \infty,$$

and the weak convergence of $X^{n,N}(\cdot)$ to $X^N(\cdot)$, we conclude $X^n(\cdot)$ converges weakly to $X(\cdot)$. This leads to the desired result. The proof of the theorem is completed.   □

6.5   Notes

Chapter 4 provides sufficient conditions for ergodicity of switching diffusions with state-dependent switching. Based on the work Yin, Mao, and Yin [171], this chapter addresses the ergodicity for the corresponding numerical algorithms. The main result here is the demonstration of convergence to the invariant measure of the Euler–Maruyama-type numerical algorithms when the invariant measure exists. To obtain the result, we have first proved weak convergence of the algorithms. Here our approach is inspired by Kushner's work [101], in which he considered convergence to invariant measures for systems driven by wideband noise. We have adopted the method in that reference to treat the numerical approximation problem. Moreover, convergence of numerical algorithms has been proved using ideas from stochastic approximation (see Kushner and Yin [104]). We have dealt with both algorithms with decreasing stepsizes and constant stepsize.

Here our approach is based on weak convergence methods and we work with the associated measure. A different approach concentrating on the associated differential equations is in Mao, Yuan, and Yin [121]. The rate of convergence of the algorithms may be studied, for example, by means of strong invariance principles. Further study may also be directed to the large deviation analysis related to convergence to invariant measures.

Part III

Stability

7

Stability

7.1 Introduction

Continuing our effort of studying positive recurrence and ergodicity of switching diffusion processes in Chapters 3 and 4, this chapter focuses on stability of the dynamic systems described by switching diffusions. For some of the recent progress in stability analysis, we refer the reader to [48, 116, 136, 182, 183] and references therein. For treating dynamic systems in science and engineering, linearization techniques are used most often. Nevertheless, the nonlinear systems and their linearizations may or may not share similar asymptotic behavior. A problem of great interest is: If a linear system is stable, what can we say about the associated nonlinear systems? This chapter provides a systematic approach for treating such problems for switching diffusions. We solve these problems using Liapunov function methods.

The rest of the chapter is arranged as follows. Section 7.2 begins with the formulation of the problem together with an auxiliary result, which is used in our stability analysis. Section 7.3 recalls various notions of stability, and presents p-stability and exponential p-stability results. Easily verifiable conditions for stability and instability of linearized systems are provided in Section 7.4. To demonstrate our results, we provide several examples in Section 7.5. Further remarks are made in Section 7.6 to conclude this chapter.

G.G. Yin and C. Zhu, *Hybrid Switching Diffusions: Properties and Applications*, Stochastic Modelling and Applied Probability 63, DOI 10.1007/978-1-4419-1105-6_7, © Springer Science + Business Media, LLC 2010

7.2 Formulation and Auxiliary Results

Recall that we use z' to denote the transpose of $z \in \mathbb{R}^{\ell_1 \times \ell_2}$ with $\ell_i \geq 1$, and $i = 1, 2$, whereas $\mathbb{R}^{\ell \times 1}$ is simply written as $\mathbb{R}^{\ell}$; $\mathbb{1} = (1, 1, \ldots, 1)' \in \mathbb{R}^{m_0}$ is a column vector with all entries being 1; the Euclidean norm for a row or column vector x is denoted by $|x|$. As usual, I denotes the identity matrix. For a matrix A, its trace norm is denoted by $|A| = \sqrt{\mathrm{tr}(A'A)}$. If B is a set, its indicator function is denoted by $I_B(\cdot)$.

Suppose that $(X(t), \alpha(t))$ is a two-component Markov process such that $X(\cdot)$ is a continuous component taking values in $\mathbb{R}^r$ and $\alpha(\cdot)$ is a jump process taking values in a finite set $\mathcal{M} = \{1, 2, \ldots, m_0\}$. The process $(X(t), \alpha(t))$ has a generator $\mathcal{L}$ given as follows. For any twice continuously differentiable function $g(\cdot, i)$, $i \in \mathcal{M}$,

$$
\begin{aligned}
\mathcal{L}g(x, i) &= \frac{1}{2} \sum_{j,k=1}^{r} a_{jk}(x, i) \frac{\partial^2 g(x, i)}{\partial x_j \partial x_k} + \sum_{j=1}^{r} b_j(x, i) \frac{\partial g(x, i)}{\partial x_j} \\
&\quad + Q(x) g(x, \cdot)(i) \\
&= \frac{1}{2} \mathrm{tr}(a(x, i) \nabla^2 g(x, i)) + b'(x, i) \nabla g(x, i) + Q(x) g(x, \cdot)(i),
\end{aligned}
\tag{7.1}
$$

where $x \in \mathbb{R}^r$, $Q(x) = (q_{ij}(x))$ is an $m_0 \times m_0$ matrix depending on x satisfying $q_{ij}(x) \geq 0$ for $i \neq j$ and $\sum_{j \in \mathcal{M}} q_{ij}(x) = 0$ for each $i \in \mathcal{M}$, and

$$
\begin{aligned}
Q(x) g(x, \cdot)(i) &= \sum_{j \in \mathcal{M}} q_{ij}(x) g(x, j) \\
&= \sum_{j \in \mathcal{M}} q_{ij}(x)(g(x, j) - g(x, i)), \quad i \in \mathcal{M},
\end{aligned}
$$

and $\nabla g(\cdot, i)$ and $\nabla^2 g(\cdot, i)$ denote the gradient and Hessian of $g(\cdot, i)$, respectively.

The process $(X(t), \alpha(t))$ can be described by

$$
\begin{aligned}
dX(t) &= b(X(t), \alpha(t))dt + \sigma(X(t), \alpha(t))dw(t), \\
X(0) &= x, \quad \alpha(0) = \alpha,
\end{aligned}
\tag{7.2}
$$

and for $i \neq j$,

$$
\begin{aligned}
\mathbf{P}\{\alpha(t + \Delta t) &= j | \alpha(t) = i, (X(s), \alpha(s)), s \leq t\} \\
&= q_{ij}(X(t))\Delta t + o(\Delta t),
\end{aligned}
\tag{7.3}
$$

where $w(t)$ is a d-dimensional standard Brownian motion, $b(\cdot, \cdot) : \mathbb{R}^r \times \mathcal{M} \mapsto \mathbb{R}^r$, and $\sigma(\cdot, \cdot) : \mathbb{R}^r \times \mathcal{M} \mapsto \mathbb{R}^{r \times d}$ satisfies $\sigma(x, i)\sigma'(x, i) = a(x, i)$.

To proceed, we need conditions on the smoothness and growth of the functions involved, and the condition that 0 is the only equilibrium point of the random dynamic system. Hence we assume that the following conditions are valid throughout this chapter.

(A7.1) The matrix-valued function $Q(\cdot)$ is bounded and continuous.

(A7.2) $b(0, \alpha) = 0$, and $\sigma(0, \alpha) = 0$ for each $\alpha \in \mathcal{M}$. Moreover, assume that $\sigma(x, \alpha)$ vanishes only at $x = 0$ for each $\alpha \in \mathcal{M}$.

(A7.3) There exists a constant $K_0 > 0$ such that for each $\alpha \in \mathcal{M}$, and for any $x, y \in \mathbb{R}^r$,

$$|b(x, \alpha) - b(y, \alpha)| + |\sigma(x, \alpha) - \sigma(y, \alpha)| \le K_0 |x - y|. \qquad (7.4)$$

Under these conditions, the system given by (7.2) and (7.3) has a unique solution; see Chapter 2 for more details. In what follows, a process starting from (x, α) is denoted by $(X^{x,\alpha}(t), \alpha^{x,\alpha}(t))$ if the emphasis on initial condition is needed. If the context is clear, we simply write $(X(t), \alpha(t))$.

To study stability of the equilibrium point $x = 0$, we first present the following "nonzero" property, which asserts that almost all the sample paths of any solution of the system given by (7.2) and (7.3) starting from a nonzero state will never reach the origin. For diffusion processes, such a result was established in [83, Section 5.2]; for Markovian regime-switching processes, similar results were obtained in [116, Lemma 2.1]. In what follows, we give a proof for switching diffusions with continuous-state-dependent switching processes. The result is useful, provides us with flexibility for choices of Liapunov functions, and enables us to build Liapunov functions in a deleted neighborhood of the origin.

Lemma 7.1. *Under conditions* (A7.1)–(A7.3), *we have*

$$\mathbf{P}\{X^{x,\alpha}(t) \ne 0, t \ge 0\} = 1, \quad \text{for any } x \ne 0, \ \alpha \in \mathcal{M}, \qquad (7.5)$$

and for any $\beta \in \mathbb{R}$ *and* $t > 0$

$$\mathbf{E}\left[|X^{x,\alpha}(t)|^\beta\right] \le |x|^\beta e^{Kt}, \quad x \ne 0, \ \alpha \in \mathcal{M}, \qquad (7.6)$$

where K *is a constant depending only on* β, m_0, *and the Lipschitz constant* K_0 *in* (7.4).

Proof. For $x \ne 0$, for each $i \in \mathcal{M}$, define $V(x, i) = |x|^\beta$ for any $\beta \in \mathbb{R} - \{0\}$. For any $\Delta > 0$ and $|x| > \Delta$,

$$\nabla V(x, i) = \beta |x|^{\beta-2} x,$$
$$\nabla^2 V(x, i) = \beta |x|^{\beta-4} \left(|x|^2 I + (\beta - 2) xx' \right).$$

Then it follows that

$$\mathcal{L}V(x, i) = \left\langle \beta |x|^{\beta-2} x, b(x, i) \right\rangle$$
$$+ \frac{1}{2} \text{tr} \left[\sigma(x, i)\sigma'(x, i)\beta |x|^{\beta-4} \left(|x|^2 I + (\beta - 2)xx' \right) \right].$$

Moreover, using conditions (A7.2) and (A7.3),

$$|\mathcal{L}V(x, i)| \le K\,|x|^\beta, \quad \text{for any } (x, i) \in \mathbb{R}^r \times \mathcal{M} \text{ with } x \neq 0. \tag{7.7}$$

Let τ_Δ be the first exit time from $\{x \in \mathbb{R}^r : |x| > \Delta\}$. Denote $(X(t), \alpha(t)) = (X^{x,\alpha}(t), \alpha^{x,\alpha}(t))$. Now applying generalized Itô's formula (see (2.7)) to V, we obtain

$$\begin{aligned}
|X(\tau_\Delta \wedge t)|^\beta &= |x|^\beta + \int_0^{\tau_\Delta \wedge t} \mathcal{L}\,|X(u)|^\beta\,du \\
&\quad + \int_0^{\tau_\Delta \wedge t} \beta\,|X(u)|^{\beta-2}\,X'(u)\sigma(X(u), \alpha(u))dw(u).
\end{aligned} \tag{7.8}$$

Note that by virtue of conditions (A7.2) and (A7.3)

$$\mathbf{E} \int_0^{\tau_\Delta \wedge t} \left| \beta\,|X(u)|^{\beta-2}\,X'(u)\sigma(X(u), \alpha(u)) \right|^2 du \le \mathbf{E} \int_0^{\tau_\Delta \wedge t} K\,|X(u)|^{2\beta}\,du.$$

Thus if $\beta > 0$, (2.12) implies that

$$\mathbf{E} \int_0^{\tau_\Delta \wedge t} K\,|X(u)|^{2\beta}\,du \le \mathbf{E} \int_0^t |X(u)|^{2\beta}\,du \le KCt < \infty,$$

where $C = C(x, t, \beta) > 0$. On the other hand, if $\beta < 0$, then by the definition of τ_Δ, we have

$$\mathbf{E} \int_0^{\tau_\Delta \wedge t} |X(u)|^{2\beta}\,du \le \mathbf{E} \int_0^{\tau_\Delta \wedge t} K\Delta^{2\beta}du \le K\Delta^{2\beta}t < \infty.$$

Therefore we have verified that the stochastic integral in (7.8) is a martingale with mean 0. Hence, by taking expectations on both sides of (7.8), and taking into account (7.7), it follows that

$$\begin{aligned}
\mathbf{E}|X(\tau_\Delta \wedge t)|^\beta &\le |x|^\beta + K\mathbf{E} \int_0^{\tau_\Delta \wedge t} |X(u)|^\beta du \\
&\le |x|^\beta + K\mathbf{E} \int_0^t |X(u \wedge \tau_\Delta)|^\beta du.
\end{aligned}$$

In the above, we have used the fact $\tau_\Delta \wedge u = u$ for $u \le \tau_\Delta \wedge t$. An application of Gronwall's inequality implies that for any $\beta \neq 0$,

$$\mathbf{E}|X(\tau_\Delta \wedge t)|^\beta \le |x|^\beta \exp(Kt). \tag{7.9}$$

Taking $\beta = -1$, we have

$$\mathbf{E}\left[|X(\tau_\Delta \wedge t)|^{-1} \right] \le |x|^{-1} \exp(Kt).$$

By Tchebeshev's inequality, for any $\Delta > 0$,

$$\mathbf{P}(\tau_\Delta < t) = \mathbf{P}(|X(\tau_\Delta \wedge t)| \leq \Delta) = \mathbf{P}\left(|X(\tau_\Delta \wedge t)|^{-1} \geq \Delta^{-1}\right)$$
$$\leq \Delta\left[\mathbf{E}\,|X(\tau_\Delta \wedge t)|^{-1}\right] \leq \Delta\left[|x|^{-1}\exp(Kt)\right].$$

Letting $\Delta \to 0$, we have $\mathbf{P}(\tau_\Delta < t) \to 0$ as $\Delta \to 0$. Therefore (7.6) follows by letting $\Delta \to 0$ in (7.9) and Fatou's lemma.

Finally, assume (7.5) were false, then there would exist some $(x_0, \alpha_0) \in \mathbb{R}^r \times \mathcal{M}$ and $T > 0$ such that

$$\mathbf{P}\left\{X^{x_0,\alpha_0}(T) = 0\right\} > 0.$$

Then we would have

$$\mathbf{E}\left[|X^{x_0,\alpha_0}(T)|^{-1}\right] = \infty,$$

which would contradict with (7.6) that we just proved. This completes the proof of the lemma. $\square$

Remark 7.2. In view of (7.5), we can work with functions $V(\cdot, i)$, $i \in \mathcal{M}$ that are twice continuously differentiable and are defined on a deleted neighborhood of 0 in what follows.

To proceed, we present an auxiliary result, namely, the solvability of a system of deterministic equations. Suppose that Q, an $m_0 \times m_0$ constant matrix, is the generator of a continuous-time Markov chain $r(t)$ and that Q is irreducible.

Remark 7.3. In the above, by the irreducibility, we mean that the system of equations

$$\begin{cases} \nu Q = 0 \\ \nu \mathbb{1} = 1, \end{cases}$$

has a unique solution such that $\nu = (\nu_1, \ldots, \nu_{m_0})$ satisfies $\nu_i > 0$; see Definition A.7 and discussions there for further details.

Note that if Q is irreducible, the rank of Q is $m_0 - 1$. Denote by $\mathcal{R}(Q)$ and $\mathcal{N}(Q)$ the range and the null space of Q, respectively. It follows that $\mathcal{N}(Q)$ is one-dimensional spanned by $\mathbb{1}$ (i.e., $\mathcal{N}(Q) = \text{span}\{\mathbb{1}\}$). As a consequence, the Markov chain $r(t)$ is ergodic; see, for example, [29]. In what follows, denote the associated stationary distribution by

$$\nu = (\nu_1, \nu_2, \ldots, \nu_{m_0}) \in \mathbb{R}^{1 \times m_0}. \tag{7.10}$$

We are interested in solving a linear system of equations

$$Qc = \eta, \tag{7.11}$$

where $Q \in \mathbb{R}^{m_0 \times m_0}$ and $\eta \in \mathbb{R}^{m_0}$ are given and $c \in \mathbb{R}^{m_0}$ is an unknown vector. Note that (7.11) is a Poisson equation. The properties of solutions of (7.11) are provided in Lemma A.12. Basically, it indicates that under the irreducibility of Q, equation (7.11) has a solution if and only if $\nu\eta = 0$. Moreover, suppose that c_1 and c_2 are two solutions of (7.11). Then $c_1 - c_2 = \alpha_0 \mathbb{1}$ for some $\alpha_0 \in \mathbb{R}$.

7.3 p-Stability

This section is concerned with stability of the equilibrium point $x = 0$ for the system given by (7.2) and (7.3). Adopting the terminologies of [83], we first present definitions of stability, p-stability, and exponential p-stability. Then general results in terms of Liapunov functions are provided.

7.3.1 Stability

Definition 7.4. The equilibrium point $x = 0$ of the system given by (7.2) and (7.3) is said to be

(i) *stable in probability*, if for any $\varepsilon > 0$ and any $\alpha \in \mathcal{M}$,

$$\lim_{x \to 0} \mathbf{P}\{\sup_{t \geq 0} |X^{x,\alpha}(t)| > \varepsilon\} = 0,$$

and $x = 0$ is said to be *unstable in probability* if it is not stable in probability;

(ii) *asymptotically stable in probability*, if it is stable in probability and satisfies

$$\lim_{x \to 0} \mathbf{P}\{\lim_{t \to \infty} X^{x,\alpha}(t) = 0\} = 1, \text{ for any } \alpha \in \mathcal{M};$$

(iii) *p-stable (for $p > 0$)*, if

$$\lim_{\delta \to 0} \sup_{|x| \leq \delta, \alpha \in \mathcal{M}, t \geq 0} \mathbf{E}|X^{x,\alpha}(t)|^p = 0;$$

(iv) *asymptotically p-stable*, if it is p-stable and satisfies $\mathbf{E}|X^{x,\alpha}(t)|^p \to 0$ as $t \to \infty$ for any $(x, \alpha) \in \mathbb{R}^r \times \mathcal{M}$;

(v) *exponentially p-stable*, if for some positive constants K and k,

$$\mathbf{E}|X^{x,\alpha}(t)|^p \leq K|x|^p \exp(-kt), \quad \text{for any } (x, \alpha) \in \mathbb{R}^r \times \mathcal{M}.$$

Using similar arguments as those of [83, Theorems 5.3.1, 5.4.1, and 5.4.2], we establish the following three lemmas. The statements are given together their proofs.

Lemma 7.5. *Let $D \subset \mathbb{R}^r$ be a neighborhood of 0. Suppose that for each $i \in \mathcal{M}$, there exists a nonnegative function $V(\cdot, i) : D \mapsto \mathbb{R}$ such that*

(i) *$V(\cdot, i)$ is continuous in D and vanishes only at $x = 0$;*

(ii) *$V(\cdot, i)$ is twice continuously differentiable in $D - \{0\}$ and satisfies*

$$\mathcal{L}V(x, i) \leq 0, \quad \text{for all } x \in D - \{0\}.$$

Then the equilibrium point $x = 0$ is stable in probability.

Proof. Let $\varsigma > 0$ be such that the ball $B_\varsigma = \{x \in \mathbb{R}^r : |x| < \varsigma\}$ and its boundary $\partial B_\varsigma = \{x \in \mathbb{R}^r : |x| = \varsigma\}$ are contained in D. Set

$$V_\varsigma := \inf \{V(y, j) : y \in D \setminus B_\varsigma, j \in \mathcal{M}\}.$$

Then $V_\varsigma > 0$ by assumption (i). Next, by virtue of assumption (ii) and Dynkin's formula, we have

$$\mathbf{E}_{x,i} V(X(t \wedge \tau_\varsigma), \alpha(t \wedge \tau_\varsigma)) = V(x, i) + \mathbf{E}_{x,i} \int_0^{\tau_\varsigma \wedge t} \mathcal{L}V(X(s), \alpha(s))ds$$
$$\leq V(x, i),$$

where $(x, i) \in B_\varsigma \times \mathcal{M}$ and τ_ς is the first exit time from B_ς, that is, $\tau_\varsigma := \{t \geq 0 : |X(t)| \geq \varsigma\}$. Because V is nonnegative, we further have

$$V_\varsigma \mathbf{P}\{\tau_\varsigma \leq t\} \leq \mathbf{E}_{x,i} \left[V(X(\tau_\varsigma), \alpha(\tau_\varsigma)) I_{\{\tau_\varsigma \leq t\}} \right] \leq V(x, i).$$

Note that $\tau_\varsigma \leq t$ if and only if $\sup_{0 \leq u \leq t} |x(u)| > \varsigma$. Therefore it follows that

$$\mathbf{P}_{x,i} \left\{ \sup_{0 \leq u \leq t} |X(u)| > \varsigma \right\} \leq \frac{V(x, i)}{V_\varsigma}.$$

Letting $t \to \infty$, we obtain

$$\mathbf{P}_{x,i} \left\{ \sup_{0 \leq t} |X(t)| > \varsigma \right\} \leq \frac{V(x, i)}{V_\varsigma}.$$

Finally, the desired conclusion follows from the assumptions that $V(0, i) = 0$ and $V(\cdot, i)$ is continuous for each $i \in \mathcal{M}$.   □

Introduce the notation

$$\tau_{\varepsilon, r_0}^{x, \alpha} := \inf\{t \geq 0 : |X^{x, \alpha}(t)| = \varepsilon \text{ or } |X^{x, \alpha}(t)| = r_0\}, \tag{7.12}$$

for any $0 < \varepsilon < r_0$ and any $(x, \alpha) \in \mathbb{R}^r \times \mathcal{M}$ with $\varepsilon < |x| < r_0$.

Lemma 7.6. *Assume the conditions of Lemma 7.5. If for any sufficiently small $0 < \varepsilon < r_0$ and any $(x, \alpha) \in \mathbb{R}^r \times \mathcal{M}$ with $\varepsilon < |x| < r_0$, we have*

$$\mathbf{P}\{\tau_{\varepsilon, r_0}^{x, \alpha} < \infty\} = 1, \tag{7.13}$$

then the equilibrium point $x = 0$ is asymptotically stable in probability.

Proof. We divide the proof into several steps.

Step 1. Let $\varsigma > 0$ and $(x, i) \in B_\varsigma \times \mathcal{M}$, and define τ_ς as in the proof of Lemma 7.5. For any $t \geq 0$, let $g(t) := V(X^{x,i}(\tau_\varsigma \wedge t), \alpha^{x,i}(\tau_\varsigma \wedge t))$. Then as in the proof of Lemma 7.5, the assumption that $\mathcal{L}V(y, j) \leq 0$ for all $(y, j) \in (D - \{0\}) \times \mathcal{M}$ implies that g is a nonnegative supermartingale. Therefore the martingale convergence theorem implies that

$$\lim_{t \to \infty} g(t) = g(\infty) \quad \text{exists a.s.} \tag{7.14}$$

Step 2. By virtue of Lemma 7.5, the equilibrium point $x = 0$ is stable in probability. Therefore for any $\varepsilon > 0$, there exists a $\delta > 0$ (we may further assume that $\delta < \varsigma$) such that

$$\mathbf{P}_{y,j} \{\tau_\varsigma < \infty\} < \varepsilon/2, \quad \text{for all} \quad (y, j) \in B_\delta \times \mathcal{M}. \tag{7.15}$$

Now choose an arbitrary point $(x, \alpha) \in B_\delta \times \mathcal{M}$. Then both (7.14) and (7.15) hold. Hence it follows from (7.13) and (7.15) that for any $\rho > 0$ with $\rho < |x|$, we have

$$\mathbf{P}_{x,\alpha} \{\tau_\rho < \infty\} \geq \mathbf{P}_{x,\alpha} \{\tau_{\rho,\varsigma} < \infty\} - \mathbf{P}_{x,\alpha} \{\tau_\varsigma < \infty\} \geq 1 - \varepsilon/2.$$

This implies that $\mathbf{P}_{x,\alpha} \{\inf_{t \geq 0} |X(t)| \leq \rho\} \geq 1 - \varepsilon/2$. Since $\rho > 0$ can be arbitrarily small,

$$\mathbf{P}_{x,\alpha} \left\{ \inf_{t \geq 0} |X(t)| = 0 \right\} \geq 1 - \varepsilon/2.$$

Now let

$$A := \left\{ \omega \in \Omega : \tau_\varsigma(\omega) = \infty, \inf_{t \geq 0} |X(t, \omega)| = 0 \right\}.$$

Then $\mathbf{P}_{x,\alpha}(A) \geq 1 - \varepsilon/2$.

Step 3. We claim that for almost all $\omega \in A$, we have

$$\liminf_{t \to \infty} |X^{x,\alpha}(t, \omega)| = 0.$$

If the claim were false, there would exist a $B \subset A$ with $\mathbf{P}_{x,\alpha}(B) > 0$ such that for all $\omega \in B$, we would have

$$\liminf_{t \to \infty} |X^{x,\alpha}(t, \omega)| \geq \theta > 0.$$

Then for any $\omega \in B$, there exists a $T = T(\omega) > 0$ such that $|X^{x,\alpha}(t, \omega)| \geq \theta$ for all $t \geq T$. Therefore for any $\omega \in B$ and $n \in \mathbb{N}$ sufficiently large, $\tau_{1/n}(\omega) \leq T$, where $\tau_{1/n}(\omega) := \inf \{t \geq 0 : |X^{x,\alpha}(t, \omega)| \leq 1/n\}$. Hence it follows that

$$\lim_{n \to \infty} \tau_{1/n}(\omega) \leq T(\omega) < \infty.$$

Then we would have

$$\mathbf{P}_{x,\alpha}\left\{\lim_{n\to\infty}\tau_{1/n}<\infty\right\}\geq\mathbf{P}_{x,\alpha}(B)>0.$$

But this would lead to a contradiction because by virtue of Lemma 7.1, the equilibrium point 0 is inaccessible with probability 1. Thus it follows that

$$\mathbf{P}_{x,\alpha}\left\{\lim_{n\to\infty}\tau_{1/n}=\infty\right\}=1,\quad\text{or}\quad\mathbf{P}_{x,\alpha}\left\{\lim_{n\to\infty}\tau_{1/n}<\infty\right\}=0.$$

Hence the claim is verified.

Step 4. Since $V(\cdot,i)$ is continuous and $V(0,i)=0$ for each $i\in\mathcal{M}$, we have from Step 3 that

$$\liminf_{t\to\infty}V(X^{x,\alpha}(t),\alpha^{x,\alpha}(t))=0,\quad\text{for almost all }\omega\in A. \tag{7.16}$$

Now by virtue of (7.14) and the definition of A, we have

$$\begin{aligned}\lim_{t\to\infty}g(t)&=\lim_{t\to\infty}V(X^{x,\alpha}(\tau_\varsigma\wedge t),\alpha^{x,\alpha}(\tau_\varsigma\wedge t))\\&=\lim_{t\to\infty}V(X^{x,\alpha}(t),\alpha^{x,\alpha}(t))=g(\infty)\quad\text{a.s.}\end{aligned} \tag{7.17}$$

Thus it follows from (7.16) and (7.17) that

$$\lim_{t\to\infty}V(X^{x,\alpha}(t),\alpha^{x,\alpha}(t))=0,\quad\text{for almost all }\omega\in A.$$

But $V(\cdot,i)$ vanishes only at $x=0$ for each $i\in\mathcal{M}$, so $\lim_{t\to\infty}X^{x,\alpha}(t)=0$ on A. Thus we have

$$\mathbf{P}_{x,\alpha}\left\{\lim_{t\to\infty}X(t)=0\right\}\geq\mathbf{P}_{x,\alpha}(A)\geq1-\varepsilon/2.$$

Note that (x,α) is an arbitrary point in $B_\delta\times\mathcal{M}$. Thus, for any $\varepsilon>0$, there exists a $\delta>0$ such that

$$\mathbf{P}_{x,\alpha}\left\{\lim_{t\to\infty}X(t)=0\right\}\geq1-\varepsilon/2,\quad\text{for all }(x,\alpha)\in B_\delta\times\mathcal{M}.$$

That is, the equilibrium point $x=0$ is asymptotically stable in probability as desired. □

Lemma 7.7. *Let $D\subset\mathbb{R}^r$ be a neighborhood of 0. Assume that the conditions of Lemma 7.6 hold and that for each $i\in\mathcal{M}$, there exists a nonnegative function $V(\cdot,i):D\mapsto\mathbb{R}$ such that $V(\cdot,i)$ is twice continuously differentiable in every deleted neighborhood of 0, and*

$$\mathcal{L}V(x,i)\leq0\quad\text{for all }x\in D-\{0\}; \tag{7.18}$$

$$\lim_{|x|\to0}V(x,i)=\infty,\quad\text{for each }i\in\mathcal{M}. \tag{7.19}$$

Then the equilibrium point $x=0$ is unstable in probability if (7.13) holds.

Proof. Let $\varsigma > 0$ and $(x, \alpha) \in B_\varsigma \times \mathcal{M}$, and define τ_ς as in the proof of Lemma 7.5. Let also $0 < \varepsilon < |x|$ and define $\tau_\varepsilon := \inf \{t \geq 0 : |X^{x,\alpha}(t)| \leq \varepsilon\}$. Then for any $t > 0$, we have

$$\mathbf{E}_{x,\alpha} V(X(t \wedge \tau_{\varepsilon,\varsigma}), \alpha(t \wedge \tau_{\varepsilon,\varsigma}))$$
$$= V(x, \alpha) + \mathbf{E}_{x,\alpha} \int_0^{t \wedge \tau_{\varepsilon,\varsigma}} \mathcal{L}V(X(s), \alpha(s)) ds \qquad (7.20)$$
$$\leq V(x, \alpha).$$

Letting $t \to \infty$ in (7.20), we obtain by virtue of Fatou's lemma and (7.13) that

$$\mathbf{E}_{x,\alpha} V(X(\tau_\varepsilon \wedge \tau_\varsigma), \alpha(\tau_\varepsilon \wedge \tau_\varsigma)) \leq V(x, \alpha).$$

Furthermore, since V is nonnegative, we have

$$V(x, \alpha) \geq \mathbf{E}_{x,\alpha} \left[V(X(\tau_\varepsilon), \alpha(\tau_\varepsilon)) I_{\{\tau_\varepsilon < \tau_\varsigma\}} \right]$$
$$\geq \inf \{V(y, j) : |y| = \varepsilon, j \in \mathcal{M}\} \mathbf{P}_{x,\alpha} \{\tau_\varepsilon < \tau_r\}$$
$$= V_\varepsilon \mathbf{P}_{x,\alpha} \left\{ \sup_{0 \leq t \leq \tau_\varepsilon} |X(t)| < \varsigma \right\},$$

where $V_\varepsilon = \inf \{V(y, j) : |y| = \varepsilon, j \in \mathcal{M}\}$. By Lemma 7.1, the equilibrium point 0 is inaccessible with probability 1 and hence $\tau_\varepsilon \to \infty$ a.s. as $\varepsilon \to 0$. Also, it follows from (7.19) that $V_\varepsilon \to \infty$ as $\varepsilon \to 0$. Therefore it follows that as $\varepsilon \to 0$, we have

$$\mathbf{P}_{x,\alpha} \left\{ \sup_{t \geq 0} |X(t)| < \varsigma \right\} = 0.$$

This shows that the equilibrium point $x = 0$ is unstable in probability. $\square$

Remark 7.8. Note that (7.13) is an essential assumption in Lemma 7.6 and Lemma 7.7. Here we present two sufficient conditions for which (7.13) holds.

(i) Let $N \subset \mathbb{R}^r$ be a neighborhood of 0. Assume that for each $i \in \mathcal{M}$, there exists a nonnegative function $V(\cdot, i) : N \mapsto \mathbb{R}$ such that $V(\cdot, i)$ is twice continuously differentiable in every deleted neighborhood of 0, and that for any sufficiently small $0 < \varepsilon < r_0$ there is a positive constant $\kappa = \kappa(\varepsilon)$ such that

$$\mathcal{L}V(x, i) \leq -\kappa, \quad \text{for all } x \in N \text{ with } \varepsilon < |x| < r_0. \qquad (7.21)$$

Then (7.13) holds.

(ii) If for any sufficiently small $0 < \varepsilon < r_0$, there exist some $\iota = 1, 2, \ldots, r$ and some constant $\kappa = \kappa(\varepsilon) > 0$ such that

$$a_{\iota\iota}(x, i) \geq \kappa, \quad \text{for all } (x, i) \in (\{x : \varepsilon < |x| < r_0\}) \times \mathcal{M}, \qquad (7.22)$$

then (7.13) holds.

Assertion (i) can be established using almost the same proof as that for [83, Theorem 3.7.1]. Also (ii) follows by observing that if (7.22) is satisfied, then we can construct some Liapunov function $V(\cdot, \cdot)$ satisfying (7.21); see also a similar argument in the proof of [83, Corollary 3.7.2]. We omit the details here for brevity.

7.3.2 Auxiliary Results

Concerning the exponential p-stability of the equilibrium point $x = 0$ of the system given by (7.2) and (7.3), sufficient conditions in terms of the existence of certain Liapunov functions being homogeneous of degree p were obtained in [116]. In what follows, we first derive a Kolmogorov-type backward equation and provide a couple of lemmas as a preparation, and then we present a necessary condition for the exponential p-stability.

Note that in Theorem 7.10, we do not assume that the operator $\mathcal{L}$ is uniformly parabolic. In other words, the operator $\mathcal{L}$ is degenerate. Nevertheless, we prove that the function $u(t, x, i)$ defined in (7.24) is a *classical* solution to the initial value problem (7.30)–(7.31).

Theorem 7.9. *Assume that for each $i \in \mathcal{M}$, the coefficients of the operator $\mathcal{L}$ defined in (7.1) satisfy $b(\cdot, i) \in C^2$ and $\sigma(\cdot, i) \in C^2$ and that $|q_{ij}(x)| \leq K$ for all $x \in \mathbb{R}^r$ and some $K > 0$. Suppose that $\phi(\cdot, i) \in C^2$ and that $D_x^\theta \phi(\cdot, i)$ is Lipschitz continuous for each $i \in \mathcal{M}$ and $|\theta| = 2$, and that*

$$\left| D_x^\beta b(x, i) \right| + \left| D_x^\beta \sigma(x, i) \right| + \left| D_x^\theta \phi(x, i) \right| \leq K(1 + |x|^\gamma), \quad i \in \mathcal{M}, \quad (7.23)$$

where K and γ are positive constants and β and θ are multi-indices with $|\beta| \leq 2$ and $|\theta| \leq 2$. Then for any $T > 0$, the function

$$u(t, x, i) := \mathbf{E}_{x,i}[\phi(X(t), \alpha(t))] = \mathbf{E}[\phi(X^{x,i}(t), \alpha^{x,i}(t))] \qquad (7.24)$$

is twice continuously differentiable with respect to the variable x and satisfies

$$\left| D_x^\beta u(t, x, i) \right| \leq K(1 + |x|^\gamma),$$

where $t \in [0, T]$, $x \in \mathbb{R}^r$, and $i \in \mathcal{M}$.

Proof. For notational simplicity, we prove the theorem when $X(t)$ is one-dimensional, the multidimensional case can be handled in a similar manner. Fix $(t, x, i) \in [0, T] \times \mathbb{R}^r \times \mathcal{M}$. Let $\widetilde{x} = x + \Delta$ with $0 < |\Delta| < 1$. As in the proof of Theorem 2.27, we denote $(X(t), \alpha(t)) = (X^{x,i}(t), \alpha^{x,i}(t))$ and $(\widetilde{X}(t), \widetilde{\alpha}(t)) = (X^{\widetilde{x},i}(t), \alpha^{\widetilde{x},i}(t))$. By virtue of Theorem 2.27, the mean square derivative $\zeta(t) = (\partial/\partial x)X^{x,i}(t)$ exists and is mean square continuous with respect to x and t.

Write

$$\frac{u(t,\widetilde{x},i) - u(t,x,i)}{\Delta} = \frac{1}{\Delta}\mathbf{E}[\phi(\widetilde{X}(t),\widetilde{\alpha}(t)) - \phi(X(t),\alpha(t))]$$

$$= \frac{1}{\Delta}\mathbf{E}[\phi(\widetilde{X}(t),\widetilde{\alpha}(t)) - \phi(\widetilde{X}(t),\alpha(t))] \qquad (7.25)$$

$$+ \frac{1}{\Delta}\mathbf{E}[\phi(\widetilde{X}(t),\alpha(t)) - \phi(X(t),\alpha(t))].$$

Similar to the proof of Lemma 2.28, we can show that

$$\frac{1}{\Delta^2}\mathbf{E}\left[\sup_{0\leq t\leq T}\left|\phi(\widetilde{X}(t),\widetilde{\alpha}(t)) - \phi(\widetilde{X}(t),\alpha(t))\right|^2\right] \to 0, \qquad (7.26)$$

as $\Delta \to 0$. To proceed, for each $i \in \mathcal{M}$, we use $\phi_x(\cdot,i)$ and $\phi_{xx}(\cdot,i)$ to denote the first and second derivatives of $\phi(\cdot,i)$ with respect to x, respectively. We obtain

$$\frac{1}{\Delta}\mathbf{E}[\phi(\widetilde{X}(t),\alpha(t)) - \phi(X(t),\alpha(t))]$$

$$= \frac{1}{\Delta}\mathbf{E}\int_0^1 \frac{d}{dv}\phi(X(t) + v(\widetilde{X}(t) - X(t)),\alpha(t))dv$$

$$= \mathbf{E}\left[Z(t)\int_0^1 \phi_x(X(t) + v(\widetilde{X}(t) - X(t)),\alpha(t))dv\right],$$

where

$$Z(t) = \frac{\widetilde{X}(t) - X(t)}{\Delta}. \qquad (7.27)$$

Thus it follows that

$$\left|\frac{1}{\Delta}\mathbf{E}[\phi(\widetilde{X}(t),\alpha(t)) - \phi(X(t),\alpha(t))] - \mathbf{E}\left[\phi_x(X(t),\alpha(t))\zeta(t)\right]\right|$$

$$\leq \mathbf{E}\left|\int_0^1 \phi_x(X(t) + v(\widetilde{X}(t) - X(t)),\alpha(t))dv\,Z(t) - \phi_x(X(t),\alpha(t))\zeta(t)\right|$$

$$\leq \mathbf{E}\left|\left[\int_0^1 \phi_x(X(t) + v(\widetilde{X}(t) - X(t)),\alpha(t))dv - \phi_x(X(t),\alpha(t))\right]Z(t)\right|$$

$$+ \mathbf{E}\left|\phi_x(X(t),\alpha(t))\left[Z(t) - \zeta(t)\right]\right|$$

$$:= e_1 + e_2,$$

It follows from (7.23), Proposition 2.3, and (2.73) that

$$e_2 = \mathbf{E}\left|\phi_x(X(t),\alpha(t))\left[Z(t) - \zeta(t)\right]\right|$$

$$\leq \mathbf{E}^{1/2}\left|\phi_x(X(t),\alpha(t))\right|^2 \mathbf{E}^{1/2}\left|Z(t) - \zeta(t)\right|^2$$

$$\leq K\mathbf{E}^{1/2}\left|Z(t) - \zeta(t)\right|^2 \to 0,$$

as $\Delta \to 0$. To estimate the term e_1, we note that (7.23) and Proposition 2.3 imply that

$$\mathbf{E}\left|\phi_x(X(t) + v(\widetilde{X}(t) - X(t)), \alpha(t)) - \phi_x(X(t), \alpha(t))\right|^2 \leq K$$

for all $0 < |\Delta| < 1$. Recall also from the proof of Theorem 2.27 that $\widetilde{X}(t) \to X(t)$ in probability for any $t \in [0, T]$. Thus it follows that

$$\mathbf{E}\left|\phi_x(X(t) + v(\widetilde{X}(t) - X(t)), \alpha(t)) - \phi_x(X(t), \alpha(t))\right|^2 \to 0,$$

as $\Delta \to 0$. Note that we proved in Corollary 2.32 that $\mathbf{E}|Z(t)|^2 \leq K$, where $Z(t)$ is the "difference quotient" defined in (7.27). Then we have from the Cauchy–Schwartz inequality that

$$
\begin{aligned}
e_1 &= \mathbf{E}\left|\left[\int_0^1 \phi_x(X(t) + v(\widetilde{X}(t) - X(t)), \alpha(t)) dv - \phi_x(X(t), \alpha(t))\right] Z(t)\right| \\
&\leq \mathbf{E}^{1/2}\left|\left[\int_0^1 \phi_x(X(t) + v(\widetilde{X}(t) - X(t)), \alpha(t)) dv - \phi_x(X(t), \alpha(t))\right]\right|^2 \\
&\quad \times \mathbf{E}^{1/2}|Z(t)|^2 \\
&\to 0 \quad \text{as} \quad \Delta \to 0.
\end{aligned}
$$

Hence we have shown that as $\Delta \to 0$,

$$\left|\frac{1}{\Delta}\mathbf{E}[\phi(\widetilde{X}(t), \alpha(t)) - \phi(X(t), \alpha(t))] - \mathbf{E}\left[\phi_x(X(t), \alpha(t))\zeta(t)\right]\right| \to 0. \quad (7.28)$$

Therefore it follows from (7.25), (7.26), and (7.28) that

$$\left|\frac{u(t, \widetilde{x}, i) - u(t, x, i)}{\Delta} - \mathbf{E}\left[\phi_x(X(t), \alpha(t))\zeta(t)\right]\right| \to 0 \quad \text{as } \Delta \to 0.$$

Thus $u(t, \cdot, i)$ is differentiable with respect to the variable x and

$$\frac{\partial u(t, x, i)}{\partial x} = \mathbf{E}\left[\phi_x(X(t), \alpha(t))\zeta(t)\right] = \mathbf{E}\left[\phi_x(X^{x,i}(t), \alpha^{x,i}(t))\frac{\partial X^{x,i}(t)}{\partial x}\right].$$

$$(7.29)$$

Moreover, (7.23), Proposition 2.3, and (2.74) imply that for some $K > 0$, we have

$$
\begin{aligned}
\left|\frac{\partial u(t, x, i)}{\partial x}\right| &\leq \mathbf{E}|\phi_x(X(t), \alpha(t))\zeta(t)| \\
&\leq \mathbf{E}^{1/2}|\phi_x(X(t), \alpha(t))|^2 \mathbf{E}^{1/2}|\zeta(t)|^2 \\
&\leq K\mathbf{E}^{1/2}(1 + |X(t)|^\gamma) \leq K(1 + |x|^{\gamma_0}).
\end{aligned}
$$

Next, we verify that $(\partial/\partial x)u(t, x, i)$ is continuous with respect to x. To this purpose, we consider

$$\left| \frac{\partial u(t, x, i)}{\partial x} - \frac{\partial u(t, \widetilde{x}, i)}{\partial x} \right| \leq \mathbf{E} \left| \phi_x(\widetilde{X}(t), \widetilde{\alpha}(t))\widetilde{\zeta}(t) - \phi_x(X(t), \alpha(t))\zeta(t) \right|$$
$$\leq \mathbf{E} \left| \phi_x(X(t), \alpha(t))(\widetilde{\zeta}(t) - \zeta(t)) \right|$$
$$+ \mathbf{E} \left| [\phi_x(\widetilde{X}(t), \widetilde{\alpha}(t)) - \phi_x(X(t), \alpha(t))]\zeta(t) \right|,$$

where

$$\widetilde{\zeta}(t) = \zeta^{\widetilde{x}, i}(t) = \frac{\partial X^{\widetilde{x}, i}(t)}{\partial x}.$$

By virtue of Theorem 2.27, $\zeta(t) = \partial X(t)/\partial x$ is mean square continuous. Hence it follows that

$$\mathbf{E} \left| \phi_x(X(t), \alpha(t))(\widetilde{\zeta}(t) - \zeta(t)) \right| \leq \mathbf{E}^{1/2} \left| \phi_x(X(t), \alpha(t)) \right|^2 \mathbf{E}^{1/2} \left| \widetilde{\zeta}(t) - \zeta(t) \right|^2$$
$$\to 0 \quad \text{as} \quad \widetilde{x} \to x.$$

In addition, detailed calculations similar to those used in deriving (7.26) lead to

$$\mathbf{E} \left| [\phi_x(\widetilde{X}(t), \widetilde{\alpha}(t)) - \phi_x(X(t), \alpha(t))]\zeta(t) \right|$$
$$\leq \mathbf{E}^{1/2} \left| \phi_x(\widetilde{X}(t), \widetilde{\alpha}(t)) - \phi_x(X(t), \alpha(t)) \right|^2 \mathbf{E}^{1/2} |\zeta(t)|^2$$
$$\leq K \mathbf{E}^{1/2} \left| \frac{\phi_x(\widetilde{X}(t), \widetilde{\alpha}(t)) - \phi_x(X(t), \alpha(t))}{\widetilde{x} - x} \right|^2 |\widetilde{x} - x|^2$$
$$\to 0 \quad \text{as} \quad \widetilde{x} \to x.$$

Hence it follows that $(\partial/\partial x)u(t, x, i)$ is continuous with respect to x and therefore $u(t, x, i)$ is continuously differentiable with respect to the variable x.

In a similar manner, we can show that $u(t, x, i)$ is twice continuously differentiable with respect to the variable x and that

$$\frac{\partial^2 u(t, x, i)}{\partial x^2} = \mathbf{E}_{x, i} \left[\phi_{xx}(X(t), \alpha(t)) \left(\frac{\partial X(t)}{\partial x} \right)^2 + \phi_x(X(t), \alpha(t)) \frac{\partial^2 X(t)}{\partial x^2} \right].$$

Consequently, we can verify that

$$\left| \frac{\partial^2 u(t, x, i)}{\partial x^2} \right| \leq K(1 + |x|^\gamma).$$

This completes the proof of the theorem. □

Theorem 7.10. *Assume the conditions of Theorem 7.9. Then the function u defined in (7.24) is continuously differentiable with respect to the variable t. Moveover, u satisfies the system of Kolmogorov backward equations*

$$\frac{\partial u(t,x,i)}{\partial t} = \mathcal{L}u(t,x,i), \quad (t,x,i) \in (0,T] \times \mathbb{R}^r \times \mathcal{M}, \qquad (7.30)$$

with initial condition

$$\lim_{t\downarrow 0} u(t,x,i) = \phi(x,i), \quad (x,i) \in \mathbb{R}^r \times \mathcal{M}, \qquad (7.31)$$

where $\mathcal{L}u(t,x,i)$ in (7.30) is to be interpreted as $\mathcal{L}$ applied to the function $(x,i) \mapsto u(t,x,i)$.

Proof. First note that by virtue of Proposition 2.4, the process $(X(t), \alpha(t))$ is càdlàg. Hence the initial condition (7.31) follows from the continuity of ϕ. We divide the rest of the proof into several steps.

Step 1. For fixed $(x,i) \in \mathbb{R}^r \times \mathcal{M}$, $u(t,x,i)$ is absolutely continuous with respect to $t \in [0,T]$. In fact, for any $0 \le s \le t \le T$, we have from Dynkin's formula that

$$
\begin{aligned}
u(t,x,i) - u(s,x,i) &= \mathbf{E}_{x,i}\phi(X(t),\alpha(t)) - \mathbf{E}_{x,i}\phi(X(s),\alpha(s)) \\
&= \mathbf{E}_{x,i}\left[\mathbf{E}_{x,i}[(\phi(X(t),\alpha(t)) - \phi(X(s),\alpha(s)))|\mathcal{F}_s]\right] \\
&= \mathbf{E}_{x,i}\mathbf{E}_{x,i}\left[\int_s^t \mathcal{L}\phi(X(v),\alpha(v))dv|\mathcal{F}_s\right] \\
&\le \mathbf{E}_{x,i}\int_s^t \mathbf{E}_{x,i}\left[|\mathcal{L}\phi(X(v),\alpha(v))|\,|\mathcal{F}_s\right]dv.
\end{aligned}
$$

Using (7.23), for some positive constants K and γ_0, we have

$$|\mathcal{L}\phi(x,i)| \le K(1 + |x|^{\gamma_0}) \quad \text{for all} \quad (x,i) \in \mathbb{R}^r \times \mathcal{M}.$$

Hence it follows from Proposition 2.3 that

$$\mathbf{E}_{x,i}\left[|\mathcal{L}\phi(X(v),\alpha(v))|\,|\mathcal{F}_s\right] \le K\mathbf{E}_{x,i}\left[(1+|X(v)|^{\gamma_0})|\mathcal{F}_s\right] \le C,$$

where C is independent of t, s, or v. Thus we have

$$|u(t,x,i) - u(s,x,i)| \le C|t-s|.$$

Thus u is absolutely continuous with respect to t and $(\partial/\partial t)u(t,x,i)$ exists a.s. on $[0,T]$ and we have

$$u(t,x,i) = u(0,x,i) + \int_0^t \frac{\partial u(v,x,i)}{\partial v}dv. \qquad (7.32)$$

Step 2. For any $h > 0$, we have from the strong Markov property that

$$
\begin{aligned}
u(t+h,x,i) &= \mathbf{E}_{x,i}\phi(X(t+h),\alpha(t+h)) \\
&= \mathbf{E}_{x,i}\left[\mathbf{E}_{x,i}[\phi(X(t+h),\alpha(t+h))|\mathcal{F}_h]\right] \\
&= \mathbf{E}_{x,i}\left[\mathbf{E}_{X(h),\alpha(h)}\phi(X(t+h),\alpha(t+h))\right] \\
&= \mathbf{E}_{x,i}u(t,X(h),\alpha(h)).
\end{aligned}
\tag{7.33}
$$

Now let $g(x,i) := u(t,x,i)$. Then Theorem 7.9 implies that $g(\cdot,i) \in C^2$ for each $i \in \mathcal{M}$ and for some $K > 0$ and $\gamma_0 > 0$,

$$
\left|D_x^\beta g(x,i)\right| \le K(1+|x|^{\gamma_0}), \quad i \in \mathcal{M}.
$$

Thus it follows from Dynkin's formula that

$$
\mathbf{E}_{x,i}g(X(h),\alpha(h)) - g(x,i) = \mathbf{E}_{x,i}\int_0^h \mathcal{L}g(X(v),\alpha(v))dv.
$$

Using the same argument as in the proof of [47, Theorem 5.6.1], we can show that

$$
\frac{1}{h}\mathbf{E}_{x,i}\int_0^h \mathcal{L}g(X(v),\alpha(v))dv \to \mathcal{L}g(x,i) \quad \text{as} \ h \downarrow 0.
\tag{7.34}
$$

Therefore,

$$
\lim_{h\downarrow 0}\frac{\mathbf{E}_{x,i}g(X(h),\alpha(h)) - g(x,i)}{h} = \mathcal{L}g(x,i).
$$

But by the definition of g, we have from (7.33) that

$$
\lim_{h\downarrow 0}\frac{u(t+h,x,i) - u(t,x,i)}{h} = \mathcal{L}g(x,i) = \mathcal{L}u(t,x,i).
\tag{7.35}
$$

Thus a combination of (7.32) and (7.35) leads to

$$
u(t,x,i) = u(0,x,i) + \int_0^t \mathcal{L}u(v,x,i)dv.
\tag{7.36}
$$

Step 3. We claim that $\mathcal{L}u(t,x,i)$ is continuous with respect to the variable t. Note that

$$
\mathcal{L}u(t,x,i) = b(x,i)\frac{\partial u(t,x,i)}{\partial x} + \frac{1}{2}\sigma^2(x,i)\frac{\partial^2 u(t,x,i)}{\partial x^2} + \sum_{j=1}^{m_0} q_{ij}(x)u(t,x,j).
$$

The claim is verified if we can show $(\partial/\partial x)u(t,x,i)$ and $(\partial^2/\partial x^2)u(t,x,i)$ are continuous with respect to t, since Step 1 above shows that $u(t,x,i)$ is continuous with respect to t. To this end, let $t, s \in [0, T]$. Then we have

$$
\left| \frac{\partial u(t, x, i)}{\partial x} - \frac{\partial u(s, x, i)}{\partial x} \right|
$$
$$
= \left| \mathbf{E}_{x,i} \left[\phi_x(X(t), \alpha(t)) \zeta(t) \right] - \mathbf{E}_{x,i} \left[\phi_x(X(s), \alpha(s)) \zeta(s) \right] \right|
$$
$$
\leq \mathbf{E}_{x,i} \left| \phi_x(X(t), \alpha(t)) \zeta(t) - \phi_x(X(s), \alpha(s)) \zeta(s) \right|
$$
$$
\leq \mathbf{E}_{x,i} \left| \left[\phi_x(X(t), \alpha(t)) - \phi_x(X(s), \alpha(s)) \right] \zeta(t) \right|
$$
$$
+ \mathbf{E}_{x,i} \left| \phi_x(X(s), \alpha(s)) \left[\zeta(t) - \zeta(s) \right] \right|
$$
$$
\leq \mathbf{E}_{x,i}^{1/2} \left| \phi_x(X(t), \alpha(t)) - \phi_x(X(s), \alpha(s)) \right|^2 \mathbf{E}_{x,i}^{1/2} \left| \zeta(t) \right|^2
$$
$$
+ \mathbf{E}_{x,i}^{1/2} \left| \phi_x(X(s), \alpha(s)) \right|^2 \mathbf{E}_{x,i}^{1/2} \left| \zeta(t) - \zeta(s) \right|^2
$$

As we demonstrated before,

$$
\mathbf{E}_{x,i}^{1/2} \left| \phi_x(X(s), \alpha(s)) \right|^2 \leq K.
$$

While Corollary 2.32 implies that $\zeta(t)$ is mean square continuous with respect to t. Hence it follows that

$$
\mathbf{E}_{x,i}^{1/2} \left| \zeta(t) - \zeta(s) \right|^2 \to 0 \quad \text{as} \quad |t - s| \to 0.
$$

Meanwhile,

$$
\mathbf{E}_{x,i} \left| \phi_x(X(t), \alpha(t)) - \phi_x(X(s), \alpha(s)) \right|^2
$$
$$
\leq K \mathbf{E}_{x,i} \left| \phi_x(X(t), \alpha(t)) - \phi_x(X(s), \alpha(t)) \right|^2
$$
$$
+ K \mathbf{E}_{x,i} \left| \phi_x(X(s), \alpha(t)) - \phi_x(X(s), \alpha(s)) \right|^2
$$
$$
:= e_1 + e_2.
$$

Using Theorem 2.13 or (2.26) and (2.74), detailed computations show that

$$
e_1 \leq K \mathbf{E}_{x,i} \left| \int_0^1 \phi_{xx}(X(s) + v(X(t) - X(s)), \alpha(t)) dv(X(t) - X(s)) \right|^2
$$
$$
\to 0 \quad \text{as} \quad |t - s| \to 0.
$$

To treat the term e_2, we assume without loss of generality that $t > s$ and compute

$$
\begin{aligned}
e_2 &= K\mathbf{E}_{x,i}\left|\phi_x(X(s),\alpha(t)) - \phi_x(X(s),\alpha(s))\right|^2 \\
&= \sum_{i=1}^{m_0}\sum_{j\neq i}\mathbf{E}_{x,i}\left|\phi_x(X(s),j) - \phi_x(X(s),i)\right|^2 I_{\{\alpha(t)=j\}}I_{\{\alpha(s)=i\}} \\
&= \sum_{i=1}^{m_0}\sum_{j\neq i}\mathbf{E}_{x,i}\left[\left|\phi_x(X(s),j) - \phi_x(X(s),i)\right|^2 I_{\{\alpha(s)=i\}}\mathbf{E}_{x,i}[I_{\{\alpha(t)=j\}}|\mathcal{F}_s]\right] \\
&= \sum_{i=1}^{m_0}\sum_{j\neq i}\mathbf{E}_{x,i}\Big[\left|\phi_x(X(s),j) - \phi_x(X(s),i)\right|^2 I_{\{\alpha(s)=i\}} \\
&\qquad \times q_{ij}(X(s))(t-s) + o(t-s)\Big] \\
&\leq K(t-s).
\end{aligned}
$$

Thus it follows that $e_2 \to 0$ as $|t - s| \to 0$. Hence we have shown that

$$
\left|\frac{\partial u(t,x,i)}{\partial x} - \frac{\partial u(s,x,i)}{\partial x}\right| \to 0 \quad \text{as} \quad |t - s| \to 0
$$

and so $(\partial/\partial x)u(t,x,i)$ is continuous with respect to the variable t. Similarly, we can show that $(\partial^2/\partial x^2)u(t,x,i)$ is also continuous with respect to the variable t. Therefore $\mathcal{L}u(t,x,i)$ is continuous with respect to the variable t.

Step 4. Finally, by virtue of (7.36) and Step 3 above, we conclude that $(\partial/\partial t)u(t,x,i)$ exists everywhere for $t \in (0,T]$ and that

$$
\frac{\partial u(t,x,i)}{\partial t} = \mathcal{L}u(t,x,i).
$$

This finishes the proof of the theorem.   □

Lemma 7.11. *Let $X^{x,\alpha}(t)$ be the solution to the system given by (7.2) and (7.3) with initial data $X(0) = x$, $\alpha(0) = \alpha$. Assume that for each $i \in \mathcal{M}$, $b(\cdot,i)$ and $\sigma(\cdot,i)$ have continuous partial derivatives with respect to the variable x up to the second order and $b(0,i) = \sigma(0,i) = 0$. Then for any $\gamma \in \mathbb{R}$, the function*

$$
u(t,x,i) := \mathbf{E}\left|X^{x,i}(t)\right|^p \tag{7.37}
$$

is twice continuously differentiable with respect to x, except possibly at $x = 0$. Moreover, we have

$$
\begin{aligned}
\left|\frac{\partial u(t,x,\alpha)}{\partial x_j}\right| &\leq K|x|^{p-1}e^{\kappa_0 t}, \quad \text{and} \\
\left|\frac{\partial^2 u(t,x,\alpha)}{\partial x_j \partial x_k}\right| &\leq K|x|^{p-2}e^{\kappa_0 t},
\end{aligned} \tag{7.38}
$$

where $j,k = 1,2,\ldots,r$, and K and κ_0 are positive constants.

Proof. Once again, for notational simplicity, we present the proof for $X(t)$ being a real-valued process. By virtue of Theorem 7.9, $u(t, x, i) := \mathbf{E}\left|X^{x,i}(t)\right|^p$ is twice continuously differentiable with respect to x, except possibly at $x = 0$. We need only show that the partial derivatives satisfy (7.38). To this end, similar to the proofs of Theorems 7.9 and 7.10, we assume x to be a scalar without loss of generality. By virtue of (7.29),

$$\frac{\partial u(t, x, \alpha)}{\partial x} = p\mathbf{E}\left(|X^{x,\alpha}(t)|^{p-1}\mathrm{sgn}(X^{x,\alpha}(t))\frac{\partial X^{x,\alpha}(t)}{\partial x}\right). \tag{7.39}$$

Then it follows from (7.6) and (2.74) that

$$\left|\frac{\partial u(t, x, \alpha)}{\partial x}\right| \leq K\mathbf{E}\left[|X^{x,\alpha}(t)|^{p-1}\left|\frac{\partial X^{x,\alpha}(t)}{\partial x}\right|\right]$$

$$\leq \mathbf{E}^{1/2}\left||X^{x,\alpha}(t)|^{2p-2}\right|\mathbf{E}^{1/2}\left|\frac{\partial X^{x,\alpha}(t)}{\partial x}\right|^2$$

$$\leq K(|x|^{2p-2}e^{Kt})^{1/2} = K|x|^{p-1}e^{\kappa_0 t}.$$

Similarly, detailed calculations lead to

$$\left|\frac{\partial^2 u(t, x, \alpha)}{\partial x^2}\right| \leq K|x|^{p-2}e^{\kappa_0 t}.$$

The lemma is thus proved. $\qquad\square$

7.3.3 Necessary and Sufficient Conditions for p-Stability

Theorem 7.12. *Suppose that the equilibrium point 0 is exponentially p-stable. Moreover assume that the coefficients b and σ have continuous bounded derivatives with respect to the variable x up to the second order. Then for each $i \in \mathcal{M}$, there exists a function $V(\cdot, i) : \mathbb{R}^r \mapsto \mathbb{R}$ such that*

$$k_1|x|^p \leq V(x, i) \leq k_2|x|^p, \quad x \in N, \tag{7.40}$$

$$\mathcal{L}V(x, i) \leq -k_3|x|^p \quad \text{for all} \quad x \in N - \{0\}, \tag{7.41}$$

$$\left|\frac{\partial V}{\partial x_j}(x, i)\right| < k_4|x|^{p-1},$$

$$\left|\frac{\partial^2 V}{\partial x_j \partial x_k}(x, i)\right| < k_4|x|^{p-2}, \tag{7.42}$$

for all $1 \leq j, k \leq r$, $x \in N - \{0\}$, and for some positive constants k_1, k_2, k_3, and k_4, where N is a neighborhood of 0.

Proof. For each $i \in \mathcal{M}$, consider the function

$$V(x, i) = \int_0^T \mathbf{E}|X^{x,i}(u)|^p du.$$

It follows from Lemma 7.11 that the functions $V(x,i)$, $i \in \mathcal{M}$ are twice continuously differentiable with respect to x except possibly at $x = 0$.

The equilibrium point 0 is exponentially p-stable, therefore by the definition of exponential p-stability, there is a $\beta > 0$ such that

$$V(x,i) \le \int_0^T K|x|^p \exp(-\beta u)du \le K|x|^p.$$

Since 0 is an equilibrium point, $|A(x,i)| \le K|x|^2$ and $b(x,i)| \le K|x|$. Consequently, $|\mathcal{L}|x|^p| \le K|x|^p$. An application of Itô's lemma to $g(x) = |x|^p$ implies that

$$
\begin{aligned}
\mathbf{E}|X^{x,i}(T)|^p - |x|^p &= \mathbf{E}\int_0^T \mathcal{L}|X^{x,i}(u)|^p du \\
&\ge -K\int_0^T \mathbf{E}|X^{x,i}(u)|^p du = -KV(x,i).
\end{aligned}
$$

Again, by the exponential p-stability, we can choose T so that

$$\mathbf{E}|X^{x,i}(T)|^p \le (1/2)|x|^p,$$

and as a result $V(x,i) \ge |x|^p/(2K)$. Thus, (7.40) is verified.

We note that

$$
\begin{aligned}
\left|\frac{\partial V(x,i)}{\partial x_\ell}\right| &= \int_0^T \frac{\partial}{\partial x_\ell}\mathbf{E}|X^{x,i}(u)|^p du \\
&\le \int_0^T K|x|^{p-1}\exp(Ku)du \le K|x|^{p-1}.
\end{aligned}
$$

Likewise, we can verify the second part of (7.42). Thus the proof is completed. $\qquad\square$

We end this section with the following results on linear systems. Assume that the evolution (7.2) is replaced by

$$dX(t) = b(\alpha(t))X(t)dt + \sum_{j=1}^d \sigma_j(\alpha(t))X(t)dw_j(t), \qquad (7.43)$$

where $b(i)$ and $\sigma_j(i)$ are $r \times r$ constant matrices, and $w_j(t)$ are independent one-dimensional standard Brownian motions for $i = 1, 2, \ldots, m_0$, and $j = 1, 2, \ldots, d$. Then we have the following two theorems.

Theorem 7.13. *The equilibrium point $x = 0$ of system (7.43) together with (7.3) is exponentially p-stable if and only if for each $i \in \mathcal{M}$, there is a function $V(\cdot,i) : \mathbb{R}^r \mapsto \mathbb{R}$ satisfying equations (7.40) and (7.42) for some constants $k_i > 0$, $i = 1, \ldots, 4$.*

Proof. The proof of sufficiency was contained in [116]. However, the necessity follows from Theorem 7.12 because the coefficients of (7.43) and (7.3) satisfy the conditions of Theorem 7.12. We omit the details here. $\qquad\square$

Theorem 7.14. *Let $Q(x) \equiv Q$ be a constant matrix. Assume also that the Markov chain $\alpha(t)$ is independent of the Brownian motion*

$$w(t) = (w_1(t), w_2(t), \dots, w_d(t))'$$

(or equivalently, $\alpha(0)$ is independent of the Brownian motion $w(\cdot)$). If the equilibrium point $x = 0$ of the system given by (7.43) and (7.3) is stable in probability, then it is p-stable for sufficiently small $p > 0$.

Proof. The proof follows from a crucial observation. In this case, because (7.43) is linear in $X(t)$, $X^{\lambda x, \alpha}(t) = \lambda X^{x, \alpha}(t)$. Using a similar argument as that for [83, Lemma 6.4.1], we can conclude the proof; a few details are omitted.   □

7.4   Stability and Instability of Linearized Systems

This section provides criteria for stability and instability. To proceed, we put an assumption.

(A7.4)   For each $i \in \mathcal{M}$, there exist $b(i), \sigma_J(i) \in \mathbb{R}^{r \times r}, J = 1, 2, \dots, d$, and a generator of a continuous-time Markov chain $\widehat{Q} = (\widehat{q}_{ij})$ such that as $x \to 0$,

$$b(x, i) = b(i)x + o(|x|),$$
$$\sigma(x, i) = (\sigma_1(i)x, \sigma_2(i)x, \dots, \sigma_d(i)x) + o(|x|), \qquad (7.44)$$
$$Q(x) = \widehat{Q} + o(1).$$

Moreover, $\widehat{Q}$ is irreducible and $\widehat{\alpha}(t)$ is a Markov chain with generator $\widehat{Q}$.

Remark 7.15. Note that condition (A7.4) is rather natural. It is equivalent to $Q(x)$ being continuous at $x = 0$, and $b(x, i)$ and $\sigma(x, i)$ continuously differentiable at $x = 0$. It follows from (A7.4) that $\widehat{\alpha}(t)$ is an ergodic Markov chain. Denote the stationary distribution of $\widehat{\alpha}(t)$ by $\pi = (\pi_1, \pi_2, \dots, \pi_{m_0}) \in \mathbb{R}^{1 \times m_0}$.

Remark 7.16. For any square matrix $A \in \mathbb{R}^{r \times r}$, A can be decomposed to the sum of a symmetric matrix A_1 and an antisymmetric matrix A_2. In fact, $A_1 = (A + A')/2$ and $A_2 = (A - A')/2$. Moreover, the quadratic form satisfies

$$x'Ax = x'A_1x = x'\frac{A + A'}{2}x. \qquad (7.45)$$

This observation is used in what follows.

Theorem 7.17. *Assume condition* (A7.4). *Then the equilibrium point* $x = 0$ *of system given by* (7.2) *and* (7.3) *is asymptotically stable in probability if*

$$\sum_{i=1}^{m_0} \pi_i \lambda_{\max}\left(b(i) + b'(i) + \sum_{j=1}^{d} \sigma_j'(i)\sigma_j(i) \right) < 0, \qquad (7.46)$$

and is unstable in probability if

$$\sum_{i=1}^{m_0} \pi_i \left(\lambda_{\min}\left(b(i) + b'(i) + \sum_{j=1}^{d} \sigma_j'(i)\sigma_j(i) \right) \right.$$
$$\left. - \frac{1}{2} \sum_{j=1}^{d} \left[\rho(\sigma_j(i) + \sigma_j'(i)) \right]^2 \right) > 0. \qquad (7.47)$$

Proof. (a) We first prove that the equilibrium point $x = 0$ of the system given by (7.2) and (7.3) is asymptotically stable in probability if (7.46) holds. For notational simplicity, define the column vector

$$\mu = (\mu_1, \mu_2, \ldots, \mu_{m_0})' \in \mathbb{R}^{m_0}$$

with

$$\mu_i = \lambda_{\max}\left(b(i) + b'(i) + \sum_{j=1}^{d} \sigma_j'(i)\sigma_j(i) \right).$$

Also let $\beta := -\pi\mu$. Note that $\beta > 0$ by (7.46). It follows from assumption (A7.4) and Lemma A.12 that the equation

$$\widehat{Q}c = \mu + \beta \mathbb{1}$$

has a solution $c = (c_1, c_2, \ldots, c_{m_0})' \in \mathbb{R}^{m_0}$. Thus we have

$$\mu_i - \sum_{j=1}^{m_0} \widehat{q}_{ij} c_j = -\beta, \quad i \in \mathcal{M}. \qquad (7.48)$$

For each $i \in \mathcal{M}$, consider the Liapunov function

$$V(x, i) = (1 - \gamma c_i)|x|^\gamma,$$

where $0 < \gamma < 1$ is sufficiently small so that $1 - \gamma c_i > 0$ for each $i \in \mathcal{M}$. It is readily seen that for each $i \in \mathcal{M}$, $V(\cdot, i)$ is continuous, nonnegative, and vanishes only at $x = 0$. Detailed calculations reveal that for $x \neq 0$, we have

$$\nabla V(x, i) = (1 - \gamma c_i)\gamma |x|^{\gamma-2} x,$$
$$\nabla^2 V(x, i) = (1 - \gamma c_i)\gamma \left[|x|^{\gamma-2} I + (\gamma - 2)|x|^{\gamma-4} xx' \right].$$

In addition, it follows from (7.44) that

$$a(x,i) = \sigma(x,i)\sigma'(x,i) = \sum_{j=1}^{d}\sigma_j(i)xx'\sigma'_j(i) + o(|x|^2).$$

Note that for any matrix $A \in \mathbb{R}^{r \times r}$, we have

$$\text{tr}(\sigma_j(i)xx'\sigma'_j(i)A) = x'\sigma'_j(i)A\sigma_j(i)x.$$

Therefore, we have that

$$
\begin{aligned}
\mathcal{L}V(x,i) &= \frac{1}{2}\text{tr}\left[\left(\sum_{j=1}^{d}\sigma_j(i)xx'\sigma'_j(i) + o(|x|^2)\right)\nabla^2 V(x,i)\right] \\
&\quad + \nabla V'(x,i)(b(i)x + o(|x|)) - \sum_{j \neq i}q_{ij}(x)|x|^{\gamma}\gamma(c_j - c_i) \\
&= \gamma(1 - \gamma c_i)|x|^{\gamma}\left\{\frac{1}{2}\sum_{j=1}^{d}\left(\frac{x'\sigma'_j(i)\sigma_j(i)x}{|x|^2} + (\gamma - 2)\frac{(x'\sigma'_j(i)x)^2}{|x|^4}\right)\right. \\
&\quad \left. + \frac{x'b(i)x}{|x|^2} - \sum_{j \neq i}q_{ij}(x)\frac{c_j - c_i}{1 - \gamma c_i} + o(1)\right\}.
\end{aligned}
$$

$$(7.49)$$

By virtue of Remark 7.16, we obtain

$$
\begin{aligned}
&\frac{x'b(i)x}{|x|^2} + \frac{1}{2}\sum_{j=1}^{d}\frac{x'\sigma'_j(i)\sigma_j(i)x}{|x|^2} \\
&\quad \leq \frac{1}{2}\lambda_{\max}\left(b(i) + b'(i) + \sum_{j=1}^{d}\sigma'_j(i)\sigma_j(i)\right) = \mu_i.
\end{aligned}
$$

$$(7.50)$$

Next, it follows from condition (A7.4) that when $|x|$ and γ are sufficiently small,

$$
\begin{aligned}
&\sum_{j \neq i}q_{ij}(x)\frac{c_j - c_i}{1 - \gamma c_i} \\
&\quad = \sum_{j=1}^{m_0}q_{ij}(x)c_j + \sum_{j \neq i}q_{ij}(x)\frac{c_i(c_j - c_i)}{1 - \gamma c_i}\gamma \\
&\quad = \sum_{j=1}^{m_0}\widehat{q}_{ij}c_j + O(\gamma) + o(1),
\end{aligned}
$$

$$(7.51)$$

where $o(1) \to 0$ as $|x| \to 0$. Hence it follows from (7.49) and (7.51) that when $|x| < r_0$ with r_0 and $0 < \gamma < 1$ sufficiently small, we have

$$\mathcal{L}V(x,i) \leq \gamma(1 - \gamma c_i)|x|^{\gamma}\left\{\mu_i - \sum_{j=1}^{m_0}\widehat{q}_{ij}c_j + o(1) + O(\gamma)\right\}.$$

Furthermore, by virtue of (7.48), we have

$$\mathcal{L}V(x,i) \leq \gamma(1-\gamma c_i)|x|^\gamma(-\beta + o(1) + O(\gamma)) \leq -\kappa(\varepsilon) < 0,$$

for any $(x,i) \in N \times \mathcal{M}$ with $\varepsilon < |x| < r_0$, where $N \subset \mathbb{R}^r$ is a small neighborhood of 0 and $\kappa(\varepsilon)$ is a positive constant. Therefore we conclude from Lemma 7.6 and Remark 7.8 that the equilibrium point $x = 0$ is asymptotically stable in probability.

(b) Now we prove that the equilibrium point $x = 0$ is unstable in probability if (7.47) holds. Define the column vector $\theta = (\theta_1, \theta_2, \ldots, \theta_{m_0})' \in \mathbb{R}^{m_0}$ by

$$\theta_i := \frac{1}{2}\lambda_{\min}\left(b(i) + b'(i) + \sum_{j=1}^d \sigma_j'(i)\sigma_j(i)\right) - \frac{1}{4}\sum_{j=1}^d [\rho(\sigma_j(i) + \sigma_j'(i))]^2,$$

and set

$$\delta := -\pi\theta = \sum_{i=1}^{m_0} \pi_i\theta_i < 0.$$

As in part (a), assumption (A7.4), the definition of δ, and Lemma A.12 imply that the equation $\widehat{Q}c = \theta + \delta\mathbb{1}$ has a solution $c = (c_1, c_2, \ldots, c_{m_0})' \in \mathbb{R}^{m_0}$ and

$$\theta_i - \sum_{j=1}^d \widehat{q}_{ij}c_j = -\delta > 0, \quad i \in \mathcal{M}. \tag{7.52}$$

For $i \in \mathcal{M}$, consider the Liapunov function

$$V(x,i) = (1-\gamma c_i)|x|^\gamma,$$

where $-1 < \gamma < 0$ is sufficiently small so that $1 - \gamma c_i > 0$ for each $i \in \mathcal{M}$. Obviously the nonnegative function $V(\cdot, i), i \in \mathcal{M}$ satisfies (7.19). Similar to the arguments in part (a), Remark 7.16 implies that

$$\frac{x'b(i)x}{|x|^2} + \frac{\displaystyle\sum_{j=1}^d x'\sigma_j'(i)\sigma_j(i)x}{2|x|^2} \geq \frac{1}{2}\lambda_{\min}\left(b(i) + b'(i) + \sum_{j=1}^d \sigma_j'(i)\sigma_j(i)\right).$$

Note that for any symmetric matrix A with real eigenvalues $\lambda_1 \geq \lambda_2 \geq \cdots \geq \lambda_n$, using the transformation $x = Uy$, where U is a real orthogonal matrix such that $U'AU = \mathrm{diag}(\lambda_1, \lambda_2, \ldots, \lambda_n)$ (see [59, Theorem 8.1.1]), we have

$$|x'Ax| = |\lambda_1 y_1^2 + \lambda_2 y_2^2 + \cdots + \lambda_n y_n^2| \leq \rho_A|y|^2 = \rho_A|x|^2. \tag{7.53}$$

Thus by applying (7.53) to the matrix $\sigma_j'(i) + \sigma_j(i)$, we obtain that

$$\frac{(x'\sigma_j'(i)x)^2}{|x|^4} = \frac{(x'(\sigma_j'(i) + \sigma_j(i))x)^2}{4|x|^4} \leq \frac{1}{4}[\rho(\sigma_j(i) + \sigma_j'(i))]^2.$$

Therefore, detailed computations as in part (a) (taking into account the extra term involving $\frac{1}{4}\left[\rho(\sigma_j(i) + \sigma_j'(i))\right]^2$) show that for any sufficiently small $0 < \varepsilon < r$, we have

$$\mathcal{L}V(x,i) \leq -\kappa(\varepsilon) < 0, \quad \text{for any } (x,i) \in N \times \mathcal{M} \text{ with } \varepsilon < |x| < r_0,$$

where $N \subset \mathbb{R}^r$ is a small neighborhood of 0 and $\kappa(\varepsilon)$ is a positive constant. Therefore Lemma 7.7 and Remark 7.8 imply that the equilibrium point $x = 0$ is unstable in probability. The proof of the theorem is concluded. $\square$

Remark 7.18. Suppose that for all $i \in \mathcal{M}$ and $j = 1, 2, \ldots, d$, the matrices $(\sigma_j'(i) + \sigma_j(i))$ are nonnegative definite. Then we have

$$\rho(\sigma_j(i) + \sigma_j'(i)) = \lambda_{\max}(\sigma_j(i) + \sigma_j'(i)) \geq \lambda_{\min}(\sigma_j(i) + \sigma_j'(i)) \geq 0.$$

Consequently a close examination of the proof of Theorem 7.17 reveals that the conditions (7.46) and (7.47) can be replaced by

$$\sum_{i=1}^{m_0} \pi_i \left(\lambda_{\max}\left(b(i) + b'(i) + \sum_{j=1}^{d} \sigma_j(i)\sigma_j'(i) \right) \right.$$
$$\left. -\frac{1}{2} \sum_{j=1}^{d} \left[\lambda_{\min}(\sigma_j(i) + \sigma_j'(i)) \right]^2 \right) < 0, \tag{7.54}$$

and

$$\sum_{i=1}^{m_0} \pi_i \left(\lambda_{\min}\left(b(i) + b'(i) + \sum_{j=1}^{d} \sigma_j(i)\sigma_j'(i) \right) \right.$$
$$\left. -\frac{1}{2} \sum_{j=1}^{d} \left[\lambda_{\max}(\sigma_j(i) + \sigma_j'(i)) \right]^2 \right) > 0, \tag{7.55}$$

respectively. In a sense, the above two inequalities, in particular (7.54), strengthen the corresponding results in Theorem 7.17.

Theorem 7.17 gives sufficient conditions in terms of the maximum and minimum eigenvalues of the matrices for stability and instability of the equilibrium point $x = 0$. Because there is a "gap" between the maximum and minimum eigenvalues, a natural question arises: Can we obtain necessary and sufficient conditions for stability? If the component $X(t)$ is one-dimensional, we have the following result.

We replace the first and second equations of (7.44) in assumption (A7.4) by

$$b(x,i) = b_i x + o(x),$$
$$\sigma(x,i) = \sigma_i x + o(|x|). \tag{7.56}$$

where $x \in \mathbb{R}$, and b_i and σ_i^2 are real constants with $\sigma_i^2 \geq 0, i \in \mathcal{M}$. Then we immediately have the following corollary from Theorem 7.17 and Remark 7.18.

Corollary 7.19. *Let assumption* (A7.4) *and* (7.56) *be valid. Then the equilibrium point $x = 0$ is asymptotically stable in probability if*

$$\sum_{i=1}^{m_0} \pi_i \left(b_i - \frac{\sigma_i^2}{2} \right) < 0,$$

and is unstable in probability if

$$\sum_{i=1}^{m_0} \pi_i \left(b_i - \frac{\sigma_i^2}{2} \right) > 0.$$

Remark 7.20. As can be seen from Corollary 7.19, if the continuous component of the system is one-dimensional, we obtain a necessary and sufficient condition for stability. One question of particular interest is: Will we be able to obtain a similar condition for a multidimensional counterpart. For linear systems of stochastic differential equations with constant coefficients without switching, such a condition was obtained in Khasminskii [83, pp. 220–224]. The main ingredient is the use of the transformations $y = x/|x|$ and $\ln|x|$. The result is a sharp necessary and sufficient condition. In Mao, Yin, and Yuan [119], inspired by the approach of [83], Markov-modulated regime-switching diffusions were considered, and necessary and sufficient conditions were obtained for exponential stability. The main ingredient is the use of a logarithm transformation technique leading to the derivation of the so-called Liapunov exponent. Such an approach can be adopted to treat switching diffusions with state-dependent switching with no essential difficulty. Because in Chapter 8, we will also examine a related problem for switched ordinary differential equations (a completely degenerate switching diffusion with the absence of the diffusion terms), we will not dwell on it here.

7.5   Examples

Example 7.21. To illustrate Theorem 7.17 and Corollary 7.19, we consider a real-valued process given by

$$\begin{cases} dX(t) = b(X(t), \alpha(t))dt + \sigma(X(t), \alpha(t))dw(t) \\ \mathbf{P}\{\alpha(t + \Delta) = j | \alpha(t) = i, X(s), \alpha(s), s \le t\} = q_{ij}(X(t))\Delta + o(\Delta), \end{cases}$$
$$(7.57)$$

for $j \ne i$, where the jump process $\alpha(t)$ has three states and is generated by

$$Q(x) = \begin{bmatrix} -3 - \sin x \cos x + \sin x^2 & 1 + \sin x \cos x & 2 - \sin x^2 \\ 2 & -2 - \dfrac{x^2}{2 + x^2} & \dfrac{x^2}{2 + x^2} \\ 4 - \sin x & \sin^2 x & -4 + \sin x - \sin^2 x \end{bmatrix},$$

and the drift and diffusion coefficients are given by

$$b(x,1) = x - x\sin x, \quad b(x,2) = x - x\sin x^2, \quad b(x,3) = 4x + x\sin x,$$
$$\sigma(x,1) = -\frac{3x}{1+x^2}, \quad \sigma(x,2) = x + \frac{1}{3}x\sin x, \quad \sigma(x,3) = x - \frac{1}{2}x\sin^2 x.$$

Associated with (7.57), there are three diffusions

$$dX(t) = (X(t) - X(t)\sin X(t))dt - \frac{3X(t)}{1+X^2(t)}dw(t), \tag{7.58}$$

$$dX(t) = \left(X(t) - X(t)\sin X^2(t)\right)dt + \left(X(t) + \frac{1}{3}X(t)\sin X(t)\right)dw(t), \tag{7.59}$$

$$dX(t) = (4X(t) + X(t)\sin X(t))dt + \left(X(t) - \frac{1}{2}X(t)\sin^2 X(t)\right)dw(t), \tag{7.60}$$

switching from one to another according to the movement of the jump process $\alpha(t)$. It is readily seen that as $x \to 0$, the constants $b_i, \sigma_i^2, i = 1,2,3$, as in (7.56) are given by

$$b_1 = 1, \quad b_2 = 1, \quad b_3 = 4$$
$$\sigma_1^2 = 9, \quad \sigma_2^2 = 1, \quad \sigma_3^2 = 1. \tag{7.61}$$

Also as $x \to 0$, $Q(x)$ tends to

$$\widehat{Q} = \begin{pmatrix} -3 & 1 & 2 \\ 2 & -2 & 0 \\ 4 & 0 & -4 \end{pmatrix}.$$

The matrix is irreducible. By solving the system of equations $\pi\widehat{Q} = 0$ and $\mathbb{1}\pi = 1$, we obtain the stationary distribution π associated with $\widehat{Q}$,

$$\pi = (0.5,\ 0.25,\ 0.25). \tag{7.62}$$

Finally, by virtue of (7.61) and (7.62) we examine that

$$\sum_{i=1}^{3} \pi_i \left(b_i - \frac{\sigma_i^2}{2}\right) = -0.75 < 0.$$

Therefore, we conclude from Corollary 7.19 that the equilibrium point $x = 0$ of (7.57) is asymptotically stable in probability.

This example is interesting and provides insight. It was proven in [83, pp. 171–172] that a one-dimensional nonlinear diffusion is stable if and only if its linear approximation is stable. Hence we can check that the equilibrium point $x = 0$ of (7.58) is stable in probability whereas (7.59) and (7.60) are unstable in probability. Therefore the jump process $\alpha(t)$ could be considered as a stabilization factor. Note that similar examples were demonstrated in [116] under the assumptions that the jump component $\alpha(\cdot)$ is generated by a constant matrix Q and that the Markov chain $\alpha(\cdot)$ is independent of the Brownian motion $w(\cdot)$. Note also Examples 4.1 and 4.2 in [48] are also concerned with stability of switching systems. Their result indicates that if the switching takes place sufficiently fast, the system will be stable even if the individual mode may be unstable. Essentially, it is related to singularly perturbed systems. Due to the fast variation, there is a limit system that is an average with respect to the stationary distribution of the Markov chain and that is stable. Then if the rate of switching is fast enough the original system will also be stable. Such an idea was also used in an earlier paper [18].

To illustrate, we plot a sample path of (7.57) in Figure 7.1. For comparison, we also demonstrate the sample paths of (7.58), (7.59), and (7.60) (without switching) in Figures 7.2–7.4, respectively.

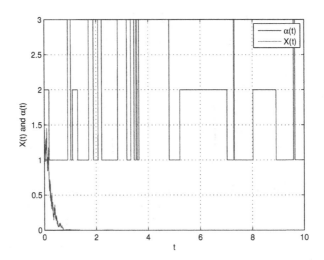

FIGURE 7.1. A sample path of (7.57) with initial condition $(x, \alpha) = (1.5, 1)$.

Example 7.22. (Lotka–Volterra model). This is a continuation of the discussion of the Lotka–Volterra model given in Example 1.1. The notation and problem setup are as in Example 1.1. Motivated by the works [34] and [118], define $V(t, x, \alpha) = e^t \log(|x|)$ for $(t, x, \alpha) \in [0, \infty) \times \mathbb{R}_+^n \times \mathcal{M}$. It

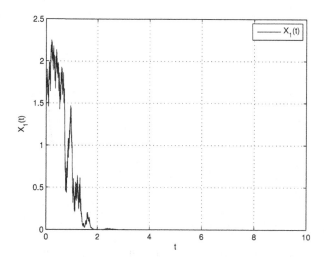

FIGURE 7.2. A sample path of (7.58) with initial condition $x = 1.5$.

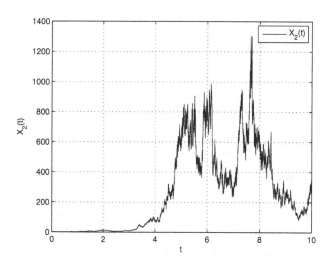

FIGURE 7.3. A sample path of (7.59) with initial condition $x = 1.5$.

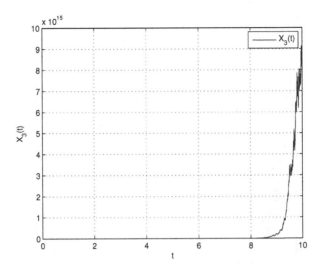

FIGURE 7.4. A sample path of (7.60) with initial condition $x = 1.5$.

follows from Itô's lemma that

$$
\begin{aligned}
& e^t \log(|x(t)|) - \log(|x(0)|) \\
& = \int_0^t e^s \sum_{i=1}^n \frac{x_i^2(s)}{|x(s)|^2} \left[r_i(\alpha(s)) - \sum_{j=1}^n a_{ij}(\alpha(s)) x_j(s) \right. \\
& \quad \left. + \frac{1}{2} \left(1 - \frac{2x_i^2(s)}{|x(s)|^2} \right) \sigma_i^2(\alpha(s)) \right] ds \\
& \quad + \int_0^t e^s \log(|x(s)|) ds + \int_0^t e^s \sum_{i=1}^n \frac{x_i^2(s)}{|x(s)|^2} \sigma_i(\alpha(s)) dw_i(s).
\end{aligned}
$$

Denote

$$
M_i(t) = \int_0^t e^s \frac{x_i^2(s)}{|x(s)|^2} \sigma_i(\alpha(s)) dw_i(s),
$$

whose quadratic variation is

$$
\langle M_i(t), M_i(t) \rangle = \int_0^t \frac{e^{2s}}{|x(s)|^4} x_i^4(s) \sigma_i^2(\alpha(s)) ds.
$$

By virtue of the exponential martingale inequality [47], for any positive constants T, δ, and β, we have

$$
\mathbf{P} \left\{ \sup_{0 \le t \le T} [M_i(t) - \frac{\delta}{2} \langle M_i(t), M_i(t) \rangle] > \beta \right\} \le e^{-\delta\beta}.
$$

Choose $T = k\gamma$, $\delta = n\varepsilon e^{-k\delta}$, and $\beta = (\theta e^{k\delta} \log(k))/(\varepsilon n)$, where $k \in \mathbb{N}$, $0 < \varepsilon < 1$, $\theta > 1$, and $\gamma > 0$ in the above equation. Then it follows that

$$\mathbf{P}\left\{ \sup_{0 \le t \le k\gamma} [M_i(t) - \frac{n\varepsilon e^{-k\gamma}}{2}\langle M_i(t), M_i(t)\rangle] > \frac{\theta e^{k\delta}\log k}{\varepsilon n} \right\} \le k^{-\theta}.$$

Because $\sum_{k=1}^{\infty} k^{-\theta} < \infty$, it follows from the Borel–Cantalli lemma that there exists some $\Omega_i \subset \Omega$ with $\mathbf{P}(\Omega_i) = 1$ such that for any $\omega \in \Omega_i$, an integer $k_i = k_i(\omega)$ such that for any $k > k_i$, we have

$$M_i(t) \le \frac{n\varepsilon e^{-k\gamma}}{2}\langle M_i(t), M_i(t)\rangle + \frac{\theta e^{k\gamma}\log k}{\varepsilon n} \quad \text{for all } 0 \le t \le k\gamma.$$

Now let $\Omega_0 := \bigcap_{i=1}^{n} \Omega_i$. Then $\mathbf{P}(\Omega_0) = 1$. Moreover, for any $\omega \in \Omega_0$, let

$$k_0(\omega) := \max\{k_i(\omega), i = 1, 2, \ldots, n\}.$$

Then for any $\omega \in \Omega_0$ and any $k \ge k_0(\omega)$, we have

$$\sum_{i=1}^{n} \int_0^t e^s \frac{x_i^2(s)}{|x(s)|^2}\sigma_i(\alpha(s))dw_i(s)$$
$$= \sum_{i=1}^{n} M_i(t) \le \frac{n\varepsilon e^{-k\gamma}}{2}\sum_{i=1}^{n}\langle M_i(t), M_i(t)\rangle + \frac{\theta e^{k\gamma}\log k}{\varepsilon},$$

where $0 \le t \le k\gamma$. Then it follows that

$$e^t \log(|x(t)|) - \log(|x(0)|)$$
$$\le \int_0^t e^s \sum_{i=1}^{n} \frac{x_i^2(s)}{|x(s)|^2}\left[r_i(\alpha(s)) - \sum_{j=1}^{n} a_{ij}(\alpha(s))x_j(s) \right.$$
$$\left. + \frac{1}{2}\left(1 - \frac{2x_i^2(s)}{|x(s)|^2}\right)\sigma_i^2(\alpha(s))\right]ds$$
$$+ \int_0^t e^s \log(|x(s)|)ds + \int_0^t \frac{n\varepsilon e^{-k\gamma}}{2}e^{2s}\sum_{i=1}^{n}\frac{x_i^4(s)}{|x(s)|^4}\sigma_i^2(\alpha(s))ds$$
$$+ \frac{\theta e^{k\gamma}\log k}{\varepsilon}$$
$$\le \int_0^t e^s \left\{ \log(|x(s)|) + \sum_{i=1}^{n}\frac{x_i^2(s)}{|x(s)|^2}\left[b_i(\alpha(s)) + \sigma_i^2(\alpha(s)) \right.\right.$$
$$\left. - \sum_{j=1}^{n} a_{ij}(\alpha(s))x_j(s) \right]$$
$$+ \sum_{i=1}^{n}\left(\frac{\varepsilon n e^{-k\gamma}}{2}e^s - 1\right)\frac{x_i^4(s)\sigma_i^2(\alpha(s))}{|x(s)|^4} \Bigg\} + \frac{\theta e^{k\gamma}\log k}{\varepsilon}.$$

Note that for any $t \in [0, k\gamma]$, $s \in [0, t]$, and $(x, \alpha) \in \mathbb{R}_+^n \times \mathcal{M}$, we have

$$\log(|x|) + \sum_{i=1}^{n} \left[\frac{x_i^2}{|x|^2} \left[b_i(\alpha) + \sigma_i^2(\alpha) - \sum_{j=1}^{n} a_{ij}(\alpha) x_j \right] \right.$$
$$\left. + \left(\frac{\varepsilon n e^{s-k\gamma}}{2} - 1 \right) \frac{x_i^4 \sigma_i^2(\alpha)}{|x|^4} \right]$$
$$\leq \log(|x|) + \kappa - \frac{1}{|x|^2} \beta \sum_{i=1}^{n} x_i^3 + K$$
$$\leq \log(|x|) + \kappa - \frac{\beta}{\sqrt{n}} |x| + K \leq K.$$

Hence it follows that for all $0 \leq t \leq k\gamma$ with $k \geq k_0(\omega)$, we have

$$e^t \log(|x(t)|) - \log(|x(0)|) \leq \int_0^t K e^s ds + \frac{\theta e^{k\gamma} \log k}{\varepsilon}$$
$$= K(e^t - 1) + \frac{\theta e^{k\gamma} \log k}{\varepsilon}.$$

Thus for $(k-1)\gamma \leq t \leq k\gamma$, we have

$$\log(|x(t)|) \leq e^{-t} \log(|x(0)|) + K(1 - e^{-t}) + \frac{\theta e^{k\gamma} \log k}{\varepsilon e^{(k-1)\gamma}}$$
$$= e^{-t} \log(|x(0)|) + K(1 - e^{-t}) + \frac{\theta e^{\gamma} \log k}{\varepsilon},$$

and hence it follows that

$$\frac{\log(|x(t)|)}{\log t} \leq \frac{\log(|x(0)|)}{e^t \log t} + \frac{K(1 - e^{-t})}{\log t} + \frac{\theta e^{\gamma} \log k}{\varepsilon \log((k-1)\gamma)}.$$

Now let $k \to \infty$ (and so $t \to \infty$) and we obtain

$$\limsup_{t \to \infty} \frac{\log(|x(t)|)}{\log t} \leq \frac{\theta e^{\gamma}}{\varepsilon}.$$

Finally, by sending $\gamma \downarrow 0$, $\varepsilon \uparrow 1$, and $\theta \downarrow 1$, we have

$$\limsup_{t \to \infty} \frac{\log(|x(t)|)}{\log t} \leq 1,$$

as desired. Thus, the solution $x(t)$ of (1.1) satisfies

$$\limsup_{T \to \infty} \frac{\log(|x(T)|)}{\log T} \leq 1. \tag{7.63}$$

Furthermore, since

$$\limsup_{T \to \infty} \frac{\log |x(T)|}{T} = \limsup_{T \to \infty} \frac{\log |x(T)|}{\log T} \limsup_{T \to \infty} \frac{\log T}{T}$$
$$\leq \limsup_{T \to \infty} \frac{\log T}{T},$$

$$\limsup_{T \to \infty} \frac{\log |x(T)|}{T} \leq 0 \quad \text{a.s.} \tag{7.64}$$

7.6 Notes

The framework of this chapter is based on the paper of Khasminskii, Zhu, and Yin [92]. Some new results are included, e.g., Theorems 7.9 and 7.10, where we derived Kolmogorov-type backward equations without assuming nondegeneracy of the diffusion part. These results are interesting in their own right.

For brevity, we are not trying to cover every angle of the stability analysis. For example, sufficient conditions for exponential p-stability can be derived similar to that of [116], and as a result the verbatim proof for the sufficiency is omitted. Nevertheless, necessary conditions for stability and stability under linearization are provided. Finally, we note that using linearization to infer the stability of the associated nonlinear systems should be interesting owing to its wide range of applications.

8
Stability of Switching ODEs

8.1 Introduction

The main motivational forces for this chapter are the work of Davis [30] on piecewise deterministic systems, and the work of Kac and Krasovskii [79] on stability of randomly switched systems. In recent years, growing attention has been drawn to deterministic dynamic systems formulated as differential equations modulated by a random switching process. This is because of the increasing demands for modeling large-scale and complex systems, designing optimal controls, and carrying out optimization tasks. In this chapter, we consider stability of such hybrid systems modulated by a random-switching process, which are "equivalent" to a number of ordinary differential equations coupled by a switching or jump process.

In this chapter, for random-switching systems, we first obtain sufficient conditions for stability and instability. Our approach leads to a necessary and sufficient condition for systems whose continuous component is one-dimensional. For multidimensional systems, our conditions involve the use of minimal and maximal eigenvalues of appropriate matrices. The difference of maximal and minimal eigenvalues results in a gap for stability and instability. To close this gap, we introduce a logarithm transformation leading to the continuous component taking values in the unit sphere. This in turn, enables us to obtain necessary and sufficient conditions for stability. The essence is the utilization of the so-called Liapunov exponent.

Because the systems we are interested in have continuous components (representing continuous dynamics) as well as discrete components (repre-

G.G. Yin and C. Zhu, *Hybrid Switching Diffusions: Properties and Applications*,
Stochastic Modelling and Applied Probability 63, DOI 10.1007/978-1-4419-1105-6_8,
© Springer Science + Business Media, LLC 2010

senting discrete events), their asymptotic behavior can be quite different from a single system of differential equations. As noted, a random-switching differential system may be considered as several differential equations coupled by a switching process. We show that even though some of the individual equations are not stable, the entire switching system may still be stable as long as certain conditions are satisfied.

For random switching systems that are linear in their continuous component, suppose that corresponding to different discrete states, some of the associated differential equations are stable and the others are not. If the jump component is ergodic, we show that as long as the stable part of the differential equations dominates the rest (in an appropriate sense), the coupled hybrid system will be stable. For nonlinear differential equations, the well-known Hartman–Grobman theorem (see [134, Section 2.8]) provides an important result concerning the local qualitative behavior. It says that near a hyperbolic equilibrium point x_0, the nonlinear system $\dot{x} = f(x)$ has the same qualitative structure as that of the linear system $\dot{x} = \nabla f(x_0)x$, although the topological equivalence may not hold for a non-hyperbolic equilibrium point (e.g., a center). Treating hybrid systems, consider the differential equations $\dot{x}(t) = f(x(t), \alpha(t))$ and $\dot{x}(t) = \nabla f(x_0, \alpha(t))x(t)$ for $\alpha(t)$ belonging in a finite set. We show although some of the linear equations have centers, as long as the spectrum of the coefficients of the differential equation corresponding to the stable node dominates that of the centers, the overall system will still be topologically equivalent to the linear (in continuous component) system. To reveal the salient features, we present a number of examples, and display the corresponding phase portraits. The results are quite revealing.

The rest of the chapter is arranged as follows. Section 8.2 begins with the formulation of the random-switching systems and provides definitions of stability, instability, and asymptotical stability of the equilibrium point of the random-switching hybrid systems and gives some preliminary results. For the purpose of our asymptotic analysis, we also present sufficient conditions for stability, instability, and asymptotical stability. Easily verifiable conditions for stability and instability of the systems are provided in Section 8.3. Section 8.4 presents a sharper condition for systems that are linear in the continuous state variable. Discussions on Liapunov exponent are given in Section 8.5. To demonstrate our results, we provide several examples in Section 8.6. Finally, we conclude the chapter with further remarks in Section 8.7.

8.2 Formulation and Preliminary Results

8.2.1 Problem Setup

Throughout the chapter, we use z' to denote the transpose of $z \in \mathbb{R}^{\ell_1 \times \ell_2}$ with $\ell_i \geq 1$, whereas $\mathbb{R}^{\ell \times 1}$ is simply written as $\mathbb{R}^\ell$; $\mathbb{1} = (1, 1, \ldots, 1)' \in \mathbb{R}^{m_0}$ is a column vector with all entries being 1; the Euclidean norm for a row or column vector x is denoted by $|x|$. As usual, I denotes the identity matrix. For a matrix A, its trace norm is denoted by $|A| = \sqrt{\mathrm{tr}(A'A)}$. When B is a set, $I_B(\cdot)$ denotes the indicator function of B. For $A \in \mathbb{R}^{r \times r}$ being a symmetric matrix, we use $\lambda_{\max}(A)$ and $\lambda_{\min}(A)$ to denote the maximum and minimum eigenvalues of A, respectively.

Consider the system with random switching

$$\dot{X}(t) = f(X(t), \alpha(t)), \quad X(0) = x \in \mathbb{R}^r, \quad \alpha(0) = \alpha \in \mathcal{M}, \qquad (8.1)$$

where $X(t)$ is the continuous state, $f(\cdot, \cdot) : \mathbb{R}^r \times \mathcal{M} \mapsto \mathbb{R}^r$, and $\alpha(\cdot)$ is a jump process taking value in a finite state space $\mathcal{M} = \{1, 2, \ldots, m_0\}$ with generator $Q(x) = q_{ij}(x)$ satisfying $q_{ij}(x) \geq 0$ for $j \neq i$ and $\sum_{j \in \mathcal{M}} q_{ij}(x) = 0$ for all $x \in \mathbb{R}^r$ and $i \in \mathcal{M}$. The evolution of the jump component is described by

$$\begin{aligned} \mathbf{P}\{\alpha(t + \Delta) &= j | \alpha(t) = i, (X(s), \alpha(s)), s \leq t\} \\ &= q_{ij}(X(t))\Delta + o(\Delta), \quad i \neq j. \end{aligned} \qquad (8.2)$$

Note that in our formulation, x-dependent $Q(x)$ is considered, whereas in [6, 79, 116, 183], the constant generator Q was used.

Associated with the process $(X(t), \alpha(t))$ defined by (8.1)–(8.2), there is an operator $\mathcal{L}$ defined as follows. For each $i \in \mathcal{M}$ and any $g(\cdot, i) \in C^1(\mathbb{R}^r)$,

$$\mathcal{L}g(x, i) = f'(x, i)\nabla g(x, i) + Q(x)g(x, \cdot)(i), \qquad (8.3)$$

where $\nabla g(x, i)$ denotes the gradient (with respect to the variable x) of $g(x, i)$, and

$$Q(x)g(x, \cdot)(i) = \sum_{j \in \mathcal{M}} q_{ij}(x)g(x, j) \quad \text{for each } i \in \mathcal{M}. \qquad (8.4)$$

For further references on the associated operator (or generator) of the hybrid system (8.1)–(8.2), we refer the reader to Chapter 2 of this book; see also [79] and [150].

To proceed, we need conditions regarding the smoothness and growth of the functions involved, and the condition that 0 is an equilibrium point of the dynamic system. We assume the following hypotheses throughout the chapter.

(A8.1) The matrix-valued function $Q(\cdot)$ is bounded and continuous.

(A8.2) For each $\alpha \in \mathcal{M}$, $f(\cdot, \alpha)$ is locally Lipschitz continuous and satisfies $f(0, \alpha) = 0$.

(A8.3) There exists a constant $K_0 > 0$ such that for each $\alpha \in \mathcal{M}$,

$$|f(x, \alpha)| \leq K_0(1 + |x|), \quad \text{for all } x \in \mathbb{R}^r. \tag{8.5}$$

It is well known that under these conditions, system (8.1)–(8.2) has a unique solution; see [150] for details. In what follows, a process starting from (x, α) is denoted by $Y^{x,\alpha}(t) = (X^{x,\alpha}(t), \alpha^{x,\alpha}(t))$ to emphasize the dependence on the initial condition. If the context is clear, we simply write $Y(t) = (X(t), \alpha(t))$.

8.2.2 Preliminary Results

In this subsection, we first recall the definitions of stability, instability, asymptotic stability, and exponential p-stability. Then we present some preparatory results of stability and instability in terms of Liapunov functions.

Definition 8.1. ([79]) The equilibrium point $x = 0$ of system (8.1)–(8.2) is said to be

(i) *stable in probability*, if for any $\alpha = 1, \ldots, m_0$ and $r_0 > 0$,

$$\lim_{x \to 0} \mathbf{P}\{\sup_{t \geq 0} |X^{x,\alpha}(t)| > r_0\} = 0,$$

and it is said to be *unstable in probability* if it is not stable in probability;

(ii) *asymptotically stable in probability*, if it is stable in probability and

$$\lim_{x \to 0} \mathbf{P}\{\lim_{t \to \infty} X^{x,\alpha}(t) = 0\} = 1, \quad \text{for each } \alpha = 1, \ldots, m_0;$$

(iii) *exponentially p-stable*, if for some positive constants K and γ,

$$\mathbf{E}|X^{x,\alpha}(t)|^p \leq K|x|^p \exp\{-\gamma t\}, \quad \text{for any } (x, \alpha) \in \mathbb{R}^r \times \mathcal{M}.$$

The definitions above should be compared to Definition 7.4. They are of the same spirit although now we have a completely degenerate case with the diffusion matrix being identically 0. To study stability of the equilibrium point $x = 0$, we first observe that almost all trajectories of the system (8.1)–(8.2) starting from a nonzero state will never reach the origin with probability one.

Proposition 8.2. *Let conditions* (A8.1)–(A8.3) *be satisfied. Then*

$$\mathbf{P}\{X^{x,\alpha}(t) \neq 0, t \geq 0\} = 1, \quad \text{for any } (x, \alpha) \in \mathbb{R}^r \times \mathcal{M} \text{ with } x \neq 0. \tag{8.6}$$

Proof. This proposition can be proved using a slight modification of the argument in Lemma 7.1. The details are omitted for brevity. □

In view of (8.6), we can work with functions $V(\cdot, i), i \in \mathcal{M}$, which are continuously differentiable in a deleted neighborhood of 0 in what follows. This turns out to be quite convenient. Another immediate consequence of (8.6) is the following L^p estimate for the solution of the system (8.1)–(8.2). The result is interesting in its own right.

Theorem 8.3. *Let conditions* (A8.1)–(A8.3) *be satisfied. Then for any* $p \geq 1$ *and any* $(x, \alpha) \in \mathbb{R}^r \times \mathcal{M}$ *with* $x \neq 0$, *we have*

$$
\begin{aligned}
\mathbf{E}|X^{x,\alpha}(t)| &\leq \left(|x|^p + \frac{1}{2}\right) \exp(2pm_0 K_0 t) - \frac{1}{2} \\
&\leq \left(|x|^p + \frac{1}{2}\right) \exp(2pm_0 K_0 t).
\end{aligned}
\tag{8.7}
$$

Proof. For each $\alpha \in \mathcal{M}$, the function $V(x, \alpha) = |x|^p$ is continuously differentiable in the domain $|x| > \delta$ for any $\delta > 0$. Let τ_δ be the first exit time of the process $X^{x,\alpha}(\cdot)$ from the set $\{x \in \mathbb{R}^r : |x| > \delta\} \times \mathcal{M}$. That is,

$$
\tau_\delta := \inf\{t \geq 0 : |X^{x,\alpha}(t)| \leq \delta\}.
$$

For any $t > 0$, set $\tau_\delta(t) := \min\{\tau_\delta, t\}$. Because $V(x, i)$ is independent of i, $\sum_{j \in \mathcal{M}} q_{ij}(x)V(x, j) = 0$. Then it follows from the Cauchy–Schwartz inequality and (8.5) that

$$
\begin{aligned}
\mathbf{E}&|X^{x,\alpha}(\tau_\delta(t))|^p \\
&= |x|^p + \mathbf{E} \int_0^{\tau_\delta(t)} p |X^{x,\alpha}(s)|^{p-2} \left\langle X^{x,\alpha}(s), f(X^{x,\alpha}(s), \alpha^{x,\alpha}(s)) \right\rangle ds \\
&\leq |x|^p + pm_0 K_0 \mathbf{E} \int_0^{\tau_\delta(t)} |X^{x,\alpha}(s)|^{p-1} \left(1 + |X^{x,\alpha}(s)|\right) ds \\
&\leq |x|^p + 2pm_0 K_0 \mathbf{E} \int_0^{\tau_\delta(t)} |X^{x,\alpha}(s)|^p ds \\
&\quad + pm_0 K_0 \mathbf{E} \int_0^{\tau_\delta(t)} |X^{x,\alpha}(s)|^{p-1} I_{\{\delta \leq |X^{x,\alpha}(s)| < 1\}} ds \\
&\leq |x|^p + 2pm_0 K_0 \mathbf{E} \int_0^{\tau_\delta(t)} \left(|X^{x,\alpha}(s)|^p + \frac{1}{2}\right) ds.
\end{aligned}
$$

Note that for all $s \leq \tau_\delta(t)$, we have $s = \tau_\delta(s)$. Hence we have

$$
\begin{aligned}
\mathbf{E}|X^{x,\alpha}(\tau_\delta(t))|^p &\leq |x|^p + 2pm_0 K_0 \mathbf{E} \int_0^{\tau_\delta(t)} \left(|X^{x,\alpha}(\tau_\delta(s))|^p + \frac{1}{2}\right) ds \\
&\leq |x|^p + 2pm_0 K_0 \int_0^t \left(\mathbf{E}|X^{x,\alpha}(\tau_\delta(s))|^p + \frac{1}{2}\right) ds.
\end{aligned}
$$

Applying Gronwall's inequality to $[\mathbf{E}\,|X^{x,\alpha}(\tau_\delta(t))|^p + (1/2)]$ leads to

$$\mathbf{E}\,|X^{x,\alpha}(\tau_\delta(t))|^p + \frac{1}{2} \le \left(|x|^p + \frac{1}{2}\right)\exp(2pm_0K_0t), \qquad (8.8)$$

or equivalently,

$$\begin{aligned}
\mathbf{E}\,|X^{x,\alpha}(\tau_\delta(t))|^p &\le \left(|x|^p + \frac{1}{2}\right)\exp(2pm_0K_0t) - \frac{1}{2} \\
&\le \left(|x|^p + \frac{1}{2}\right)\exp(2pm_0K_0t).
\end{aligned} \qquad (8.9)$$

Note that we have from (8.6) that

$$\tau_\delta(t) \to t \text{ as } \delta \to 0 \text{ with probability 1 for any } t > 0.$$

Finally, letting $\delta \to 0$ in (8.9), by Fatou's lemma, we obtain (8.7).   □

Remark 8.4. If condition (8.5) is replaced by

$$|f(x,\alpha)| \le K|x|, \quad \text{for all } (x,\alpha) \in \mathbb{R}^r \times \mathcal{M}, \qquad (8.10)$$

where K is some positive constant, then the conclusion Theorem 8.3 can be strengthened to the following. For any $\beta \in \mathbb{R}$ and any $(x,\alpha) \in \mathbb{R}^r \times \mathcal{M}$ with $x \ne 0$, we have

$$\mathbf{E}|X^{x,\alpha}(t)|^\beta \le |x|^\beta e^{\rho t}, \qquad (8.11)$$

where ρ is a constant depending only on β, m_0, and the constant K given in (8.10).

In fact, by virtue of (8.10), we obtain by a slight modification of the argument in the proof of Theorem 8.3 that

$$\mathbf{E}|X^{x,\alpha}(\tau_\delta(t))|^\beta \le |x|^\beta e^{\rho t},$$

where ρ is a constant depending only on β, m_0, and the constant K in (8.10). Then similar to the proof of Theorem 8.3, (8.11) follows from (8.6) and Fatou's lemma.

We finally note that if $f(x,\alpha)$ is Lipschitzian with a global Lipschitz constant L_0, and $f(0,\alpha) = 0$, then (8.10) is verified.

Next, concerning stability and asymptotical stability of the equilibrium point $x = 0$ of the system (8.1)–(8.2), we have the following results.

Proposition 8.5. *Let $D \subset \mathbb{R}^r$ be a neighborhood of 0. Suppose that for each $i \in \mathcal{M}$, there exists a nonnegative function $V(\cdot,i) : D \mapsto \mathbb{R}$ such that*

(i) *$V(\cdot,i)$ is continuous in D and vanishes only at $x = 0$;*

(ii) *$V(\cdot,i)$ is continuously differentiable in $D - \{0\}$ and satisfies*

$$\mathcal{L}V(x,i) \le 0, \quad \text{for all } x \in D - \{0\}. \qquad (8.12)$$

Then the equilibrium point $x = 0$ is stable in probability.

Proof. Choose $r_0 > 0$ such that the ball $B_{r_0} = \{x \in \mathbb{R}^r : |x| < r_0\}$ and its boundary $\partial B_{r_0} = \{x \in \mathbb{R}^r : |x| = r_0\}$ are contained in D. Set $V_{r_0} := \inf \{V(x,i) : x \in D \setminus B_{r_0}, i \in \mathcal{M}\}$. Then $V_{r_0} > 0$ by assumption (i). Next, assumption (ii) leads to

$$\mathbf{E}_{x,i} V(X(t \wedge \tau_{r_0}), \alpha(t \wedge \tau_{r_0})) = V(x,i) + \mathbf{E}_{x,i} \int_0^{\tau_{r_0} \wedge t} \mathcal{L}V(X(s), \alpha(s)) ds$$

$$\leq V(x,i),$$

where $(x,i) \in B_{r_0} \times \mathcal{M}$ and τ_{r_0} is the first exit time from B_{r_0}; that is, $\tau_{r_0} := \{t \geq 0 : |x(t)| \geq r_0\}$. Because V is nonnegative, we further have

$$V_{r_0} \mathbf{P} \{\tau_{r_0} \leq t\} \leq \mathbf{E}_{x,i} \left[V(X(\tau_{r_0}), \alpha(\tau_{r_0})) I_{\{\tau_{r_0} \leq t\}} \right] \leq V(x,i).$$

Note that $\tau_{r_0} \leq t$ if and only if $\sup_{0 \leq u \leq t} |x(u)| > r_0$. Therefore it follows that

$$\mathbf{P}_{x,i} \left\{ \sup_{0 \leq u \leq t} |X(u)| > r_0 \right\} \leq \frac{V(x,i)}{V_{r_0}}.$$

Letting $t \to \infty$, we obtain from assumption (i) that

$$\mathbf{P}_{x,i} \left\{ \sup_{0 \leq t} |X(t)| > r_0 \right\} \leq \frac{V(x,i)}{V_{r_0}}.$$

Finally, the desired conclusion follows from the assumptions that $V(0,i) = 0$ and $V(\cdot,i)$ is continuous for each $i \in \mathcal{M}$.   □

Proposition 8.6. *Assume the conditions of Proposition 8.5. Suppose also that for each $i \in \mathcal{M}$, the function $V(\cdot,i)$ satisfies*

$$\mathcal{L}V(x,i) \leq -\kappa(\varrho) < 0, \quad \text{for all } x \in D - \{x \in \mathbb{R}^r : |x| > \varrho\}, \quad (8.13)$$

where $\varrho > 0$ and $\kappa(\varrho)$ is a positive constant. Then the equilibrium point $x = 0$ is asymptotically stable in probability.

Proof. By virtue of Proposition 8.5, the equilibrium point $x = 0$ is stable in probability. It remains to show that

$$\lim_{x \to 0} \mathbf{P}_{x,i} \left\{ \lim_{t \to \infty} X(t) = 0 \right\} = 1.$$

The equilibrium point $x = 0$ is stable in probability, therefore it follows that for any $\varepsilon > 0$ and $r_0 > 0$, there exists some $\delta > 0$ (without loss of generality, we may assume that $\delta < r_0$) such that

$$\mathbf{P}_{x,i} \left\{ \sup_{t \geq 0} |X(t)| < r_0 \right\} \geq 1 - \frac{\varepsilon}{2}, \quad \text{for any } (x,i) \in B_\delta \times \mathcal{M}, \quad (8.14)$$

where $B_\delta := \{x \in \mathbb{R}^r : |x| < \delta\}$. Now fix some $(x, \alpha) \in (B_\delta - \{0\}) \times \mathcal{M}$ and let $\varrho_1 > \varrho > 0$ be arbitrary satisfying $\varrho_1 < |x|$. Define

$$\tau_\varrho := \{t \geq 0 : |X(t)| \leq \varrho\},$$
$$\tau_{r_0} := \{t \geq 0 : |X(t)| \geq r_0\}.$$

Then it follows that for any $t > 0$,

$$\mathbf{E}_{x,\alpha} V(X(t \wedge \tau_\varrho \wedge \tau_{r_0}), \alpha(t \wedge \tau_\varrho \wedge \tau_{r_0})) - V(x, \alpha)$$
$$\leq \mathbf{E}_{x,\alpha} \int_0^{t \wedge \tau_\varrho \wedge \tau_{r_0}} \mathcal{L}V(X(s), \alpha(s)) ds$$
$$\leq -\kappa(\varrho) \mathbf{E}_{x,\alpha}[t \wedge \tau_\varrho \wedge \tau_{r_0}].$$

Because V is nonnegative, we have $\mathbf{E}_{x,\alpha}[t \wedge \tau_\varrho \wedge \tau_{r_0}] \leq V(x, \alpha)/(\kappa(\varrho))$ and hence $t\mathbf{P}_{x,\alpha}\{\tau_\varrho \wedge \tau_{r_0} > t\} \leq V(x, \alpha)/(\kappa(\varrho))$. Letting $t \to \infty$, we obtain

$$\mathbf{P}_{x,\alpha}\{\tau_\varrho \wedge \tau_{r_0} = \infty\} = 0 \quad \text{or} \quad \mathbf{P}_{x,\alpha}\{\tau_\varrho \wedge \tau_{r_0} < \infty\} = 1. \tag{8.15}$$

Note that (8.14) implies that $\mathbf{P}_{x,\alpha}\{\tau_{r_0} < \infty\} \leq \varepsilon/2$. Hence it follows that

$$\mathbf{P}_{x,\alpha}\{\tau_\varrho < \infty\} \geq \mathbf{P}_{x,\alpha}\{\tau_\varrho \wedge \tau_{r_0} < \infty\} - \mathbf{P}_{x,\alpha}\{\tau_{r_0} < \infty\}$$
$$\geq 1 - \frac{\varepsilon}{2}. \tag{8.16}$$

Now let

$$\tau_{\varrho_1} := \{t \geq \tau_\varrho : |X(t)| \geq \varrho_1\}.$$

We use the convention that $\inf \emptyset = \infty$. Then for any $t > 0$, we have

$$\mathbf{E}_{x,\alpha} V(X(t \wedge \tau_{\varrho_1}), \alpha(t \wedge \tau_{\varrho_1}))$$
$$= \mathbf{E}_{x,\alpha} V(X(t \wedge \tau_\varrho), \alpha(t \wedge \tau_\varrho)) + \mathbf{E}_{x,\alpha} \int_{t \wedge \tau_\varrho}^{t \wedge \tau_{\varrho_1}} \mathcal{L}V(X(s), \alpha(s)) ds$$
$$\leq \mathbf{E}_{x,\alpha} V(X(t \wedge \tau_\varrho), \alpha(t \wedge \tau_\varrho)).$$

$$\tag{8.17}$$

Note that $\tau_\varrho \leq \tau_{\varrho_1}$ by definition and hence $\tau_\varrho \geq t$ implies that $\tau_{\varrho_1} \geq t$. Therefore it follows that

$$\mathbf{E}_{x,\alpha} \left[I_{\{\tau_\varrho \geq t\}} V(X(\tau_\varrho \wedge t), \alpha(\tau_\varrho \wedge t)) \right]$$
$$= \mathbf{E}_{x,\alpha} \left[I_{\{\tau_\varrho \geq t\}} V(X(t), \alpha(t)) \right] \tag{8.18}$$
$$= \mathbf{E}_{x,\alpha} \left[I_{\{\tau_\varrho \geq t\}} V(X(\tau_{\varrho_1} \wedge t), \alpha(\tau_{\varrho_1} \wedge t)) \right].$$

Then we have by virtue of (8.17) and (8.18) that

$$
\begin{aligned}
\mathbf{E}_{x,\alpha}&\left[I_{\{\tau_\varrho<t\}}V(X(\tau_{\varrho_1}\wedge t),\alpha(\tau_{\varrho_1}\wedge t))\right]\\
&\leq \mathbf{E}_{x,\alpha}\left[I_{\{\tau_\varrho<t\}}V(X(\tau_\varrho\wedge t),\alpha(\tau_\varrho\wedge t))\right]\\
&= \mathbf{E}_{x,\alpha}\left[I_{\{\tau_\varrho<t\}}V(X(\tau_\varrho),\alpha(\tau_\varrho))\right]\\
&\leq \widehat{V}_\varrho,
\end{aligned}
$$

where $\widehat{V}_\varrho := \sup\{V(y,j) : |y| = \varrho, j \in \mathcal{M}\}$. Furthermore,

$$
\begin{aligned}
\widehat{V}_\varrho &\geq \mathbf{E}_{x,\alpha}\left[I_{\{\tau_\varrho<t\}}I_{\{\tau_{\varrho_1}<t\}}V(X(\tau_{\varrho_1}\wedge t),\alpha(\tau_{\varrho_1}\wedge t))\right]\\
&= \mathbf{E}_{x,\alpha}\left[I_{\{\tau_{\varrho_1}<t\}}V(X(\tau_{\varrho_1}\wedge t),\alpha(\tau_{\varrho_1}\wedge t))\right]\\
&= \mathbf{E}_{x,\alpha}\left[I_{\{\tau_{\varrho_1}<t\}}V(X(\tau_{\varrho_1}),\alpha(\tau_{\varrho_1}))\right]\\
&\geq V_{\varrho_1}\mathbf{P}_{x,\alpha}\{\tau_{\varrho_1}<t\},
\end{aligned}
$$

where $V_{\varrho_1} := \inf\{V(y,j) : |y| = \varrho_1, j \in \mathcal{M}\}$. Because for each $i \in \mathcal{M}$, $V(\cdot,i)$ vanishes only at $x = 0$, $V_{\varrho_1} > 0$. Since V is continuous, we can choose ϱ sufficiently small so that

$$
\mathbf{P}_{x,\alpha}\{\tau_{\varrho_1}<t\}\leq \frac{\widehat{V}_\varrho}{V_{\varrho_1}}\leq \frac{\varepsilon}{2}.
$$

Letting $t \to \infty$, we obtain

$$
\mathbf{P}_{x,\alpha}\{\tau_{\varrho_1}<\infty\}\leq \frac{\varepsilon}{2}. \tag{8.19}
$$

Finally, it follows from (8.16) and (8.19) that

$$
\begin{aligned}
\mathbf{P}_{x,\alpha}\{\tau_\varrho<\infty,\tau_{\varrho_1}=\infty\} &\geq \mathbf{P}_{x,\alpha}\{\tau_\varrho<\infty\}-\mathbf{P}_{x,\alpha}\{\tau_{\varrho_1}<\infty\}\\
&\geq 1-\frac{\varepsilon}{2}-\frac{\varepsilon}{2}=1-\varepsilon.
\end{aligned}
$$

This implies that

$$
\mathbf{P}_{x,\alpha}\left\{\limsup_{t\to\infty}|X(t)|\leq \varrho_1\right\}\geq 1-\varepsilon.
$$

Because $\varrho_1 > 0$ can be chosen to be arbitrarily small, we have

$$
\mathbf{P}_{x,\alpha}\left\{\lim_{t\to\infty}X(t)=0\right\}\geq 1-\varepsilon.
$$

This finishes the proof of the proposition.   □

Proposition 8.7. *Let $D \subset \mathbb{R}^r$ be a neighborhood of 0. Assume that for each $i \in \mathcal{M}$, there exists a nonnegative function $V(\cdot, i) : D \mapsto \mathbb{R}$ such that $V(\cdot, i)$ is continuously differentiable in every deleted neighborhood of 0,*

$$\mathcal{L}V(x, i) \leq -\kappa(\varepsilon) < 0, \quad \text{for all } x \in D - \{x \in \mathbb{R}^r : |x| > \varepsilon\}, \qquad (8.20)$$

where $\varepsilon > 0$ and $\kappa(\varepsilon)$ is a positive constant, and

$$\lim_{|x| \to 0} V(x, i) = \infty, \quad \text{for each } i \in \mathcal{M}. \qquad (8.21)$$

Then the equilibrium point $x = 0$ is unstable in probability.

Proof. Let r_0 be a positive real number so that $B_{r_0} := \{x \in \mathbb{R}^r : |x| < r_0\}$ and its boundary $\partial B_{r_0} := \{x \in \mathbb{R}^r : |x| = r_0\}$ are contained in D. Fix some $(x, \alpha) \in B_{r_0} \times \mathcal{M}$ and let $0 < \varepsilon < |x|$. Then for any $t > 0$, we have

$$\begin{aligned}
\mathbf{E}_{x,\alpha} &V(X(t \wedge \tau_\varepsilon \wedge \tau_{r_0}), \alpha(t \wedge \tau_\varepsilon \wedge \tau_{r_0})) \\
&= V(x, \alpha) + \mathbf{E}_{x,\alpha} \int_0^{t \wedge \tau_\varepsilon \wedge \tau_{r_0}} \mathcal{L}V(X(s), \alpha(s)) ds \qquad (8.22) \\
&\leq V(x, \alpha).
\end{aligned}$$

As in the proof of Proposition 8.6, (8.20) implies that

$$\mathbf{P}_{x,\alpha} \{\tau_\varepsilon \wedge \tau_{r_0} < \infty\} = 1.$$

Hence letting $t \to \infty$ in (8.22), we obtain by virtue of Fatou's lemma that

$$\mathbf{E}_{x,\alpha} V(X(\tau_\varepsilon \wedge \tau_{r_0}), \alpha(\tau_\varepsilon \wedge \tau_{r_0})) \leq V(x, \alpha).$$

Furthermore, because V is nonnegative, we have

$$\begin{aligned}
V(x, \alpha) &\geq \mathbf{E}_{x,\alpha} \left[V(X(\tau_\varepsilon), \alpha(\tau_\varepsilon)) I_{\{\tau_\varepsilon < \tau_{r_0}\}} \right] \\
&\geq \inf \{V(y, j) : |y| = \varepsilon, j \in \mathcal{M}\} \mathbf{P}_{x,\alpha} \{\tau_\varepsilon < \tau_r\} \\
&= V_\varepsilon \mathbf{P}_{x,\alpha} \left\{ \sup_{0 \leq t \leq \tau_\varepsilon} |X(t)| < r_0 \right\},
\end{aligned}$$

where $V_\varepsilon = \inf \{V(y, j) : |y| = \varepsilon, j \in \mathcal{M}\}$. By virtue of Proposition 8.2, $\tau_\varepsilon \to \infty$ a.s. as $\varepsilon \to 0$. Also, it follows from (8.21) that $V_\varepsilon \to \infty$ as $\varepsilon \to 0$. Therefore it follows that as $\varepsilon \to 0$, we have

$$\mathbf{P}_{x,\alpha} \left\{ \sup_{t \geq 0} |X(t)| < r_0 \right\} = 0.$$

This demonstrates that the equilibrium point $x = 0$ is unstable in probability. $\qquad \square$

8.3 Stability and Instability: Sufficient Conditions

In the previous section, we obtained sufficient conditions for stability, instability, and asymptotic stability, using a Liapunov function argument. Because the results are based on the existence of Liapunov functions, to apply them, it is necessary to find appropriate Liapunov functions. Nevertheless, finding suitable Liapunov functions is more often than not a very challenging task. In many applications, it is often more convenient to be able to analyze the stability through conditions on the coefficients of the corresponding stochastic differential equations. Thus in this section we continue our study by providing easily verifiable conditions on the coefficients of the system (8.1)–(8.2).

Note that if the generator Q is irreducible, then all the states of the Markov chain belong to the same ergodic class. For multiple ergodic class cases, one may use the idea of two-time-scale formulation and singular perturbation methods as in [176] and [179]. To proceed, we assume the following condition holds throughout the rest of the section.

(A8.4) For each $i \in \mathcal{M}$, there exist $A_i \in \mathbb{R}^{r \times r}$ and $\widehat{Q} = (\widehat{q}_{ij}) \in \mathbb{R}^{m_0 \times m_0}$, a generator of a continuous-time Markov chain $\widehat{\alpha}(t)$ such that as $x \to 0$,

$$
\begin{aligned}
f(x, i) &= A_i x + o(|x|), \\
Q(x) &= \widehat{Q} + o(1).
\end{aligned}
\tag{8.23}
$$

Moreover, $\widehat{Q}$ is irreducible. Denote the unique stationary distribution of the associated Markov chain $\widehat{\alpha}(t)$ by

$$
\pi = (\pi_1, \pi_2, \ldots, \pi_{m_0}) \in \mathbb{R}^{1 \times m_0}.
$$

Theorem 8.8. *Assume conditions (A8.1)–(A8.4). Then the following assertions hold.*

(i) *If there exists a symmetric and positive definite matrix G such that*

$$
\sum_{i=1}^{m_0} \pi_i \lambda_{\max}(GA_iG^{-1} + G^{-1}A_i'G) < 0,
\tag{8.24}
$$

then the equilibrium point $x = 0$ of the system (8.1)–(8.2) is asymptotically stable in probability.

(ii) *If there exists a symmetric and positive definite matrix G such that*

$$
\sum_{i=1}^{m_0} \pi_i \lambda_{\min}(GA_iG^{-1} + G^{-1}A_i'G) > 0,
\tag{8.25}
$$

then the equilibrium point $x = 0$ of the system (8.1)– (8.2) is unstable in probability.

Proof. (a) We first prove that the equilibrium point $x = 0$ of system (8.1)–(8.2) is asymptotically stable in probability if (8.24) holds for some symmetric and positive definite matrix G. For notational simplicity, define the column vector $\mu = (\mu_1, \mu_2, \ldots, \mu_{m_0})' \in \mathbb{R}^{m_0}$ with

$$\mu_i = \frac{1}{2}\lambda_{\max}(GA_iG^{-1} + G^{-1}A_i'G),$$

where G is as in (8.24). Also let

$$\beta := -\pi\mu = -\sum_{i=1}^{m_0} \pi_i\mu_i.$$

Note that $\beta > 0$ by (8.24). By virtue of condition (A8.4) and Lemma A.12, the equation

$$\widehat{Q}c = \mu + \beta\mathbb{1}$$

has a solution $c = (c_1, c_2, \ldots, c_{m_0})' \in \mathbb{R}^{m_0}$. Thus we have

$$\mu_i - \sum_{j=1}^{m_0} \widehat{q}_{ij}c_j = -\beta, \quad i \in \mathcal{M}. \tag{8.26}$$

For each $i \in \mathcal{M}$, consider the Liapunov function

$$V(x, i) = (1 - \gamma c_i)(x'G^2x)^{\gamma/2},$$

where $0 < \gamma < 1$ is sufficiently small so that $1 - \gamma c_i > 0$ for each $i \in \mathcal{M}$. It is readily seen that for each $i \in \mathcal{M}$, $V(\cdot, i)$ is continuous, nonnegative, and vanishes only at $x = 0$. In addition, since $\gamma > 0$ and $1 - \gamma c_i > 0$, we have

$$\lim_{|x|\to\infty} V(x, i) \geq \lim_{|x|\to\infty} (1 - \gamma c_i)(\lambda_{\min}(G^2))^{\gamma/2}|x|^\gamma = \infty. \tag{8.27}$$

Detailed calculation reveals that for $x \neq 0$, we have

$$\nabla V(x, i) = (1 - \gamma c_i)\gamma(x'G^2x)^{\gamma/2-1}G^2x.$$

It yields that for $x \neq 0$,

$$\begin{aligned}
\mathcal{L}V(x, i) &= (1 - \gamma c_i)\gamma(x'G^2x)^{\gamma/2-1}x'G^2(A_ix + o(|x|)) \\
&\quad - \sum_{j\neq i} q_{ij}(x)(x'G^2x)^{\gamma/2}\gamma(c_j - c_i) \\
&= (1 - \gamma c_i)\gamma(x'G^2x)^{\gamma/2}\left[\frac{x'G^2A_ix}{x'G^2x} + o(1) - \sum_{j\neq i} q_{ij}(x)\frac{c_j - c_i}{1 - \gamma c_i}\right].
\end{aligned} \tag{8.28}$$

It follows from condition (A8.4) that for sufficiently small $|x|$,

$$
\begin{aligned}
\sum_{j\neq i} q_{ij}(x)\frac{c_j - c_i}{1 - \gamma c_i} & \\
= \sum_{j=1}^{m_0} q_{ij}(x)c_j + \sum_{j\neq i} q_{ij}(x)\frac{c_i(c_j - c_i)}{1 - \gamma c_i}\gamma & \\
= \sum_{j=1}^{m_0} \hat{q}_{ij}c_j + O(\gamma) + o(1), &
\end{aligned}
\tag{8.29}
$$

where $o(1) \to 0$ as $|x| \to 0$ and $O(\gamma) \to 0$ as $\gamma \to 0$. Meanwhile, using the transformation $y = Gx$ we have

$$
\begin{aligned}
\frac{x'G^2 A_i x}{x'G^2 x} &= \frac{x'(G^2 A_i + A_i'G^2)x}{2x'G^2 x} = \frac{y'G^{-1}(G^2 A_i + A_i'G^2)G^{-1}y}{2y'y} \\
&\leq \frac{1}{2}\lambda_{\max}(G^{-1}(G^2 A_i + A_i'G^2)G^{-1}) \\
&= \frac{1}{2}\lambda_{\max}(GA_i G^{-1} + G^{-1}A_i'G) = \mu_i.
\end{aligned}
\tag{8.30}
$$

Moreover, note that

$$
(\lambda_{\min}(G^2))^{\gamma/2}|x|^\gamma \leq (x'G^2 x)^{\gamma/2} \leq (\lambda_{\max}(G^2))^{\gamma/2}|x|^\gamma.
\tag{8.31}
$$

When $|x| < \delta$ with δ and $0 < \gamma < 1$ sufficiently small, (8.28)–(8.31) lead to

$$
\mathcal{L}V(x, i) \leq \gamma(1 - \gamma c_i)(\lambda_{\min}(G^2))^{\gamma/2}|x|^\gamma \left\{ \mu_i - \sum_{j=1}^{m_0} \hat{q}_{ij}c_j + o(1) + O(\gamma) \right\}.
$$

Furthermore, by virtue of (8.26), we have

$$
\begin{aligned}
\mathcal{L}V(x, i) &\leq \gamma(1 - \gamma c_i)(\lambda_{\min}(G^2))^{\gamma/2}|x|^\gamma(-\beta + o(1) + O(\gamma)) \\
&\leq -\kappa(\varepsilon) < 0,
\end{aligned}
$$

for any $(x, i) \in N_0 \times \mathcal{M}$ with $|x| > \varepsilon$, where $N_0 \subset \mathbb{R}^r$ is a small neighborhood of 0 and $\kappa(\varepsilon)$ is a positive constant. Therefore we conclude from Proposition 8.6 that the equilibrium point $x = 0$ is asymptotically stable in probability.

(b) Now we prove that the equilibrium point $x = 0$ is unstable in probability if (8.25) holds for some symmetric and positive definite matrix G. Define the column vector $\theta = (\theta_1, \theta_2, \ldots, \theta_{m_0})' \in \mathbb{R}^{m_0}$ by

$$
\theta_i := \frac{1}{2}\lambda_{\min}(GA_i G^{-1} + G^{-1}A_i'G),
$$

and set $\delta = -\pi\theta$. Note that

$$
\delta = -\sum_{i=1}^{m_0} \pi_i \theta_i < 0.
$$

As in part (a), assumption (A8.4), the definition of δ, and Lemma A.12 imply that the equation $\widehat{Q}c = \theta + \delta \mathbb{1}$ has a solution $c = (c_1, c_2, \ldots, c_{m_0})' \in \mathbb{R}^{m_0}$ and

$$\theta_i - \sum_{j=1}^{m_0} \widehat{q}_{ij} c_j = -\delta > 0, \quad i \in \mathcal{M}. \tag{8.32}$$

For $i \in \mathcal{M}$, consider the Liapunov function

$$V(x,i) = (1 - \gamma c_i)(x'G^2 x)^{\gamma/2},$$

where $-1 < \gamma < 0$ is sufficiently small so that $1 - \gamma c_i > 0$ for each $i \in \mathcal{M}$. Similar to the argument in (8.27), we can verify that $V(\cdot, i)$ satisfies (8.21) for each $i \in \mathcal{M}$. Detailed computations as in part (a) show that for any sufficiently small $0 < \varepsilon < r_0$,

$$\mathcal{L}V(x,i) \le -\kappa(\varepsilon) < 0, \quad \text{for any } (x,i) \in N_0 \times \mathcal{M} \text{ with } |x| > \varepsilon,$$

where $N_0 \subset \mathbb{R}^r$ is a small neighborhood of 0 and $\kappa(\varepsilon)$ is a positive constant. Therefore Proposition 8.7 implies that the equilibrium point $x = 0$ is unstable in probability. This completes the proof of the theorem. $\square$

Corollary 8.9. *Assume conditions (A8.1)–(A8.4). Then the following assertions hold.*

(i) *The equilibrium point $x = 0$ is asymptotically stable in probability if*

$$\sum_{i=1}^{m_0} \pi_i \lambda_{\max}(A_i + A_i') < 0. \tag{8.33}$$

(ii) *The equilibrium point $x = 0$ is unstable in probability if*

$$\sum_{i=1}^{m_0} \pi_i \lambda_{\min}(A_i + A_i') > 0. \tag{8.34}$$

Proof. This corollary follows from Theorem 8.8 immediately by choosing the symmetric and positive definite matrix G in (8.24) and (8.25) to be the identity matrix I. $\square$

Theorem 8.8 and Corollary 8.9 give sufficient conditions in terms of the maximum and minimum eigenvalues of the matrices for stability and instability of the equilibrium point $x = 0$. Because there is a "gap" between the maximum and minimum eigenvalues, a natural question arises: Can we obtain a necessary and sufficient condition for stability. If the component $X(t)$ is one-dimensional, we have the following result from Theorem 8.8, which is a necessary and sufficient condition.

Corollary 8.10. *Assume conditions* (A8.1)–(A8.4). *Let the continuous component* $X(t)$ *of the hybrid process* $(X(t), \alpha(t))$ *given by* (8.1) *and* (8.2) *be one-dimensional. Then the equilibrium point* $x = 0$ *is asymptotically stable in probability if*

$$\sum_{i=1}^{m_0} \pi_i A_i < 0,$$

and is unstable in probability if

$$\sum_{i=1}^{m_0} \pi_i A_i > 0.$$

8.4   A Sharper Result

This section deals with systems that are linear in the continuous state variable x. The system we are interested in is given by

$$\dot{X}(t) = A(\alpha(t))X(t), \tag{8.35}$$

where $A(i) = A_i \in \mathbb{R}^{r \times r}$, $X(t) \in \mathbb{R}^r$, and $\alpha(t)$ is a continuous-time Markov chain with a generator Q independent of t. Assume moreover that the Markov chain $\alpha(t)$ is irreducible. Denote the corresponding stationary distribution by $\pi = (\pi_1, \pi_2, \ldots, \pi_{m_0})$. Our main objective is to obtain a necessary and sufficient condition, or, in other words, to close the gap due to the presence of minimal and maximal eigenvalues as we mentioned in the previous section. Denote the solution of (8.35) by $X(t)$ with initial condition $(X(0), \alpha(0))$. Assume that $X(0) \neq 0$. Then according to Corollary 8.2, $X(t) \neq 0$ for any $t \geq 0$ with probability 1. As a result, $Y(t) = X(t)/|X(t)|$ is well defined and takes values on the unit sphere

$$\mathbb{S}_r = \{y \in \mathbb{R}^r : |y| = 1\}.$$

Define $z = \ln |x|$. It is readily seen that

$$\dot{z}(t) = \frac{X'(t)\dot{X}(t)}{|X(t)|^2}. \tag{8.36}$$

In what follows, we deal with a case that the Markov chain is fast varying. It acts as a "noise," whereas the $X(t)$ is slowly varying. In the end, the noise is averaged out and replaced by its stationary distribution. To put this in a mathematical form, we suppose that there is a small parameter $\varepsilon > 0$ and

$$Q = Q^\varepsilon = \frac{Q_0}{\varepsilon},$$

where Q_0 is the generator of an irreducible Markov chain. Note that in this case, $\alpha(t)$ should really be written as $\alpha^\varepsilon(t)$. In what follows, we adopt this notation.

Using (8.36), for any $T > 0$, we have

$$
\begin{aligned}
\frac{z(T) - z(0)}{T} &= \frac{1}{T} \int_0^T \frac{X'(t) A(\alpha^\varepsilon(t)) X(t)}{|X(t)|^2} dt \\
&= \frac{1}{T} \sum_{i=1}^{m_0} \int_0^T \frac{X'(t) A_i X(t)}{|X(t)|^2} [I_{\{\alpha^\varepsilon(t)=i\}} - \pi_i] dt \\
&\quad + \frac{1}{T} \sum_{i=1}^{m_0} \int_0^T \frac{X'(t) A_i X(t)}{|X(t)|^2} \pi_i dt.
\end{aligned}
\tag{8.37}
$$

For arbitrary $T > 0$, we partition the interval $[0, T]$ by use of $T^{-\Delta} > 0$ where $\Delta > 0$ is a parameter. Then we choose ε as a function of $T^{-\Delta}$. Such a choice will lead to the desired result. To proceed, choose a real number $\Delta > 1/2$. Denote $\delta_\Delta = T^{-\Delta}$ and $N = \lfloor T/\delta_\Delta \rfloor$, where $\lfloor y \rfloor$ is the usual floor function notation for a real number y. Note that $N = O(T^{1+\Delta})$ and $N\delta_\Delta = O(T)$. Next, let $0 = t_0 < t_1 < \cdots < t_N = T$ be a partition of $[0, T]$ such that $t_k = k\delta_\Delta$ for $k = 0, 1, \ldots, N$. For notational simplicity, for $i = 1, \ldots, m_0$, denote

$$
\zeta_i(t) = \frac{X'(t) A_i X(t)}{|X(t)|^2},
$$

$$
\widetilde{\zeta}_i(t) = \begin{cases} \zeta_i(0), & \text{if } 0 \le t < t_2, \\ \zeta_i(t_{k-1}), & \text{if } t_k \le t < t_{k+1}, \ k = 2, \ldots, N-1, \\ \zeta_i(t_{N-1}), & \text{if } t_N \le t \le T. \end{cases}
$$

$$
I_i^\varepsilon(t) = I_{\{\alpha^\varepsilon(t)=i\}} - \pi_i.
$$

Note that in the above, $\widetilde{\zeta}_i(t)$ is a piecewise constant approximation of $\zeta_i(t)$ with the interpolation intervals $[t_k, t_{k+1})$, $k = 0, 1, \ldots, N$.

Lemma 8.11. *Suppose that Q_0 is irreducible and that*

$$
\varepsilon = o(T^{-\Delta}) \quad \text{as } T \to \infty.
\tag{8.38}
$$

Then

$$
\frac{1}{T} \sum_{i=1}^{m_0} \int_0^T \frac{X'(t) A_i X(t)}{|X(t)|^2} [I_{\{\alpha^\varepsilon(t)=i\}} - \pi_i] dt \to 0 \quad \text{in probability as } T \to \infty.
$$

Proof. The assertion of Lemma 8.11 will follow immediately if we can show that for each $i \in \mathcal{M}$,

$$
\mathbf{E} \left| \frac{1}{T} \int_0^T \zeta_i(t) I_i^\varepsilon(t) dt \right|^2 \to 0 \quad \text{as } T \to \infty.
$$

Using the triangle inequality and the Cauchy–Schwarz inequality, we can verify that for each $i \in \mathcal{M}$,

$$
\mathbf{E} \left| \frac{1}{T} \int_0^T \zeta_i(t) I_i^\varepsilon(t) dt \right|^2
$$

$$
\leq \frac{2}{T^2} \mathbf{E} \left| \int_0^T [\zeta_i(t) - \widetilde{\zeta}_i(t)] I_i^\varepsilon(t) dt \right|^2 + \frac{2}{T^2} \mathbf{E} \left| \int_0^T \widetilde{\zeta}_i(t) I_i^\varepsilon(t) dt \right|^2 \quad (8.39)
$$

$$
\leq \frac{K}{T} \int_0^T \mathbf{E}[\zeta_i(t) - \widetilde{\zeta}_i(t)]^2 dt + \frac{2}{T^2} \mathbf{E} \left| \int_0^T \widetilde{\zeta}_i(t) I_i^\varepsilon(t) dt \right|^2.
$$

In the above and hereafter, K is used as a generic positive constant, whose values may change for different appearances. Clearly, for each $i \in \mathcal{M}$, $\zeta_i(\cdot)$ is uniformly bounded by $|A_i|$, the norm of A_i. Thus

$$
\sup_{0 \leq t \leq T} \mathbf{E} |\zeta_i(t)|^2 < \infty.
$$

Because $X(t)/|X(t)|$ takes values on the unit sphere, it is readily verified that

$$
\sup_{0 \leq t \leq T} \left| \frac{d}{dt} \frac{X'(t) A_i X(t)}{|X(t)|^2} \right| \leq K, \quad (8.40)
$$

where K is independent of T. As a result, $\zeta_i(\cdot)$ is Lipschitz continuous, uniformly on $[0, T]$. Consequently,

$$
\int_0^T \mathbf{E}[\zeta_i(t) - \widetilde{\zeta}_i(t)]^2 dt = \sum_{k=0}^{N-1} \int_{t_k}^{t_{k+1}} \mathbf{E}[\zeta_i(t) - \zeta_i(t_{k-1})]^2 dt
$$

$$
\leq K \sum_{k=0}^{N-1} \int_{t_k}^{t_{k+1}} [t - t_{k-1}]^2 dt \quad (8.41)
$$

$$
= O(N\delta_\Delta^3) = O(T\delta_\Delta^2) \to 0 \quad \text{as} \quad T \to \infty,
$$

by the choice of δ_Δ.

By means of (8.39) and (8.41), it remains to show that the last term of (8.39) converges to 0 as $T \to \infty$. For any $0 \leq t \leq T$, define

$$
h_i(t) = \mathbf{E} \left[\int_0^t \widetilde{\zeta}_i(s) I_i^\varepsilon(s) ds \right]^2. \quad (8.42)
$$

Then

$$
\frac{dh_i(t)}{dt} = 2 \int_0^t \mathbf{E}[\widetilde{\zeta}_i(s) I_i^\varepsilon(s) \widetilde{\zeta}_i(t) I_i^\varepsilon(t)] ds. \quad (8.43)
$$

For $0 \leq t \leq t_2$,

$$
\left| \int_0^t \mathbf{E}[\widetilde{\zeta}_i(s) I_i^\varepsilon(s) \widetilde{\zeta}_i(t) I_i^\varepsilon(t)] ds \right|
$$

$$
\leq \int_0^{t_2} \mathbf{E}^{1/2} |\widetilde{\zeta}_i(s)|^2 \mathbf{E}^{1/2} |\widetilde{\zeta}_i(t)|^2 ds \leq K t_2 = O(T^{-\Delta}).
$$

For $t_k \leq t < t_{k+1}$ with $k = 2, \ldots, N$, we have

$$\frac{dh_i(t)}{dt} = 2 \left(\int_0^{t_{k-1}} + \int_{t_{k-1}}^t \right) \mathbf{E} \left[\widetilde{\zeta}_i(s) I_i^\varepsilon(s) \widetilde{\zeta}_i(t) I_i^\varepsilon(t) \right] ds.$$

Furthermore,

$$\left| \int_{t_{k-1}}^t \mathbf{E}[\widetilde{\zeta}_i(s) I_i^\varepsilon(s) \widetilde{\zeta}_i(t) I_i^\varepsilon(t)] ds \right|$$
$$\leq \int_{t_{k-1}}^t \mathbf{E}^{1/2} |\widetilde{\zeta}_i(s)|^2 \mathbf{E}^{1/2} |\widetilde{\zeta}_i(t)|^2 ds$$
$$\leq K(t - t_{k-1}) = O(T^{-\Delta}).$$

It follows that

$$\frac{dh_i(t)}{dt} = 2 \int_0^{t_{k-1}} \mathbf{E}[\widetilde{\zeta}_i(s) I_i^\varepsilon(s) \widetilde{\zeta}_i(t) I_i^\varepsilon(t)] ds + O(T^{-\Delta}). \tag{8.44}$$

For $s \leq t_{k-1} \leq t < t_{k+1}$, by use of (8.38),

$$\left| \mathbf{E}[\widetilde{\zeta}_i(s) I_i^\varepsilon(s) \widetilde{\zeta}_i(t) I_i^\varepsilon(t)] \right|$$
$$= |\mathbf{E}[\widetilde{\zeta}_i(s) I_i^\varepsilon(s) \mathbf{E}(\widetilde{\zeta}_i(t) I_i^\varepsilon(t) | \mathcal{F}_{t_{k-1}})]|$$
$$= |\mathbf{E}[\widetilde{\zeta}_i(s) I_i^\varepsilon(s) \zeta_i(t_{k-1}) \mathbf{E}(I_i^\varepsilon(t) | \mathcal{F}_{t_{k-1}})]| \tag{8.45}$$
$$\leq \mathbf{E}^{1/2} |\widetilde{\zeta}_i(s)|^2 \mathbf{E}^{1/2} |\zeta_i(t_{k-1})|^2 O(\varepsilon + \exp(-(t - t_{k-1})/\varepsilon))$$
$$= O(\varepsilon + \exp(-(t_k - t_{k-1})/\varepsilon))$$
$$= O(\varepsilon + \exp(-T^{-\Delta}/\varepsilon)).$$

The next to the last inequality follows from the asymptotic expansions obtained in, for example, [176, Lemma 5.1].

Using (8.44) and (8.45), we obtain that

$$\sup_{0 \leq t \leq T} \left| \frac{dh_i(t)}{dt} \right| \leq O(T) O\left(\varepsilon + \exp\left(-\frac{T^{-\Delta}}{\varepsilon} \right) \right) \quad \text{for each} \quad i \in \mathcal{M}.$$

Because $h_i(0) = 0$, we conclude

$$|h_i(T)| \leq O(T^2) O\left(\varepsilon + \exp\left(-\frac{T^{-\Delta}}{\varepsilon} \right) \right).$$

Thus $h_i(T)/T^2 \to 0$ as $T \to \infty$ as long as $\varepsilon = o(T^{-\Delta})$. The lemma then follows. $\square$

By virtue of Lemma 8.11, we need only consider the last term in (8.37). To this end, for each $i \in \mathcal{M}$, there is a $\lambda_i \in \mathbb{R}$ such that

$$\frac{1}{T} \int_0^T \frac{X'(t) A_i X(t)}{|X(t)|^2} dt = \frac{1}{T} \int_0^T Y'(t) A_i Y(t) dt \tag{8.46}$$
$$\to \lambda_i \quad \text{in probability as} \quad T \to \infty,$$

where the existence of the limit follows from a slight modification of [61, p. 344, Theorem 6]. The number λ_i above is precisely the average of $Y'(t)A_iY(t)$, known as the mean value of $Y'(\cdot)A_iY(\cdot)$ and denoted by $M[Y'A_iY]$ in the literature of almost periodic functions [61, Appendix]. Define

$$\lambda = \sum_{i=1}^{m_0} \pi_i\lambda_i. \tag{8.47}$$

It follows from Lemma 8.11, equations (8.37), (8.46), and (8.47) that

$$\frac{1}{T}\ln\frac{|X(T)|}{|X(0)|} = \frac{1}{T}\int_0^T \frac{X'(t)A(\alpha^\varepsilon(t))X(t)}{|X(t)|^2}dt \to \lambda \quad \text{as } T \to \infty. \tag{8.48}$$

To proceed, note that (8.48) can be rewritten as

$$\frac{1}{T}\ln\frac{|X(T)|}{|X(0)|} = \lambda + o(1), \tag{8.49}$$

where $o(1) \to 0$ in probability as $T \to \infty$. If $\lambda < 0$, denote $\lambda = -\lambda_0$ with $\lambda_0 > 0$. We can make $-\lambda_0 + o(1) \le -\lambda_1$ where $0 < \lambda_1 < \lambda_0$. Thus it follows from (8.49) that

$$|X(T)| \le |X(0)|\exp(-\lambda_1 T) \to 0 \quad \text{in probability as } T \to \infty.$$

Likewise, if $\lambda > 0$, we can find $0 < \lambda_2 < \lambda$ such that $\lambda + o(1) \ge \lambda_2$, which in turn, implies that

$$|X(T)| \ge |X(0)|\exp(\lambda_2 T) \to \infty \quad \text{in probability as } T \to \infty. \tag{8.50}$$

We summarize the result obtained thus far in the following theorem.

Theorem 8.12. *The equilibrium point $x = 0$ of (8.35) is asymptotically stable in probability if $\lambda < 0$ and asymptotically unstable in probability in the sense of (8.50) if $\lambda > 0$.*

Note that the results in Theorem 8.12 present a dichotomy for the property of the equilibrium point $x = 0$ according to $\lambda < 0$ or $\lambda > 0$. One naturally asks the question: What can we say about the equilibrium point $x = 0$ if $\lambda = 0$. To answer this question, we need the following lemma.

Lemma 8.13. *If the equilibrium point $x = 0$ of (8.35) is asymptotically stable in probability, then it is exponentially p-stable for all sufficiently small positive p.*

Proof. This lemma can be established using a similar argument as that of [83, Theorem 6.4.1]; we omit the details here for brevity. $\square$

With Lemma 8.13 at our hand, we are now able to describe the local qualitative behavior of the equilibrium point $x = 0$ when $\lambda = 0$.

Theorem 8.14. *Suppose that $\lambda = 0$. Then the equilibrium point $x = 0$ is neither asymptotically stable nor asymptotically unstable in probability.*

Proof. Assume first that the equilibrium point $x = 0$ is asymptotically stable in probability. Then Lemma 8.13 implies that it is exponentially p-stable for all sufficiently small positive p. Now applying Jensen's inequality to the convex function $\varphi(x) = e^x$, from equation (8.37) and Definition 8.1 (iii), we obtain that

$$|X(0)|^p \exp\left\{ p\mathbf{E} \int_0^T \frac{X'(t)A(\alpha^\varepsilon(t))X(t)}{|X(t)|^2} dt \right\}$$
$$= |X(0)|^p \exp\left\{ p\mathbf{E}(\ln|X(T)| - \ln|X(0)|) \right\}$$
$$\leq \mathbf{E}|X(T)|^p \leq K|X(0)|^p e^{-kT}.$$

Consequently, with positive probability,

$$\lim_{T \to \infty} \frac{1}{T} \int_0^T \frac{X'(t)A(\alpha^\varepsilon(t))X(t)}{|X(t)|^2} dt < 0.$$

According to (8.48), this contradicts the assumption that $\lambda = 0$. Thus the equilibrium point $x = 0$ is not asymptotically stable. Similarly we can prove that the equilibrium point $x = 0$ is not asymptotically unstable in probability if $\lambda = 0$.   □

8.5   Remarks on Liapunov Exponent

This section is divided into two parts. In the first part, we consider a more general case with the process given by (8.35), where $\alpha(t)$ is a continuous-time Markov chain whose generator Q does not involve a small parameter $\varepsilon > 0$. Here, the result is presented by the use of the Liapunov exponent and the stationary distributions. Then, in the second part, we consider the problems of finding the stationary density of the random process.

8.5.1   Stability under General Setup

To date, a method used frequently for analyzing stability of stochastic systems is based on the work set forth in Khasminskii's book [83]. It is particularly effective in treating linear stochastic differential equations, using transformation techniques. Such an approach depends on the calculation of a limit quantity later commonly referred to as the Liapunov exponent.

Assume that the Markov chain $\alpha(t)$ is irreducible. Our main objective is to obtain a necessary and sufficient condition for stability. It follows that $Y(t) = X(t)/|X(t)|$ is well-defined and takes values on the unit sphere $\mathbb{S}_r =$

$\{y \in \mathbb{R}^r : |y| = 1\}$. Differentiating $Y(t)$ with respect to t and expressing the result in terms of y again, we obtain

$$\dot{Y}(t) = -[Y'(t)A(\alpha(t))y(t)]Y(t) + A(\alpha(t))Y(t). \tag{8.51}$$

Defining a process $z(t) = \ln|X(t)|$ as before, we obtain

$$\dot{z}(t) = Y'(t)A(\alpha(t))Y(t). \tag{8.52}$$

Thus

$$\ln|X(t)| = \ln|X(0)| + \int_0^t Y'(u)A(\alpha(u))Y(u)du. \tag{8.53}$$

In view of the discussion in Chapter 4 and the well-known results for Markov processes, denote the transition density of $(Y(t), \alpha(t))$ by $p(y, i, t)$ for $i \in \mathcal{M}$. (Note that in general, we would use $p(y_0, i_0; y, i, t)$ for each $i \in \mathcal{M}$ to denote the transition probability density function, where (y_0, i_0) denotes the initial position of the process $(Y(t), \alpha(t))$. Nevertheless, in view of (8.35), the process is time homogeneous. Using the convention in Markov processes, we can simply write the transition density by $p(y, i, t)$.) Then $p(y, i, t)$ satisfies the system of forward equations

$$\frac{\partial p(y, i, t)}{\partial t} = \mathcal{L}^* p(y, i, t)$$

$$= -\sum_{l=1}^r \frac{\partial}{\partial y_l}((A(i)y)_l p(y, i, t)) + Q'p(y, \cdot, t)(i), \quad i \in \mathcal{M},$$

where $(A(i)y)_l$ denotes the lth component of $A(i)y$ and Q' is the transpose of Q. Moreover, for any $\Gamma \subset \mathbb{R}^r$ and $i \in \mathcal{M}$,

$$\mathbf{P}(y(t) \in \Gamma, \alpha(t) = i) = \int_\Gamma p(y, i, t)dx.$$

Note that the process $(Y(t), \alpha(t))$ is on the compact set $\mathbb{S}_r \times \mathcal{M}$. The existence of the invariant density of $(Y(t), \alpha(t))$ is thus guaranteed owing to the compactness. We further assume the uniqueness of the invariant density, and denote it by $(\mu(y, i) : i = 1, \ldots, m_0)$. Then $\mu(y, i)$ satisfies

$$\mathcal{L}^* \mu(y, i) = 0, \quad \sum_{i=1}^{m_0} \int_{\mathbb{R}^r} \mu(y, i)dy = 1.$$

Under this condition, as in the proof similar to [74, Theorem 1], it can be shown that there is a $\kappa_0 > 0$ such that

$$|p(y, i, t) - \mu(y, i)| \leq K\exp(-\kappa_0 t). \tag{8.54}$$

Similar to (8.37), we obtain

$$\frac{z(T) - z(0)}{T} = \frac{1}{T}\int_0^T Y'(t)A(\alpha(t))Y(t)dt$$

$$= \frac{1}{T}\int_0^T Y'(t)A(\alpha(t))Y(t)dt.$$

Redefine λ as

$$\lambda = \sum_{i=1}^{m_0} \int_{\mathbb{R}^r} y' A(i) y \mu(y, i) dy,$$

where $\mu(y, i)$ is the stationary density of $(Y(t), \alpha(t))$. Then we have

$$
\begin{aligned}
\mathbf{E} &\left| \frac{1}{T} \int_0^T Y'(t) A(\alpha(t)) Y(t) dt - \lambda \right|^2 \\
&= \left| \mathbf{E} \frac{1}{T^2} \int_0^T \int_0^T (Y'(t) A(\alpha(t)) Y(t) - \lambda)' \right. \\
&\quad \times \left. (Y'(s) A(\alpha(s)) Y(s) - \lambda) \, dt ds \right|.
\end{aligned}
\tag{8.55}
$$

By virtue of (8.54), for $t \geq s$,

$$
\begin{aligned}
\Big| \mathbf{E} \left(Y'(t) A(\alpha(t)) Y(t) - \lambda \right)' &\left(Y'(s) A(\alpha(s)) Y(s) - \lambda \right) \Big| \\
&\leq K \exp(-\kappa_0(t - s)).
\end{aligned}
$$

This together with (8.55) then yields that

$$\mathbf{E} \left| \frac{1}{T} \int_0^T Y'(t) A(\alpha(t)) Y(t) dt - \lambda \right|^2 \leq \frac{K}{T} \to 0 \quad \text{as} \quad T \to \infty.$$

Because $z(0)/T \to 0$ as $T \to \infty$, we conclude that $z(T)/T \to \lambda$ in probability as $T \to \infty$ as desired. The rest of the arguments are the same as in the previous section. In fact, we can also obtain the limit in the sense of convergence w.p.1. That is,

$$\lim_{t \to \infty} \frac{\ln |X(t)|}{t} = \lambda = \sum_{i=1}^{m_0} \int_{\mathbb{S}_r} y' A(i) y \mu(y, i) dy \quad \text{w.p.1.}
\tag{8.56}$$

The limit λ is precisely the Liapunov exponent.

8.5.2 Invariant Density

The last section was devoted to calculation of the Liapunov exponent. In obtaining the Liapunov exponent, it is crucial to find the associated stationary density or invariant density. For ordinary differential equations with Markovian random switching, when the continuous component is two dimensional, using Liapunov exponents and the framework of [83], stability analysis was carried out in [8, 9, 112] and interesting results were obtained for special classes of problems such as certain forms of Markov modulated harmonic oscillators. Nevertheless, general results on sufficient conditions guaranteeing the existence and uniqueness of invariant density are still scarce.

In this section, we discuss the problem of finding the associated invariant densities of $(X(t), \alpha(t))$. In the literature, considerable attention has been devoted to the case that $X(t) \in \mathbb{R}^2$; see [8, 9, 112]. First, second-order differential equations with random switching are frequently used in mechanical systems to describe oscillatory motions, as well as in mathematical physics to study problems involving Schrödinger equations with random potential. Linearization of the regime-switching Liénard equations (in particular, Van der Pol equations) also leads to switching systems that are linear in the x component. In addition, for two-dimensional systems, polar coordinate transformation can be used to facilitate the study. To give better insight and for simplicity, we let $\alpha(t) \in \mathcal{M} = \{1, 2\}$. Conditions for existence and uniqueness of solutions are provided. The equations satisfied by the invariant density are nonlinear and complex. It appears that closed-form solutions are difficult to obtain.

Assume that the Markov chain $\alpha(t)$ is irreducible. To proceed, for simplicity, we consider the case that $x \in \mathbb{R}^2$ and

$$A(i) = \begin{pmatrix} a_{11}(i) & a_{12}(i) \\ a_{21}(i) & a_{22}(i) \end{pmatrix} \in \mathbb{R}^{2 \times 2}, \quad \mathcal{M} = \{1, 2\},$$

and the generator

$$Q = \begin{pmatrix} -q_1 & q_1 \\ q_2 & -q_2 \end{pmatrix} \in \mathbb{R}^{2 \times 2}.$$

For $x = (x_1, x_2)' \in \mathbb{R}^2$, it is more convenient to use coordinate transformation. Introduce the polar coordinate system by defining $\varphi = \varphi(x) = \tan^{-1}(x_2/x_1)$. Conditioning on $\alpha(t) = i$, with the use of variable φ, we obtain

$$\gamma(\varphi, i) = \frac{d\varphi}{dt} = -[\sin \varphi \cos \varphi a_{11}(i) + \sin^2 \varphi a_{12}(i)]$$
$$+ [\cos^2 \varphi a_{21}(i) + a_{22}(i) \cos \varphi \sin \varphi]. \tag{8.57}$$

To proceed, for each $i \in \mathcal{M}$, consider a Liapunov function $V(\cdot, i) : [0, 2\pi] \mapsto \mathbb{R}$, where $\mathbb{S}$ is the unit circle. It can be verified that

$$\mathcal{L}V(\varphi, i) = \frac{\partial V}{\partial \varphi} \gamma(\varphi, i) + QV(\varphi, \cdot)(i), \quad i \in \mathcal{M}. \tag{8.58}$$

We slightly change the notation and write the vector-valued function $\mu(\varphi) = (\mu_1(\varphi), \mu_2(\varphi))' \in \mathbb{R}^{2 \times 1}$ in what follows. This invariant density then is a solution of the system of forward equations

$$
\begin{cases}
-\dfrac{\partial}{\partial \varphi}\begin{pmatrix} \gamma_1(\varphi)\mu_1(\varphi) \\ \gamma_2(\varphi)\mu_2(\varphi) \end{pmatrix} + Q'\begin{pmatrix} \mu_1(\varphi) \\ \mu_2(\varphi) \end{pmatrix} = 0, \\[2mm]
\displaystyle\sum_{i=1}^{2}\int_0^{2\pi}\mu_i(\varphi)d\varphi = 1, \\[2mm]
\mu_i(\cdot) \text{ is } 2\pi \text{ periodic for } i \in \mathcal{M}.
\end{cases}
\tag{8.59}
$$

Recall that Q' denotes the transpose of Q. To ensure the existence of the solution of (8.59), we propose the following conditions.

(A8.5) For each $i \in \mathcal{M}$,

$$
[a_{22}(i) - a_{11}(i)]^2 + 4a_{12}(i)a_{21}(i) < 0.
\tag{8.60}
$$

Next, we proceed with solving the system of equations (8.59). Suppress the argument when there is no confusion. Then the first part of (8.59) leads to

$$
\begin{cases}
-\dfrac{d}{d\varphi}(\gamma_1\mu_1) - q_1\mu_1 + q_2\mu_2 = 0, \\[2mm]
-\dfrac{d}{d\varphi}(\gamma_2\mu_2) + q_1\mu_1 - q_2\mu_2 = 0.
\end{cases}
\tag{8.61}
$$

Thus, we have

$$
\frac{d}{d\varphi}(\gamma_1\mu_1 + \gamma_2\mu_2) = 0.
\tag{8.62}
$$

As a result, $\gamma_1(\varphi)\mu_1(\varphi) + \gamma_2(\varphi)\mu_2(\varphi) = c$ for all $\varphi \in [0, 2\pi]$, where c is a constant.

Let $\overline{\gamma}_i(\varphi) = \gamma_i(\varphi)\mu_i(\varphi)$ for $i = 1, 2$. Then (8.61) and (8.62) can be rewritten as

$$
\begin{cases}
\dfrac{d\overline{\gamma}_1}{d\varphi} + q_1\dfrac{\overline{\gamma}_1}{\gamma_1} - q_2\dfrac{\overline{\gamma}_2}{\gamma_2} = 0 \\[2mm]
\dfrac{d\overline{\gamma}_2}{d\varphi} - q_1\dfrac{\overline{\gamma}_1}{\gamma_1} + q_2\dfrac{\overline{\gamma}_2}{\gamma_2} = 0 \\[2mm]
\overline{\gamma}_1 + \overline{\gamma}_2 = c.
\end{cases}
\tag{8.63}
$$

Denote

$$
h(\varphi) = \exp\left(\int_0^\varphi \left(\frac{q_1}{\gamma_1(s)} + \frac{q_2}{\gamma_2(s)}\right)ds\right).
$$

By solving the above system of linear differential equations, we have

$$
\begin{cases}
\overline{\gamma}_1 = \dfrac{\displaystyle\int_0^\varphi q_2 c\dfrac{h(s)}{\gamma_2(s)}ds + \overline{c}}{h(\varphi)} \\[4mm]
\overline{\gamma}_2 = \dfrac{\displaystyle\int_0^\varphi q_1 c\dfrac{h(s)}{\gamma_1(s)}ds + c - \overline{c}}{h(\varphi)},
\end{cases}
\tag{8.64}
$$

where $\bar{c} = \bar{\gamma}_1(0)$. Note that in deriving the second equation in (8.63), we have used $\bar{\gamma}_2(0) = c - \bar{\gamma}_1(0)$. Thus,

$$
\begin{cases}
\mu_1(\varphi) = \dfrac{\displaystyle\int_0^\varphi q_2 c \dfrac{h(s)}{\gamma_2(s)} ds + \bar{c}}{h(\varphi)\gamma_1(\varphi)} \\[4ex]
\mu_2(\varphi) = \dfrac{\displaystyle\int_0^\varphi q_1 c \dfrac{h(s)}{\gamma_1(s)} ds + c - \bar{c}}{h(\varphi)\gamma_2(\varphi)}.
\end{cases}
\tag{8.65}
$$

We also know that $\mu_i(\varphi)$ must be a 2π-periodic function of φ for $i \in \mathcal{M}$, so $\mu_1(0) = \mu_1(2\pi)$. We thus have a system of linear equations for c and $\bar{c}$, namely,

$$
D \begin{pmatrix} c \\ \bar{c} \end{pmatrix} = \begin{pmatrix} 0 \\ 1 \end{pmatrix},
\tag{8.66}
$$

where $D = (d_{ij}) \in \mathbb{R}^{2\times 2}$, with

$$
d_{11} = \int_0^{2\pi} \frac{q_2 h(s)}{\gamma_2(s)} ds
$$

$$
d_{12} = 1 - h(2\pi)
$$

$$
d_{21} = \int_0^{2\pi} \frac{\gamma_2(\varphi) \displaystyle\int_0^\varphi q_2 \dfrac{h(s)}{\gamma_2(s)} ds + \gamma_1(\varphi) \displaystyle\int_0^\varphi q_1 \dfrac{h(s)}{\gamma_1(s)} ds + \gamma_1(\varphi)}{h(\varphi)\gamma_1(\varphi)\gamma_2(\varphi)} d\varphi
$$

$$
d_{22} = \int_0^{2\pi} \frac{\gamma_2(\varphi) - \gamma_1(\varphi)}{h(\varphi)\gamma_1(\varphi)\gamma_2(\varphi)} d\varphi.
$$

(A8.6) The matrix D is nonsingular, or equivalently, $\det(D) \neq 0$.

The uniqueness of (8.66) leads to the uniqueness of $\mu(y)$. Thus, we arrive at the following conclusion: Suppose conditions (A8.5) and (A8.6) hold. Then (8.59) has a unique solution.

8.6 Examples

For simplicity, by a slightly abuse of notation, we call a system that is linear with respect to the continuous state a linear system or hybrid linear system. To illustrate, we provide several examples in this section. In addition, we demonstrate results that are associated with the well-known Hartman–Grobman theorem for random switching systems.

Example 8.15. Consider a system (8.35) (linear in the variable x) with the following specifications. The Markov chain $\alpha(t)$ has two states and is generated by

$$
Q = \begin{pmatrix} -3 & 3 \\ 1 & -1 \end{pmatrix},
$$

and

$$A_1 = A(1) = \begin{pmatrix} -1 & 2 \\ 0 & 2 \end{pmatrix}, \quad A_2 = A(2) = \begin{pmatrix} -3 & -1 \\ 1 & -2 \end{pmatrix}.$$

Thus associated with the hybrid system (8.35), there are two ordinary differential equations

$$\dot{X}(t) = A_1 X(t), \tag{8.67}$$

and

$$\dot{X}(t) = A_2 X(t), \tag{8.68}$$

switching back and forth from one to another according to the movement of the jump process $\alpha(t)$. It is readily seen that the eigenvalues of A_1 are -1 and 2. Hence the motion of (8.67) is unstable. Similarly, by computing the eigenvalues of A_2, the motion of (8.68) is asymptotically stable.

Next we use Corollary 8.9 to show that the hybrid system (8.35) is asymptotically stable owing to the presence of the Markov chain $\alpha(t)$. That is, the Markov chain becomes a stabilizing factor. The stationary distribution of the Markov chain $\alpha(t)$ is $\pi = (0.25, \; 0.75)$, which is obtained by solving the system of equations $\pi Q = 0$ and $\pi 1\!\!1 = 1$. The maximum eigenvalues of $A_1 + A_1'$ and $A_2 + A_2'$ are

$$\lambda_{\max}(A_1 + A_1') = 4.6056, \quad \lambda_{\max}(A_2 + A_2') = -4,$$

respectively. This yields that

$$\pi_1 \lambda_{\max}(A_1 + A_1') + \pi_2 \lambda_{\max}(A_2 + A_2') = -1.8486 < 0.$$

Therefore, we conclude from Corollary 8.9 that the hybrid system (8.35) is asymptotically stable. The phase portrait Figure 8.1(a) confirms our findings. It is interesting to note the dynamic movements and the interactions of the continuous and discrete components. To see the difference between hybrid systems and ordinary differential equations, we also present the phase portraits of (8.67) and (8.68) in Figure 8.1(b).

Example 8.16. Consider a system (linear in the x variable)

$$\dot{X}(t) = A(\alpha(t))X(t) \tag{8.69}$$

with the following specifications. Here we consider an x-dependent generator $Q(x)$. The discrete event process $\alpha(t)$ has two states and is generated by

$$Q(x) = Q(x_1, x_2)$$
$$= \begin{pmatrix} -2 - \sin x_1 \cos x_1^2 - \sin x_2^2 & 2 + \sin x_1 \cos(x_1^2) + \sin(x_2^2) \\ 1 - 0.5 \sin(x_1 x_2) & -1 + 0.5 \sin(x_1 x_2) \end{pmatrix},$$

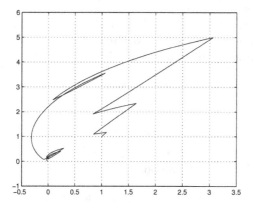

(a) Hybrid linear system: Phase portrait of (8.35) with initial condition $(x, \alpha) = ([1, 1]', 1)$.

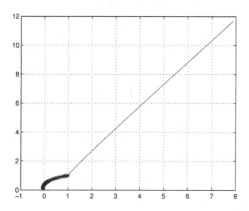

(b) Phase portraits of (8.67) (solid line) and (8.68) (starred line) with the same initial condition $x = [1, 1]'$.

FIGURE 8.1. Comparisons of switching system and the associated ordinary differential equations.

and

$$A_1 = A(1) = \begin{pmatrix} 0 & 0 \\ 3 & 0 \end{pmatrix}, \qquad A_2 = A(2) = \begin{pmatrix} -1 & 2 \\ -2 & -1 \end{pmatrix}.$$

Note that the distinct feature of this example compared with the last one is that the Q matrix is x dependent, which satisfies the approximation condition posed in Section 8.3. Associated with the hybrid system (8.35), there are two ordinary differential equations

$$\dot{X}(t) = A_1 X(t), \tag{8.70}$$

and

$$\dot{X}(t) = A_2 X(t) \tag{8.71}$$

switching from one to another according to the movement of the jump process $\alpha(t)$. Solving (8.70), we obtain

$$x_1(t) = c_1, \quad x_2(t) = 3c_1 t + c_2$$

for some constants c_1 and c_2. Hence the motion of (8.70) is unstable if the initial point (c_1, c_2) is not in the y-axis. In addition, by computing the eigenvalues of A_2, we obtain that the motion of (8.71) is asymptotically stable.

By virtue of Corollary 8.9, the hybrid system (8.69) is asymptotically stable due to the stabilizing jump process $\alpha(t)$. We first note that as $x \to 0$,

$$Q(x) \to \widehat{Q} = \begin{pmatrix} -2 & 2 \\ 1 & -1 \end{pmatrix},$$

and hence the stationary distribution of the Markov chain $\widehat{\alpha}(t)$ is $\pi = (1/3, \, 2/3)$. The maximum eigenvalues of $A_1 + A_1'$ and $A_2 + A_2'$ are

$$\lambda_{\max}(A_1 + A_1') = 3, \quad \lambda_{\max}(A_2 + A_2') = -2,$$

respectively. It follows that

$$\pi_1 \lambda_{\max}(A_1 + A_1') + \pi_2 \lambda_{\max}(A_2 + A_2') = -1/3 < 0.$$

Therefore, we conclude from Corollary 8.9 that the hybrid system (8.69) is asymptotically stable. The phase portrait in Figure 8.2(a) confirms our results. To delineate the difference between hybrid systems and ordinary differential equations, we also present the phase portraits of (8.70) and (8.71) in Figure 8.2(b). The phase portraits reveal the interface of continuous and discrete components. They illustrate the hybrid characteristics in an illuminating way.

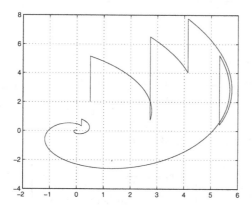

(a) Hybrid linear system: Phase portrait of (8.69) with initial condition $(x, \alpha) = ([0.5, 2]', 1)$.

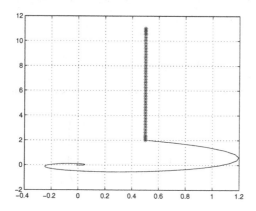

(b) Phase portraits of (8.70) (starred line) and (8.71) (solid line) with the same initial condition $x = [0.5, 2]'$.

FIGURE 8.2. Comparisons of switching system and the associated ordinary differential equations.

Example 8.17. In this example, we consider a nonlinear hybrid system

$$\dot{X}(t) = f(X(t), \alpha(t)), \qquad (8.72)$$

where $X(t)$ is a two-dimensional state trajectory, and $\alpha(t)$ is a jump process taking value in $\mathcal{M} = \{1, 2, 3\}$ with generator

$$Q(x_1, x_2) =$$

$$\begin{bmatrix} \sin x_2 - 2 - \cos x_1 & 1 - \sin x_2 & 1 + \cos x_1 \\ 1 - \frac{\sin(x_1 x_2)}{2} & \frac{\sin(x_1 x_2)}{2} - 1 - \cos^2(x_1 x_2) & \cos^2(x_1 x_2) \\ 0 & 3 + \sin x_1 + \sin x_2 \cos x_2 & -3 - \sin x_1 - \sin x_2 \cos x_2 \end{bmatrix}.$$

The functions $f(x, i), i = 1, 2, 3$ are defined as

$$f(x, 1) = \begin{bmatrix} x_1 + \frac{x_1^2}{1 + x_1^2 + x_2^2} \\ -2 x_2 + \sin(x_1 x_2) \cos x_2 \end{bmatrix},$$

$$f(x, 2) = \begin{bmatrix} -2 x_1 + x_2 - 2 x_1 \sin x_2 \\ -x_1 - x_2 + x_1 \cos x_1 \sin x_2 \end{bmatrix},$$

$$f(x, 3) = \begin{bmatrix} x_2 + 2 x_1 \cos x_1 \sin x_2 \\ -x_1 + x_2 \sin x_1 \end{bmatrix}.$$

Note that the matrices $\widehat{Q}$, and $A_i, i \in \mathcal{M}$ in assumption (A8.4) can be obtained as follows

$$\widehat{Q} = \begin{pmatrix} -3 & 1 & 2 \\ 1 & -1 & 0 \\ 0 & 3 & -3 \end{pmatrix} \qquad (8.73)$$

and

$$A_1 = \begin{pmatrix} 1 & 0 \\ 0 & -2 \end{pmatrix}, \quad A_2 = \begin{pmatrix} -2 & 1 \\ -1 & -1 \end{pmatrix}, \quad \text{and } A_3 = \begin{pmatrix} 0 & 1 \\ -1 & 0 \end{pmatrix}.$$

$$(8.74)$$

For a system of differential equations without switching, the well-known Hartman–Grobman theorem holds. It indicates that for hyperbolic equilibria, a linear system arising from approximation is topologically equivalent to the associated nonlinear system, whereas a system with a center is not. Here we demonstrate that this phenomenon changes a little. We show that the nonlinear system and its approximation could be equivalent even for equilibria with a center in one or more of its components as long as the component with hyperbolic equilibrium dominates the rest of them.

We proceed to use Theorem 8.8 and Corollary 8.9 to verify that the hybrid system (8.72) is asymptotically stable. The stationary distribution of the Markov chain $\widehat{\alpha}(t)$ is $\pi = (3/14,\ 9/14,\ 1/7)$. Now consider the symmetric and positive definite matrix

$$G = \begin{pmatrix} 3 & -1 \\ -1 & 3 \end{pmatrix}$$

and compute the maximum eigenvalues of the matrices $GA_iG^{-1}+G^{-1}A_i'G$ with $i = 1, 2, 3$,

$$\lambda_{\max}(GA_1G^{-1} + G^{-1}A_1'G) = \frac{11}{4},$$
$$\lambda_{\max}(GA_2G^{-1} + G^{-1}A_2'G) = -\frac{11}{4},$$

and

$$\lambda_{\max}(GA_3G^{-1} + G^{-1}A_3'G) = \frac{3}{2}.$$

Thus we obtain

$$\sum_{i=1}^{3} \pi_i \lambda_{\max}(GA_iG^{-1} + G^{-1}A_i'G) = -\frac{27}{28} < 0. \qquad (8.75)$$

Hence Theorem 8.8 implies that the hybrid system (8.72) is asymptotically stable.

It is interesting to note that the linear approximation

$$\dot{X}(t) = A(\widehat{\alpha}(t))X(t), \qquad (8.76)$$

is also asymptotically stable, where the matrices $A(i) = A_i, i \in \mathcal{M}$ are as in (8.74), and the Markov chain $\widehat{\alpha}(t)$ is generated by $\widehat{Q}$ in (8.73). To demonstrate, we present the phase portrait of (8.72) in Figure 8.3(a), whereas Figure 8.3(b) presents the phase portrait of its first–order linear approximation (8.76).

8.7 Notes

Consideration of switching ordinary differential equations of the form (8.1) stems from a wide variety of applications including control, optimization, estimation, and tracking. For example, in [168], with motivation of using stochastic recursive algorithms for tracking Markovian parameters such

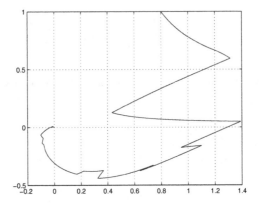

(a) Hybrid nonlinear system: Phase portrait of (8.72) with initial condition $(x, \alpha) = ([0.8, 1]', 1)$.

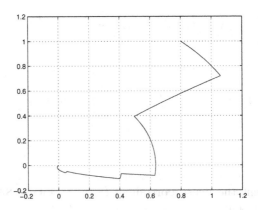

(b) Hybrid linear system: Phase portrait of (8.76) with initial condition $(x, \alpha) = ([0.8, 1]', 1)$.

FIGURE 8.3. Comparisons of nonlinear system and its linear approximation.

as those in spreading code optimization in CDMA (Code Division Multiple Access) wireless communication, we used an adaptive algorithm with constant stepsize to construct a sequence of estimates of the time-varying distribution. It was shown that under simple conditions, a continuous-time interpolation of the iteration converges weakly not to an ODE as is widely known in the literature of stochastic approximation [104], but to a system of ODEs with regime switching. Subsequently, treating least-squares-type algorithms involving Markovian jump processes in [169], random-switching ODEs were also obtained. Thus, not only is the study of systems given by (8.1) and (8.2) of mathematical interest, but also it provides practical guidance for many applications.

Taking into consideration that many real-world systems are in operation for a long period of time, longtime behavior of such systems is of foremost importance. Recently, much attention has been drawn to the study of stability of such systems; see [6, 78, 116, 183] among others. Much of the contemporary study of stochastic stability of dynamic systems can be traced back to the original work [79] by Kac and Krasovskii, in which a systematic approach was developed for stability of systems with Markovian switching using Liapunov function methods. This important work stimulated much of the subsequent development.

This chapter focuses on stability and instability of random switching systems of differential equations. Sufficient conditions have been derived. These conditions are easily verifiable and are based on coefficients of the systems. For systems that are linear in the continuous component, certain necessary and sufficient conditions are derived using a transformation technique, which close the gap in using maximal and minimal eigenvalues of certain matrices. Somewhat remarkable, a particularly interesting discovery is: Different from a single system of differential equations, in which the Hartman–Grobman theorem is in force, for switching systems of differential equations modulated by a random process, even if some of the equilibria are not hyperbolic (e.g., center), the original system and that of the "linearized" system (with respect to the continuous variable) could still have the same asymptotic behavior. This is demonstrated by our analytical results as well as computations with the use of phase portraits. The stability study is based on the results from Zhu, Yin, and Song [190]; the discussion on Liapunov exponent is based on He and Yin [65].

Owing to the increasing needs of random environment models, stability of hybrid systems has received resurgent attention lately. Effort has been placed on deriving more easily verifiable conditions for stability and instability. This work also aims to contribute in this direction. One of the main features of [79] is to use quadratic Liapunov functions to obtain verifiable conditions for stability of the system $\dot{X}(t) = A(\alpha(t))X(t)$, where $\alpha(t)$ is a continuous-time Markov chain with a constant generator Q. Their conditions amount to solving a system of linear nonhomogeneous equations, which is generally complicated. Here, we presented easily verifiable conditions for stability and instability with the aid of nonquadratic Liapunov functions. Compared with the conditions in [79], our conditions in Theorem 8.8, Corollary 8.9, and Corollary 8.10 are simpler and easier to verify. Moreover, our results can be applied to more general models.

9

Invariance Principles

9.1 Introduction

In the previous two chapters, we have studied stability of switching diffusions and random switching ordinary differential equations. Continuing our effort, this chapter is concerned with invariance principles of switching diffusion processes. This chapter together with the previous two chapters delineates long-time behavior and gives a complete picture of the switching-diffusion processes under consideration.

The rest of the chapter is arranged as follows. Section 9.2 begins with the formulation of the problem. Section 9.3 is devoted to the invariance principles using sample paths and kernels of the associated Liapunov functions. Here, Liapunov function-type criteria are obtained first. Then linear (in x) systems are treated. Section 9.4 switches gears to examine the invariance using the associated measures. Finally, a few more remarks are made in Section 9.5 to conclude the chapter.

9.2 Formulation

As in the previous chapters, we use z' to denote the transpose of $z \in \mathbb{R}^{\ell_1 \times \ell_2}$ with $\ell_i \geq 1$, whereas $\mathbb{R}^{\ell \times 1}$ is simply written as $\mathbb{R}^\ell$; $\mathbb{1} = (1, 1, \ldots, 1)' \in \mathbb{R}^{m_0}$ is a column vector with all entries being 1; the Euclidean norm for a row or a column vector x is denoted by $|x|$. As usual, I denotes the identity matrix with suitable dimension. For a matrix A, its trace norm is denoted by

G.G. Yin and C. Zhu, *Hybrid Switching Diffusions: Properties and Applications*,
Stochastic Modelling and Applied Probability 63, DOI 10.1007/978-1-4419-1105-6_9,
© Springer Science + Business Media, LLC 2010

$|A| = \sqrt{\operatorname{tr}(A'A)}$. If a matrix A is real and symmetric, we use $\lambda_{\max}(A)$ and $\lambda_{\min}(A)$ to denote the maximal and minimal eigenvalues of A, respectively, and set $\rho(A) := \max\{|\lambda_{\max}(A)|, |\lambda_{\min}(A)|\}$. When B is a set, $I_B(\cdot)$ denotes the indicator function of B.

We work with $(\Omega, \mathcal{F}, \mathbf{P})$, a complete probability space, and consider a two-component Markov process $(X(t), \alpha(t))$, where $X(\cdot)$ is a continuous component taking values in $\mathbb{R}^r$ and $\alpha(\cdot)$ is a jump component taking values in a finite set $\mathcal{M} = \{1, 2, \dots, m_0\}$. The process $(X(t), \alpha(t))$ has a generator $\mathcal{L}$ given as follows. For each $i \in \mathcal{M}$ and any twice continuously differentiable function $g(\cdot, i)$,

$$
\begin{aligned}
\mathcal{L}g(x, i) &= \frac{1}{2} \sum_{j,k=1}^{r} a_{jk}(x, i) \frac{\partial^2 g(x, i)}{\partial x_j \partial x_k} + \sum_{j=1}^{r} b_j(x, i) \frac{\partial g(x, i)}{\partial x_j} + Q(x)g(x, \cdot)(i), \\
&= \frac{1}{2} \operatorname{tr}(a(x, i)\nabla^2 g(x, i)) + b'(x, i)\nabla g(x, i) + Q(x)g(x, \cdot)(i),
\end{aligned}
\tag{9.1}
$$

where $x \in \mathbb{R}^r$, and $Q(x) = (q_{ij}(x))$ is an $m_0 \times m_0$ matrix depending on x satisfying $q_{ij}(x) \geq 0$ for $i \neq j$ and $\sum_{j \in \mathcal{M}} q_{ij}(x) = 0$ for each $i \in \mathcal{M}$,

$$
\begin{aligned}
Q(x)g(x, \cdot)(i) &= \sum_{j \in \mathcal{M}} q_{ij}(x)g(x, j) \\
&= \sum_{j \in \mathcal{M}, j \neq i} q_{ij}(x)(g(x, j) - g(x, i)), \quad i \in \mathcal{M},
\end{aligned}
$$

and $\nabla g(\cdot, i)$ and $\nabla^2 g(\cdot, i)$ denote the gradient and Hessian of $g(\cdot, i)$, respectively.

The process $(X(t), \alpha(t))$ can be described by

$$
\begin{aligned}
dX(t) &= b(X(t), \alpha(t))dt + \sigma(X(t), \alpha(t))dw(t), \\
X(0) &= x, \ \alpha(0) = \alpha,
\end{aligned}
\tag{9.2}
$$

and

$$
\begin{aligned}
\mathbf{P}\{\alpha(t + \Delta t) = j | \alpha(t) = i, (X(s), \alpha(s)), s \leq t\} \\
= q_{ij}(X(t))\Delta t + o(\Delta t), \quad i \neq j,
\end{aligned}
\tag{9.3}
$$

where $w(t)$ is a d-dimensional standard Brownian motion, $b(\cdot, \cdot) : \mathbb{R}^r \times \mathcal{M} \mapsto \mathbb{R}^r$, and $\sigma(\cdot, \cdot) : \mathbb{R}^r \times \mathcal{M} \mapsto \mathbb{R}^{r \times d}$ satisfies $\sigma(x, i)\sigma'(x, i) = a(x, i)$.

Throughout the chapter, we assume that both $b(\cdot, i)$ and $\sigma(\cdot, i)$ satisfy the usual Lipschitz condition and linear growth condition for each $i \in \mathcal{M}$ and that $Q(\cdot)$ is bounded and continuous. Under these conditions, the system (9.2)–(9.3) has a unique strong solution; see Chapter 2 and also [77] or [172] for details. From time to time, we often wish to emphasize the initial data $(X(0), \alpha(0)) = (x, \alpha)$ dependence of the solution of (9.2)–(9.3), which is denoted by $(X^{x,\alpha}(t), \alpha^{x,\alpha}(t))$.

9.3   Invariance (I): A Sample Path Approach

Recall that the evolution of the discrete component $\alpha(\cdot)$ can be represented as a stochastic integral with respect to a Poisson random measure; see Chapter 2, and also, for example, [52, 150]. Indeed, for $x \in \mathbb{R}^r$ and $i, j \in \mathcal{M}$ with $j \neq i$, let $\Delta_{ij}(x)$ be consecutive (with respect to the lexicographic ordering on $\mathcal{M} \times \mathcal{M}$), left-closed, right-open intervals of the real line, each having length $q_{ij}(x)$. Define a function $h : \mathbb{R}^r \times \mathcal{M} \times \mathbb{R} \mapsto \mathbb{R}$ by

$$h(x, i, z) = \sum_{j=1}^{m_0} (j - i) I_{\{z \in \Delta_{ij}(x)\}}. \tag{9.4}$$

Then (9.3) is equivalent to

$$d\alpha(t) = \int_{\mathbb{R}} h(X(t), \alpha(t-), z) \mathfrak{p}(dt, dz), \tag{9.5}$$

where $\mathfrak{p}(dt, dz)$ is a Poisson random measure with intensity $dt \times m(dz)$, and m is the Lebesgue measure on $\mathbb{R}$. The Poisson random measure $\mathfrak{p}(\cdot, \cdot)$ is independent of the Brownian motion $w(\cdot)$. Denote the natural filtration by $\mathcal{F}_t := \sigma\{(X(s), \alpha(s)), s \leq t\}$. Without loss of generality, assume the filtration $\{\mathcal{F}_t\}_{t \geq 0}$ satisfies the usual condition. That is, it is right continuous with $\mathcal{F}_0$ containing all $\mathbf{P}$-null sets.

For any $(x, i), (y, j) \in \mathbb{R}^r \times \mathcal{M}$, define

$$d((x, i), (y, j)) = \begin{cases} |x - y|, & \text{if } i = j, \\ |x - y| + 1, & \text{if } i \neq j. \end{cases} \tag{9.6}$$

It is easy to verify that for any $(x, i), (y, j)$, and (z, l),

(i) $d((x, i), (y, j)) \geq 0$ and $d((x, i), (y, j)) = 0$ if and only if $(x, i) = (y, j)$,

(ii) $d((x, i), (y, j)) = d((y, j), (x, i))$, and

(iii) $d((x, i), (y, j)) \leq d((x, i), (z, l)) + d((z, l), (y, j))$.

Thus d is a distance function of $\mathbb{R}^r \times \mathcal{M}$. Also if U is a subset of $\mathbb{R}^r \times \mathcal{M}$, we define

$$d((x, i), U) = \inf\{d((x, i), (y, j)) : (y, j) \in U\}. \tag{9.7}$$

Let $\mathcal{M}$ be endowed with the trivial topology. As usual, we denote by $d(x, D)$ the distance between $x \in \mathbb{R}^r$ and $D \subset \mathbb{R}^r$; that is,

$$d(x, D) = \inf\{|x - y| : y \in D\}. \tag{9.8}$$

In addition, we use $\mathbf{P}_{x,\alpha}$ and $\mathbf{E}_{x,\alpha}$ to denote the probability and expectation with $(X(0), \alpha(0)) = (x, \alpha)$, respectively. Then for a fixed $U \in \mathbb{R}^r \times \mathcal{M}$, the function $d(\cdot, U)$ is continuous.

9.3.1   Invariant Sets

Inspired by the study in [47], we define the invariant set as follows.

Definition 9.1. A Borel measurable set $U \subset \mathbb{R}^r \times \mathcal{M}$ is said to be *invariant* with respect to the solutions of (9.2)–(9.5) or simply, U is *invariant* with respect to the process $(X(t), \alpha(t))$ if

$$\mathbf{P}_{x,i}\{(X(t), \alpha(t)) \in U, \ \text{for all} \ t \geq 0\} = 1, \ \text{for any} \ (x, i) \in U.$$

That is, a process starting from U will remain in U a.s.

As shown in Lemma 7.1, when the coefficients of (9.2)–(9.5) satisfy

$$b(0, \alpha) = \sigma(0, \alpha) = 0, \ \text{for each} \ \alpha \in \mathcal{M},$$

then any solution with initial condition (x, i) satisfying $x \neq 0$ will never reach the origin almost surely, in other words, the set $(\mathbb{R}^r - \{0\}) \times \mathcal{M}$ is invariant with respect to the solutions of (9.2)–(9.5).

Using the terminologies in [47, 83], we recall the definitions of stability and asymptotic stability of a set. Then general results in terms of the Liapunov function are provided.

Definition 9.2. A closed and bounded set $K \subset \mathbb{R}^r \times \mathcal{M}$ is said to be

(i) *stable in probability* if for any $\varepsilon > 0$ and $\rho > 0$, there is a $\delta > 0$ such that

$$\mathbf{P}_{x,i}\left\{\sup_{t \geq 0} d((X(t), \alpha(t)), K) < \rho\right\} \geq 1 - \varepsilon, \ \text{whenever} \ d((x, i), K) < \delta;$$

(ii) *asymptotically stable in probability* if it is stable in probability, and moreover

$$\mathbf{P}_{x,i}\left\{\lim_{t \to \infty} d((X(t), \alpha(t)), K) = 0\right\} \to 1, \ \text{as} \ d((x, i), K) \to 0;$$

(iii) *stochastically asymptotically stable in the large* if it is stable in probability, and

$$\mathbf{P}_{x,i}\left\{\lim_{t \to \infty} d((X(t), \alpha(t)), K) = 0\right\} = 1, \ \text{for any} \ (x, i) \in \mathbb{R}^r \times \mathcal{M};$$

(iv) *asymptotically stable with probability one* if

$$\lim_{t \to \infty} d((X(t), \alpha(t)), K) = 0, \ \text{a.s.}$$

Theorem 9.3. *Assume that there exists a nonnegative function $V(\cdot,\cdot)$:*
$\mathbb{R}^r \times \mathcal{M} \mapsto \mathbb{R}_+$ *such that*

$$\mathrm{Ker}(V) := \{(x,i) \in \mathbb{R}^r \times \mathcal{M} : V(x,i) = 0\} \qquad (9.9)$$

is nonempty and bounded, and that for each $\alpha \in \mathcal{M}$, $V(\cdot,\alpha)$ is twice con-tinuously differentiable with respect to x, and

$$\mathcal{L}V(x,i) \le 0, \quad \text{for all } (x,i) \in \mathbb{R}^r \times \mathcal{M}. \qquad (9.10)$$

Then

(i) $\mathrm{Ker}(V)$ *is an invariant set for the process $(X(t), \alpha(t))$, and*

(ii) $\mathrm{Ker}(V)$ *is stable in probability.*

Proof. Let $(x_0, i_0) \in \mathrm{Ker}(V)$. By virtue of generalized Itô's lemma [150], we have for any $t \ge 0$,

$$V(X(t), \alpha(t)) = V(x_0, i_0) + \int_0^t \mathcal{L}V(X(s), \alpha(s))ds + M(t), \qquad (9.11)$$

where $M(t) = M_1(t) + M_2(t)$ is a local martingale with

$$
\begin{aligned}
M_1(t) &= \int_0^t \langle \nabla V(X(s), \alpha(s)), \sigma(X(s), \alpha(s))dw(s) \rangle, \\
M_2(t) &= \int_0^t \int_{\mathbb{R}} \big[V(X(s), i_0 + h(X(s), \alpha(s-), z)) \\
&\quad -V(X(s), \alpha(s))\big] \mu(ds, dz),
\end{aligned}
$$

where $\langle \cdot, \cdot \rangle$ denotes the usual inner product, and

$$\mu(ds, dz) = \mathfrak{p}(ds, dz) - ds \times m(dz)$$

is a martingale measure with $\mathfrak{p}(dt, dz)$ being the Poisson random measure with intensity $dt \times m(dz)$ as in (9.5). Taking expectations on both sides of (9.11) (using a sequence of stopping times and Fatou's lemma, if necessary as the argument in Theorem 3.14), it follows from (9.10) that

$$\mathbf{E}_{x_0, i_0}[V(X(t), \alpha(t))] \le V(x_0, i_0) = 0.$$

The last equality above holds because $(x_0, i_0) \in \mathrm{Ker}(V)$. But V is non-negative, so we must have $V(X(t), \alpha(t)) = 0$ a.s. for any $t \ge 0$. Then we have

$$\mathbf{P}_{x_0, i_0} \left\{ \sup_{t_n \in \mathbb{Q}^+} V(X(t_n), \alpha(t_n)) = 0 \right\} = 1,$$

where $\mathbb{Q}^+$ denotes the set of nonnegative rational numbers. Now by virtue of Proposition 2.4, the process $(X(t), \alpha(t))$ is cádlág (sample paths being right continuous and having left limits). Thus we obtain

$$\mathbf{P}_{x_0, i_0}\left\{\sup_{t \geq 0} V(X(t), \alpha(t)) = 0\right\} = 1.$$

That is,

$$\mathbf{P}_{x_0, i_0}\left\{(X(t), \alpha(t)) \in \mathrm{Ker}(V), \quad \text{for all } t \geq 0\right\} = 1.$$

This proves the first assertion of the theorem.

We proceed to prove the second assertion. For any $\delta > 0$, let U_δ be a neighborhood of $\mathrm{Ker}(V)$ such that

$$U_\delta := \{(x, i) \in \mathbb{R}^r \times \mathcal{M} : d((x, i), \mathrm{Ker}(V)) < \delta\}. \tag{9.12}$$

Let the initial condition $(x, i) \in U_\delta - \mathrm{Ker}(V)$ and τ be the first exit time of the process from U_δ. That is,

$$\tau = \inf\{t : (X(t), \alpha(t)) \notin U_\delta\}.$$

Then for any $t \geq 0$, by virtue of generalized Itô's lemma,

$$V(X(t \wedge \tau), \alpha(t \wedge \tau)) = V(x, i) + \int_0^{t \wedge \tau} \mathcal{L}V(X(s), \alpha(s))ds + M(t \wedge \tau),$$

where

$$\begin{aligned}
M(t \wedge \tau) &= \int_0^{t \wedge \tau} \left\langle \nabla V(X(s), \alpha(s)), \sigma(X(s), \alpha(s))dw(s)\right\rangle \\
&\quad + \int_0^{t \wedge \tau} \int_{\mathbb{R}} \big[V(X(s), i + h(X(s), \alpha(s-), z)) \\
&\quad - V(X(s), \alpha(s))\big]\mu(ds, dz).
\end{aligned}$$

As argued in the previous paragraph, by virtue of (9.10), we can use a sequence of stopping times and Fatou's lemma, if necessary, to obtain

$$\begin{aligned}
\mathbf{E}_{x, i}[V(X(t \wedge \tau), \alpha(t \wedge \tau))] &\leq V(x, i) + \mathbf{E}_{x, i}\int_0^{t \wedge \tau} \mathcal{L}V(X(s), \alpha(s))ds \\
&\leq V(x, i).
\end{aligned}$$

Because V is nonnegative, we further have

$$\begin{aligned}
V(x, i) &\geq \mathbf{E}_{x, i}[V(X(\tau), \alpha(\tau))I_{\{\tau < t\}}] + \mathbf{E}_{x, i}[V(X(t), \alpha(t))I_{\{t \leq \tau\}}] \\
&\geq \mathbf{E}_{x, i}[V(X(\tau), \alpha(\tau))I_{\{\tau < t\}}].
\end{aligned}$$

$$\tag{9.13}$$

For notational simplicity, denote $(\xi, \ell) = (X(\tau), \alpha(\tau))$. We claim that

$$V(\xi, \ell) > \rho, \quad \text{for some constant } \rho > 0. \tag{9.14}$$

To this end, write $\mathrm{Ker}(V) = \bigcup_{l=1}^{k}(N_{j_l} \times \{j_l\})$, where $k \leq m_0$, $N_{j_l} \subset \mathbb{R}^r$ and $j_l \in \mathcal{M}$, for $j_l = 1, \ldots, k$. We denote further $J = \{j_1, \ldots, j_k\} \subset \mathcal{M}$. Let us first consider the case when $\ell \notin J$. Note that $\xi \in D$, where D is a bounded neighborhood of $\bigcup_{l=1}^{k} N_l$ (such a neighborhood D exists because $\mathrm{Ker}(V)$ is bounded by the assumption of the theorem). Then we have

$$\inf \{V(x, \ell) : x \in D\} \geq \rho_1 > 0. \tag{9.15}$$

Suppose (9.15) were not true. Then there would exist a sequence $\{x_n\} \subset D$ such that $\lim_{n \to \infty} V(x_n, \ell) = 0$. Because $\{x_n\}$ is bounded, there exists a subsequence $\{x_{n_k}\}$ such that $x_{n_k} \to \tilde{x}$. Thus by the continuity of $V(\cdot, \ell)$, we have

$$V(\tilde{x}, \ell) = \lim_{k \to \infty} V(x_{n_k}, \ell) = 0.$$

That is, $(\tilde{x}, \ell) \in \mathrm{Ker}(V)$. This is a contradiction to the assumption that $\ell \notin J$. Thus (9.15) is true and hence $V(\xi, \ell) \geq \rho_1$.

Now let us consider the case $\ell \in J$. It follows that $\delta \leq d(\xi, N_\ell) \leq \tilde{\delta} < \infty$. A similar argument using contradiction as in the previous case shows that

$$\inf \left\{ V(x, \ell) : \delta \leq d(x, N_\ell) \leq \tilde{\delta} \right\} \geq \rho_2 > 0.$$

Thus it follows that $V(\xi, \ell) \geq \rho_2$. A combination of the two cases gives us $V(\xi, \ell) \geq \rho$, where $\rho = \rho_1 \wedge \rho_2$. Hence the claim follows.

Finally, we have from (9.13) and (9.14) that

$$\mathbf{P}_{x,i} \{\tau < t\} \leq \frac{1}{\rho} V(x, i).$$

Letting $t \to \infty$,

$$\mathbf{P}_{x,i} \{\tau < \infty\} \leq \frac{1}{\rho} V(x, i).$$

Note that

$$\{\tau < \infty\} = \left\{ \sup_{0 \leq t < \infty} d((X(t), \alpha(t)), \mathrm{Ker}(V)) \geq \delta \right\}.$$

Therefore, it follows that

$$\mathbf{P}_{x,i} \left\{ \sup_{0 \leq t < \infty} d((X(t), \alpha(t)), \mathrm{Ker}(V)) \geq \delta \right\} \leq \frac{1}{\rho} V(x, i) \to 0,$$

as $d((x, i), \mathrm{Ker}(V)) \to 0$. This finishes the proof of the theorem. $\quad\square$

Next we consider asymptotic stability. To this end, we need the following lemma.

Lemma 9.4. *Assume that there exists a nonnegative function* $V : \mathbb{R}^r \times \mathcal{M} \mapsto \mathbb{R}_+$ *with nonempty and bounded* $\text{Ker}(V)$ *such that for each* $\alpha \in \mathcal{M}$, $V(\cdot, \alpha)$ *is twice continuously differentiable with respect to* x, *and that for any* $\varepsilon > 0$,

$$\mathcal{L}V(x, i) \leq -\kappa_\varepsilon < 0, \quad \text{for any } (x, i) \in (\mathbb{R}^r \times \mathcal{M}) - \overline{U}_\varepsilon, \tag{9.16}$$

where κ_ε *is a positive constant depending on* ε, U_ε *is a neighborhood of* $\text{Ker}(V)$ *as defined in* (9.12), *and* $\overline{U}_\varepsilon$ *denotes the closure of* U_ε. *Then for any* $0 < \varepsilon < r_0$, *we have*

$$\mathbf{P}_{x,i}\{\tau_{\varepsilon,r_0} < \infty\} = 1, \quad \text{for any } (x, i) \in U_{\varepsilon,r_0},$$

where

$$U_{\varepsilon,r_0} = \{(y, j) \in \mathbb{R}^r \times \mathcal{M} : \varepsilon < d((y, j), \text{Ker}(V)) < r_0\},$$

and τ_{ε,r_0} *is the first exit time from* U_{ε,r_0}; *that is,*

$$\tau_{\varepsilon,r_0} := \inf\{t \geq 0 : (X(t), \alpha(t)) \notin U_{\varepsilon,r_0}\}.$$

Proof. Fix any $(x, i) \in U_{\varepsilon,r_0}$. By virtue of generalized Itô's lemma, we have that for any $t \geq 0$,

$$V(X(t \wedge \tau_{\varepsilon,r_0}), \alpha(t \wedge \tau_{\varepsilon,r_0}))$$
$$= V(x, i) + \int_0^{t \wedge \tau_{\varepsilon,r_0}} \mathcal{L}V(X(s), \alpha(s))ds + M(t \wedge \tau_{\varepsilon,r_0}),$$

where $M(t \wedge \tau_{\varepsilon,r_0})$ is a martingale with mean zero. Thus by taking expectations on both sides, and using (9.16), we obtain

$$\mathbf{E}_{x,i}\left[V(X(t \wedge \tau_{\varepsilon,r_0}), \alpha(t \wedge \tau_{\varepsilon,r_0}))\right] \leq V(x, i) - \mathbf{E}_{x,i} \int_0^{t \wedge \tau_{\varepsilon,r_0}} \kappa_\varepsilon ds$$
$$= V(x, i) - \kappa_\varepsilon \mathbf{E}_{x,i}[t \wedge \tau_{\varepsilon,r_0}].$$

Note that V is nonnegative; hence we have

$$\mathbf{E}_{x,i}[t \wedge \tau_{\varepsilon,r_0}] \leq \frac{1}{\kappa_\varepsilon}V(x, i).$$

But

$$\mathbf{E}_{x,i}[t \wedge \tau_{\varepsilon,r_0}] = t\mathbf{P}_{x,i}\{\tau_{\varepsilon,r_0} > t\} + \mathbf{E}_{x,i}[\tau_{\varepsilon,r_0}I_{\{\tau_{\varepsilon,r_0} \leq t\}}]$$
$$\geq t\mathbf{P}_{x,i}\{\tau_{\varepsilon,r_0} > t\}.$$

Thus it follows that

$$t\mathbf{P}_{x,i}\{\tau_{\varepsilon,r_0} > t\} \leq \frac{1}{\kappa_\varepsilon}V(x, i).$$

Now letting $t \to \infty$, we have

$$\mathbf{P}_{x,i}\left\{\tau_{\varepsilon,r_0} = \infty\right\} = 0 \quad \text{or} \quad \mathbf{P}_{x,i}\left\{\tau_{\varepsilon,r_0} < \infty\right\} = 1.$$

The assertion thus follows. $\qquad\square$

Theorem 9.5. *Assume that there exists a function V satisfying the conditions of Lemma 9.4. Then $\mathrm{Ker}(V)$ is an invariant set for the process $(X(t), \alpha(t))$ and $\mathrm{Ker}(V)$ is asymptotically stable in probability.*

Proof. Motivated by Mao and Yuan [120, Theorem 5.36], we use similar ideas and proceed as follows. By virtue of Theorem 9.3, we know that $\mathrm{Ker}(V)$ is an invariant set for the process $(X(t), \alpha(t))$ and that $\mathrm{Ker}(V)$ is stable in probability. Hence it remains to show that

$$\mathbf{P}_{x,i}\left\{\lim_{t \to \infty} d((X(t), \alpha(t)), \mathrm{Ker}(V)) = 0\right\} \to 1 \quad \text{as} \quad d((x,i), \mathrm{Ker}(V)) \to 0.$$

Because $\mathrm{Ker}(V)$ is stable in probability, for any $\varepsilon > 0$ and any $\theta > 0$, there exists some $\delta > 0$ (without loss of generality, we may assume that $\delta < \theta$) such that

$$\mathbf{P}_{x,i}\left\{\sup_{t \geq 0} d((X(t), \alpha(t)), \mathrm{Ker}(V)) < \theta\right\} \geq 1 - \frac{\varepsilon}{2}, \tag{9.17}$$

for any $(x,i) \in U_\delta$, where U_δ is defined in (9.12). Now fix any $(x, \alpha) \in U_\delta - \mathrm{Ker}(V)$ and let $\rho > 0$ be arbitrary satisfying $0 < \rho < d((x,i), \mathrm{Ker}(V))$ and choose some $\varrho \in (0, \rho)$. Define

$$\tau_\varrho := \inf\left\{t \geq 0 : d((X(t), \alpha(t)), \mathrm{Ker}(V)) \leq \varrho\right\},$$
$$\tau_\theta := \inf\left\{t \geq 0 : d((X(t), \alpha(t)), \mathrm{Ker}(V)) \geq \theta\right\}.$$

Then it follows from Lemma 9.4 that

$$\mathbf{P}_{x,\alpha}\left\{\tau_\varrho \wedge \tau_\theta < \infty\right\} = \mathbf{P}_{x,\alpha}\left\{\tau_{\varrho,\theta} < \infty\right\} = 1, \tag{9.18}$$

where $\tau_{\varrho,\theta}$ is the first exit time from $U_{\varrho,\theta}$ that is defined as

$$U_{\varrho,\theta} := \left\{(y,j) \in \mathbb{R}^r \times \mathcal{M} : \varrho < d((X(t), \alpha(t)), \mathrm{Ker}(V)) < \theta\right\}.$$

But (9.17) implies that $\mathbf{P}_{x,\alpha}\left\{\tau_\theta < \infty\right\} \leq \varepsilon/2$. Note also

$$\mathbf{P}_{x,\alpha}\left\{\tau_\varrho \wedge \tau_\theta < \infty\right\} \leq \mathbf{P}_{x,\alpha}\left\{\tau_\varrho < \infty\right\} + \mathbf{P}_{x,\alpha}\left\{\tau_\theta < \infty\right\}.$$

Thus it follows that

$$\mathbf{P}_{x,\alpha}\left\{\tau_\varrho < \infty\right\} \geq \mathbf{P}_{x,\alpha}\left\{\tau_\varrho \wedge \tau_\theta < \infty\right\} - \mathbf{P}_{x,\alpha}\left\{\tau_\theta < \infty\right\} \geq 1 - \frac{\varepsilon}{2}. \tag{9.19}$$

Now let
$$\tau_\rho := \inf\left\{t \geq \tau_\varrho : d((X(t), \alpha(t)), \mathrm{Ker}(V)) \geq \rho\right\}.$$

We have used the convention that $\inf\{\emptyset\} = \infty$. For any $t \geq 0$, we apply generalized Itô's lemma and (9.16) to obtain

$$
\begin{aligned}
\mathbf{E}_{x,\alpha} V(X(\tau_\rho \wedge t), \alpha(\tau_\rho \wedge t)) &\leq \mathbf{E}_{x,\alpha} V(X(\tau_\varrho \wedge t), \alpha(\tau_\varrho \wedge t)) \\
&\quad + \mathbf{E}_{x,\alpha} \int_{\tau_\varrho \wedge t}^{\tau_\rho \wedge t} \mathcal{L}V(X(s), \alpha(s)) ds \quad (9.20) \\
&\leq \mathbf{E}_{x,\alpha} V(X(\tau_\varrho \wedge t), \alpha(\tau_\varrho \wedge t)).
\end{aligned}
$$

Note that $\tau_\varrho \geq t$ implies $\tau_\rho \geq t$ since $\varrho \in (0, \rho)$. As a result, on the set $\{\omega \in \Omega : \tau_\varrho(\omega) \geq t\}$, we have

$$
\begin{aligned}
\mathbf{E}_{x,\alpha} V(X(\tau_\rho \wedge t), \alpha(\tau_\rho \wedge t)) \\
= \mathbf{E}_{x,\alpha} V(X(t), \alpha(t)) \quad (9.21) \\
= \mathbf{E}_{x,\alpha} V(X(\tau_\varrho \wedge t), \alpha(\tau_\varrho \wedge t)).
\end{aligned}
$$

Therefore, it follows from (9.20) and (9.21) that

$$
\begin{aligned}
\mathbf{E}_{x,\alpha} &\left[I_{\{\tau_\varrho < t\}} V(X(\tau_\rho \wedge t), \alpha(\tau_\rho \wedge t)) \right] \\
&\leq \mathbf{E}_{x,\alpha} \left[I_{\{\tau_\varrho < t\}} V(X(\tau_\varrho \wedge t), \alpha(\tau_\varrho \wedge t)) \right] \\
&= \mathbf{E}_{x,\alpha} \left[I_{\{\tau_\varrho < t\}} V(X(\tau_\varrho), \alpha(\tau_\varrho)) \right] \\
&\leq \widehat{V}_\varrho,
\end{aligned}
$$

where $\widehat{V}_\varrho := \sup\left\{V(y, j) : d((y, j), \mathrm{Ker}(V)) = \varrho\right\}$. Note that $\tau_\rho < t$ implies $\tau_\varrho < t$. Hence we further have

$$
\begin{aligned}
\widehat{V}_\varrho &\geq \mathbf{E}_{x,\alpha} \left[I_{\{\tau_\varrho < t\}} I_{\{\tau_\rho < t\}} V(X(\tau_\rho \wedge t), \alpha(\tau_\rho \wedge t)) \right] \\
&= \mathbf{E}_{x,\alpha} \left[I_{\{\tau_\rho < t\}} V(X(\tau_\rho \wedge t), \alpha(\tau_\rho \wedge t)) \right] \\
&= \mathbf{E}_{x,\alpha} \left[I_{\{\tau_\rho < t\}} V(X(\tau_\rho), \alpha(\tau_\rho)) \right] \\
&\geq V_\rho \mathbf{P}_{x,\alpha} \{\tau_\rho < t\},
\end{aligned}
$$

where $V_\rho := \inf\left\{V(y, j) : \rho \leq d((y, j), \mathrm{Ker}(V)) \leq \widetilde{\rho}\right\}$, with $\widetilde{\rho} > 0$ being some constant. Recall that we showed in the proof of Theorem 9.3 that $V_\rho > 0$. Because V is continuous, we may choose ϱ sufficiently small so that

$$\mathbf{P}_{x,\alpha} \{\tau_\rho < t\} \leq \frac{\widehat{V}_\varrho}{V_\rho} \leq \frac{\varepsilon}{2}.$$

Letting $t \to \infty$, we obtain

$$\mathbf{P}_{x,\alpha} \{\tau_\rho < \infty\} \leq \frac{\varepsilon}{2}. \tag{9.22}$$

Finally, it follows from (9.19) and (9.22) that

$$\mathbf{P}_{x,\alpha} \{\tau_\varrho < \infty, \tau_\rho = \infty\} \geq \mathbf{P}_{x,\alpha} \{\tau_\varrho < \infty\} - \mathbf{P}_{x,\alpha} \{\tau_\rho < \infty\}$$
$$\geq 1 - \frac{\varepsilon}{2} - \frac{\varepsilon}{2} = 1 - \varepsilon.$$

This implies that

$$\mathbf{P}_{x,\alpha} \left\{ \limsup_{t\to\infty} d((X(t), \alpha(t)), \mathrm{Ker}(V)) \leq \rho \right\} \geq 1 - \varepsilon.$$

But $\rho > 0$ can be chosen to be arbitrarily small. Therefore we have

$$\mathbf{P}_{x,\alpha} \left\{ \lim_{t\to\infty} d((X(t), \alpha(t)), \mathrm{Ker}(V)) = 0 \right\} \geq 1 - \varepsilon.$$

This finishes the proof of the theorem.   □

Theorem 9.6. *Assume there exists a function V satisfying the conditions of Lemma 9.4. If V also satisfies*

$$\lim_{|x|\to\infty} \inf_{\alpha\in\mathcal{M}} V(x,\alpha) = \infty, \tag{9.23}$$

Then $\mathrm{Ker}(V)$ is asymptotically stable in probability in the large; that is, $\mathrm{Ker}(V)$ is stable in probability and

$$\mathbf{P}_{x,\alpha} \left\{ \lim_{t\to\infty} d((X(t), \alpha(t)), \mathrm{Ker}(V)) = 0 \right\} = 1, \tag{9.24}$$

for any $(x,\alpha) \in \mathbb{R}^r \times \mathcal{M}$.

Proof. By virtue of Theorem 9.3, $\mathrm{Ker}(V)$ is stable in probability. Thus it remains to verify (9.24). To this end, as in the proof of Theorem 9.3, write $\mathrm{Ker}(V) = \bigcup_{l=1}^{k} (N_{j_l} \times \{j_l\})$, where $k \leq m_0$, $N_{j_l} \subset \mathbb{R}^r$, and $j_l \in \mathcal{M}$. Since $\mathrm{Ker}(V)$ is bounded by assumption, in particular, $\bigcup_{l=1}^{k} N_{j_l}$ is bounded, there exists some $R > 0$ such that

$$\sup \left\{ |y| : y \in \bigcup_{l=1}^{k} N_{j_l} \right\} \leq R. \tag{9.25}$$

Let $\varepsilon > 0$ and fix any $(x,\alpha) \in \mathbb{R}^r \times \mathcal{M}$. Then (9.23) implies that there exists some positive constant $\beta > (R+2) \vee d((x,\alpha), \mathrm{Ker}(V))$ such that

$$\inf \{V(y,j) : |y| \geq \beta, j \in \mathcal{M}\} \geq \frac{2V(x,\alpha)}{\varepsilon}. \tag{9.26}$$

Define
$$\tau_\beta := \inf\{t \geq 0 : d((X(t), \alpha(t)), \mathrm{Ker}(V)) \geq 2\beta\}.$$

For any $t \geq 0$, we have by virtue of generalized Itô's lemma and (9.16) that

$$\mathbf{E}_{x,\alpha} V(X(t \wedge \tau_\beta), \alpha(t \wedge \tau_\beta)) \leq V(x, \alpha). \tag{9.27}$$

We claim that $|X(\tau_\beta)| \geq \beta$. If this were not true, it would follow from (9.25) that for any $(y, j) \in \mathrm{Ker}(V)$,

$$d((X(\tau_\beta), \alpha(\tau_\beta)), (y, j)) \leq |X(\tau_\beta) - y| + 1$$
$$\leq |X(\tau_\beta)| + |y| + 1 < \beta + R + 1 < 2\beta - 1,$$

where in the last inequality above, we used the fact that $\beta > R + 2$. Then we have $d((X(\tau_\beta), \alpha(\tau_\beta)), \mathrm{Ker}(V)) \leq 2\beta - 1 < 2\beta$. This contradicts the definition of τ_β. Thus we must have $|X(\tau_\beta)| \geq \beta$. Then it follows from (9.27) that

$$V(x, \alpha) \geq \mathbf{E}_{x,\alpha}\left[V(X(\tau_\beta), \alpha(\tau_\beta))I_{\{\tau_\beta < t\}}\right]$$
$$\geq \inf\{V(y, j) : |y| \geq \beta, j \in \mathcal{M}\} \mathbf{P}_{x,\alpha}\{\tau_\beta < t\},$$

and hence (9.26) implies that

$$\mathbf{P}_{x,\alpha}\{\tau_\beta < t\} \leq \frac{\varepsilon}{2}.$$

By letting $t \to \infty$, we have

$$\mathbf{P}_{x,\alpha}\{\tau_\beta < \infty\} \leq \frac{\varepsilon}{2}.$$

Then we can finish the proof using the same argument in the proof of Theorem 9.5. □

The following theorem provides a criterion for asymptotic stability with probability 1.

Theorem 9.7. *If there exists a nonnegative function $V : U \mapsto \mathbb{R}_+$ such that for each $\alpha \in \mathcal{M}$, $V(\cdot, \alpha)$ is twice continuously differentiable with respect to x and that there exists a continuous function $\widehat{W} : \mathbb{R}^r \times \mathcal{M} \mapsto \mathbb{R}_+$ satisfying*

$$\mathcal{L}V(x, i) \leq -\widehat{W}(x, i), \quad \text{for any } (x, i) \in U, \tag{9.28}$$

where $U \subset \mathbb{R}^r \times \mathcal{M}$ is an invariance set for the process $(X(t), \alpha(t))$, assume also that either U is bounded or

$$\lim_{|x| \to \infty,\ (x,\alpha) \in U} V(x, \alpha) = \infty. \tag{9.29}$$

Then for any initial condition $(x, \alpha) \in \mathbb{R}^r \times \mathcal{M}$, the following assertions hold.

(i) $\limsup\limits_{t\to\infty} V(X^{x,\alpha}(t), \alpha^{x,\alpha}(t)) < \infty$ $a.s.$,

(ii) $\text{Ker}(\widehat{W}) \neq \emptyset$,

(iii) $\lim\limits_{t\to\infty} d((X^{x,\alpha}(t), \alpha^{x,\alpha}(t)), \text{Ker}(\widehat{W})) = 0$ $a.s.$, and

(iv) *if moreover,* $\text{Ker}(\widehat{W}) = \{0\} \times \mathcal{M}$, *then*

$$\lim_{t\to\infty} X^{x,\alpha}(t) = 0 \quad a.s.$$

Proof. This theorem can be proved using the arguments in [117, Theorem 2.1] although some modifications are needed. $\square$

9.3.2 Linear Systems

We end this section with the following results on linear systems. Again, by linear systems we mean systems that are linear in the x variable. Consider

$$dX(t) = b(\alpha(t))X(t)dt + \sum_{j=1}^{d} \sigma_j(\alpha(t))X(t)dw_j(t), \qquad (9.30)$$

where $b(i), \sigma_j(i)$ are $r \times r$ constant matrices and $w_j(t)$ are independent one-dimensional standard Brownian motions for $i = 1, 2, \ldots, m_0$ and $j = 1, 2, \ldots, d$.

Note that 0 is an equilibrium point for the system given by (9.30) and (9.3). As we indicated earlier, it was shown in [92, 116] that the set $\mathbb{R}^r \times \mathcal{M}$ is invariant with respect to the process $(X(t), \alpha(t))$.

Theorem 9.8. *Assume that the discrete component* $\alpha(\cdot)$ *is ergodic with constant generator* $Q = (q_{ij})$ *and invariant distribution* $\pi = (\pi_1, \ldots, \pi_{m_0}) \in \mathbb{R}^{1 \times m_0}$. *Then the equilibrium point* $x = 0$ *of the system given by (9.30) and (9.3)*

(i) *is asymptotically stable with probability one if*

$$\sum_{i=1}^{m} \pi_i \lambda_{\max}\left(b(i) + b'(i) + \sum_{j=1}^{d} \sigma_j(i)\sigma_j'(i)\right) < 0, \qquad (9.31)$$

(ii) *is unstable in probability if*

$$\sum_{i=1}^{m} \pi_i \left[\lambda_{\min}\left(b(i) + b'(i) + \sum_{j=1}^{d} \sigma_j(i)\sigma_j'(i)\right) - \frac{1}{2}\left(\rho(\sigma_j(i) + \sigma_j'(i))\right)^2\right] > 0. \qquad (9.32)$$

Proof. We need only prove assertion (i), because assertion (ii) was considered in Theorem 7.17 in Chapter 7. For notational simplicity, define the column vector

$$\mu = (\mu_1, \mu_2, \ldots, \mu_{m_0})' \in \mathbb{R}^{m_0}$$

with

$$\mu_i = \frac{1}{2}\lambda_{\max}\left(b(i) + b'(i) + \sum_{j=1}^{d}\sigma_j(i)\sigma_j'(i)\right).$$

Also let $\beta := -\pi\mu$. Note that $\beta > 0$ by (9.31). As in Chapter 7, it follows that the equation

$$Qc = \mu + \beta \mathbb{1}$$

has a solution $c = (c_1, c_2, \ldots, c_{m_0})' \in \mathbb{R}^{m_0}$. Thus we have

$$\mu_i - \sum_{j=1}^{m_0} q_{ij}c_j = -\beta, \quad i \in \mathcal{M}. \tag{9.33}$$

For each $i \in \mathcal{M}$, consider the Liapunov function

$$V(x, i) = (1 - \gamma c_i)|x|^{\gamma},$$

where $0 < \gamma < 1$ is sufficiently small so that $1 - \gamma c_i > 0$ for each $i \in \mathcal{M}$. It is readily seen that for each $i \in \mathcal{M}$, $V(\cdot, i)$ is continuous, nonnegative, vanishes only at $x = 0$, and satisfies (9.29). Detailed calculations as in the proof of Theorem 7.17 in Chapter 7 reveal that for $x \neq 0$, we have

$$\mathcal{L}V(x, i) = \gamma(1 - \gamma c_i)|x|^{\gamma}\left\{\frac{x'b(i)x}{|x|^2} - \sum_{j\neq i}q_{ij}\frac{c_j - c_i}{1 - \gamma c_i}\right.$$
$$\left. + \frac{1}{2}\sum_{j=1}^{d}\left(\frac{x'\sigma_j'(i)\sigma_j(i)x}{|x|^2} + (\gamma - 2)\frac{(x'\sigma_j'(i)x)^2}{|x|^4}\right)\right\}. \tag{9.34}$$

Note that

$$\frac{x'b(i)x}{|x|^2} + \frac{1}{2}\sum_{j=1}^{d}\frac{x'\sigma_j'(i)\sigma_j(i)x}{|x|^2}$$

$$= \frac{x'(b'(i) + b(i))x}{2|x|^2} + \frac{1}{2}\sum_{j=1}^{d}\frac{x'\sigma_j'(i)\sigma_j(i)x}{|x|^2} \tag{9.35}$$

$$\leq \frac{1}{2}\lambda_{\max}\left(b(i) + b(i)' + \sum_{j=1}^{d}\sigma_j'(i)\sigma_j(i)\right) = \mu_i.$$

Next, it follows that when γ is sufficiently small,

$$\sum_{j\neq i}q_{ij}\frac{c_j - c_i}{1 - \gamma c_i} = \sum_{j=1}^{m}q_{ij}c_j + \sum_{j\neq i}q_{ij}\frac{c_i(c_j - c_i)}{1 - \gamma c_i}\gamma = \sum_{j=1}^{m}\widehat{q}_{ij}c_j + O(\gamma). \tag{9.36}$$

Hence it follows from (9.33)–(9.36) that when $0 < \gamma < 1$ sufficiently small, we have

$$
\begin{aligned}
\mathcal{L}V(x,i) &\leq \gamma(1 - \gamma c_i)|x|^\gamma \Big\{ \mu_i - \sum_{j=1}^{m} \widehat{q}_{ij} c_j + O(\gamma) \Big\} \\
&= \gamma(1 - \gamma c_i)|x|^\gamma \Big(-\beta + O(\gamma) \Big) \\
&\leq -\frac{\beta}{2}\gamma(1 - \gamma c_i)\,|x|^\gamma := -\widehat{W}(x,i),
\end{aligned}
$$

Note that $\mathrm{Ker}(\widehat{W}) = \{0\} \times \mathcal{M}$. Thus, we conclude from Theorem 9.7 that the equilibrium point $x = 0$ is asymptotically stable with probability 1. $\square$

9.4   Invariance (II): A Measure-Theoretic Approach

This section aims to study the invariance principles by adopting the idea of Kushner [98]; see also the related work [99, 100]. The motivation stems from the study of deterministic system $\dot{x} = b(x)$. In this case, for a suitable Liapunov function $V(x)$ and a suitable function $k(x)$, let $O_\lambda = \{x : V(x) < \lambda\}$ and denote $Z = \{x \in O_\lambda : k(x) = 0\}$. In his well-known treatment of the invariance theory, LaSalle [106] showed that if $b(\cdot)$ is continuous and $V(\cdot)$ is a continuously differentiable function being nonnegative in the bounded open set O_λ satisfying $\dot{V}(x) = -k(x) \leq 0$, where $k(\cdot)$ is continuous in O_λ, then $x(t)$ tends to Z_I, the largest invariant set contained in Z as $t \to \infty$. Note that for the invariance, we often work with the time interval $(-\infty, \infty)$ rather than the usual $[0, \infty)$. As commented by Hale and Infante in [62], working on $[0, \infty)$ "at first sight to be a reasonable definition; however, this definition does not impart any special significance to the limit set of an orbit and appears unreasonable since it generally occurs that trajectories having limits can be used to define functions on $(-\infty, \infty)$."

In deterministic systems, one usually assumes the boundedness of the trajectories, whereas in the stochastic setup, this boundeness can no longer be assumed. It is replaced by certain weak boundedness (a precise definition is provided later). The stochastic counterpart of the invariance principle is based upon a set of measures that have the semigroup property. For our case, consider the two-component Markov process $Y(t) = (X(t), \alpha(t))$ in which we are interested. Roughly, the idea of invariant set from a measure-theoretic view point can be described as follows. A set of measures S is an invariant set, if for any $\psi \in S$, there is a process $Y(t)$ for $t \in (-\infty, \infty)$ with measure $\psi(t)$ so that $\psi(0) = \psi$ and $\psi(t) \in S$. Thus if $\psi \in S$, so is an entire trajectory of measures over $(-\infty, \infty)$. All of these are made more precise in what follows. The main point is that the state of the flow of the process is the measure, analogous to the state of the deterministic model. To proceed, we first state a result, whose proof is essentially in

[97]. We recall that the system is stable with respect to (O_1, O_2, ρ) (or stable relative to (O_1, O_2, ρ)) if for each $i \in \mathcal{M}$ and $x \in O_1$ implies that $\mathbf{P}_{x,i}(X(t) \in O_2 \text{ for all } t < \infty) \leq \rho$.

Proposition 9.9. *Let* $Y(t) = (X(t), \alpha(t))$ *be a switching diffusion process on* $\mathbb{R}^r \times \mathcal{M}$. *For each* $i \in \mathcal{M}$, *let* $V(x, i)$ *be a nonnegative continuous function,* $O_m = \{x : V(x, i) < m\}$ *be a bounded set, and*

$$\tau = \inf\{t : X(t) \notin O_m\}. \tag{9.37}$$

Suppose that for each $i \in \mathcal{M}$, $\mathcal{L}V(x, i) \leq -k(x) \leq 0$, *where* $k(\cdot)$ *is continuous in* O_m. *Denote* $\widehat{O}_m = O_m \cap \{x : k(x) = 0\}$. *Suppose that*

$$\mathbf{P}_{x,i}\left(\sup_{0 \leq s \leq t} |X(s) - x| \geq \eta\right) \to 0 \quad as \quad t \to \infty$$

for any $\eta > 0$ *and for each* $x \in \overline{O}_m$. *Then* $X(t) \to \widehat{O}_m$ *with probability at least* $1 - V(x, i)/m$.

Remark 9.10. Recall that an η-neighborhood of a set $\widetilde{M}$ relative to an open set O is an open set $N_\eta(\widetilde{M}) \subset O$ such that

$$N_\eta(\widetilde{M}) = \{x \in O : |x - y| < \eta \text{ for some } y \in \widetilde{M}\}.$$

In addition, $\overline{N}_\eta(\widetilde{M}) = N_\eta(\widetilde{M}) + \partial N_\eta(\widetilde{M})$ is the closure of $N_\eta(\widetilde{M})$. In our case, $k(x)$ is uniformly continuous on $\widehat{O}_m$. If $k(x) > 0$ for some $x \in O_m$, then for some $d_0 > 0$ and for each $0 < d < d_0$, there is an $\eta_d > 0$ such that for the η_d-neighborhood $N_{\eta_d}(\widehat{O}_m)$ of $\widehat{O}_m$ relative to O_m, $k(x) \geq d > 0$ on $O_m - N_{\eta_d}(\widehat{O}_m)$. Before proceeding to the proof of Proposition 9.9, we establish a lemma below.

Lemma 9.11. *In addition to the conditions of Proposition 9.9, assume also that* $V(0, i) = 0$ *for each* $i \in \mathcal{M}$, *for some* $m > 0$, *and for any* $d_1 > 0$ *satisfying* $d_1 \leq V(x, i) \leq m$. *Then the switching diffusion is stable with respect to* $(O_{d_1}, O_m, 1 - d_1/m)$. *In addition, for almost all* $\omega \in \Omega_m = \{\omega : X(t) \in O_m \text{ for all } t < \infty\}$, *there is a* $c(\omega)$ *satisfying* $0 \leq c(\omega) \leq m$ *such that* $V(X(t \wedge \tau_m), \alpha(t \wedge \tau_m)) \to c(\omega)$, *where* $\tau_m = \inf\{t : X(t) \notin O_m\}$ *is the first exit time from* O_m.

Idea of Proof of Lemma 9.11. By Dynkin's formula,

$$\mathbf{E}_{x,i}V(X(t \wedge \tau_m), \alpha(t \wedge \tau_m)) - V(x, i) = \int_0^{t \wedge \tau_m} \mathcal{L}V(X(s), \alpha(s))ds \leq 0.$$

Then $\mathbf{E}_{x,i}V(X(t \wedge \tau_m), \alpha(t \wedge \tau_m)) \leq V(x, i)$. Thus the stopped process is a supermartingale. We have

$$\mathbf{P}_{x,i}\left(\sup_{0 \leq t < \infty} V(X(t \wedge \tau_m), \alpha(t \wedge \tau_m)) \geq \lambda\right) \leq \frac{V(x, i)}{\lambda}.$$

Thus, $\mathbf{P}_{x,i}(\omega \in \Omega_m) \geq 1 - V(x,i)/m$. By virtue of the supermartingale convergence theorem, there exists a $c(\omega)$ such that $V(X(t \wedge \tau_m), \alpha(t \wedge \tau_m)) \to c(\omega)$ a.s. Moreover, the structure of the sets O_m, $\widehat{O}_m$, and $\widetilde{O}_m$ implies that $0 \leq c(\omega) \leq m$.

Sketch of Proof of Proposition 9.9. The proof in fact, is almost the same as that of [97, Theorem 2]. If $k(x) \equiv 0$ in O_m, the result is evident as a consequence of Lemma 9.11. Moreover, by Lemma 9.11, $\mathbf{P}(X(t) \in O_m^0) \geq 1 - V(x,i)/m$, where O_m^0 denotes the interior of O_m.

For each d_1 and $d_2 > 0$, and d_0 given in Remark 9.10 satisfying $d_0 > d_1 > d_2$, let η_ℓ be such that for $x \in O_m - N_{\eta_\ell}(\widehat{O}_m)$, $k(x) \geq d_\ell$. Without loss of generality, $N_{\eta_2}(\widehat{O}_m)$ is a proper subset of $N_{\eta_1}(\widehat{O}_m)$. Define

$$\chi(s) = I_{x,i}(s, \omega, \eta_\ell) = \begin{cases} 1, & \text{if } X(t) \in O_m - N_{\eta_\ell}(\widehat{O}_m), \\ 0, & \text{otherwise.} \end{cases}$$

Define

$$T(t, \eta_\ell) = T_{x,i}(t, \eta_\ell) = \int_{t \wedge \tau_m}^{\tau_m} \chi(s)ds.$$

That is, $T(t, \eta_\ell)$ is the total time that the process spent in $O_m - N_{\eta_\ell}(\widehat{O}_m)$ with $t < T(t, \eta_\ell)$, and $T(t, \eta_\ell) < \min(\infty, \tau_m)$; $T(t, \eta_\ell) = 0$ if $\tau_m < t$. Using Lemma 9.11,

$$\mathbf{P}_{x,i}\,(X(t) \text{ leaves } O_m \text{ at least once before } t = \infty)$$
$$= 1 - \mathbf{P}_{x,i}(\Omega_m) \leq \frac{V(x,i)}{m}.$$

Again, by Dynkin's formula, for any $x \in O_m - N_{\eta_\ell}(\widehat{O}_m)$,

$$V(x,i) - \mathbf{E}_{x,i}V(X(t \wedge \tau_m), \alpha(t \wedge \tau_m))$$
$$= -\mathbf{E}_{x,i}\int_0^{t \wedge \tau_m} \mathcal{L}V(X(s), \alpha(s))ds$$
$$\geq \gamma \mathbf{E}_{x,i}\int_0^{t \wedge \tau_m} I_{x,i}(s)ds = \gamma \mathbf{E}_{x,i}(t \wedge \tau_m),$$

where $I_{x,i}(s)$ is the indicator of the set of (s, ω)s where $\mathcal{L}V(x,i) \leq -\gamma$. Using the nonnegativity of $V(\cdot)$ and taking the limit as $t \to \infty$, we obtain $\mathbf{E}_{x,i}\tau_m < V(x,i)/\gamma$. Thus, $T_{x,i}(t, \eta_\ell) < \infty$ a.s. and $T_{x,i}(t, \eta_\ell) \to 0$ as $t \to \infty$.

There are only two possibilities. (i) There is a random variable $\tau(\eta_1) < \infty$ a.s. such that for all $t > \tau(\eta_1)$, $X(t) \in N_{\eta_1}(\widehat{O}_m)$ with probability at least $1 - V(x,i)/m$ (this can also be represented as $X(t) \in N_{\eta_1}(\widehat{O}_m)$ a.s. relative to Ω_m); (ii) for $\omega \in \Omega_m$, $X(t)$ moves from $N_{\eta_2}(\widehat{O}_m)$ back and forth to $O_m - N_{\eta_1}(\widehat{O}_m)$ infinitely often in any interval $[t, \infty)$. We demonstrate that the second alternative cannot happen.

Consider (ii). Since $T_{x,i}(t,\eta_\ell) \to 0$ a.s. as $t \to \infty$, there are infinitely many movements from $N_{\eta_2}(\widehat{O}_m)$ to $O_m - N_{\eta_1}(\widehat{O}_m)$ and back to $N_{\eta_2}(\widehat{O}_m)$ in a total time being arbitrarily small. We claim that the probability of the second alternative is 0.

For any $\Delta_1 > 0$, choose $\widetilde{t} > 0$ such that

$$\sup_{x \in O_m - N_{\eta_2}(\widehat{O}_m)} \mathbf{P}_{x,i}(\sup_{0 \le s \le \widetilde{t}} |X(s) - x| \ge \eta_1 - \eta_2) < \Delta_1. \tag{9.38}$$

Owing to the stochastic continuity and compactness of $\overline{O}_m$ (the closure of O_m), for each $\Delta_2 > 0$, there is a $\widetilde{t} < \infty$ such that $\mathbf{P}_{x,i}(T_{x,i}(t,\eta_2) > \widetilde{t}) < \Delta_2$. This is a contradiction to (9.38). Thus

$$\mathbf{P}_{x,i}(X(s) \in O_m - N_{\eta_1}(\widehat{O}_m) \text{ for some } s \in (t,\infty)) \to 0 \text{ as } t \to \infty,$$

for any $x \in O_m$. The arbitrariness of η_ℓ then implies the result. $\square$

Remark 9.12. Denote $D_{\eta_\ell} = \{\omega : \omega \in \Omega_m \text{ and } \int_0^\infty \chi_\ell(s)ds < \infty\}$, without assuming the boundedness of O_m and $\widehat{O}_m$, but require that for any $\varepsilon > 0$, $\mathbf{P}_{x,i}\{|X^{x,i}(t)| \to \infty, \ X(t) \in O_m, D_{\eta_\ell}\} = 0$. Then the conclusions of Proposition 9.9 continue to hold. The proof of this can be carried out as in [97, Theorem 3]. We omit the details for brevity.

To continue, we first set up some notation and conventions. Consider the two-component Markov process $Y(t) = (X(t), \alpha(t))$ with the generator given by (9.1) or equivalently with the dynamics specified by (9.2) and (9.3) with initial condition $(X(0), \alpha(0)) = (X_0, \alpha_0) \in \mathbb{R}^r \times \mathcal{M}$ being possibly random. The process takes values in $\mathbb{R}^r \times \mathcal{M}$.

Suppose that $m(t, \varphi, dx \times i)$ is the measure induced on $\mathcal{B}(\mathbb{R}^r \times \mathcal{M})$, the Borel sets of $\mathbb{R}^r \times \mathcal{M}$, at time t such that

$$\varphi = m(0, \varphi, dx \times i) \text{ is the measure induced at } t = 0,$$
$$m(t+s, \varphi, dx \times i) = m(t, m(s, \varphi), dx \times i), \ t,s \ge 0. \tag{9.39}$$

It is the measure of the process at time t, given "initial data" φ. The last line of (9.39) is just the semigroup property. In the above, we continue to use the convention as mentioned in Remark 6.6. That is, $m(t, \varphi, dx \times i)$ is meant to be $m(t, \varphi, dx \times \{i\})$.

Let C_b be the space of bounded and continuous functions on $\mathbb{R}^r \times \mathcal{M}$. Let $\overline{\mathbf{M}}$ be the collection of all measures on $(\mathbb{R}^r \times \mathcal{M}, \mathcal{B}(\mathbb{R}^r \times \mathcal{M}))$. A sequence of measures $\{\psi_n\}$ converges to ψ weakly in $\overline{\mathbf{M}}$ if as $n \to \infty$,

$$\sum_{i=1}^{m_0} \int_{\mathbb{R}^r} h(x,i)\psi_n(dx \times i) \to \sum_{i=1}^{m_0} \int_{\mathbb{R}^r} h(x,i)\psi(dx \times i), \text{ for all } h \in C_b.$$

A set $S \subset \overline{\mathbf{M}}$ is weakly bounded or tight, if for any $\eta > 0$, and each $i \in \mathcal{M}$, there is a compact set $K_\eta \subset \mathbb{R}^r$ such that

$$\psi((\mathbb{R}^r - K_\eta) \times \{i\}) < \eta \text{ for all } \psi \in S.$$

Remark 9.13. In the above and henceforth, for simplicity, we use the phrase "$f \in C_b$" for a f function defined on $\mathbb{R}^r \times \mathcal{M}$. In fact, since $\alpha(t)$ takes values in a finite set $\mathcal{M}$, for each x, $f(x, \cdot)$ is trivially continuous. Thus, the requirement of $f \in C_b$ can be rephrased as "for each $i \in \mathcal{M}$, $f(\cdot, i) \in C_b$." In all subsequent development, when we say $f \in C_b$, it is understood to be in this sense.

9.4.1 ω-Limit Sets and Invariant Sets

Definition 9.14. A measure $\psi \in \overline{\mathbf{M}}$ is in the ω-limit set $\widetilde{\Omega}(\varphi)$ (with φ given by (9.39)), if there is a sequence of real numbers $\{t_n\}$ satisfying $t_n \to \infty$ such that

$$\sum_{j \in \mathcal{M}} \int_{\mathbb{R}^r} f(y, j)[m(t_n, \varphi, dy \times j) - \psi(dy \times j)] \to 0 \quad \text{as} \quad n \to \infty$$

for each $f \in C_b$.

Definition 9.15. A set $S \subset \overline{\mathbf{M}}$ is an invariant set if for each $\psi \in S$, there is a function $\widetilde{m}(s, \psi, dx \times j)$ defined for all $s \in (-\infty, \infty)$ and taking values in S such that

(1) $\widetilde{m}(0, \psi, dx \times j) = \psi(dx \times j)$,

(2) $\widetilde{m}(t + s, \psi, dx \times j) = m(t, \widetilde{m}(s, \psi), dx \times j)$ for any $s \in (-\infty, \infty)$ and $t \in [0, \infty)$.

Note that the $\widetilde{m}(t, \psi, dx)$ is the measure of the process. The variable s taking values in $(-\infty, \infty)$ is consistent with the definitions used in dynamic systems for the flows of deterministic problems. The idea is that, for any measure ψ to be an element of the invariant set, there must be a flow for $s \in (-\infty, \infty)$ that takes value ψ at time zero. As commented upon in the beginning of this section, the time interval is two-sided. We aim to obtain the invariance principle using measure concepts. The result is spelled out in the following theorem.

Theorem 9.16. *Suppose that*

(a) *for each $i \in \mathcal{M}$, $\cup_{t \in [0, \infty)} m(t, \varphi, dx \times i)$ is weakly bounded;*

(b) *on any finite time interval,*

$$\sum_{j=1}^{m_0} \int f(x, j) m(t, \varphi, dx \times j) \text{ is a continuous function of } t$$

uniformly in φ, for φ in any weakly bounded set of $\overline{\mathbf{M}}$;

(c) *for each $f(\cdot) \in C_b$ and some weakly bounded sequence $\{\varphi_n\}$ converging weakly to φ, we have*

$$\sum_{j=1}^{m_0} \int f(x,j) m(\varphi_n, dx \times j) \to \sum_{j=1}^{m_0} \int f(x,j) m(\varphi, dx \times j).$$

Then (i) $\widetilde{\Omega}(\varphi) \neq \emptyset$, (ii) $\widetilde{\Omega}(\varphi)$ *is weakly bounded,* (iii) $\widetilde{\Omega}(\varphi)$ *is an invariant set, and* (iv)

$$\sum_{j=1}^{m_0} \int f(x,j)[m(t,\varphi, dx \times j) - \varphi(dx \times j)] \to 0, \quad as \ t \to \infty.$$

Proof. We divide the proof into several steps.

Step 1. It is well known that on a completely separable metric space, weak boundedness is equivalent to sequential compactness; see [16, p. 37], [43, p. 103] and so on. Consequently, the ω-limit set $\widetilde{\Omega}(\varphi) \neq \emptyset$.

Step 2. For some sequence $\{t_n\}$ with $t_n \to \infty$ as $n \to \infty$, some $\psi \in \overline{\mathbf{M}}$, and all $f(\cdot,j) \in C_b$ for each $j \in \mathcal{M}$, suppose that

$$\sum_{j=1}^{m_0} \int f(x,j) m(t_n,\varphi, dx \times j) \to \sum_{j=1}^{m_0} \int f(x,j)\psi(dx \times j) \quad \text{as} \ n \to \infty.$$

Let K be a compact set in $\mathbb{R}^r$. Denote by C_K the restriction of C_b to K. Then C_K is a complete separable metric space equipped with the sup-norm restricted to K. Let $\mathcal{T}_K$ be a countable dense set in C_K. For each $i \in \mathcal{M}$ and each $f(\cdot,i) \in \mathcal{T}_K$, define

$$\widetilde{f}_n(t;f) = \sum_{j \in \mathcal{M}} \int_K f(x,j) m(t_n + t,\varphi, dx \times j).$$

For some $T > 0$, consider the family of functions

$$\left\{ \widetilde{f}_n(t;f) : f \in \mathcal{T}_K, \ t \in [-T,T], \ t_n \geq T \right\},$$

which is uniformly bounded. By virtue of condition (b) of the theorem, this is an equicontinuous family. The well-known Ascoli–Arzela lemma implies that we can extract a subsequence $\{\widetilde{f}_{n_k}(t;f)\}$ that converges uniformly to $\widetilde{f}(t;f)$ on $[-T,T]$. By using the diagonal process and repeatedly selecting subsequences, we can show that we can extract a further subsequence $\{\widetilde{f}_{n_\ell}(t;f)\}$ that converges uniformly to $\widetilde{f}(t;f)$ on any finite interval in $(-\infty,\infty)$. Moreover, because $\mathcal{T}_K$ is dense in C_K, the aforementioned convergence takes place for any $f \in C_K$. The semigroup property of

$m(t, \varphi, dx \times i)$ and weak convergence of

$$\tilde{f}_n(t; f) = \sum_{j \in \mathcal{M}} \int_K f(x, dx \times j) m(t_n + t, \varphi, dx \times j)$$

$$= \sum_{j \in \mathcal{M}} \int_K f(x, dx \times j) m(t, m(t_n, \varphi), dx \times j)$$

for each $t \in (-\infty, \infty)$ and each $f(\cdot, i) \in C_K$ for each $i \in \mathcal{M}$ together with the weak boundedness of $\{m(t_n + t, \varphi, dx \times i)\}$ implies that the restriction of $\{m(t_n + t, \varphi)\}$ to $(K, \mathcal{B}((\mathbb{R}^r \cap K) \times \mathcal{M}))$ converges weakly to a measure $\psi_K(t)$ for each $t \in (-\infty, \infty)$.

Step 3. By the weak boundedness of $\{m(t, \varphi)\}$, there exists a sequence $\eta_i \to 0$, and a sequence of compact sets $\{K_i\}$ in $\mathbb{R}^r$ satisfying $K_i \subset K_{i+1}$ such that for each $j \in \mathcal{M}$,

$$m(t, \varphi, K_i^c \times \{j\}) \leq \eta_i, \tag{9.40}$$

where $K_i^c \times \{j\} = (\mathbb{R}^r - K_i) \times \{j\}$. By means of a diagonal process, we can find a subsequence $\{t_n\}$ and a sequence of measures $\{\psi_i(t, \cdot \times \cdot) : t \in (-\infty, \infty)\}$ such that for each $f(\cdot) \in C_b$,

$$\sum_{j \in \mathcal{M}} \int_{K_i} f(x, j) m(t_n + t, \varphi, dx \times j) \to \sum_{j \in \mathcal{M}} \int_{K_i} f(x, j) \psi_{K_i}(t, dx \times j). \tag{9.41}$$

For $k > i$,

$$\sum_{j \in \mathcal{M}} \int_{K_i} f(x, j) \psi_{K_i}(t, dx \times j) = \sum_{j \in \mathcal{M}} \int_{K_i} f(x, j) \psi_{K_k}(t, dx \times j). \tag{9.42}$$

By (9.40)–(9.42), there is a measure $\psi(\cdot)$ such that for all $t \in (-\infty, \infty)$ and all $f(\cdot) \in C_b$,

$$\sum_{j=1}^{m_0} \int f(x, j) m(t_n + t, \varphi, dx \times j) \to \sum_{j=1}^{m_0} \int f(x, j) \psi(t, dx \times j).$$

Noting $\psi(0) = \psi$ and by (c) in the assumption of the theorem, for each $f \in C_b$, $s \geq 0$, and $t \in (-\infty, \infty)$,

$$\sum_{j=1}^{m_0} \int f(x, j) m(s, m(t_n + t, \varphi), dx \times j)$$

$$\to \sum_{j=1}^{m_0} \int f(x, j) m(s, \psi(t), dx \times j) \text{ as } n \to \infty,$$

and

$$\sum_{j=1}^{m_0} \int f(x,j)m(s,m(t_n + t,\varphi,dx \times j),dx \times j)$$

$$= \sum_{j=1}^{m_0} \int f(x,j)m(0,m(t_n + t + s,\varphi),dx \times j)$$

$$\rightarrow \sum_{j=1}^{m_0} \int f(x,j)m(0,\psi(t + s),dx \times j)$$

$$= \sum_{j=1}^{m_0} \int f(x,j)\psi(t + s,dx \times j).$$

This shows that $\{\psi(t,dx \times j)\}$ is an invariant set.

Step 4. Suppose that there were a sequence $\{t_n\}$ satisfying for any subsequence $\{t_{n_k}\}$ and some $f(\cdot) \in C_b$,

$$\lim_{t \rightarrow \infty} \sup_{k} \inf_{\psi \in \tilde{\Omega}(\varphi)} \left| \sum_{j=1}^{m_0} \int f(x,j)m(t_{n_k},\varphi,dx \times j) \right.$$
$$\left. - \sum_{j=1}^{m_0} \int f(x,j)\psi(dx \times j) \right| > 0.$$

The weak boundedness of $\{m(t_{n_k},\varphi)\}$ implies that there is a subsequence that converges weakly to some $\psi \in \overline{\mathbf{M}}$. However, $\psi \in \tilde{\Omega}(\varphi)$, which is a contradiction. Thus the desired result follows.   □

Proposition 9.17. *If $(X(t),\alpha(t))$ is Feller, then condition* (c) *of Theorem 9.16 holds. That is, for any $t \in [0,T]$ with $T > 0$ and any weakly bounded sequence $\{\varphi_n\}$ such that φ_n converges weakly to φ as $n \rightarrow \infty$, we have*

$$\sum_{j=1}^{m_0} \int_{\mathbb{R}^r} f(x,j)m(t,\varphi_n,dx \times j) \rightarrow \sum_{j=1}^{m_0} \int_{\mathbb{R}^r} f(x,j)m(t,\varphi,dx \times j),$$

for any $f \in C_b$ as $n \rightarrow \infty$.

Proof. Fix any $t \in [0,T]$. For any $f \in C_b$, let $h(x,i) = \mathbf{E}_{x,i}f(X(t),\alpha(t))$. Then by virtue of the Feller property, h is continuous and bounded. Note that

$$\sum_{j=1}^{m_0} \int_{\mathbb{R}^r} f(x,j)m(t,\varphi_n,dx \times j) = \sum_{i=1}^{m_0} \int_{\mathbb{R}^r} \mathbf{E}_{x,i}f(X(t),\alpha(t))\varphi_n(dx \times j)$$

$$= \sum_{i=1}^{m_0} \int_{\mathbb{R}^r} h(x,i)\varphi_n(dx \times i).$$

Because φ_n converges weakly to φ and $h \in C_b$, it follows that

$$\lim_{n \to \infty} \sum_{i=1}^{m_0} \int_{\mathbb{R}^r} h(x,i) \varphi_n(dx \times i) = \sum_{i=1}^{m_0} \int_{\mathbb{R}^r} h(x,i) \varphi(dx \times i).$$

But

$$\sum_{j=1}^{m_0} \int_{\mathbb{R}^r} f(x,j) m(t,\varphi, dx \times j) = \sum_{i=1}^{m_0} \int_{\mathbb{R}^r} \mathbf{E}_{x,i} f(X(t), \alpha(t)) \varphi(dx \times i)$$

$$= \sum_{i=1}^{m_0} \int_{\mathbb{R}^r} h(x,i) \varphi(dx \times i).$$

Thus we conclude that

$$\lim_{n \to \infty} \sum_{j=1}^{m_0} \int_{\mathbb{R}^r} f(x,j) m(t,\varphi_n, dx \times j) = \sum_{j=1}^{m_0} \int_{\mathbb{R}^r} f(x,j) m(t,\varphi, dx \times j).$$

This finishes the proof.   □

Proposition 9.18. *Assume that*

(i) $(X(t), \alpha(t))$ *is Feller,*

(ii) $(X(t), \alpha(t))$ *is continuous in probability uniformly in $t \in [0,T]$, where $T > 0$,*

(iii) $\bigcup_{t \geq 0} m(t,\varphi)$ *is weakly bounded.*

Then condition (b) of Theorem 9.16 holds. That is, for any $f \in C_b$, the function

$$t \mapsto \sum_{j=1}^{m_0} \int_{\mathbb{R}^r} f(x,j) m(t,\varphi, dx \times j)$$

is continuous, uniformly in φ for φ in any weakly bounded set of $\overline{\mathbf{M}}$.

Proof. The proof is divided into several steps.

Step 1. Let $f \in C_b$. Then the weak boundedness, the continuity of f, and the stochastic continuity of $(X(t), \alpha(t))$ imply that $f(X(t), \alpha(t))$ is continuous in probability in t.

Step 2. Next we show that the function $t \mapsto \mathbf{E}_{x,i} f(X(t), \alpha(t))$ is continuous at any fixed finite t for any $(x,i) \in \mathbb{R}^r \times \mathcal{M}$. In fact, for any fixed finite t, any $(x,i) \in \mathbb{R}^r \times \mathcal{M}$, and any $\varepsilon > 0$, by virtue of Step 1, we can find some $\Delta = \Delta(x,i) > 0$ such that

$$\mathbf{P}_{x,i}\{|f(X(t+s), \alpha(t+s)) - f(X(t), \alpha(t))| > \varepsilon/6\} < \frac{\varepsilon}{12\|f\|}$$

for $|s| \leq \Delta$, where $\|f\|$ denotes that essential sup-norm of f. Then it follows that for $|s| \leq \Delta$,

$$
\begin{aligned}
&|\mathbf{E}_{x,i} f(X(t+s), \alpha(t+s)) - \mathbf{E}_{x,i} f(X(t), \alpha(t))| \\
&\leq \mathbf{E}_{x,i} |f(X(t+s), \alpha(t+s)) - f(X(t), \alpha(t))| \\
&= \mathbf{E}_{x,i} [|f(X(t+s), \alpha(t+s)) - f(X(t), \alpha(t))| (I_1 + I_2)] \qquad (9.43) \\
&\leq \frac{\varepsilon}{6} + 2 \|f\| \frac{\varepsilon}{12 \|f\|} \\
&\leq \frac{\varepsilon}{3},
\end{aligned}
$$

where $I_1 = I_{\{|f(X(t+s), \alpha(t+s)) - f(X(t), \alpha(t))| \leq \varepsilon/6\}}$ and $I_2 = 1 - I_1$. Thus the function $\mathbf{E}_{x,i} f(X(t), \alpha(t))$ is continuous at t.

Step 3. We claim that the function $t \mapsto \mathbf{E}_{x,i} f(X(t), \alpha(t))$ is continuous at any fixed finite t, uniformly for $(x, i) \in K \times \mathcal{M}$, where K is any compact subset of $\mathbb{R}^r$. In fact, since the process $(X(t), \alpha(t))$ is Feller, the function $x \mapsto \mathbf{E}_{x,i} f(X(t), \alpha(t))$ is continuous, and hence uniformly continuous in K, where K is any compact subset of $\mathbb{R}^r$. Moreover, Proposition 2.30 enables us to conclude that the function $x \mapsto \mathbf{E}_{x,i} f(X(t), \alpha(t))$ is continuous, uniformly in any finite t interval. Thus for any $\varepsilon > 0$, there exists a $\delta > 0$ such that

$$
|\mathbf{E}_{x_1,i} f(X(t), \alpha(t)) - \mathbf{E}_{x_2,i} f(X(t), \alpha(t))| < \frac{\varepsilon}{3}, \qquad (9.44)
$$

for any $x_1, x_2 \in K$ with $|x_1 - x_2| < \delta$. Because K is compact, we can cover K by a finite union of balls; that is, there exists a positive integer N and $x_1, \ldots, x_N \in K$ such that $K \subset \cup_{k=1}^{N} B(x_k, \delta)$. It follows from (9.43) that for each $k = 1, 2, \ldots, N$ and each $i \in \mathcal{M}$, there exists a $\Delta_{k,i} = \Delta(x_k, i)$ such that

$$
|\mathbf{E}_{x_k,i} f(X(t+s), \alpha(t+s)) - \mathbf{E}_{x_k,i} f(X(t), \alpha(t))| < \frac{\varepsilon}{3}, \qquad (9.45)
$$

for all $|s| \leq \Delta_{k,i}$. Let $\Delta = \min \{\Delta_{k,i}, k = 1, \ldots, N, i \in \mathcal{M}\}$. Note that for any $x \in K$, $x \in B(x_k, \delta)$ for some $k = 1, \ldots, N$. Hence for any $i \in \mathcal{M}$ and any $|s| \leq \Delta$, it follows from (9.44) and (9.45) that

$$
\begin{aligned}
&|\mathbf{E}_{x,i} f(X(t+s), \alpha(t+s)) - \mathbf{E}_{x,i} f(X(t), \alpha(t))| \\
&\leq |\mathbf{E}_{x,i} f(X(t+s), \alpha(t+s)) - \mathbf{E}_{x_k,i} f(X(t+s), \alpha(t+s))| \\
&\quad + |\mathbf{E}_{x_k,i} f(X(t+s), \alpha(t+s)) - \mathbf{E}_{x_k,i} f(X(t), \alpha(t))| \\
&\quad + |\mathbf{E}_{x_k,i} f(X(t), \alpha(t)) - \mathbf{E}_{x,i} f(X(t), \alpha(t))| \\
&\leq \frac{\varepsilon}{3} + \frac{\varepsilon}{3} + \frac{\varepsilon}{3} = \varepsilon.
\end{aligned}
$$

The claim thus follows.

Step 4. Now for any compact set $K \subset \mathbb{R}^r$ and any $f \in C_b$, we have

$$
\left| \sum_{j=1}^{m_0} \int_{\mathbb{R}^r} f(y,j) m(t+s, \varphi, dy \times j) - \sum_{j=1}^{m_0} \int_{\mathbb{R}^r} f(y,j) m(t, \varphi, dy \times j) \right|
$$

$$
= \left| \sum_{j=1}^{m_0} \int_{\mathbb{R}^r} \left(\mathbf{E}_{x,i} f(X(t+s), \alpha(t+s)) - \mathbf{E}_{x,i} f(X(t), \alpha(t)) \right) \varphi(dx \times i) \right|
$$

$$
\leq \sum_{i=1}^{m_0} \int_K |\mathbf{E}_{x,i} f(X(t+s), \alpha(t+s)) - \mathbf{E}_{x,i} f(X(t), \alpha(t))| \, \varphi(dx \times i)
$$

$$
+ \sum_{i=1}^{m_0} \int_{K^c} |\mathbf{E}_{x,i} f(X(t+s), \alpha(t+s)) - \mathbf{E}_{x,i} f(X(t), \alpha(t))| \, \varphi(dx \times i).
$$

Let $S \subset \overline{\mathbf{M}}$ be weakly bounded. Fix $t \geq 0$ and $\varepsilon > 0$. There is a compact $K_0 \subset \mathbb{R}^r$ such that $\varphi(K_0^c \times \mathcal{M}) < \varepsilon/(4\|f\|)$ for all $\varphi \in S$. By virtue of Step 3, there is a $\Delta > 0$ such that

$$
|\mathbf{E}_{x,i} f(X(t+s), \alpha(t+s)) - \mathbf{E}_{x,i} f(X(t), \alpha(t))| < \frac{\varepsilon}{2},
$$

for all $|s| \leq \Delta$ and all $(x,i) \in K_0 \times \mathcal{M}$. Hence it follows that for any $\varphi \in S$ and any $|s| \leq \Delta$, we have

$$
\left| \sum_{j=1}^{m_0} \int_{\mathbb{R}^r} f(y,j) m(t+s, \varphi, dy \times j) - \sum_{j=1}^{m_0} \int_{\mathbb{R}^r} f(y,j) m(t, \varphi, dy \times j) \right|
$$

$$
\leq \sum_{i=1}^{m_0} \int_{K_0} |\mathbf{E}_{x,i} f(X(t+s), \alpha(t+s)) - \mathbf{E}_{x,i} f(X(t), \alpha(t))| \, \varphi(dx \times i)
$$

$$
+ \sum_{i=1}^{m_0} \int_{K_0^c} |\mathbf{E}_{x,i} f(X(t+s), \alpha(t+s)) - \mathbf{E}_{x,i} f(X(t), \alpha(t))| \, \varphi(dx \times i)
$$

$$
\leq \sum_{i=1}^{m_0} \int_{K_0} \frac{\varepsilon}{2} \varphi(dx \times i) + \sum_{i=1}^{m_0} \int_{K_0} 2\|f\| \, \varphi(dx \times i)
$$

$$
\leq \frac{\varepsilon}{2} + 2\|f\| \frac{\varepsilon}{4\|f\|}
$$

$$
= \varepsilon.
$$

This shows that the function $t \mapsto \sum_{j=1}^{m_0} \int_{\mathbb{R}^r} f(y,j) m(t, \varphi, dy \times j)$ is continuous, uniformly in $\varphi \in S$, a weakly bounded set of $\overline{\mathbf{M}}$. $\square$

9.4.2   Switching Diffusions

Let us recall the definitions of the support of a measure and a family of measures here. Let μ be a measure on $(\mathbb{R}, \mathcal{B}(\mathbb{R}^r))$, where $\mathcal{B}(\mathbb{R}^r)$ denotes the collection of Borel sets on $\mathbb{R}^r$. Then the support of μ is defined to be the set of all points $x \in \mathbb{R}^r \times \mathcal{M}$ for which every neighborhood $N(x)$ of x has positive measure. That is, if μ is a measure, then the support of μ, denoted by $\mathrm{supp}(\mu)$, is defined as

$$\mathrm{supp}(\mu) := \{x : \mu(N(x)) > 0 \text{ for all open neighborhood } N(x) \text{ of } x\}.$$

If S is a family of measures, then

$$\mathrm{supp}(S) := \bigcup_{\mu \in S} \mathrm{supp}(\mu).$$

In what follows, for a measure defined on $\mathbb{R}^r \times \mathcal{M}$, by its x-section (or x-component), we mean the measure defined on $\mathbb{R}^r$.

Theorem 9.19. *Assume the conditions of Theorem 9.16. Then the following assertions hold.*

(a) *Suppose that $F = \mathrm{supp}_x(\widetilde{\Omega}(\varphi))$ is the support of the x-section of set $\widetilde{\Omega}(\varphi)$. Then the process $X(t)$ converges to F in probability as $t \to \infty$. That is,*

$$\mathbf{P}_\varphi(d(X(t), F) > \eta) \to 0 \quad as \quad t \to \infty \quad for \ any \quad \eta > 0.$$

(b) *Suppose $X(t)$ converges in probability to a set G. Let L_I be the largest invariant set whose support is contained in $\overline{G}$ (i.e., $\mathrm{supp}(L_I) \subset \overline{G}$). Then $X(t)$ converges to $\mathrm{supp}(L_I)$ in probability.*

Proof. To prove (a), let O be an arbitrary open set such that $\overline{F} \subset O$. We claim

$$\limsup_{t \to \infty} \mathbf{P}_\varphi \left\{ X(t) \in \mathbb{R}^r - \overline{O} \right\} = 0. \tag{9.46}$$

We verify (9.46) by contradiction. Suppose it were not true. There would be an open set $\widetilde{O}$ containing $\overline{O}$, and a nonnegative function $\widetilde{f}(\cdot, i) \in C_b$ (for each $i \in \mathcal{M}$) satisfying

$$\widetilde{f}(y, i) \begin{cases} = 0, & \text{if } y \in \overline{O} \\ \geq 1, & \text{if } y \in \mathbb{R}^r - \widetilde{O} \end{cases}$$

such that

$$\limsup_{t \to \infty} \sum_{i=1}^{m_0} \int_{\mathbb{R}^r} \widetilde{f}(x, i) m(t, \varphi, dx \times i) > 0.$$

It follows that there is a sequence $\{t_n\}$ and a $\gamma > 0$ such that

$$\sum_{i=1}^{m_0} \int \widetilde{f}(x, i) m(t_n, \varphi, dx \times i) \to \gamma > 0.$$

By Theorem 9.16, $\{m(t_n, \varphi)\}$ is weakly bounded. So there is a subsequence $\{t_{n_k}\}$ and a $\psi \in \widetilde{\Omega}(\varphi)$ such that

$$\sum_{i=1}^{m_0} \int f(x,i) m(t_{n_k}, \varphi, dx \times i) \to \sum_{i=1}^{m_0} \int f(x,i) \psi(dx \times i)$$

for all $f(\cdot, i) \in C_b$. Thus it also holds for

$$\sum_{i=1}^{m_0} \int \widetilde{f}(x,i) \psi(dx \times i) = \gamma > 0.$$

However, this implies that there is some point $x \in \mathbb{R}^r - \overline{O}$ such that $\psi(N \times i) > 0$ for all open set N containing x. Hence it follows that $(x, i) \in \text{supp}(\psi)$, which leads to a contradiction, because $\text{supp}(\psi) \subset \text{supp}(\Omega(\varphi)) = F \subset O$. Therefore (9.46) must be true and hence the assertion of (a) is established.

To prove (b), by virtue of part (a), $X(t)$ converges in probability to F as $t \to \infty$. In addition, by the hypothesis, $X(t)$ converges to G in probability as $t \to \infty$. Thus it follows that $F \subset \overline{G}$. Meanwhile, Theorem 9.16 implies that $\widetilde{\Omega}(\varphi)$ is an invariance set. Hence it follows that $F = \text{supp}_x(\widetilde{\Omega}(\varphi)) \subset \text{supp}(L_I)$. As a result, $X(t)$ converges to $\text{supp}(L_I)$ in probability as $t \to \infty$.

Remark 9.20. Let the conditions of Theorem 2.1 be satisfied. Then by virtue of Theorems 2.13 and 2.18, the process $(X(t), \alpha(t))$ is continuous in probability and Feller. Suppose that for each $i \in \mathcal{M}$, there are nonnegative functions $V(\cdot, i) \in C^2$ and $k(x) \in C$ satisfying

$$\mathcal{L}V(x, i) \leq -k(x) \leq 0.$$

Assume also that

(i) $\lim_{|x| \to \infty} \inf_{i \in \mathcal{M}} V(x, i) = \infty$, and

(ii) $\mathbf{E}V(X(0), \alpha(0)) = \sum_{i \in \mathcal{M}} \int_{\mathbb{R}^r} V(x, i) \varphi(dx \times i) < \infty.$

Then using Dynkin's formula, we can verify that

$$\mathbf{E}_\varphi V(X(t), \alpha(t)) \leq \mathbf{E}V(X(0), \alpha(0)) < \infty$$

for any $t \geq 0$. Thus the set of measures induced by $\{V(X(t), \alpha(t)), t \geq 0\}$ is weakly bounded. Hence for any $\varepsilon > 0$, there is a compact $K \subset \mathbb{R}$ such that

$$\mathbf{P}_\varphi \{V(X(t), \alpha(t)) \in K \times \mathcal{M}\} \geq 1 - \varepsilon, \text{ for any } t \geq 0.$$

Now let $\widetilde{K} := \{x \in \mathbb{R}^r : V(x, i) \in K, \text{ for some } i \in \mathcal{M}\}$. Because K is compact and $V(\cdot, i)$ is continuous, $\widetilde{K}$ is closed. In addition, a simple contradiction argument and condition (i) imply that $\widetilde{K}$ is bounded. Therefore $\widetilde{K}$ is compact. Note that

$$V(X(t), \alpha(t)) \in K \times \mathcal{M} \text{ implies } (X(t), \alpha(t)) \in \widetilde{K} \times \mathcal{M}$$

for any $t \geq 0$. Thus it follows that

$$\mathbf{P}_\varphi \left\{ (X(t), \alpha(t)) \in \widetilde{K} \times \mathcal{M} \right\} \geq 1 - \varepsilon.$$

This shows that the collection of measures induced by $\{(X(t), \alpha(t)), t \geq 0\}$ or $\cup_{t \geq 0} m(t, \varphi)$ is weakly bounded. Therefore we conclude from Propositions 9.17 and 9.18 that all conditions of Theorem 9.16 are satisfied.

Suppose further that $k(X^{x,i}(t)) \to 0$ in probability for each $i \in \mathcal{M}$, and there is a sequence of compact sets $\{K_n\}$ with $K_n \times \mathcal{M} \subset \mathbb{R}^r \times \mathcal{M}$ satisfying

$$m(t, \varphi, K_n^c \times \mathcal{M}) < \frac{1}{n}.$$

Then for any open set O in $\mathbb{R}^r$ containing the set $K_n \cap \{x : k(x) = 0\}$, we have

$$\lim_{t \to \infty} \mathbf{P}_\varphi(X(t) \notin O) < \frac{1}{n}.$$

So we conclude from part (b) of Theorem 9.19 that $X(t)$ converges in probability to the largest support of an invariant set that is contained in $\lim_n K_n \cap \{x : k(x) = 0\}$.

Example 9.21. We consider a randomly switching Liénard equation, a real-valued equation of the second order of the following form

$$\frac{dX^2(t)}{dt^2} + f(X(t), \alpha(t)) \frac{dX(t)}{dt} + g(X(t)) = 0,$$

where $\alpha(t)$ is a continuous-time Markov chain taking values in $\mathcal{M} = \{1, 2\}$, and for each $i \in \mathcal{M}$, $f(\cdot, i) : \mathbb{R} \mapsto \mathbb{R}$ and $g(\cdot) : \mathbb{R} \mapsto \mathbb{R}$ are continuously differentiable functions satisfying for each $i \in \mathcal{M}$,

$$f(x, i) > 0 \quad \text{for all} \quad x \in \mathbb{R},$$

$$g(0) = 0, \ xg(x) > 0 \quad \text{for all} \quad x \neq 0,$$

$$\int_0^x g(u)du \to \infty \quad \text{as} \quad x \to \infty.$$

The above equation may be converted to a system of equations

$$\begin{cases} \dfrac{dX_1}{dt} = X_2(t) \\ \dfrac{dX_2}{dt} = -g(X_1(t)) - f(X_1(t), \alpha(t))X_2(t). \end{cases}$$

For each $i \in \mathcal{M}$, we can then define a Liapunov function

$$V(x_1, x_2, i) = \frac{x_2^2}{2} + \int_0^{x_1} g(u)du,$$

where as in the case of ordinary differential equations, the first term in the Liapunov function has the meaning of kinetic energy and the term involving the integral is potential energy. Note that the Liapunov function constructed is independent of $i \in \mathcal{M}$. Thus,

$$\sum_{j=1}^{2} q_{ij} V(x_1, x_2, j) = 0 \text{ for each } i = 1, 2.$$

Denote by

$$O_\lambda = \{(x_1, x_2, i) : V(x_1, x_2, i) < \lambda\}.$$

It is easily checked that

$$\mathcal{L}V(x_1, x_2, i) = -x_2^2 f(x_1, i) \leq 0$$

since $f(x_1, i) > 0$ for all $x_1 \in \mathbb{R}$. With $(X_1(0), X_2(0), \alpha(0)) = (x_1, x_2, \alpha)$, by Dynkin's formula,

$$\mathbf{E}_{x_1,x_2,\alpha} V(X_1(t), X_2(t), \alpha(t)) - V(x_1, x_2, \alpha)$$
$$= \mathbf{E}_{x_1,x_2,\alpha} \int_0^t \mathcal{L}V(X_1(u), X_2(u), \alpha(u))du \leq 0.$$

So $\mathbf{E}_{x_1,x_2,\alpha} V(X_1(t), X_2(t), \alpha(t)) \leq V(x_1, x_2, \alpha)$, and it is a supermartingale. It follows that for any $t_1 \geq 0$,

$$\mathbf{P}_{x_1,x_2,\alpha} \left(\sup_{t_1 \leq t < \infty} V(X_1(t), X_2(t), \alpha(t)) \geq \lambda \right)$$
$$\leq \frac{\mathbf{E}_{x_1,x_2,\alpha} V(X_1(t_1), X_2(t_1), \alpha(t))}{\lambda}.$$

By virtue of Proposition 9.9, it can be shown that with probability 1, relative to

$$\Omega_\lambda = \{\omega : \sup V(X_1(t), X_2(t), \alpha(t)) < \lambda\},$$

and

$$\mathbf{P}(\Omega_\lambda) \geq 1 - V(x_1, x_2, i)/\lambda,$$

$$(X_1(t), X_2(t)) \to G = \{(x_1, x_2) : x_2 = 0, V(x_1, x_2, i) < \lambda; i \in \mathcal{M}\}$$
$$= \{(x_1, x_2, i) : x_2 = 0, \int_0^{x_1} g(u)du < \lambda; i \in \mathcal{M}\}.$$

Another way to rephrase the notion "convergence in probability relative to Ω_λ" is $X(t) \to G$ with probability at least $1 - V(x_1, x_2, i)/\lambda$.

Referring to Theorem 9.19, the $k(x)$ there now takes the form

$$k(x) = -x_2^2 f(x_1, i) \leq 0 \text{ for each } i \in \mathcal{M}.$$

Thus by Theorem 9.19, $X(t) = (X_1(t), X_2(t))$ converges in probability to the largest invariance set whose support is contained in $\overline{G}$. For $\omega \in \Omega - \Omega_\lambda$,

$(X_1(t), X_2(t)) \to \partial O_\lambda$. It follows that $(X_1(t), X_2(t)) \to (0,0)$ in probability relative to Ω_λ. Therefore, for each $\eta > 0$, there is a $T < \infty$, such that for all $t \geq T$, $\mathbf{P}(V(X_1(t), X_2(t)) \geq \lambda_0) \leq \eta$. In particular,

$$\mathbf{P}(V(X_1(t), X_2(t)) \geq \lambda_0) \leq \eta. \tag{9.47}$$

By the supermartingale inequality,

$$\mathbf{P}_{x_1, x_2, \alpha} \left(\sup_{T \leq t < \infty} V(X_1(t), X_2(t), \alpha(t)) \geq \lambda_0 \right)$$
$$\leq \frac{\mathbf{E}_{x_1, x_2, \alpha} V(X_1(T), X_2(T), \alpha(T))}{\lambda_0}. \tag{9.48}$$

Then using (9.47),

$$\mathbf{E}_{x_1, x_2, \alpha} V(X_1(T), X_2(T), \alpha(T)) \leq \eta(1 - \eta) + \eta\lambda.$$

Thus,

$$\mathbf{P}_{x_1, x_2, \alpha} \left(\sup_{T \leq t < \infty} V(X_1(t), X_2(t), \alpha(t)) \geq \lambda_0 \right) \leq \frac{\eta(1 + \lambda)}{\lambda_0}.$$

Because η and λ_0 are arbitrary, $(X_1(t), X_2(t)) \to (0,0)$ almost surely relative to Ω_λ. However, O_λ is also arbitrary and bounded for any $0 < \lambda$. Since $\mathbf{P}(\Omega_\lambda) \to 1$ as $\lambda \to \infty$, $(X_1(t), X_2(t)) \to (0,0)$ almost surely.

9.5 Notes

For systems running for a long time, it is crucial to learn their long-run behavior; see [92, 116, 183] for recent progress on stability of such systems. The rapid progress in natural science, life science, engineering, as well as in social science demands the consideration of stability of such systems. In fact, the advent of switching diffusions is largely because of the practical needs in modeling complex dynamic systems; see [7, 52, 74, 92, 116, 168, 180, 183, 187, 188] for some of the recent studies.

Most works to date has been concentrated on Markov-modulated diffusions, in which the Brownian motion and the switching force are independent, whereas less is known for systems with continuous-component dependent switching processes. As demonstrated in Chapter 2 (see also [188]), when x-dependent switching diffusions are encountered, even such properties as continuous and smooth dependence on the initial data are nontrivial and fairly difficult to establish. Nevertheless, studying such systems is both practically useful and theoretically interesting. In our recent work, basic properties such as recurrence, positive recurrence, and ergodicity are studied in [187]; stability is treated in [92]; stability of randomly switching ordinary differential equations is treated in [190].

This chapter has taken up the issue of examination of the invariance principle akin to LaSalle's theorem for deterministic systems [61, 62, 106]. Previous study of invariance principles for stochastic systems can be found in [98, 117].

In this chapter, two different approaches are used to study the invariance. The first one is inspired by the work of Mao [117] using kernels of Liapunov functions. The second one uses the approach of the measure-theoretic viewpoint of Kushner [98]. The results obtained can also be adopted to treat random-switching ordinary differential equations.

Part IV

Two-Time-Scale Modeling and Applications

10
Positive Recurrence: Weakly Connected Ergodic Classes

10.1   Introduction

To study the positive recurrence and ergodicity, one of the conditions used in Chapters 3 and 4 is that the states of the switching process belong to only one ergodic class. In this chapter, we further our study by treating a more general class of problems. We consider the case that the states of the discrete event process belong to several "ergodic" classes that are weakly connected. This notion is made more precise in what follows. A key idea is the use of two-time-scale formulation; see [176, 177] and many references therein.

The rest of the chapter is arranged as follows. Section 10.2 begins with the formulation. Section 10.3 focuses on hybrid diffusions whose discrete component lives in weakly connected "ergodic" (irreducible) classes. Finally, the chapter is concluded with additional remarks in Section 10.4.

10.2   Problem Setup and Notation

Let $x \in \mathbb{R}^r$, $\mathcal{M} = \{1, \ldots, m_0\}$, and $Q(x) = (q_{ij}(x))$ an $m_0 \times m_0$ matrix depending on x satisfying that for any $x \in \mathbb{R}^r$ and $i \in \mathcal{M}$, $q_{ij}(x) \geq 0$ for $i \neq j$ and $\sum_{j=1}^{m_0} q_{ij}(x) = 0$. Consider a switching diffusion process $Y(t) = (X(t), \alpha(t))$, which has two components, an r-dimensional diffusion component $X(t)$ and a jump component $\alpha(t)$ taking value in $\mathcal{M} = \{1, \ldots, m_0\}$ representing discrete events. The hybrid diffusion process

G.G. Yin and C. Zhu, *Hybrid Switching Diffusions: Properties and Applications*,
Stochastic Modelling and Applied Probability 63, DOI 10.1007/978-1-4419-1105-6_10,
© Springer Science + Business Media, LLC 2010

$Y(t) = (X(t), \alpha(t))$ satisfies

$$dX(t) = b(X(t), \alpha(t))dt + \sigma(X(t), \alpha(t))dw(t),$$
$$X(0) = x, \ \alpha(0) = \alpha, \tag{10.1}$$

and

$$\mathbf{P}(\alpha(t + \Delta t) = j | \alpha(t) = i, X(s), \alpha(s), s \le t)$$
$$= q_{ij}(X(t))\Delta t + o(\Delta t), \ i \ne j, \tag{10.2}$$

where $w(t)$ is a d-dimensional standard Brownian motion, $b(\cdot, \cdot) : \mathbb{R}^r \times \mathcal{M} \mapsto \mathbb{R}^r$, and $\sigma(\cdot, \cdot) : \mathbb{R}^r \times \mathcal{M} \mapsto \mathbb{R}^{r \times d}$ satisfying

$$b(x, i) = (b_j(x, i)) \in \mathbb{R}^r, \ \sigma(x, i)\sigma'(x, i) = a(x, i) = (a_{jk}(x, i)) \in \mathbb{R}^{r \times r},$$

with z' denoting the transpose of z for $z \in \mathbb{R}^{\iota_1 \times \iota_2}$ and $\iota_1, \iota_2 \ge 1$.

Associated with the process given in (10.1) and (10.2), there is a generator $\mathcal{L}_0$ defined as follows. For each $i \in \mathcal{M}$ and for any twice continuously differentiable function $g(\cdot, i)$, let

$$\mathcal{L}_0 g(x, i) = \frac{1}{2}\text{tr}(a(x, i)\nabla^2 g(x, i)) + b'(x, i)\nabla g(x, i) + Q(x)g(x, \cdot)(i), \tag{10.3}$$

where $\nabla g(\cdot, i)$ and $\nabla^2 g(\cdot, i)$ denote the gradient and Hessian of $g(\cdot, i)$, respectively, and

$$Q(x)g(x, \cdot)(i) = \sum_{j=1}^{m_0} q_{ij}(x)g(x, j)$$
$$= \sum_{j \ne i, j \in \mathcal{M}} q_{ij}(x)(g(x, j) - g(x, i)), \ i \in \mathcal{M}. \tag{10.4}$$

For further references on stochastic differential equations involving Poisson measures describing the evolution of the jump processes, we refer the reader to Chapter 2 of this book; see also Skorohod [150].

Throughout the chapter, we assume that for each $i \in \mathcal{M}$, both $b(\cdot, i)$ and $\sigma(\cdot, i)$ satisfy the usual local Lipschitz and linear growth conditions. It is well known that under these conditions, the system (10.1)–(10.2) has a unique solution; see Chapter 2 of this book and also [150] for details. In what follows, denote the solution by $Y^{x, \alpha}(t) = (X^{x, \alpha}(t), \alpha^{x, \alpha}(t))$ to emphasize the dependence on the initial data when needed.

10.3 Weakly Connected, Multiergodic-Class Switching Processes

As mentioned, one of the conditions used in Chapters 3 and 4 is that the states of the jump component are in a single ergodic class. This section deals with the situation that weakly connected, multiple ergodic classes are included. We assume that the discrete jump component is generated by

$$Q^\varepsilon = \frac{1}{\varepsilon}\widetilde{Q} + \widehat{Q}, \qquad (10.5)$$

where $0 < \varepsilon \ll 1$, and $\widetilde{Q}$ and $\widehat{Q}$ are themselves generators of certain Markov chains. Corresponding to (10.5), the states of the switching process live in a number of ergodic classes that are weakly connected through the generator $\widehat{Q}$. We say switching processes with such a structure have multiergodic classes that are weakly connected. This section is divided into two parts. We first concern ourselves with the case that the jump components are divided into l recurrent classes. Later, we consider the case that transient states are also included.

10.3.1 Preliminary

Before proceeding further, let us give the motivation for using such models. First, we note that Q^ε is a constant matrix independent of x. The rationale is similar to those considered in the previous chapters, which can be considered as linearizing $Q(x)$ at "point of ∞." We may begin with an x-dependent matrix, say $Q^\varepsilon(x)$. Then for x large enough, it can be replaced by a constant Q^ε. To be more precise, assume

$$Q^\varepsilon(x) = Q^\varepsilon + o(1), \quad \text{as } |x| \to \infty, \qquad (10.6)$$

where Q^ε has the form (10.5). Note that in the previous chapters, a condition similar to (10.6) was used without the ε-dependence, but the corresponding constant matrix (the limit at $|x| \to \infty$) is a generator of an ergodic Markov chain. Here, we are mostly concerned with the case that the generator could possibly be reducible with several ergodic classes. Nevertheless, the states belonging to different ergodic classes are not completely separable. They are linked together through weak interaction due to the presence of the slow part of the generator $\widetilde{Q}$.

The formulation of Q^ε being a generator of an ε-dependent Markov chain for a small parameter ε, stems from an effort of using two-time-scale models to reduce the complexity of the underlying systems. It has been observed in [149] that there are natural hierarchical structures in many large-scale and complex systems. The formulation in (10.5) is an effort to highlight the different parts of subsystems varying at different rates. It is often possible to partition the system states into a number of groups so that within each group, the state transitions take place rapidly, whereas among different groups, the changes are relatively infrequent. Such scenarios, in fact, appear in many applications. Thus an effective way is to treat the systems through decompositions and aggregations. Loosely, one can use the natural scales shown in the system to aggregate the states in each ergodic class into one state. In this way, the total number of states is much reduced for the aggregated system. This point of view was exclusively discussed in Yin and Zhang [176]. To begin, one may not have an ε in the system, but it is brought into the formulation to highlight the different rates of change so as to separate the fast and slow motions.

10.3.2 Weakly Connected, Multiple Ergodic Classes

Suppose

$$\widetilde{Q} = \mathrm{diag}(\widetilde{Q}^1, \ldots, \widetilde{Q}^l). \tag{10.7}$$

In view of (10.7), the state space $\mathcal{M}$ of the underlying Markov chain is decomposable into l subspaces. That is, we can relabel the states so that

$$\mathcal{M} = \mathcal{M}_1 \cup \mathcal{M}_2 \cup \cdots \cup \mathcal{M}_l, \tag{10.8}$$

with $\mathcal{M}_i = \{s_{i1}, \ldots, s_{im_i}\}$ and $m_0 = m_1 + m_2 + \cdots + m_l$ such that $\widetilde{Q}^i$, the generator associated with the subspace $\mathcal{M}_i$ for each $i = 1, \ldots, l$ is irreducible. Thus the corresponding $\mathcal{M}_i$ for $i = 1, \ldots, l$ consist of recurrent states belonging to l ergodic classes.

To signal the ε-dependence, we index the process by ε and write it as $Y^\varepsilon(t) = (X^\varepsilon(t), \alpha^\varepsilon(t))$. Then (10.1) and (10.2) become

$$dX^\varepsilon(t) = b(X^\varepsilon(t), \alpha^\varepsilon(t))dt + \sigma(X^\varepsilon(t), \alpha^\varepsilon(t))dw(t),$$
$$X^\varepsilon(0) = x, \ \alpha(0) = \alpha, \tag{10.9}$$

and

$$\mathbf{P}(\alpha^\varepsilon(t + \Delta) = j | \alpha^\varepsilon(t) = i, X^\varepsilon(s), \alpha^\varepsilon(s), s \le t) = q_{ij}^\varepsilon \Delta + o(\Delta), \quad i \neq j. \tag{10.10}$$

The associated operator for the switching diffusion is given by

$$\mathcal{L}^\varepsilon g(x, \iota) = \frac{1}{2}\mathrm{tr}(a(x, \iota)\nabla^2 g(x, \iota)) + b'(x, \iota)\nabla g(x, \iota) + Q^\varepsilon g(x, \cdot)(\iota), \ \iota \in \mathcal{M}. \tag{10.11}$$

To proceed, lump the states of the jump component in each $\mathcal{M}_i$ into a single state and define

$$\overline{\alpha}^\varepsilon(t) = i \text{ if } \alpha^\varepsilon(t) \in \mathcal{M}_i. \tag{10.12}$$

Denote the state space of $\overline{\alpha}^\varepsilon(\cdot)$ by $\overline{\mathcal{M}} = \{1, \ldots, l\}$, and $\tilde{\nu} = \mathrm{diag}(\nu^1, \ldots, \nu^l)$, where ν^i is the stationary distribution corresponding to $\widetilde{Q}^i$. Define

$$\overline{Q} = \tilde{\nu}\widehat{Q}\mathbb{1} \tag{10.13}$$

with $\mathbb{1} = \mathrm{diag}(\mathbb{1}_{m_1}, \ldots, \mathbb{1}_{m_l})$ and $\mathbb{1}_\ell = (1, \ldots, 1)' \in \mathbb{R}^\ell$. The essence of the aggregated process is to treat all the states in $\mathcal{M}_i$ as one state, so the total number of states in the "effective" state space is much reduced. We need the following assumption about the generator of $\alpha^\varepsilon(\cdot)$.

(A10.1) For each $i = 1, \ldots, l$, $\widetilde{Q}^i$ is irreducible.

Lemma 10.1. *Under* (A10.1), *the following assertions hold.*

(a) *The probability vector* $p^\varepsilon(t) \in \mathbb{R}^{1 \times m_0}$ *with*

$$p^\varepsilon(t) = (\mathbf{P}(\alpha^\varepsilon(t) = s_{ij}), i = 1, \ldots, l, j = 1, \ldots, m_i),$$

satisfies

$$p^\varepsilon(t) = \theta(t)\tilde{\nu} + O(\varepsilon(t+1) + e^{-\kappa_0 t/\varepsilon})$$

for some $\kappa_0 > 0$, *and* $\theta(t) = (\theta_1(t), \ldots, \theta_l(t)) \in \mathbb{R}^{1 \times l}$ *satisfies*

$$\frac{d\theta(t)}{dt} = \theta(t)\overline{Q}, \ \theta(0) = p(0)\mathbb{1}.$$

(b) *The transition matrix satisfies*

$$P^\varepsilon(t) = P^{(0)}(t) + O(\varepsilon(t+1) + e^{-\kappa_0 t/\varepsilon}),$$

where $P^{(0)}(t) = \mathbb{1}\Theta(t)\tilde{\nu}$ *and*

$$\frac{d\Theta(t)}{dt} = \Theta(t)\overline{Q}, \ \Theta(0) = I.$$

(c) *The aggregated process* $\overline{\alpha}^\varepsilon(\cdot)$ *converges weakly to* $\overline{\alpha}(\cdot)$ *as* $\varepsilon \to 0$, *where* $\overline{\alpha}(\cdot)$ *is a Markov chain generated by* $\overline{Q}$.

(d) *For* $i = 1, \ldots, l$, $j = 1, \ldots, m_i$,

$$\mathbf{E}\left[\int_0^\infty e^{-t}(I_{\{\alpha^\varepsilon(t)=s_{ij}\}} - \nu_j^i I_{\{\overline{\alpha}^\varepsilon(t)=i\}})dt\right]^2 = O(\varepsilon),$$

where ν_j^i *denotes the* j*th component of* ν^i *for* $i = 1, \ldots, l$ *and* $j = 1, \ldots, m_i$.

Proof. The proofs of (a) and (b) in Lemma 10.1 can be found in [176, Corollary 6.12, p. 130]. The proof of (c) is based on the martingale averaging method [176, Theorem 7.4, p. 172]; an outline of the idea is in [4], and a discrete version of the approximation may be found in [177].

As for (d), using (a) and (b), direct calculation reveals that

$$\mathbf{E}\left[\int_0^\infty e^{-t}(I_{\{\alpha^\varepsilon(t)=s_{ij}\}} - \nu_j^i I_{\{\overline{\alpha}^\varepsilon(t)=i\}})dt\right]^2$$

$$= \int_0^\infty \int_0^t e^{-t-s} O\left(\varepsilon(t+1) + e^{-\kappa_0(t-s)/\varepsilon}\right) ds dt$$

$$+ \int_0^\infty \int_0^s e^{-t-s} O\left(\varepsilon(t+1) + e^{-\kappa_0(s-t)/\varepsilon}\right) dt ds.$$

Detailed calculations then yield the desired result.   □

Due to the aggregation, certain averages take place. When ε goes to 0, we obtain a limit system. The following lemma records this fact, whose proof can be found in, for example, Yin and Zhang [176], and Yin [165]. The basic idea is to use martingale averaging; we omit the details here.

Lemma 10.2. *Assume* (A10.1). *Then* $(X^\varepsilon(\cdot), \overline{\alpha}^\varepsilon(\cdot))$ *converges weakly to* $(X(\cdot), \overline{\alpha}(\cdot))$, *whose operator is given by*

$$\mathcal{L}g(x,i) = \frac{1}{2}\mathrm{tr}(\overline{a}(x,i)\nabla^2 g(x,i)) + \overline{b}'(x,i)\nabla g(x,i) + \overline{Q}g(x,\cdot)(i), \quad (10.14)$$

where

$$\overline{b}(x,i) = \sum_{j=1}^{m_i} \nu_j^i b(x, s_{ij}),$$

$$\overline{a}(x,i) = \sum_{j=1}^{m_i} \nu_j^i a(x, s_{ij}), \quad i = 1, \dots, l. \quad (10.15)$$

Remark 10.3. In view of the weak convergence result, the limit stochastic differential equation for (10.9) is

$$dx = \overline{b}(X(t), \overline{\alpha}(t))dt + \overline{\sigma}(X(t), \overline{\alpha}(t))dw, \quad (10.16)$$

where $\overline{\sigma}(x, i)$ is defined in terms of the average in (10.15); that is,

$$\overline{\sigma}(x,i)\overline{\sigma}'(x,i) = \overline{a}(x,i).$$

(A10.2) $\overline{Q}$ is irreducible. For each $i \in \overline{\mathcal{M}}$, $\overline{a}(x, i)$ satisfies

$$\kappa_1|\xi|^2 \le \xi'\overline{a}(x,i)\xi, \quad \text{for all } \xi \in \mathbb{R}^r, \quad (10.17)$$

with some constant $\kappa_1 \in (0, 1]$ for all $x \in \mathbb{R}^r$.

With the conditions given, using the techniques of Chapter 3, to be more specific, by virtue of Theorem 3.26, we establish the following assertion.

Proposition 10.4. *Under* (A10.1) *and* (A10.2), *the switching diffusion with generator given by* (10.14) *is positive recurrent.*

We now present the result on positive recurrence of the underlying process. The main idea is that although the discrete events, described by a continuous-time Markov chain may have several weakly connected ergodic classes, when ε is sufficiently small, we still have positive recurrence for the process $(X^\varepsilon(\cdot), \alpha^\varepsilon(\cdot))$. The proof rests upon the use of perturbed Liapunov function methods, which were first used to treat diffusion approximations in [131] by Papanicolaou, Stroock, and Varadhan, and later on have been successfully used in stochastic systems theory (see Kushner [102]), and stochastic approximation (see Kushner and Yin [104]), among others. The basic idea is to introduce perturbations of Liapunov functions that are small in magnitude and that result in the desired cancellation of unwanted terms.

Theorem 10.5. *Assume that conditions* (A10.1) *and* (A10.2) *hold. Then for sufficiently small* $\varepsilon > 0$, *the process* $(X^\varepsilon(\cdot), \alpha^\varepsilon(\cdot))$ *is positive recurrent.*

Remark 10.6. First note that by Proposition 10.5, the process $(X(\cdot), \overline{\alpha}(\cdot))$ is positive recurrent. Then it follows that there are Liapunov functions $V(x, i)$ for $i = 1, \ldots, l$ for the limit system (10.16) such that

$$\mathcal{L}V(x, i) \leq -\gamma \quad \text{for some} \quad \gamma > 0. \tag{10.18}$$

In Theorem 10.5, the meaning of the property holds for sufficiently small ε. That is, there exists an $\varepsilon_0 > 0$ such that for all $0 < \varepsilon \leq \varepsilon_0$, the property holds.

Proof. To prove this result, we begin with the Liapunov function in (10.18) for the limit system. Choose n_0 to be an integer large enough so that D is contained in the ball $\{|x| < n_0\}$. For any $(x, i) \in D^c \times \mathcal{M}$, any $t > 0$, any positive integer $n > n_0$, define

$$\tau_D = \inf\{t : X^\varepsilon(t) \in D\}, \quad \text{and} \quad \tau_{D,n}(t) = t \wedge \tau_D \wedge \beta_n, \tag{10.19}$$

where β_n comes from the regularity consideration and is the first exit time of the process $(X(t), \overline{\alpha}(t))$ from the set $\{\widetilde{x} : |\widetilde{x}| < n\} \times \mathcal{M}$; this is, $\beta_n = \inf\{t : |X(t)| = n\}$. For $i \in \overline{\mathcal{M}}$, use the Liapunov function $V(x, i)$ in (10.18) to define

$$\overline{V}(x, \alpha) = \sum_{i=1}^{l} V(x, i) I_{\{\alpha \in \mathcal{M}_i\}} = V(x, i) \quad \text{if} \quad \alpha \in \mathcal{M}_i. \tag{10.20}$$

Note that

$$\overline{V}(X^\varepsilon(t), \alpha^\varepsilon(t)) = V(X^\varepsilon(t), \overline{\alpha}^\varepsilon(t)), \tag{10.21}$$

so these two expressions are to be used interchangeably in what follows. Observe that $\overline{V}(x, \alpha)$ is orthogonal to $\widetilde{Q}$; that is, $\widetilde{Q}\overline{V}(x, \cdot)(\alpha) = 0$. Then

$$\begin{aligned}
\mathcal{L}^\varepsilon \overline{V}(X^\varepsilon(t), \alpha^\varepsilon(t)) &= \overline{V}_x(X^\varepsilon(t), \alpha^\varepsilon(t)) b(X^\varepsilon(t), \alpha^\varepsilon(t)) \\
&+ \frac{1}{2} \mathrm{tr}[\overline{V}_{xx}(X^\varepsilon(t), \alpha^\varepsilon(t)) a(X^\varepsilon(t), \alpha^\varepsilon(t))] \\
&+ \widehat{Q}\overline{V}(X^\varepsilon(t), \cdot)(\alpha^\varepsilon(t)).
\end{aligned} \tag{10.22}$$

By virtue of Dynkin's formula,

$$\begin{aligned}
\mathbf{E}_{x,i}\overline{V}(X^\varepsilon(\tau_{D,n}(t)), \alpha^\varepsilon(\tau_{D,n}(t))) &- \overline{V}(x, i) \\
&= \mathbf{E}_{x,i} \int_0^{\tau_{D,n}(t)} \mathcal{L}^\varepsilon \overline{V}(X^\varepsilon(s), \alpha^\varepsilon(s)) ds,
\end{aligned} \tag{10.23}$$

which involves unwanted terms on the right-hand side. To get rid of these terms, we use methods of perturbed Liapunov functions [104, 176] to average out the "bad" terms. Note that the process $(X^\varepsilon(t), \alpha^\varepsilon(t))$ is Markov. Thus, for a suitable function $\xi(\cdot)$, $\mathcal{L}^\varepsilon \xi(t)$ can be calculated by

$$\mathcal{L}^\varepsilon \xi(t) = \lim_{\delta \to 0} \mathbf{E}_t^\varepsilon \frac{\xi(t+\delta) - \xi(t)}{\delta}, \qquad (10.24)$$

where $\mathbf{E}_t^\varepsilon$ denotes the conditional expectation with respect to the σ-algebra

$$\mathcal{F}_t^\varepsilon = \sigma\{(X^\varepsilon(s), \alpha^\varepsilon(s)) : s \le t\}.$$

To obtain the desired results, we introduce three perturbations. The integrand of each of these perturbations is formed by the difference of two terms, an original term and its "average." The goal is to use the averages in the final form for the evaluation of the Liapunov function. To ensure the integrability in an infinite horizon, exponential discounting is used in the integrals.

Define

$$V_1^\varepsilon(x, \alpha, t) = \mathbf{E}_t^\varepsilon \int_t^\infty e^{t-u} \overline{V}_x(x, \alpha) \left[b(x, \alpha^\varepsilon(u)) - \overline{b}(x, \overline{\alpha}^\varepsilon(u)) \right] du,$$

$$V_2^\varepsilon(x, \alpha, t) = \mathbf{E}_t^\varepsilon \int_t^\infty e^{t-u} \frac{1}{2} \mathrm{tr}[\overline{V}_{xx}(x, \alpha)(a(x, \alpha^\varepsilon(u)) - \overline{a}(x, \overline{\alpha}^\varepsilon(u)))] du,$$

$$V_3^\varepsilon(x, t) = \mathbf{E}_t^\varepsilon \int_t^\infty e^{t-u} [\widehat{QV}(x, \cdot)(\alpha^\varepsilon(u)) - \overline{Q}V(x, \cdot)(\overline{\alpha}^\varepsilon(u))] du.$$

$$(10.25)$$

To proceed, we first state a lemma.

Lemma 10.7. *Assume the conditions of Theorem* 10.5. *Then the following assertions hold.*

(a) *For V_i^ε with $i = 1, 2, 3$, we have the following estimates:*

$$V_1^\varepsilon(X^\varepsilon(t), \alpha^\varepsilon(t), t) = O(\varepsilon)(\overline{V}(X^\varepsilon(t), \alpha^\varepsilon(t)) + 1),$$

$$V_2^\varepsilon(X^\varepsilon(t), \alpha^\varepsilon(t), t) = O(\varepsilon)(\overline{V}(X^\varepsilon(t), \alpha^\varepsilon(t)) + 1), \qquad (10.26)$$

$$V_3^\varepsilon(X^\varepsilon(t), t) = O(\varepsilon)(\overline{V}(X^\varepsilon(t), \alpha^\varepsilon(t)) + 1).$$

(b) *Moreover,*

$$\mathcal{L}^\varepsilon V_1^\varepsilon(X^\varepsilon(t), \alpha^\varepsilon(t), t)$$
$$= -\overline{V}_x(X^\varepsilon(t), \alpha^\varepsilon(t))[b(X^\varepsilon(t), \alpha^\varepsilon(t)) - \overline{b}(X^\varepsilon(t), \overline{\alpha}^\varepsilon(t))]$$
$$+ O(\varepsilon)(\overline{V}(X^\varepsilon(t), \alpha^\varepsilon(t)) + 1),$$

$$\mathcal{L}^\varepsilon V_2^\varepsilon(X^\varepsilon(t), \alpha^\varepsilon(t), t)$$
$$= -\frac{1}{2}\text{tr}[\overline{V}_{xx}(X^\varepsilon(t), \alpha^\varepsilon(t))(a(X^\varepsilon(t), \alpha^\varepsilon(t)) - \overline{a}(X^\varepsilon(t), \overline{\alpha}^\varepsilon(t))]$$
$$+ O(\varepsilon)(\overline{V}(X^\varepsilon(t), \alpha^\varepsilon(t)) + 1),$$

$$\mathcal{L}^\varepsilon V_3^\varepsilon(X^\varepsilon(t), t) = -\widehat{QV}(X^\varepsilon(t), \cdot)(\alpha^\varepsilon(t)) + \overline{Q}V(X^\varepsilon(t), \cdot)(\overline{\alpha}^\varepsilon(t))$$
$$+ O(\varepsilon)(\overline{V}(X^\varepsilon(t), \alpha^\varepsilon(t)) + 1). \tag{10.27}$$

Proof. The proof is inspired by the technique of perturbed Liapunov function methods of Kushner and Yin in [104, Chapter 6], and the construction of the Liapunov functions is along the line of Badowski and Yin [4]. By the definition of $\overline{b}(\cdot)$,

$$b(X^\varepsilon(t), \alpha^\varepsilon(u)) - \overline{b}(X^\varepsilon(t), \overline{\alpha}^\varepsilon(u))$$
$$= \sum_{i=1}^{l}\sum_{j=1}^{m_i} b(X^\varepsilon(t), s_{ij})[I_{\{\alpha^\varepsilon(u)=s_{ij}\}} - \nu_j^i I_{\{\alpha^\varepsilon(u)\in\mathcal{M}_i\}}]. \tag{10.28}$$

It can be argued by using the Markov property and the two-time-scale structure (see [176, p. 187]) that for $u \geq t$,

$$\mathbf{E}_t^\varepsilon[I_{\{\alpha^\varepsilon(u)=s_{ij}\}} - \nu_j^i I_{\{\overline{\alpha}^\varepsilon(u)=i\}}] = O\left(\varepsilon + e^{-\kappa_0(u-t)/\varepsilon}\right). \tag{10.29}$$

Thus, using (10.21), (10.28), and (10.29) in (10.25), we obtain

$$|V_1^\varepsilon(X^\varepsilon(t), \alpha^\varepsilon(t), t)| \leq \sum_{i=1}^{l}\sum_{j=1}^{m_i} |\overline{V}_x(X^\varepsilon(t), \alpha^\varepsilon(t))b(X^\varepsilon(t), s_{ij})|$$
$$\times \left|\int_t^\infty e^{t-u}O\left(\varepsilon + e^{-\kappa_0(u-t)/\varepsilon}\right) du\right| \tag{10.30}$$
$$= O(\varepsilon)(\overline{V}(X^\varepsilon(t), \alpha^\varepsilon(t)) + 1).$$

Likewise, we obtain the estimates for $V_2^\varepsilon(X^\varepsilon(t), \alpha^\varepsilon(t), t)$ and $V_3^\varepsilon(X^\varepsilon(t), t)$ in (10.26). This establishes statement (a).

Next, we prove (b). For convenience, introduce a notation

$$\Gamma(x, \alpha, \alpha_1) = \overline{V}_x(x, \alpha)[b(x, \alpha_1) - \overline{b}(x, \overline{\alpha}_1)]. \tag{10.31}$$

By virtue of the definition of $\mathcal{L}^\varepsilon$, we have

$$
\begin{aligned}
\mathcal{L}^\varepsilon & V_1^\varepsilon(X^\varepsilon(t), \alpha^\varepsilon(t), t) \\
&= -\lim_{\delta \to 0} \frac{1}{\delta} \int_t^{t+\delta} \mathbf{E}_t^\varepsilon e^{t-u} \Gamma(X^\varepsilon(t), \alpha^\varepsilon(t), \alpha^\varepsilon(u)) du \\
&\quad + \lim_{\delta \to 0} \frac{1}{\delta} \int_{t+\delta}^\infty \mathbf{E}_t^\varepsilon [e^{t+\delta-u} - e^{t-u}] \Gamma(X^\varepsilon(t), \alpha^\varepsilon(t), \alpha^\varepsilon(u)) du \\
&\quad + \lim_{\delta \to 0} \int_{t+\delta}^\infty e^{t+\delta-u} \mathbf{E}_t^\varepsilon [\Gamma(X^\varepsilon(t+\delta), \alpha^\varepsilon(t), \alpha^\varepsilon(u)) \\
&\qquad - \Gamma(X^\varepsilon(t), \alpha^\varepsilon(t), \alpha^\varepsilon(u))] du \\
&\quad + O(\varepsilon)(\overline{V}(X^\varepsilon(t), \alpha^\varepsilon(t)) + 1).
\end{aligned}
$$

Moreover,

$$
\begin{aligned}
-\lim_{\delta \to 0} \frac{1}{\delta} \int_t^{t+\delta} & \mathbf{E}_t^\varepsilon e^{t-u} \Gamma(X^\varepsilon(t), \alpha^\varepsilon(t), \alpha^\varepsilon(u)) du \\
&= -\overline{V}_x(X^\varepsilon(t), \alpha^\varepsilon(t)) \left[b(X^\varepsilon(t), \alpha^\varepsilon(t)) - \overline{b}(X^\varepsilon(t), \overline{\alpha}^\varepsilon(t)) \right],
\end{aligned}
\tag{10.32}
$$

and

$$
\begin{aligned}
\lim_{\delta \to 0} \frac{1}{\delta} \int_{t+\delta}^\infty & \mathbf{E}_t^\varepsilon [e^{t+\delta-u} - e^{t-u}] \Gamma(X^\varepsilon(t), \alpha^\varepsilon(t), \alpha^\varepsilon(u)) du \\
&= \int_t^\infty e^{t-u} \mathbf{E}_t^\varepsilon \Gamma(X^\varepsilon(t), \alpha^\varepsilon(t), \alpha^\varepsilon(u)) du = V_1^\varepsilon(X^\varepsilon(t), \alpha^\varepsilon(t)).
\end{aligned}
\tag{10.33}
$$

The independence of $\alpha^\varepsilon(\cdot)$ and $w(\cdot)$, (10.28), and (10.31) lead to that for $u \geq t$,

$$
\begin{aligned}
\mathbf{E}_t^\varepsilon & [\Gamma(X^\varepsilon(t+\delta), \alpha^\varepsilon(t), \alpha^\varepsilon(u)) - \Gamma(X^\varepsilon(t), \alpha^\varepsilon(t), \alpha^\varepsilon(u))] \\
&= \sum_{i=1}^l \sum_{j=1}^{m_i} \mathbf{E}_t^\varepsilon [\overline{V}_x(X^\varepsilon(t+\delta), \alpha^\varepsilon(t)) b(X^\varepsilon(t+\delta), s_{ij}) \\
&\quad - \overline{V}_x(X^\varepsilon(t), \alpha^\varepsilon(t)) b(X^\varepsilon(t), s_{ij})] \mathbf{E}_{t+\delta}^\varepsilon [I_{\{\alpha^\varepsilon(u)=s_{ij}\}} - \nu_j^i I_{\{\overline{\alpha}^\varepsilon(u) \in \mathcal{M}_i\}}].
\end{aligned}
\tag{10.34}
$$

Furthermore, we have

$$
\begin{aligned}
\lim_{\delta \to 0} \frac{1}{\delta} \int_{t+\delta}^\infty & e^{t+\delta-u} \mathbf{E}_t^\varepsilon [\Gamma(X^\varepsilon(t+\delta), \alpha^\varepsilon(t), \alpha^\varepsilon(u)) \\
&\quad - \Gamma(X^\varepsilon(t), \alpha^\varepsilon(t), \alpha^\varepsilon(u))] du \\
&= \lim_{\delta \to 0} \frac{1}{\delta} \int_{t+\delta}^\infty e^{t-u} \mathbf{E}_t^\varepsilon [\Gamma(X^\varepsilon(t+\delta), \alpha^\varepsilon(t), \alpha^\varepsilon(u)) \\
&\quad - \Gamma(X^\varepsilon(t), \alpha^\varepsilon(t), \alpha^\varepsilon(u))] du + o(1),
\end{aligned}
$$

where $o(1) \to 0$ in probability uniformly in t. Using (10.29), (10.31), and (10.34),

$$
\begin{aligned}
\sum_{i=1}^{l} \sum_{j=1}^{m_i} & \left(\overline{V}_x(X^\varepsilon(t), \alpha^\varepsilon(t)) b(X^\varepsilon(t), s_{ij}) \right)_x b(X^\varepsilon(t), \alpha^\varepsilon(t)) \\
& \times \mathbf{E}_t^\varepsilon \int_t^\infty e^{t-u} \left[I_{\{\alpha^\varepsilon(u) = s_{ij}\}} - \nu_j^i I_{\{\overline{\alpha}^\varepsilon(u)=i\}} \right] du \\
& = O(\varepsilon)(\overline{V}(X^\varepsilon(t), \alpha^\varepsilon(t)) + 1),
\end{aligned}
\tag{10.35}
$$

and

$$
\begin{aligned}
\sum_{i=1}^{l} \sum_{j=1}^{m_i} & \frac{1}{2} \mathrm{tr}[(V_x(X^\varepsilon(t), \alpha^\varepsilon(t)) b(X^\varepsilon(t), s_{ij}))_{xx} a(X^\varepsilon(t), \alpha^\varepsilon(t))] \\
& \times \mathbf{E}_t^\varepsilon \int_t^\infty e^{t-u} \left[I_{\{\alpha^\varepsilon(u) = s_{ij}\}} - \nu_j^i I_{\{\overline{\alpha}^\varepsilon(u)=i\}} \right] du \\
& = O(\varepsilon)(\overline{V}(X^\varepsilon(t), \alpha^\varepsilon(t)) + 1).
\end{aligned}
\tag{10.36}
$$

Thus, (10.32), (10.33), (10.35), and (10.36) lead to

$$
\begin{aligned}
\mathcal{L}^\varepsilon V_1^\varepsilon & (X^\varepsilon(t), \alpha^\varepsilon(t), t) \\
& = -\overline{V}_x(X^\varepsilon(t), \alpha^\varepsilon(t))[b(X^\varepsilon(t), \alpha^\varepsilon(t)) - \overline{b}(X^\varepsilon(t), \overline{\alpha}^\varepsilon(t))] \\
& \quad + O(\varepsilon)(\overline{V}(X^\varepsilon(t), \alpha^\varepsilon(t)) + 1).
\end{aligned}
$$

Similar calculations enable us to obtain the rest of the estimates in part (b) of the lemma. The proof of the lemma is concluded. □

Using Lemma 10.7, we proceed to eliminate the unwanted terms and to obtain the detailed estimates. Define

$$
\begin{aligned}
V^\varepsilon(t) = \; & \overline{V}(X^\varepsilon(t), \alpha^\varepsilon(t)) + V_1^\varepsilon(X^\varepsilon(t), \alpha^\varepsilon(t), t) \\
& + V_2^\varepsilon(X^\varepsilon(t), \alpha^\varepsilon(t), t) + V_3^\varepsilon(X^\varepsilon(t), t).
\end{aligned}
\tag{10.37}
$$

It follows that

$$
V^\varepsilon(t) = V(X^\varepsilon(t), \overline{\alpha}^\varepsilon(t)) + O(\varepsilon)(\overline{V}(X^\varepsilon(t), \alpha^\varepsilon(t)) + 1),
\tag{10.38}
$$

and

$$
\begin{aligned}
\mathcal{L}^\varepsilon V^\varepsilon(t) = \; & V_x(X^\varepsilon(t), \overline{\alpha}^\varepsilon(t)) \overline{b}(X^\varepsilon(t), \overline{\alpha}^\varepsilon(t)) \\
& + \frac{1}{2} \mathrm{tr}\,[V_{xx}(X^\varepsilon(t), \overline{\alpha}^\varepsilon(t)) \overline{a}(X^\varepsilon(t), \overline{\alpha}^\varepsilon(t))] \\
& + \overline{Q} V(X^\varepsilon(t), \cdot)(\overline{\alpha}^\varepsilon(t)) \\
& + O(\varepsilon)(\overline{V}(X^\varepsilon(t), \alpha^\varepsilon(t)) + 1).
\end{aligned}
\tag{10.39}
$$

Therefore,

$$\mathcal{L}^\varepsilon V^\varepsilon(t) = \mathcal{L}V(X^\varepsilon(t), \overline{\alpha}^\varepsilon(t)) + O(\varepsilon)(\overline{V}(X^\varepsilon(t), \alpha^\varepsilon(t)) + 1). \qquad (10.40)$$

Note that through the use of perturbations, the first term on the right-hand side above involves the limit operator and the Liapunov function of the limit system, which is crucially important.

For fixed but arbitrary $T > 0$, we then obtain

$$
\begin{aligned}
\mathbf{E}_{x,i} & V^\varepsilon(\tau_{D,n}(t) \wedge T) \\
&= \overline{V}(x,i) + \mathbf{E}_{x,i} V_1^\varepsilon(x, i, 0) + \mathbf{E}_{x,i} \int_0^{\tau_{D,n}(t) \wedge T} \mathcal{L}^\varepsilon V^\varepsilon(s) ds \\
&= \overline{V}(x,i) + \mathbf{E}_{x,i} V_1^\varepsilon(x, i, 0) + \mathbf{E}_{x,i} \int_0^{\tau_{D,n}(t) \wedge T} [\mathcal{L}V(X^\varepsilon(s), \overline{\alpha}^\varepsilon(s)) \\
&\quad + O(\varepsilon)(\overline{V}(X^\varepsilon(s), \alpha^\varepsilon(s)) + 1)] ds \\
&= \overline{V}(x,i) + \mathbf{E}_{x,i} V_1^\varepsilon(x, i, 0) \\
&\quad + \mathbf{E}_{x,i} \int_0^{\tau_{D,n}(t) \wedge T} [\mathcal{L}V(X^\varepsilon(s), \overline{\alpha}^\varepsilon(s)) + O(\varepsilon)(V^\varepsilon(s) + 1)] ds \\
&\leq \overline{V}(x,i) + \mathbf{E}_{x,i} V_1^\varepsilon(x, i, 0) \\
&\quad + [O(\varepsilon) - \gamma] \mathbf{E}_{x,i} [\tau_{D,n}(t) \wedge T] + O(\varepsilon) \int_0^{\tau_{D,n}(t) \wedge T} V^\varepsilon(s) ds.
\end{aligned}
$$
$$(10.41)$$

The expression after the third equality sign follows from the estimates in (10.26), and the expression after the last inequality sign is a consequence of (10.18).

Because $\varepsilon > 0$ is sufficiently small, there is a γ_1 such that $-\gamma_1 \geq -\gamma + O(\varepsilon)$. It is clear that for the fixed T, $\tau_{D,n}(t) \wedge T \leq T$. Thus, the foregoing together with [114, Theorem 2.6.1, p. 68] and [61, p. 36] yields that

$$
\begin{aligned}
\mathbf{E}_{x,i} & V^\varepsilon(\tau_{D,n}(t) \wedge T) \\
&\leq \Big[\overline{V}(x,i) + \mathbf{E}_{x,i} \sum_{\iota=1}^3 V_\iota^\varepsilon(x, i, 0) - \gamma_1 \mathbf{E}_{x,i}(\tau_{D,n}(t) \wedge T) \Big] \exp(\varepsilon T).
\end{aligned}
$$
$$(10.42)$$

Consequently,

$$
\begin{aligned}
\exp(-\varepsilon T) & \mathbf{E}_{x,i} V^\varepsilon(\tau_{D,n}(t) \wedge T) + \gamma_1 \mathbf{E}_{x,i} [\tau_{D,n}(t) \wedge T] \\
&\leq \overline{V}(x,i) + \mathbf{E}_{x,i} \sum_{\iota=1}^3 V_\iota^\varepsilon(x, i, 0).
\end{aligned}
$$
$$(10.43)$$

Taking limit $T \to \infty$ in (10.43) leads to

$$\gamma_1 \mathbf{E}_{x,i} \tau_{D,n}(t) \leq \Big[\overline{V}(x,i) + \mathbf{E}_{x,i} \sum_{\iota=1}^3 V_\iota^\varepsilon(x, i, 0) \Big]. \qquad (10.44)$$

By means of regularity of the underlying process, passing to the limit as $n \to \infty$, (10.44) together with Fatou's lemma yields that

$$\mathbf{E}_{x,i}[t \wedge \tau_D] \leq \overline{V}(x,i) + \mathbf{E}_{x,i} \sum_{\iota=1}^{3} V_\iota^\varepsilon(x,i,0).$$

Finally, it is easily seen that

$$\mathbf{E}_{x,i} \sum_{\iota=1}^{3} V_\iota^\varepsilon(x,i,0) < \infty.$$

In addition,

$$\mathbf{E}_{x,i}[t \wedge \tau_D] = \mathbf{E}_{x,i} \tau_D I_{\{\tau_D \leq t\}} + \mathbf{E}_{x,i} t I_{\{\tau_D > t\}}. \tag{10.45}$$

Because

$$\mathbf{P}_{x,i}(\tau_D > t) \leq \frac{1}{\alpha_1 t}[\overline{V}(x,i) + \mathbf{E}_{x,i} \sum_{\iota=1}^{3} V_\iota^\varepsilon(x,i,0)] \to 0 \quad \text{as} \quad t \to \infty,$$

we obtain $\mathbf{E}_{x,i} \tau_D < \infty$ as desired. Thus the switching diffusion is positive recurrent. $\qquad \square$

10.3.3   Inclusion of Transient Discrete Events

Here, we extend the result to the case that $\alpha^\varepsilon(t)$ includes transient states in addition to the states of ergodic classes. Let $\alpha^\varepsilon(\cdot)$ be a Markov chain with generator given by (10.5) with

$$\widetilde{Q} = \begin{pmatrix} \widetilde{Q}^1 & & & & \\ & \widetilde{Q}^2 & & & \\ & & \ddots & & \\ & & & \widetilde{Q}^l & \\ \widetilde{Q}_*^1 & \widetilde{Q}_*^2 & \cdots & \widetilde{Q}_*^l & \widetilde{Q}_* \end{pmatrix}. \tag{10.46}$$

Now the jump process is again nearly completely decomposable. However, in addition to the l recurrent classes, there are a number of transient states. That is, $\mathcal{M} = \mathcal{M}_1 \cup \cdots \cup \mathcal{M}_l \cup \mathcal{M}_*$, where $\mathcal{M}_* = \{s_{*1}, \ldots, s_{*m_*}\}$ contains a collection of transient states. We replace (A10.1) by (A10.3) in this section.

(A10.3) For $i = 1, \ldots, l$, $\widetilde{Q}^i$ are irreducible, and $\widetilde{Q}_*$ is Hurwitz (i.e., all of its eigenvalues have negative real parts).

The Hurwitz condition implies that the states in $\mathcal{M}_*$ are transient. Within a short period of time, they will enter one of the recurrent classes. To proceed, we define the aggregate process. Note, however, we only lump the states in each recurrent class, not in the transient class. Define

$$
\begin{aligned}
\nu_* &= \mathrm{diag}(\tilde{\nu}, 0_{m_* \times m_*}) \in \mathbb{R}^{(l+m_*) \times m_0}, \\
\tilde{a}_i &= -\widetilde{Q}_*^{-1} \widetilde{Q}_*^i \mathbb{1}_{m_i} \in \mathbb{R}^{m_* \times 1}, \quad \text{for } i = 1, \ldots, l, \\
\widetilde{A} &= (\tilde{a}_1, \ldots, \tilde{a}_l) \in \mathbb{R}^{m_* \times l}, \\
\mathbb{1}_* &= \begin{pmatrix} \mathbb{1} & 0_{(m_0 - m_*) \times m_*} \\ \widetilde{A} & 0_{m_* \times m_*} \end{pmatrix} \in \mathbb{R}^{m_0 \times (l+m_*)}.
\end{aligned}
\tag{10.47}
$$

Write

$$
\widehat{Q} = \begin{pmatrix} \widehat{Q}^{11} & \widehat{Q}^{12} \\ \widehat{Q}^{21} & \widehat{Q}^{22} \end{pmatrix}
$$

so that $\widehat{Q}^{11} \in \mathbb{R}^{(m_0 - m_*) \times (m_0 - m_*)}$, $\widehat{Q}^{22} \in \mathbb{R}^{m_* \times m_*}$, and $\widehat{Q}^{12}$ and $\widehat{Q}^{21}$ have appropriate dimensions. Denote

$$
\overline{Q} = \tilde{\nu}(\widehat{Q}^{11}\mathbb{1} + \widehat{Q}^{12}\widetilde{A}), \quad \overline{Q}_* = \mathrm{diag}(\overline{Q}, 0_{m_* \times m_*}).
\tag{10.48}
$$

Define the aggregated process $\overline{\alpha}^\varepsilon(\cdot)$ by

$$
\overline{\alpha}^\varepsilon(t) = \begin{cases} i, & \text{if } \alpha^\varepsilon(t) \in \mathcal{M}_i, \\ U_j, & \text{if } \alpha^\varepsilon(t) = s_{*j}, \end{cases}
\tag{10.49}
$$

where

$$
U_j = \sum_{i=1}^{l} i I_{\{\sum_{j_0=1}^{i-1} \tilde{a}_{j_0 j} < U \le \sum_{j_0=1}^{i} \tilde{a}_{j_0 j}\}},
$$

and U is a random variable uniformly distributed on $[0,1]$, independent of $\alpha^\varepsilon(\cdot)$. We only lump the states in each irreducible class, thus the state space of the aggregated process $\overline{\alpha}^\varepsilon(\cdot)$ is again $\overline{\mathcal{M}} = \{1, \ldots, l\}$. The detailed proofs are omitted; see [176] and the references therein.

Lemma 10.8. *Assume condition* (A10.3). *Then the following assertions hold.*

(a) *The probability vector* $p^\varepsilon(t) = (p^{\varepsilon,1}(t), p^{\varepsilon,2}(t), \ldots, p^{\varepsilon,l}(t), p^{\varepsilon,*}(t))$, *with* $p^{\varepsilon,i}(t) \in \mathbb{R}^{1 \times m_i}$ *and*

$$
p^{\varepsilon,*}(t) = (\mathbf{P}(\alpha^\varepsilon(t) = s_{*1}), \ldots, \mathbf{P}(\alpha^\varepsilon(t) = s_{*m_*})) \in \mathbb{R}^{1 \times m_*},
$$

satisfies

$$
p^\varepsilon(t) = (\theta(t)\tilde{\nu}, 0_{m_*}) + O\left(\varepsilon(t+1) + \exp(-\kappa_0 t/\varepsilon)\right)
$$

for some $\kappa_0 > 0$, *where*

$$\frac{d\theta(t)}{dt} = \theta(t)\overline{Q},$$

$$\theta(0) = (p^1(0)\mathbb{1}_{m_i} + p^*(0)\widetilde{a}_1, \ldots, p^l(0)\mathbb{1}_{m_l} + p^*(0)\widetilde{a}_{m_l}).$$

(b) *The transition probability matrix* $P^\varepsilon(t)$ *satisfies*

$$P^\varepsilon(t) = P^{(0)}(t) + O(\varepsilon(t+1) + e^{-\kappa_0 t/\varepsilon}),$$

where $P^{(0)}(t) = \mathbb{1}_*\Theta_*(t)\nu_*$ *with* $\Theta_*(t) = \mathrm{diag}(\Theta(t), I_{m_* \times m_*})$, *where* $\Theta(t)$ *satisfies*

$$\frac{d\Theta(t)}{dt} = \Theta(t)\overline{Q}, \quad \Theta(0) = I.$$

(c) $\overline{\alpha}^\varepsilon(\cdot)$ *converges weakly to* $\overline{\alpha}(\cdot)$, *a Markov chain generated by* $\overline{Q}$.

(d) *For* $i = 1, \ldots, l$, $j = 1, \ldots, m_i$,

$$\mathbf{E}\left[\int_0^\infty e^{-t}(I_{\{\alpha^\varepsilon(t)=s_{ij}\}} - \nu_j^i I_{\{\overline{\alpha}(t)=i\}})dt\right]^2 = O(\varepsilon),$$

and for $i = *$, $j = 1, \ldots, m_*$,

$$\mathbf{E}\left[\int_0^\infty e^{-t}I_{\{\alpha^\varepsilon(t)=s_{*j}\}}dt\right]^2 = O(\varepsilon).$$

(e) $(X^\varepsilon(\cdot), \overline{\alpha}^\varepsilon(\cdot))$ *converges weakly to* $(X(\cdot), \overline{\alpha}(\cdot))$ *such that the limit operator is as given in Lemma* 10.2.

Remark 10.9. Although a collection of transient states of the discrete events are included, the limit system is still an average with respect to the stationary measures of those ergodic classes only. Asymptotically, the transient states can be discarded because the probabilities go to 0 rapidly.

Theorem 10.10. *Assume* (A10.2) *and* (A10.3), *and for each* $i \in \overline{\mathcal{M}}$, *there is a Liapunov function* $V(x, i)$ *such that* $\mathcal{L}V(x, i) \leq -\gamma$ *for some* $\gamma > 0$. *Then for sufficiently small* $\varepsilon > 0$, *the process* $(X^\varepsilon(t), \alpha^\varepsilon(t))$ *with inclusion of transient discrete events is still positive recurrent.*

Idea of Proof. Since the proof is along the line of that of Theorem 10.5, we only note the differences compared with that theorem. For $i = 1, \ldots, l$, let $V(x, i)$ be the Liapunov function associated with the limit process given by Lemma 10.8(v). The perturbed Liapunov function method is used again. This time, redefine

$$\overline{V}(x, \alpha) = \sum_{i=1}^l V(x, i)I_{\{\alpha \in \mathcal{M}_i\}} + \sum_{i=1}^{m_*} V(x, i)\widetilde{a}_{i,j}I_{\{\alpha=s_{*j}\}}. \tag{10.50}$$

It is easy to check that $\widetilde{Q}\overline{V}(x,\cdot)(\alpha) = 0$, where $\widetilde{Q}$ is defined in (10.46). Moreover,

$$
\begin{aligned}
&b(x, \alpha^\varepsilon(t)) - \overline{b}(x, \overline{\alpha}^\varepsilon(t)) \\
&= \sum_{i=1}^{i} \sum_{j=1}^{m_i} b(x, s_{ij})[I_{\{\alpha^\varepsilon(t)=s_{ij}\}} - \nu_j^i I_{\{\overline{\alpha}^\varepsilon(t)=i\}}] + \sum_{j=1}^{m_*} b(x, s_{*j})I_{\{\alpha^\varepsilon(t)=s_{*j}\}}.
\end{aligned}
$$

Therefore, we can carry out the proof in a similar manner to that of Theorem 10.5. The details are omitted.

10.4   Notes

This chapter continues our study of positive recurrence for switching diffusions. One of the crucial assumptions in the ergodicity study in the previous chapters is that the switching component has a single ergodic class. This chapter takes up the issue that the discrete component may have multiple ergodic classes that are weakly connected. The main idea is the use of the two-time-scale approach. Roughly, if the discrete events or certain components of the discrete events change sufficiently quickly, the positive recurrence can still be guaranteed. The main ingredient is the use of perturbed Liapunov function methods.

11
Stochastic Volatility Using Regime-Switching Diffusions

11.1 Introduction

This chapter aims to model stochastic volatility using regime-switching diffusions. Effort is devoted to developing asymptotic expansions of a system of coupled differential equations with applications to option pricing under regime-switching diffusions. By focusing on fast mean reversion, we aim at finding the "effective volatility." The main techniques used are singular perturbation methods. Under simple conditions, asymptotic expansions are developed with uniform asymptotic error bounds. The leading term in the asymptotic expansions satisfies a Black–Scholes equation in which the mean return rate and volatility are averaged out with respect to the stationary measure of the switching process. In addition, the full asymptotic series is developed. The asymptotic series helps us to gain insight on the behavior of the option price when the time approaches maturity. The asymptotic expansions obtained in this chapter are interesting in their own right and can be used for other problems in control optimization of systems involving fast-varying switching processes.

Nowadays, sophisticated financial derivatives such as options are used widely. The Nobel prize winning Black–Scholes formula provides an important tool for pricing options on a basic equity. It has encouraged and facilitated the union of mathematics, finance, computational sciences, and economics. On the other hand, it has been recognized, especially by practitioners in the financial market, that the assumption of constant volatility, which is essential in the Black–Scholes formula, is a less-than-perfect de-

G.G. Yin and C. Zhu, *Hybrid Switching Diffusions: Properties and Applications*, Stochastic Modelling and Applied Probability 63, DOI 10.1007/978-1-4419-1105-6_11, © Springer Science + Business Media, LLC 2010

scription of the real world. To capture the behavior of stock prices and other derivatives, there has been much effort in taking into account frequent volatility changes. It has been recognized that it is more suitable to use a stochastic process to model volatility variations. In [70], instead of the usual GBM (geometric Brownian motion) model, a second stochastic differential equation is introduced to describe the random environments. Such a formulation is known as a stochastic volatility model.

What happens if the stochastic volatility undergoes fast mean reversion? To answer this question, in [49] and the subsequent papers [50, 51], a class of volatility models has recently been studied in details. Under the setup of mean reversion, two-time-scale methods are used. The rationale is to identify the important groupings of market parameters. It also reveals that the Black–Scholes formula is a "first approximation" to such fast-varying volatility models. Assume that the volatility is a function $f(\cdot)$ of a fast-varying diffusion that is mean reverting (or ergodic). The mean reversion implies that although rapidly varying, the volatility does not blow up. By exploiting the time-scale separation, it was shown in [49, 50] that the "slow" component (the leading term or the zeroth-order outer expansion term in the approximation) of the option prices can be approximated by a Black–Scholes differential equation with constant volatility $\overline{f}$ where $\overline{f}^2$ is the average of $f^2(\cdot)$ with respect to the stationary measure of the "fast" component. Moreover, using a singular perturbation approach, the next term (the first-order outer expansion term) in the asymptotic expansion was also found. For convenience, the volatility was assumed to be driven by an Ornstein–Uhlenbeck (OU) process in [49]–[51]. The fast mean reversion has been further examined in [90] with more general models. A full asymptotic series with uniform error bounds was obtained; see also related diffusion approximation in [131] and asymptotic expansions for diffusion processes [88, 89].

Along another line, increasing attention has been drawn to modeling, analysis, and computing using regime-switching models [164], which are alternatives to the stochastic volatility models mentioned above. They present an effective way to model stochastic volatility with simpler structures. The use of the Markov chains is much simpler than the use of a second stochastic differential equation. This is motivated by the desire to use regime switching to describe uncertainty and stochastic volatility. Nowadays, it has been well recognized that due to stochastic volatility, a phenomena known as the volatility smile arises. There has been an effort to provide better models to replicate the "smile." In [164], the authors reproduced the volatility smile successfully by using regime-switching models easily. Earlier efforts in modeling and analysis of regime-switching models can be found in [5, 32, 184] among others.

To some large extent, this chapter is motivated by [49] for extremely fast mean reversion processes in the driving force of stochastic volatility. Nevertheless, rather than using an additional SDE to represent the stochastic volatility, we use the same setup of a regime-switching model as that of [164]. Therefore, it accommodates the desired goal from a different angle. Here the fast mean reversion is captured by a fast-varying continuous-time Markov chain. We demonstrate that the model under consideration leads to effective volatility, which is an average with respect to the stationary distribution of the fast-changing jump process. In mathematical terms, the fast reversion corresponds to the Markov chain being ergodic with fast variations. Using a two-time-scale formulation, the problem itself centers around fast-varying switching-diffusion processes. The approach that we are using is analytical. We focus on developing asymptotic expansions that are approximation of solutions of systems of parabolic equations, and aim to obtain such expansions with uniform asymptotic error bounds on a compact set for the continuous component. It should be noted that if one is only interested in convergence in the pointwise sense (termed pointwise asymptotic expansions henceforth) rather than obtaining uniform error bounds, then one can proceed similarly to the diffusion counterpart in [49, 50]). In any event, the "effective volatility" can be obtained. It should also be noted that the asymptotic expansions presented in this chapter give some new insight on the construction of approximation of solutions of backward type systems of PDEs, which is different from [176] where probability distributions were considered. The methods presented are constructive. The asymptotic analysis techniques are interesting in their own right.

The rest of the chapter is arranged as follows. Section 11.2 presents the formulation of the problem. Section 11.3 constructs asymptotic expansions. Section 11.4 proceeds with validation of the expansions. Finally Section 11.5 issues a few more remarks.

11.2 Formulation

In the rest of the book, r is used as the dimension of the continuous state variables (i.e., $\mathbb{R}^r$ is used as the space for the continuous state variable). In this chapter, however, to adopt the traditional convention, we use r as the interest rate throughout. We consider a basic equity, a stock whose price is given by $S(t)$. Different from the traditional geometric Brownian motion setup, we assume that the price follows a switching-diffusion model. Suppose that $\alpha(t)$ is a continuous-time Markov chain with generator $Q(t)$ and a finite state space $\mathcal{M} = \{1, \ldots, m_0\}$. The price of the stock is a solution of the stochastic differential equation

$$dS(t) = \mu(\alpha(t))S(t)dt + \sigma(\alpha(t))S(t)dw(t), \qquad (11.1)$$

where $w(\cdot)$ is a standard Brownian motion independent of $\alpha(\cdot)$, and $\mu(\cdot)$ and $\sigma(\cdot)$ are the appreciation rate and the volatility rate, respectively. Such a model is frequently referred to as a regime-switching asset model. Note that in the above, both the appreciation rate and volatility rate are functions of the Markov chain.

Consider a European type of call option. The payoff at time T is given by $H(S)$, a nonnegative function. Denote $h(S) = H(e^S)$. Suppose that the associated risk-free interest rate is given by r. Nowadays, a standard approach in option pricing is risk-neutral valuation. The rationale is to derive a suitable probability space on which the expected rate of return of all securities is equal to the risk-free interest rate. The mathematical requirement is that the discounted asset price is a martingale. The associated probability space is referred to as the risk-neutral world. The price of the option on the asset is then the expected value, with respect to this martingale measure, of the discounted option payoff. In what follows, we use the so-called risk-neutral probability measure; see, e.g., [49] and [69] among others for diffusion processes and [164] for switching diffusion processes. To proceed, we first present a lemma, which is essentially a generalized Girsanov theorem for Markov-modulated processes. The results of this type are generally known, and a proof can be found in [164].

Lemma 11.1. *The following assertions hold.*

(a) *Suppose that $\sigma(i) > 0$ for each $i \in \mathcal{M}$, and let*

$$\widetilde{w}(t) := w(t) - \int_0^t \frac{r - \mu(\alpha(u))}{\sigma(\alpha(u))} du.$$

Then, $\widetilde{w}(\cdot)$ is a $\widetilde{\mathbf{P}}$-Brownian motion, where $\widetilde{\mathbf{P}}$ is known as the risk-neutral probability measure; see [49, 164] among others.

(b) *$S(0)$, $\alpha(\cdot)$, and $\widetilde{w}(\cdot)$ are mutually independent under $\widetilde{\mathbf{P}}$;*

(c) *(Itô's lemma or Dynkin's formula) For each $i \in \mathcal{M}$ and each $g(\cdot, \cdot, i) \in C^{2,1}$, we have*

$$g(S(s), s, \alpha(s)) = g(S(t), t, \alpha(t)) + \int_t^s \mathcal{L}g(S(u), u, \alpha(u))du$$
$$+ M(s) - M(t),$$

where $M(\cdot)$ is a $\widetilde{\mathbf{P}}$ local martingale and $\mathcal{L}$ is a generator given by

$$\mathcal{L}g(S, t, i) = \frac{\partial}{\partial t}g(S, t, i) + \frac{1}{2}S^2\sigma^2(i)\frac{\partial^2}{\partial S^2}g(S, t, i)$$
$$+ rS\frac{\partial}{\partial S}g(S, t, i) + Q(t)g(S, t, \cdot)(i), \tag{11.2}$$

with

$$Q(t)g(S, t, \cdot)(i) = \sum_{j=1}^{m_0} q_{ij}(t)g(S, t, j). \tag{11.3}$$

Note that $(S(t), \alpha(t))$ is a Markov process with generator $\mathcal{L}$. To proceed, as alluded to in the introduction, in reference to [49], we also consider a fast mean reverting driving process, but the driving process is a continuous-time Markov chain not a diffusion.

By introducing a small parameter $\varepsilon > 0$, we aim to show that the regime-switching model is also a good approximation of the Black–Scholes model. To this end, suppose that the generator of the Markov chain is given by $Q^\varepsilon(t) = (q_{ij}^\varepsilon(t)) \in \mathbb{R}^{m_0 \times m_0}$.

$$Q^\varepsilon(t) = \frac{Q(t)}{\varepsilon}, \qquad (11.4)$$

where $Q(t)$ is the generator of a continuous-time Markov chain. To highlight the ε-dependence, we denote the Markov chain by $\alpha(t) = \alpha^\varepsilon(t)$.

We are mainly interested in uniform asymptotic expansions for the option price. To obtain such uniform asymptotic expansions, we need to have the continuous component of the switching diffusion be in a compact set. For simplicity, we take the compact set to be $[0, 1]$; see [90] for a comment on the corresponding stochastic volatility model without switching. We use this as a standing assumption throughout this chapter. Next, we state a couple of additional conditions.

(A11.1) There is a $T > 0$ and for all $t \in [0, T]$, the generator $Q(t)$ is weakly irreducible. There is an $n \geq 1$ such that $Q(\cdot) \in C^{n+2}[0, T]$.

(A11.2) The function $h(\cdot)$ is sufficiently smooth and vanishes outside of a compact set.

Remark 11.2. Condition (A11.1) indicates that the weak irreducibility of the Markov chain implies the existence of the unique quasi-stationary distribution $\nu(t) = (\nu_1(t), \dots, \nu_{m_0}(t)) \in \mathbb{R}^{1 \times m_0}$. Our assumption on the operator $\mathcal{L}^\varepsilon$ then leads to the equity undergoing fast mean reverting switching.

Following [49, §5.4], we have assumed $h(S)$ to be a bounded and smooth function in (A11.2). For European options, if it is a call option, $h(S) = (S - K_0)^+$, and if it is a put option, $h(S) = (K_0 - S)^+$, where K_0 is the exercise price. Thus the function $h(\cdot)$ is not smooth and is unbounded. In [50], using regularization or smoothing techniques, asymptotic expansions up to the order $O(\varepsilon)$ were obtained without assuming smoothness and boundedness of the payoff function $h(S)$. Note that the accuracy of the approximation was obtained only for a fixed state variable x and time variable t with $t < T$ in [50], however. In this chapter, as in [49], we confine our attention to the smooth and bounded payoff $h(S)$, and prove the *uniform* in S accuracy of approximation. It appears that the unbounded and nonsmooth payoff function $h(S)$ can be handled and the formal asymptotic expansions can be obtained, but such expansions are not uniform in S. Moreover, different methods and ideas must be used to justify the expansions as commented on in [90].

Recall that by weak irreducibility, we mean that the system of equations

$$\begin{cases} \nu(t)Q(t) = 0 \\ \nu(t)\mathbb{1} = 1 \end{cases} \tag{11.5}$$

has a unique solution $\nu(t) = (\nu_1(t), \ldots, \nu_{m_0}(t)) \in \mathbb{R}^{1 \times m_0}$ satisfying $\nu_i(t) \geq 0$ for each $i \in \mathcal{M}$. Such a nonnegative solution is termed a quasi-stationary distribution; see [176].

Let

$$\begin{aligned} V^\varepsilon(S, t, i) &= \mathbf{E}_{S,i}[\exp(-r(T-t))h(S^\varepsilon(T))] \\ &= \mathbf{E}[\exp(-r(T-t))h(S^\varepsilon(T))|S^\varepsilon(t) = S, \alpha^\varepsilon(t) = i]. \end{aligned} \tag{11.6}$$

The option price can be characterized by the following system of partial differential equations,

$$\begin{aligned} \frac{\partial V^\varepsilon(S, t, i)}{\partial t} &+ \frac{1}{2}\sigma^2(i)S^2 \frac{\partial^2 V^\varepsilon(S, t, i)}{\partial S^2} \\ &+ rS \frac{\partial V^\varepsilon(S, t, i)}{\partial S} - rV^\varepsilon(S, t, i) \\ &+ Q^\varepsilon(t)V^\varepsilon(S, t, \cdot)(i) = 0, \ i \in \mathcal{M}, \end{aligned} \tag{11.7}$$

which is a generalization of the usual Black–Scholes PDE. Associated with the above system of PDEs, we define an operator $\mathcal{L}^\varepsilon$. For $i \in \mathcal{M}$, and each $g(\cdot, \cdot, i) \in C^{2,1}$, let

$$\begin{aligned} \mathcal{L}^\varepsilon g(S, t, i) = \ &\frac{\partial g(S, t, i)}{\partial t} + \frac{1}{2}\sigma^2(i)S^2 \frac{\partial^2 g(S, t, i)}{\partial S^2} \\ &+ rS \frac{\partial g(S, t, i)}{\partial S} - rg(S, t, i) \\ &+ Q^\varepsilon(t)g(S, t, \cdot)(i). \end{aligned} \tag{11.8}$$

Now the setup of the problem is complete. We proceed to obtain the approximation of the option price by means of asymptotic expansions.

11.3 Asymptotic Expansions

What can one say about the effect of $\alpha^\varepsilon(\cdot)$? Is there an "effective volatility?" Is the Black–Scholes formula still a reasonable approximation to such regime-switching models? To answer these questions, we develop an asymptotic series using analytic techniques. We seek asymptotic expansions of $V^\varepsilon(S, t, i)$ of the form

$$\Phi_n^\varepsilon(S, t, i) + \Psi_n^\varepsilon(S, \tau, i) = \sum_{k=0}^{n} \varepsilon^k \varphi_k(S, t, i) + \sum_{k=0}^{n} \varepsilon^k \psi_k(S, \tau, i), \ i \in \mathcal{M}, \tag{11.9}$$

where τ is a stretched-time variable defined by

$$\tau = \frac{T - t}{\varepsilon}.$$

The $\varphi_k(\cdot)$ are called regular terms or outer expansion terms, and the $\psi_k(\cdot)$ are the boundary layer correction terms (or to be more precise, terminal layer corrections). In this problem, the terminal layer correction terms are particularly useful for behavior of the option price near the time of maturity.

We aim to obtain asymptotic expansions of the order n, and derive the uniform error bounds. For the purposes of error estimates, we need to calculate a couple of more terms for analysis reasons.

First let us look at $\Phi_n^\varepsilon(S, t, i)$, the regular part of the asymptotic expansions. Substituting it into (11.7), and comparing coefficients of like powers of ε^k for $0 \le k \le n + 1$, we obtain

$$
\begin{aligned}
&Q(t)\varphi_0(S, t, \cdot)(i) = 0, \\
&\frac{\partial \varphi_0(S, t, i)}{\partial t} + \frac{1}{2}\sigma^2(i)S^2 \frac{\partial^2 \varphi_0(S, t, i)}{\partial S^2} + rS \frac{\partial \varphi_0(S, t, i)}{\partial S} \\
&\quad - r\varphi_0(S, t, i) + Q(t)\varphi_1(S, t, \cdot)(i) = 0, \\
&\qquad \cdots \\
&\frac{\partial \varphi_k(S, t, i)}{\partial t} + \frac{1}{2}\sigma^2(i)S^2 \frac{\partial^2 \varphi_k(S, t, i)}{\partial S^2} + rS \frac{\partial \varphi_k(S, t, i)}{\partial S} \\
&\quad - r\varphi_k(S, t, i) + Q(t)\varphi_{k+1}(S, t, \cdot)(i) = 0.
\end{aligned}
\tag{11.10}
$$

To ensure the match of the terminal (or boundary) conditions, we choose

$$
\begin{aligned}
\varphi_0(S, T, i) + \psi_0(S, 0, i) &= h(S), \quad i \in \mathcal{M}, \\
\varphi_k(S, T, i) + \psi_k(S, 0, i) &= 0, \quad i \in \mathcal{M} \text{ and } k \ge 1.
\end{aligned}
\tag{11.11}
$$

Taking a Taylor expansion of $Q(\cdot)$ around T, we obtain

$$
Q(t) = Q(T - \varepsilon\tau) = \sum_{k=0}^{n+1} \frac{(-1)^k (\varepsilon\tau)^k}{k!} \frac{d^k Q(T)}{dt^k} + R_{n+1}(T - \varepsilon\tau), \tag{11.12}
$$

where by assumption (A11.1), it can be shown that $R_{n+1}(t) = O(t^{n+1})$ uniformly in $t \in [0, T]$.

Similar to the equations in (11.10) for $\Phi_{n+1}^\varepsilon(S, t, i)$, we can substitute

$\Psi_{n+1}^{\varepsilon}(S, \tau, i)$ into (11.7). This results in

$$
\begin{aligned}
&-\frac{\partial \psi_0(S, \tau, i)}{\partial \tau} + Q(T)\psi_0(S, \tau, i) = 0, \\
&-\frac{\partial \psi_1(S, \tau, i)}{\partial \tau} + Q(T)\psi_1(S, \tau, i) + \left[-\tau Q^{(1)}(T)\psi_0(S, \tau, i) \right. \\
&+ \frac{1}{2}\sigma^2(i)S^2 \frac{\partial^2 \psi_0(S, \tau, i)}{\partial S^2} \\
&\left. + rS\frac{\partial \psi_0(S, \tau, i)}{\partial S} - r\psi_0(S, \tau, i) \right] = 0, \\
&\cdots \quad \cdots \\
&-\frac{\partial \psi_k(S, \tau, i)}{\partial \tau} + Q(T)\psi_k(S, \tau, i) + \widetilde{R}_k(S, \tau, i) = 0,
\end{aligned}
\tag{11.13}
$$

where the remainder term is given by

$$
\begin{aligned}
\widetilde{R}_k(S, \tau, i) = &\sum_{j=1}^{k} \frac{(-1)^j \tau^j}{j!} \frac{d^j Q(T)}{dt^j} \psi_{k-j}(S, \tau, i) \\
&+ \frac{1}{2}\sigma^2(i)S^2 \frac{\partial^2 \psi_{k-1}(S, \tau, i)}{\partial S^2} + rS\frac{\partial \psi_{k-1}(S, \tau, i)}{\partial S} \\
&- r\psi_{k-1}(S, \tau, i).
\end{aligned}
\tag{11.14}
$$

Our task to follow is to construct the sequences $\{\varphi_k(S, t)\}$ and $\{\psi_k(S, \tau)\}$.

11.3.1   Construction of $\varphi_0(S, t, i)$ and $\psi_0(S, \tau, i)$

We claim that $\varphi_0(S, t, i)$ must be independent of i. To see this, write

$$
\varphi_0(S, t) = (\varphi_0(S, t, 1), \dots, \varphi_0(S, t, m_0))' \in \mathbb{R}^{m_0 \times 1}.
$$

Then the first equation in (11.10) may be written as $Q(t)\varphi_0(S, t) = 0$. This equation in turn implies that $\varphi_0(S, t)$ is in the null space of $Q(t)$. Because $Q(t)$ is weakly irreducible, the rank of $Q(t) = m_0 - 1$, so the null space is one-dimensional. Consequently, the null space is spanned by $\mathbb{1} = (1, \dots, 1)' \in \mathbb{R}^{m \times 1}$. This yields that $\varphi_0(S, t) = \gamma_0(S, t)\mathbb{1}$, where $\gamma_0(S, t)$ is a real-valued function and hence $\varphi_0(S, t, i)$ is a function independent of i.

Using the above argument in the second equation of (11.10), and multiplying from the left through the equation by $\nu_i(t)$, where $(\nu_1(t), \dots, \nu_{m_0}(t))$ is the quasi-stationary distribution associated with $Q(t)$, and summing over $i \in \mathcal{M}$, we obtain

$$
\frac{\partial \gamma_0(S, t)}{\partial t} + \frac{1}{2}\overline{\sigma}^2(t)S^2 \frac{\partial^2 \gamma_0(S, t)}{\partial S^2} + rS\frac{\partial \gamma_0(S, t)}{\partial S} - r\gamma_0(S, t) = 0, \tag{11.15}
$$

where

$$
\overline{\sigma}^2(t) = \sum_{i=1}^{m_0} \nu_i(t)\sigma^2(i).
$$

The terminal condition is given by

$$\gamma_0(S, T) = h(S). \tag{11.16}$$

Then (11.15) together with the terminal condition (11.16) has a unique solution.

Remark 11.3. Note that $\gamma_0(S, t)$ satisfies the Black–Scholes partial differential equation, in which the coefficients are averaged out with respect to the stationary distributions of the Markov chain. The result reveals that the Black–Scholes formulation, indeed, is a first approximation to the option model under regime switching. Thus, the regime-switching model can be thought of as another stochastic volatility model.

In view of (11.11) and (11.13), $\psi_0(S, \tau, i)$ is obtained from the first equation in (11.13) together with the condition $\psi_0(S, 0, i) = 0$. It follows that $\psi_0(S, \tau, i) = 0$ for each $i \in \mathcal{M}$.

11.3.2 Construction of $\varphi_1(S, t, i)$ and $\psi_1(S, \tau, i)$

Now, we proceed to obtain $\varphi_1(S, t, i)$ and $\psi_1(S, \tau, i)$. Similar to $\varphi_0(S, t)$, for $i \leq n + 1$, define

$$\varphi_i(S, t) = (\varphi_i(S, t, 1), \dots, \varphi_i(S, t, m_0))' \in \mathbb{R}^{m_0},$$

$$\psi_i(S, t) = (\psi_i(S, t, 1), \dots, \psi_i(S, t, m_0))' \in \mathbb{R}^{m_0}.$$

Consider $\varphi_1(S, t)$ in (11.10). It is a solution of a nonhomogeneous algebraic equation as in (A.10) in the appendix of this book. By virtue of Lemma A.12, we can write $\varphi_1(S, t, i)$ as the sum of a general solution of the associated homogeneous equation and the unique solution of the inhomogeneous equation that is orthogonal to the stationary distribution $\nu(t)$. That is,

$$\varphi_1(S, t, i) = \gamma_1(S, t) + \varphi_1^0(S, t, i), \quad i \in \mathcal{M},$$

where $\{\varphi_1^0(S, t, i) : i \in \mathcal{M}\}$ is the unique solution of (11.10) satisfying

$$\sum_{i=1}^{m_0} \nu_i(t) \varphi_1^0(S, t, i) = 0.$$

Equivalently, denote

$$\varphi_1^0(S, t) = (\varphi_1^0(S, t, 1), \dots, \varphi_1^0(S, t, m_0))' \in \mathbb{R}^{m_0}.$$

Then

$$\varphi_1(S, t) = \gamma_1(S, t)\mathbb{1} + \varphi_1^0(S, t), \tag{11.17}$$

where $\varphi_1^0(S,t)$ is the unique solution of the system of equations

$$Q(t)\varphi_1^0(S,t) = F_0(S,t)$$
$$\nu(t)\varphi_1^0(S,t) = 0, \tag{11.18}$$

where

$$F_0(S,t) = (F_0(S,t,1), \ldots, F_0(S,t,m_0))' \in \mathbb{R}^{m_0},$$

with

$$F_0(S,t,i) = -\left[\frac{\partial \gamma_0(S,t)}{\partial t} + \frac{\sigma^2(i)S^2}{2}\frac{\partial \gamma_0(S,t)}{\partial S^2} + rS\frac{\partial \gamma_0(S,t)}{\partial S} - r\gamma_0(S,t)\right], \tag{11.19}$$

by Lemma A.12. That is, $\varphi_1(S,t)$ is the sum of a general solution of the homogeneous equation plus a particular solution verifying the orthogonality condition.

Note that $\gamma_1(S,t)$ has not been determined yet. We proceed to obtain it from the next equation in (11.10). By substituting (11.17) into the next equation in (11.10) and premultiplying it by ν, we obtain

$$\frac{\partial \gamma_1(S,t)}{\partial t} + \frac{1}{2}\bar{\sigma}^2(t)S^2\frac{\partial^2 \gamma_1(S,t)}{\partial S^2} + rS\frac{\partial \gamma_1(S,t)}{\partial S} - r\gamma_1(S,t) = \overline{F}_1(S,t), \tag{11.20}$$

where

$$F_1(S,t) = (F_1(S,t,1), \ldots, F_1(S,t,m_0))' \in \mathbb{R}^{m_0},$$

with

$$F_1(S,t,i) = -\left[\frac{\partial \varphi_1^0(S,t,i)}{\partial t}\frac{\sigma^2(i)S^2}{2}\frac{\partial \varphi_1^0(S,t,i)}{\partial S^2} + rS\frac{\partial \varphi_1^0(S,t,i)}{\partial S} - r\varphi_1^0(S,t,i)\right], \tag{11.21}$$

and

$$\overline{F}_1(S,t) = \nu(t)F_1(S,t).$$

It is easily seen that (11.20) is a uniquely solvable Cauchy problem if the terminal condition $\gamma_1(S,T) \in \mathbb{R}$ is specified. We determine this by matching the terminal condition with the terminal layer term $\psi_1(S,0)$.

Since $\psi_0(S,\tau) = 0$, from the second equation in (11.13), $\psi_1(S,\tau)$ satisfies

$$\frac{\partial \psi_1(S,\tau)}{\partial \tau} = Q(T)\psi_1(S,\tau). \tag{11.22}$$

Choose $\psi_1(S,0) = \psi_1^0(S)$ such that $\overline{\psi}_1^0(S) = \nu\psi_1^0(S) = 0$ where $\nu = \nu(T)$. Here and hereafter, we always use ν to denote $\nu(T)$ for notational simplicity.

Multiplying from the left by ν, the stationary distribution associated with $Q(T)$, we obtain

$$\frac{\partial \overline{\psi}_1(S,\tau)}{\partial \tau} = 0 \in \mathbb{R},$$
$$\overline{\psi}_1(S,0) = \overline{\psi}_1^0(S) = 0, \qquad (11.23)$$

where $\overline{\psi}_1(S,\tau) = \nu \psi_1(S,\tau)$. The solution of (11.23) is thus given by

$$\overline{\psi}_1(S,\tau) = 0.$$

Remark 11.4. The above may also be seen as

$$\psi_1(S,\tau) = \exp(Q(T)\tau)\psi_1^0(S)$$

for a chosen initial $\psi_1^0(S)$. Using

$$\exp(Q(T)\tau) = \sum_{j=0}^{\infty} (Q(T)\tau)^j / j!,$$

the orthogonality of $\nu Q(T) = 0$ leads to

$$\nu \exp(Q(T)\tau)\psi_1^0(S) = \nu I \psi_1^0(S) = \nu \psi_1^0(S) = \overline{\psi}_1^0(S)$$

for each $\tau \geq 0$.

Note that once $\psi_1^0(S)$ is chosen, the $\varphi_1(S,t)$ and $\psi_1(S,\tau)$ will be determined. The choice of $\psi_1^0(S,0)$ enables us to obtain the exponential decay of $\psi_1(S,\tau)$ easily.

In view of the defining equation for $\varphi_1(S,t)$, equations (11.17)–(11.20), and the terminal condition (11.11), equation (11.20) together with the terminal condition $\gamma_1(S,T) = -\overline{\psi}_1^0(S) = 0$ has a unique solution. Up to now, $\varphi_1(S,t)$ has been completely determined. Next, we proceed to find $\psi_1(S,\tau)$. Condition (11.11) implies that

$$\psi_1(S,0) = -\varphi_1(S,T) = \overline{\psi}_1^0(S)\mathbb{1} - \varphi_1^0(S,T) = -\varphi_1^0(S,T). \qquad (11.24)$$

It follows that the Cauchy problem given by (11.22) and (11.24) has a unique solution. We are in a position to derive the exponential decay property of $\psi_1(S,\tau)$.

Lemma 11.5. *For $\psi_1(S,\tau)$ obtained from the solution of (11.22) and (11.24), we have that for some $K > 0$ and $\kappa_0 > 0$,*

$$\sup_{S\in[0,1]} |\psi_1(S,\tau)| \leq K \exp(-\kappa_0 \tau). \qquad (11.25)$$

Proof. The weak irreducibility of $Q(T)$ implies that $\exp(Q(T)\tau) \to \mathbb{1}\nu$ as $\tau \to \infty$, and

$$|\exp(Q(T)\tau) - \mathbb{1}\nu| \le \exp(-\kappa_0\tau) \text{ for some } \kappa_0 > 0.$$

The solution of (11.22) and (11.24) yields that for each $S \in [0,1]$ and for some $K > 0$,

$$
\begin{aligned}
|\psi_1(S,\tau)| &= |\psi_1(S,\tau) - \mathbb{1}\nu\psi_1(S,0)| \\
&= |[\exp(Q(T)\tau) - \mathbb{1}\nu]\psi_1(S,0)| \\
&\le |\exp(Q(T)\tau) - \mathbb{1}\nu||\psi_1(S,0)| \\
&\le K\exp(-\kappa_0\tau).
\end{aligned}
$$

Furthermore, it is readily seen that the above estimate holds uniformly for $S \in [0,1]$. The desired result thus follows. □

Remark 11.6. In what follows, for notational simplicity, K and κ_0 are generic positive constants. Their values may change for different appearances.

We are now in a position to obtain a priori bounds for the derivatives of $\psi_1(S,\tau)$. This is presented in the next lemma.

Lemma 11.7. *The function $\psi_1(S,\tau)$ satisfies*

$$
\sup_{S \in [0,1]} \left|\frac{\partial\psi_1(S,\tau)}{\partial S}\right| \le K\exp(-\kappa_0\tau),
$$

$$
\sup_{S \in [0,1]} \left|\frac{\partial^2\psi_1(S,\tau)}{\partial S^2}\right| \le K\exp(-\kappa_0\tau).
$$

Proof. Consider

$$
U(S,\tau) = \frac{\partial\psi_1(S,\tau)}{\partial S} \text{ and}
$$
$$
V(S,\tau) = \frac{\partial^2\psi_1(S,\tau)}{\partial S^2}.
$$

Then we have U and V satisfy

$$
\begin{aligned}
\frac{\partial U(S,\tau)}{\partial\tau} &= Q(T)U(S,\tau), \quad U(S,0) = \frac{\partial\psi_1(S,0)}{\partial S}, \\
\frac{\partial V(S,\tau)}{\partial\tau} &= Q(T)V(S,\tau), \quad V(S,0) = \frac{\partial^2\psi_1(S,0)}{\partial S^2}, \\
\frac{\partial\overline{U}(S,\tau)}{\partial\tau} &= 0, \quad \overline{U}(S,0) = 0, \\
\frac{\partial\overline{V}(S,\tau)}{\partial\tau} &= 0, \quad \overline{V}(S,0) = 0,
\end{aligned}
\qquad (11.26)
$$

respectively, where $\overline{U}(S,\tau) = \nu U(S,\tau)$ and $\overline{V}(S,\tau) = \nu V(S,\tau)$. Thus, we have

$$\overline{U}(S,\tau) = 0, \quad \overline{V}(S,\tau) = 0,$$

and

$$U(S,\tau) = \exp(Q(T)\tau)U(S,0),$$
$$V(S,\tau) = \exp(Q(T)\tau)V(S,0).$$

It follows that

$$|U(S,\tau)| = |\exp(Q(T)\tau)U(S,0) - \mathbb{1}\nu U(S,0)|$$
$$\leq K\exp(-\kappa_0\tau).$$

Likewise, $|V(S,\tau)| \leq K\exp(-\kappa_0\tau)$ as desired. □

11.3.3   Construction of $\varphi_k(S,t)$ and $\psi_k(S,\tau)$

Denote $\varphi_k(S,t) \in \mathbb{R}^{m_0}$, $\varphi_k^0(S,t) \in \mathbb{R}^{m_0}$, $\gamma_k(S,t) \in \mathbb{R}$, and $\psi_k(S,\tau) \in \mathbb{R}^{m_0}$ as that of $\varphi_1(S,t)$, $\varphi_1^0(S,t)$, $\gamma_1(S,t)$, and $\psi_1(S,\tau)$, respectively. By induction, we proceed to obtain the terms $\varphi_k(S,t)$ and $\psi_k(S,\tau)$ for $1 < k \leq n+1$. Similar to the last section, for $1 < k \leq n+1$, denote

$$F_k(S,t) = (F_k(S,t,1),\ldots,F_k(S,t,m_0))' \in \mathbb{R}^{m_0},$$
$$\begin{aligned} F_k(S,t,i) = -\Big[&\frac{\partial\varphi_k^0(S,t,i)}{\partial t} + \frac{\sigma^2(i)S^2}{2}\frac{\partial\varphi_k^0(S,t,i)}{\partial S^2} \\ &+ rS\frac{\partial\varphi_k^0(S,t,i)}{\partial S} - r\varphi_k^0(S,t,i)\Big], \end{aligned} \quad (11.27)$$
$$\overline{F}_k(S,t) = \nu(t)F_k(S,t).$$

Suppose that we have constructed $\varphi_{k-1}(S,t)$ and $\psi_{k-1}(S,\tau)$ for $1 < k \leq n+1$. Then similar to $\varphi_1(S,t)$ and $\psi_1(S,\tau)$, we can define $\varphi_k(S,t)$ and $\psi_k(S,\tau)$. Next, write $\varphi_k(S,t)$ as

$$\varphi_k(S,t) = \gamma_k(S,t)\mathbb{1} + \varphi_k^0(S,t), \quad (11.28)$$

where $\varphi_k^0(S,t)$ is the unique solution of the system of equations

$$\begin{aligned} Q(t)\varphi_k^0(S,t) &= F_{k-1}(S,t) \\ \nu(t)\varphi_k^0(S,t) &= 0, \end{aligned} \quad (11.29)$$

$\nu(t)$ is the quasi-stationary distribution associated with $Q(t)$, and $\gamma_k(S,t)$ satisfies

$$\frac{\partial\gamma_k(S,t)}{\partial t} + \frac{1}{2}\overline{\sigma}^2(t)S^2\frac{\partial^2\gamma_k(S,t)}{\partial S^2} + rS\frac{\partial\gamma_k(S,t)}{\partial S} - r\gamma_k(S,t) = \overline{F}_k(S,t). \quad (11.30)$$

By virtue of Lemma A.12, there is a unique solution to (11.29). To determine the solution of (11.30) as a solution of the Cauchy problem, we select the terminal condition from the terminal layer $\psi_k(S, \tau)$ via the matching condition (11.11). Note that $\psi_k(S, \tau)$ satisfies

$$-\frac{\partial \psi_k(S, \tau)}{\partial \tau} + Q(T)\psi_k(S, \tau) + \tilde{R}_k(S, \tau) = 0, \qquad (11.31)$$

where $\tilde{R}_k(S, \tau) = (\tilde{R}_k(S, \tau, 1), \dots, \tilde{R}_k(S, \tau, m_0))' \in \mathbb{R}^{m_0}$ with $\tilde{R}_k(S, \tau, i)$ given by (11.14).

Since the construction of $\psi_2(S, \tau)$ is somewhat different from $\psi_1(S, \tau)$, we single it out first. In view of (11.13) and noting $\psi_0(S, \tau) = 0$, $\psi_2(S, \tau)$ satisfies

$$
\begin{aligned}
\frac{\partial \psi_2(S, \tau)}{\partial \tau} &= Q(T)\psi_2(S, \tau) + \tilde{R}_2(S, \tau) \\
&= Q(T)\psi_2(S, \tau) + \Big[-\tau Q^{(1)}(T)\psi_1(S, \tau) + \frac{1}{2}S^2 \Sigma \frac{\partial^2 \psi_1(S, \tau)}{\partial S^2} \\
&\quad + rS\frac{\partial \psi_1(S, \tau)}{\partial S} - r\psi_1(S, \tau) \Big],
\end{aligned}
$$

$$(11.32)$$

where

$$\Sigma = \mathrm{diag}(\sigma(1), \dots, \sigma(m_0)) \in \mathbb{R}^{m_0 \times m_0}. \qquad (11.33)$$

With given initial data $\psi_2(S, 0)$, the solution of (11.32) is given by

$$\psi_2(S, \tau) = \exp(Q(T)\tau)\psi_2(S, 0) + \int_0^\tau \exp(Q(T)(\tau - s))\tilde{R}_2(S, s)ds. \qquad (11.34)$$

The solution of (11.34) can be further expanded as

$$
\begin{aligned}
\psi_2(S, \tau) = {}& \mathbb{1}\nu\psi_2(S, 0) + \mathbb{1}\int_0^\infty \nu\tilde{R}_2(S, s)ds \\
& + [\exp(Q(T)\tau) - \mathbb{1}\nu]\psi_2(S, 0) - \mathbb{1}\nu\int_\tau^\infty \tilde{R}_2(S, s)ds \qquad (11.35) \\
& + \int_0^\tau [\exp(Q(T)(\tau - s)) - \mathbb{1}\nu]\tilde{R}_2(S, s)ds.
\end{aligned}
$$

By virtue of Lemmas 11.5 and 11.7, $\tilde{R}_2(S, \tau)$ decays exponentially fast to 0 so the integral $\int_0^\infty \tilde{R}_2(S, s)ds$ is well defined. Furthermore,

$$\left| \int_\tau^\infty \tilde{R}_2(S, s)ds \right| \le K\exp(-\kappa_0\tau).$$

A similar argument to that of Lemma 11.5 shows that

$$|[\exp(Q(T)\tau) - \mathbb{1}\nu]\psi_2(S, 0)| \le K\exp(-\kappa_0\tau).$$

In addition,

$$\left| \int_0^\tau [\exp(Q(T)(\tau - s)) - \mathbb{1}\nu] \widetilde{R}_2(S, s) ds \right|$$

$$\leq K \int_0^\tau | \exp(Q(T)(\tau - s)) - \mathbb{1}\nu | | \widetilde{R}_2(S, s) | ds$$

$$\leq K \int^\tau \exp(-\kappa_0(\tau - s)) \exp(-\kappa_0 s) ds$$

$$\leq K\tau \exp(-\kappa_0 \tau).$$

Thus, $\psi_2(S, \tau)$ will decay to 0 exponentially fast if we choose

$$\mathbb{1}\overline{\psi}_2^0(S) = \mathbb{1}\nu\psi_2(S, 0) = -\mathbb{1} \int_0^\infty \nu \widetilde{R}_2(S, s) ds. \tag{11.36}$$

Note that there is only one unknown in (11.36), namely, $\overline{\psi}_2(S, 0) = \overline{\psi}_2^0(S)$. That is, (11.36) enables us to obtain $\overline{\psi}_2(S, 0)$ uniquely. Using $\gamma_2(S, T) = -\overline{\psi}_2(S, 0)$ together with (11.30) enables us to find the unique solution for the Cauchy problem (11.30) that satisfies $\overline{\psi}_2(S, 0)$. Therefore, $\varphi_2(S, t)$ and $\psi_2(S, \tau)$ are completely determined. Moreover, the construction ensures that

$$\sup_{S \in [0,1]} |\psi_2(S, \tau)| \leq K \exp(-\kappa_0 \tau).$$

In addition, it can be shown that

$$\sup_{S \in [0,1]} \left| \frac{\partial^i \psi_2(S, \tau)}{\partial S^i} \right| \leq K \exp(-\kappa_0 \tau), \quad i = 1, 2.$$

Recall that for simplicity, we have used the convention that K and κ_0 are some positive real numbers. Their values may vary. Their precise values are not important but only the exponential decay property is crucial.

Proceeding in a similar way, we obtain $\psi_k(S, \tau)$ as follows. Suppose that $\psi_1(S, \tau), \ldots, \psi_{k-1}(S, \tau)$ have been constructed so that $\psi_j(S, \tau)$ for $j = 1, \ldots, k - 1$ decay exponentially fast together with their first and second derivatives $(\partial/\partial S)\psi_j(S, \tau)$ and $(\partial^2/\partial S^2)\psi_j(S, \tau)$, respectively.

Consider

$$\frac{\partial \psi_k(S, \tau)}{\partial \tau} = Q(T)\psi_k(S, \tau) + \widetilde{R}_k(S, \tau). \tag{11.37}$$

The solution is then given by

$$\psi_k(S, \tau) = \exp(Q(T)\tau)\psi_k(S, 0) + \int_0^\tau \exp(Q(T)(\tau - s)) \widetilde{R}_k(S, s) ds$$

$$= \mathbb{1}\nu\psi_k(S, 0) + \mathbb{1} \int_0^\infty \nu \widetilde{R}_k(S, s) ds$$

$$+ [\exp(Q(T)\tau) - \mathbb{1}\nu]\psi_k(S, 0) - \mathbb{1}\nu \int_\tau^\infty \widetilde{R}_k(S, s) ds$$

$$+ \int_0^\tau [\exp(Q(T)(\tau - s)) - \mathbb{1}\nu]\widetilde{R}_k(S, s) ds.$$

$$\tag{11.38}$$

Choose

$$\overline{\psi}_k^0(S) := \mathbb{1}\nu\psi_k(S,0) = -\mathbb{1}\int_0^\infty \nu\widetilde{R}_k(S,s)ds,$$

$$\gamma_k(S,T) = -\overline{\psi}_k^0(S), \tag{11.39}$$

$$\psi_k(S,0) = -\varphi_k(S,T) = \overline{\psi}_k^0(S)\mathbb{1} - \varphi_k(S,T).$$

Then, $\varphi_k(S,t)$ and $\psi_k(S,\tau)$ are completely specified. Moreover, we can establish the following lemmas.

Lemma 11.8. *The following assertions hold.*

- $\psi_k(S,\tau)$ *can be constructed by using the first line of* (11.38) *with terminal data* (11.39);

- ψ_k *decay exponentially fast in the sense*

$$\sup_{S\in[0,1]} |\psi_k(S,\tau)| \le K\exp(-\kappa_0\tau) \tag{11.40}$$

for some $K > 0$ and $\kappa_0 > 0$.

Lemma 11.9. *For $k > 1$, the functions $\psi_k(S,\tau)$ satisfy*

$$\sup_{S\in[0,1]} \left|\frac{\partial\psi_k(S,\tau)}{\partial S}\right| \le K\exp(-\kappa_0\tau),$$

$$\sup_{S\in[0,1]} \left|\frac{\partial^2\psi_k(S,\tau)}{\partial S^2}\right| \le K\exp(-\kappa_0\tau).$$

Proof. Redefine

$$U(S,\tau) = \frac{\partial\psi_k(S,\tau)}{\partial S} \quad \text{and}$$

$$V(S,\tau) = \frac{\partial^2\psi_k(S,\tau)}{\partial S^2}.$$

Then U and V satisfy

$$\frac{\partial U(S,\tau)}{\partial\tau} = Q(T)U(S,\tau), \quad U(S,0) = \frac{\partial\psi_1(S,0)}{\partial S},$$

$$\frac{\partial V(S,\tau)}{\partial\tau} = Q(T)V(S,\tau), \quad V(S,0) = \frac{\partial^2\psi_1(S,0)}{\partial S^2}, \tag{11.41}$$

respectively. The rest of the proof is similar to that of Lemma 11.7. $\square$

We summarize what we have obtained thus far and put it in the following theorem.

Theorem 11.10. *Under conditions* (A11.1) *and* (A11.2), *we can construct sequences $\{\varphi_k(S,t,i) : i \in \mathcal{M}; k = 0,\ldots,n\}$ and $\{\psi_k(S,\tau,i) : i \in \mathcal{M}; k = 0,\ldots,n\}$ such that*

- $\varphi_0(S,t) = \gamma_0(S,t)\mathbb{1}$ with $\gamma_0(S,t)$ being the solution of (11.15) satisfying the terminal condition (11.16); $\psi_0(S,\tau,i) = 0$ for each $i \in \mathcal{M}$;

- $\varphi_1(S,t,i)$ is given by (11.17) with $\varphi_1^0(S,t)$ being the unique solution of (11.18) and $\gamma_1(S,t)$ given by (11.20) with $\gamma_1(S,T) = 0$ and the terminal layer term $\psi_1(S,\tau,i)$ specified in Lemma 11.5;

- $\varphi_k(S,t)$ is given by (11.28) with $\varphi_k^0(S,t)$ being the unique solution of (11.29) and $\gamma_k(S,t)$ being the solution of (11.30) with $\gamma(S,T)$ given in Lemma 11.8 and the terminal layer term $\psi_k(S,\tau,i)$ specified in Lemma 11.8.

11.4 Asymptotic Error Bounds

We have constructed the formal asymptotic expansions of the option price. Here we validate the expansions by deriving

$$\max_{i \in \mathcal{M}} \sup_{(S,t) \in [0,1] \times [0,T]} |\sum_{k=0}^{n} \varepsilon^k \varphi_k(S,t,i) + \sum_{k=0}^{n} \varepsilon^k \psi_k(S,\tau,i) - V^\varepsilon(S,t,i)|$$

$$= O(\varepsilon^{n+1}).$$

We first deduce a couple of lemmas.

Lemma 11.11. For each $i \in \mathcal{M}$, let $u(\cdot,\cdot,i) \in C^{2,1}([0,1] \times [0,T]; \mathbb{R})$ such that $v(S,t) = (v(S,t,1), \ldots, v(S,t,m_0))'$ and

$$\mathcal{L}^\varepsilon u(S,t,i) = v(S,t,i), \ t < s \leq T,$$

$$u(S,s,i) = 0, \ S \in [0,1],$$

where $\mathcal{L}^\varepsilon$ is defined in (11.8).
  Assume (A11.1) and (A11.2). Then

$$u(S,s,i) = -E \int_t^s v(X^{\varepsilon,S}(\xi), \alpha^{\varepsilon,i}(\xi), i) d\xi, \qquad (11.42)$$

where $(X^{\varepsilon,S}(t), \alpha^{\varepsilon,i}(t)) = (S,i)$.

Proof. Note that $S \in [0,1]$ and $t \in [0,T]$ for some finite $T > 0$ so $v(\cdot)$ is bounded together with its derivatives with respect to S up to the second order and its derivative with respect to t. In view of (11.2) and (11.3), $u(\cdot)$ is integrable. Moreover, the local martingale in Lemma 11.1 is in fact a martingale now. The desired result then follows from Dynkin's formula. The proof may also be worked out by means of martingale problem formulation. $\square$

Lemma 11.12. *Suppose that* (A11.1) *and* (A11.2) *hold and for each* $i \in \mathcal{M}$, $e^\varepsilon(\cdot, \cdot, i)$ *is a suitable function such that*

$$\sup_{(S,t)\in[0,1]\times[0,T]} |e^\varepsilon(S,t,i)| = O(\varepsilon^\ell) \quad for \;\; \ell \leq n+1. \tag{11.43}$$

Then for each $i \in \mathcal{M}$, *the solution of*

$$\mathcal{L}^\varepsilon u^\varepsilon(S,t,i) = e^\varepsilon(S,t,i), \;\; u^\varepsilon(S,T,i) = 0 \tag{11.44}$$

satisfies

$$\sup_{(S,t)\in[0,1]\times[0,T]} |u^\varepsilon(S,t,i)| = O(\varepsilon^\ell).$$

Proof. In view of Lemma 11.11, the solution of (11.44) can be written as

$$u^\varepsilon(S,t,i) = -E \int_0^t e^\varepsilon(S,\xi,i)d\xi.$$

Thus, (11.43) leads to

$$\sup_{(S,t)\in[0,1]\times[0,T]} |u^\varepsilon(S,t,i)| \leq K \int_0^T O(\varepsilon^\ell) \leq O(\varepsilon^\ell).$$

The desired result thus follows.   □

With the preparation above, we proceed to obtain the desired upper bounds on the approximation errors. For $k = 0, \ldots, n+1$, define a sequence of approximation errors

$$e_k^\varepsilon(S,t,i) = \Phi_k^\varepsilon(S,t,i) + \Psi_k^\varepsilon(S,\tau,i) - V^\varepsilon(S,t,i), \;\; i \in \mathcal{M},$$

where for each $i \in \mathcal{M}$, $V^\varepsilon(S,t,i)$ is the option price given by (11.6), and $\Phi_k^\varepsilon(S,t,i) + \Psi_k^\varepsilon(S,\tau,i)$ is the kth-order approximation to the option price. We proceed to obtain the order of magnitude estimates of $e_n^\varepsilon(S,t,i)$.

Theorem 11.13. *Assume* (A11.1) *and* (A11.2). *Then for the asymptotic expansions constructed in Theorem* 11.10,

$$\max_{i\in\mathcal{M}} \sup_{(S,t)\in[0,1]\times[0,T]} |e_n^\varepsilon(S,t,i)| = O(\varepsilon^{n+1}). \tag{11.45}$$

Proof. The proof is divided in two steps. In the first step, we obtain an estimate on $\mathcal{L}^\varepsilon e_{n+1}^\varepsilon(S,t,i)$, and in the second step, we derive the desired order estimate.

Step 1. Claim: $\mathcal{L}^\varepsilon e_{n+1}^\varepsilon(S,t,i) = O(\varepsilon^{n+1})$. To obtain this, first note that $\mathcal{L}^\varepsilon V^\varepsilon(S,t,i) = 0$. Thus

$$\begin{aligned}
\mathcal{L}^\varepsilon e_{n+1}^\varepsilon(S,t,i) &= \mathcal{L}^\varepsilon \Phi_{n+1}^\varepsilon(S,t,i) + \mathcal{L}^\varepsilon \Psi_{n+1}^\varepsilon(S,t,i) \\
&= \sum_{k=0}^{n+1} \varepsilon^k \mathcal{L}^\varepsilon \varphi_k(S,t,i) + \sum_{j=1}^{n+1} \varepsilon^k \mathcal{L}^\varepsilon \psi_k(S,\tau,i),
\end{aligned}$$

where in the last line above, we have used $\psi_0(S, \tau, i) = 0$ for each $i \in \mathcal{M}$. For the outer expansions, we have

$$
\mathcal{L}^\varepsilon \sum_{k=0}^{n+1} \varepsilon^k \varphi_k(S, t, i)
$$

$$
= \sum_{k=0}^{n+1} \varepsilon^k \Big[\frac{\partial \varphi_k(S, t, i)}{\partial t} + \frac{1}{2} \sigma^2(i) S^2 \frac{\partial^2 \varphi_k(S, t, i)}{\partial S^2}
$$

$$
+ rS \frac{\partial \varphi_k(S, t, i)}{\partial S} - r\varphi_k(S, t, i) + \frac{Q(t)}{\varepsilon} \varphi_k(S, t, \cdot)(i) \Big]
$$

$$
= \sum_{k=0}^{n} \varepsilon^k \Big[\frac{\partial \varphi_k(S, t, i)}{\partial t} + \frac{1}{2} \sigma^2(i) S^2 \frac{\partial^2 \varphi_k(S, t, i)}{\partial S^2}
$$

$$
+ rS \frac{\partial \varphi_k(S, t, i)}{\partial S} - r\varphi_k(S, t, i) + Q(t)\varphi_{k+1}(S, t, \cdot)(i) \Big]
$$

$$
+ \frac{Q(t)}{\varepsilon} \varphi_0(S, t, \cdot)(i) + \varepsilon^{n+1} \Big[\frac{\partial \varphi_{n+1}(S, t, i)}{\partial t}
$$

$$
+ \frac{1}{2} \sigma^2(i) S^2 \frac{\partial^2 \varphi_{n+1}(S, t, i)}{\partial S^2} + rS \frac{\partial \varphi_{n+1}(S, t, i)}{\partial S} - r\varphi_{n+1}(S, t, i) \Big]
$$

$$
= \varepsilon^{n+1} \Big[\frac{\partial \varphi_{n+1}(S, t, i)}{\partial t} + \frac{1}{2} \sigma^2(i) S^2 \frac{\partial^2 \varphi_{n+1}(S, t, i)}{\partial S^2}
$$

$$
+ rS \frac{\partial \varphi_{n+1}(S, t, i)}{\partial S} - r\varphi_{n+1}(S, t, i) \Big].
$$

$$(11.46)$$

The boundedness of $\varphi_{n+1}(S, t, i)$ and their derivatives up to the order 2 together with (11.46) then lead to

$$
\max_{i \in \mathcal{M}} \sup_{(S,t) \in [0,1] \times [0,T]} |\mathcal{L}^\varepsilon \Phi_{n+1}^\varepsilon(S, t, i)| = O(\varepsilon^{n+1}). \tag{11.47}
$$

As for the terminal layer terms, we have

$$
\mathcal{L}^\varepsilon \sum_{k=1}^{n+1} \varepsilon^k \psi_k(S, \tau, i)
$$

$$
= \sum_{k=1}^{n+1} \varepsilon^{k-1} \Big[\frac{\partial \psi_k(S, \tau, i)}{\partial \tau} + \frac{1}{2} \varepsilon \sigma^2(i) S^2 \frac{\partial^2 \psi_k(S, \tau, i)}{\partial S^2} + \varepsilon rS \frac{\partial \psi_k(S, \tau, i)}{\partial S}
$$

$$
- \varepsilon r \psi_k(S, \tau, i) + Q(t)\psi_k(S, \tau, \cdot)(i) \Big]
$$

$$
= \sum_{k=1}^{n+1} \varepsilon^{k-1} \Big\{ \Big[\frac{\partial \psi_k(S, \tau, i)}{\partial \tau} + \frac{1}{2} \varepsilon \sigma^2(i) S^2 \frac{\partial^2 \psi_k(S, \tau, i)}{\partial S^2} + \varepsilon rS \frac{\partial \psi_k(S, \tau, i)}{\partial S}
$$

$$
- \varepsilon r \psi_k(S, \tau, i) + Q(T)\psi_k(S, \tau, \cdot)(i)
$$

$$
+ \sum_{j=1}^{k} \frac{(-1)^j \tau^j}{j!} \frac{d^j Q(T)}{dt^j} \psi_k(S, \tau, i)
$$

$$
+ \Big[Q(t) - \sum_{j=1}^{k} \frac{(-1)^j \tau^j}{j!} \frac{d^j Q(T)}{dt^j} \Big] \psi_k(S, \tau, \cdot)(i) \Big\}.
$$

$$(11.48)$$

Note that for $k = 1, \ldots, n+1$,

$$\left| Q(t) - \sum_{j=1}^{k} \frac{(-1)^j \tau^j}{j!} \frac{d^j Q(T)}{dt^j} \right| = |R_k(T - \varepsilon\tau)| \le Kt^{k+1},$$

and that

$$\left| \sum_{j=1}^{n+1} \varepsilon^{j-1} \psi_j(S, \tau, i) O(t^{n+1-j}) \right|$$

$$\le K \sum_{j=1}^{n+1} \varepsilon^{n+1-j} t^j \exp(-\kappa_0 \tau) \le K\varepsilon^{n+1}.$$

The above observation together with (11.48) and

$$-\frac{\partial \psi_1(S, \tau, i)}{\partial \tau} + Q(T)\psi_1(S, \tau, i) = 0$$

gives us

$$\mathcal{L}^{\varepsilon} \sum_{k=1}^{n+1} \varepsilon^k \psi_k(S, \tau, i)$$

$$= \sum_{k=2}^{n+1} \varepsilon^{k-1} \left[-\frac{\partial \psi_k(S, \tau, i)}{\partial \tau} + Q(T)\psi_k(S, \tau, i) + \widetilde{R}_k \right] + O(\varepsilon^{n+1})$$

$$= O(\varepsilon^{n+1}),$$

$$(11.49)$$

which holds uniformly in $(S, t) \in [0, 1] \times [0, T]$. Thus we obtain

$$\max_{i \in \mathcal{M}} \sup_{(S,t) \in [0,1] \times [0,T]} |\mathcal{L}^{\varepsilon}[\Phi_{n+1}(S, t, i) + \Psi_{n+1}(S, \tau, i)]| = O(\varepsilon^{n+1}). \quad (11.50)$$

Step 2. Obtain estimate (11.45). Note that the definition of $e_{n+1}^{\varepsilon}(S, t, i)$ and (11.11) yield that $e_{n+1}^{\varepsilon}(S, T, i) = 0$. Because (11.50) holds, it follows from Lemma 11.12 that

$$\max_{i \in \mathcal{M}} \sup_{(S,t) \in [0,1] \times [0,T]} |e_{n+1}^{\varepsilon}(S, t, i)| = O(\varepsilon^{n+1}).$$

Note that

$$e_{n+1}^{\varepsilon}(S, t, i) = e_n^{\varepsilon}(S, t, i) + \varepsilon^{n+1} \varphi_{n+1}(S, t, i) + \varepsilon^{n+1} \psi_{n+1}(S, \tau, i).$$

$$(11.51)$$

The smoothness of $\varphi_{n+1}(S, t, i)$ and the exponential decay of $\psi_{n+1}(S, \tau, i)$ imply that

$$\max_{i \in \mathcal{M}} \sup_{(S,t) \in [0,1] \times [0,T]} |\varepsilon^{n+1} \varphi_{n+1}(S, t, i) + \varepsilon^{n+1} \psi_{n+1}(S, \tau, i)| = O(\varepsilon^{n+1}).$$

Substituting the above into (11.51), we obtain (11.45) as desired.   □

11.5   Notes

In this chapter, we have developed asymptotic expansions for a European-type option price. The essence is the use of two-time-scale formulation to deal with solutions of systems of parabolic PDEs. The result is based on the recent work of Yin [166]. The approach we are using is constructive. Thus it sheds more light on how these approximations can be carried out. Full asymptotic expansions have been obtained with uniform asymptotic error bounds for the continuous component belonging to a compact set. If one is only interested in getting asymptotic expansions with certain fixed state variables (as in [49, 50]), then one can work with the entire space $\mathbb{R}$ rather than a compact set. In lieu of $Q^\varepsilon(t)$ considered thus far, we may treat a slightly more complex model with

$$Q^\varepsilon(t) = \frac{Q_0(t)}{\varepsilon} + Q_1(t),$$

where both $Q_0(\cdot)$ and $Q_1(\cdot)$ are generators of continuous-time Markov chains such that $Q_0(t)$ is weakly irreducible. Then we can still obtain asymptotic expansions. The notation, however, will be a bit more complex due to the addition of $Q_1(t)$. In view of the work by Il'in, Khasminskii, and Yin [73, 74], the results of this chapter can be extended to switching diffusions in which the switching process has generator $Q^\varepsilon(x,t)$ that depends on x as well.

For risk-neutral valuation, it is natural to let r be a constant. Nevertheless, the techniques presented here carry over for the more general α-dependent process; that is $r(t) = r(\alpha(t))$. Although the main motivation is from mathematical finance, the techniques developed here can be used for other problems involving systems of coupled differential equations where a fast-varying switching process is a driving force.

12

Two-Time-Scale Switching Jump Diffusions

12.1 Introduction

This chapter is concerned with jump diffusions involving Markovian switching regimes. In the models, there are a finite set of regimes or configurations and a switching process that dictates which regime to take at any given instance. At each time t, once the configuration is determined by the switching process, the dynamics of the system follow a jump-diffusion process. It evolves until the next jump takes place. Then the post-jump location is determined and the process sojourns in the new location following the evolution of another jump-diffusion process and so on. The entire system consists of random switches and jump-diffusive motions.

One of our motivations stems from insurance risk theory. To capture the features of insurance policies that are subject to economic or political environment changes, generalized hybrid risk models may be considered. To reduce the complexity of the systems, time-scale separation may be used. Under the classical insurance risk model, the surplus $U(t)$ of an insurance company at $t \geq 0$ is given by

$$U(t) = u + ct - S(t),$$

where u is the initial surplus, $c > 0$ is the rate at which the premiums are received, and $S(t)$, a compound Poisson process, is the total claim in the duration $[0, t]$. In [35], Dufresne and Gerber extended the classical risk model by adding an independent diffusion process so that the surplus is given by

$$U(t) = u + ct - S(t) + \sigma w(t),$$

G.G. Yin and C. Zhu, *Hybrid Switching Diffusions: Properties and Applications*,
Stochastic Modelling and Applied Probability 63, DOI 10.1007/978-1-4419-1105-6_12,
© Springer Science + Business Media, LLC 2010

where $w(t)$ is a standard real-valued Brownian motion that represents uncertainty (often referred to as oscillations) of premium incomes and claims. Subsequently, much work has been devoted to such jump-diffusion models; see also variants of the models in [128, 140] and the references therein.

Recently, growing attention has been drawn to the use of switching models in finance and the insurance industry. For instance, taking the opportunity provided by using a switching process to represent the underlying economy switching among a finite number of discrete states, the European options under the Black–Scholes formulation of the stock market were considered in [32]; the American options were dealt with in [20]; algorithms for liquidation of a stock were constructed in [170]. Using a random pure jump process to represent a random environment, we proposed a Markovian regime-switching formulation in [163] to model the insurance surplus process. Suppose that there is a finite set $\mathcal{M} = \{1, \ldots, m_0\}$, representing the possible regimes (configurations) of the environment. At each $i \in \mathcal{M}$, assume that the premium is payable at the rate $c(i)$ continuously. Let $U(t, i)$ be the surplus process given the initial surplus $u > 0$ and initial state $\alpha(0) = i$:

$$U(t, i) = u + \int_0^t c(\alpha(s))ds - S(t),$$

where $S(t)$, as in the classical risk model, is a compound Poisson process and $\alpha(t)$ is a continuous-time Markov chain with state space $\mathcal{M}$ representing the random environment. Under suitable conditions, we obtained Lunderberg-type upper bounds and nonexponential upper bounds for the ruin probability, and treated a renewal-type system of equations for ruin probability when the claim sizes are exponentially distributed. One of the main features of [163] is that there is an additional Markov chain, which enables the underlying surplus to vary in accordance with different regimes.

Here we are concerned with a class of jump diffusions with regime switching to prepare us for treating applications involving more general risk models. In this chapter, we consider jump diffusions modulated by a continuous-time Markov chain. Because the dynamic systems are complex, it is of foremost importance to reduce the complexity. Taking into consideration the inherent hierarchy in a complex system [149], and different rates of variations of subsystems and components, we use the two-time-scale method leading to systems in which the fast and slow rates of change are in sharp contrast. Then we proceed to reduce the system complexity by aggregation/decomposition and averaging methods. We demonstrate that under broad conditions, associated with the original systems, there are limit or reduced systems, which are averages with respect to certain invariant measures. Using weak convergence methods [102, 103], we obtain the limit system (jump diffusion with regime switching) via martingale problem formulation.

Let $\Gamma \subset \mathbb{R}^r - \{0\}$ be the range space of the impulsive jumps, $w(\cdot)$ be a real-valued standard Brownian motion, and $N(\cdot, \cdot)$ be a Poisson measure such that $N(t, H)$ counts the number of impulses on $[0, t]$ with values in the set H. Let $f(\cdot, \cdot, \cdot) : [0, T] \times \mathbb{R} \times \mathcal{M} \mapsto \mathbb{R}$, $\sigma(\cdot, \cdot, \cdot) : [0, T] \times \mathbb{R} \times \mathcal{M} \mapsto \mathbb{R}$, $g(\cdot, \cdot, \cdot) : \Gamma \times \mathbb{R} \times \mathcal{M} \mapsto \mathbb{R}$, and $\alpha(\cdot)$ be a continuous-time Markov chain having a state space $\mathcal{M}$. A brief description of the jump-diffusion process with a modulating Markov chain can be found in Section A.6 of this book. Consider the following jump-diffusion process with regime switching

$$
\begin{aligned}
X(t) = x & + \int_0^t f(s, X(s), \alpha(s)) ds + \int_0^t \sigma(s, X(s), \alpha(s)) dw(s) \\
& + \int_0^t \int_\Gamma g(\gamma, X(s^-), \alpha(s^-)) N(ds, d\gamma).
\end{aligned}
\tag{12.1}
$$

Throughout the chapter, we assume that $w(\cdot)$, $N(\cdot)$, and $\alpha(\cdot)$ are mutually independent. Compared with the traditional jump-diffusion processes, the coefficients involved in (12.1) all depend on an additional switching process, namely, the Markov chain $\alpha(t)$.

In the context of risk theory, $X(t)$ can be considered as the surplus of the insurance company at time t, x is the initial surplus, $f(t, X(t), \alpha(t))$ represents the premium rate (assumed to be ≥ 0), $g(\gamma, X(t), \alpha(t))$ is the amount of the claim whenever there is one (assumed to be ≤ 0), and the diffusion is used to model additional uncertainty of the claims and/or premium incomes. Similar to the volatility in stock market models, $\sigma(\cdot, \cdot, i)$ represents the amount of oscillations or volatility in an appropriate sense. The model is sufficiently general to cover the traditional compound Poisson models as well as the diffusion perturbed ruin models. It may also be used to represent security price in finance (see [124, Chapter 3]). The process $\alpha(t)$ may be viewed as an environment variable dictating the regime. The use of the Markov chain results from consideration of the general trend of the market environment as well as other economic factors. The economic and/or political environment changes lead to the changes of regime of the surplus, resulting in markedly different behavior of the system across regimes.

Defining a centered (or compensated) Poisson measure and applying generalized Itô's rule, we can obtain the generator of the jump-diffusion process with regime switching, and formulate a related martingale problem. Instead of a single process, we have to deal with a collection of jump-diffusion processes that are modulated by a continuous-time Markov chain. Suppose that λ is positive such that $\lambda \Delta + o(\Delta)$ represents the probability of a switch of regime in the interval $[t, t + \Delta)$, and $\pi(\cdot)$ is the distribution of the jump. Then the generator of the underlying process can be written as

$$
\begin{aligned}
\mathcal{G}F(t, x, \iota) = & \left(\frac{\partial}{\partial t} + \mathcal{L} \right) F(t, x, \iota) \\
& + \int_\Gamma \lambda [F(t, x + g(\gamma, x, \iota), \iota) - F(t, x, \iota)] \pi(d\gamma) \\
& + Q(t) F(t, x, \cdot)(\iota), \quad \text{for each } \iota \in \mathcal{M},
\end{aligned}
\tag{12.2}
$$

where

$$\mathcal{L}F(t,x,\iota) = \frac{1}{2}\sigma^2(t,x,\iota)\frac{\partial^2}{\partial x^2}F(t,x,\iota) + f(t,x,\iota)\frac{\partial}{\partial x}F(t,x,\iota),$$

$$Q(t)F(t,x\cdot)(\iota) = \sum_{\ell=1}^{m_0} a_{\iota\ell}(t)F(t,x,\ell) = \sum_{\ell\neq\iota} a_{\iota\ell}(t)[F(t,x,\ell) - F(t,x,\iota)].$$

(12.3)

By concentrating on time-scale separations, in this chapter, we treat two cases. In the first one, the regime switching is significantly faster than the dynamics of the jump diffusions, whereas in the second case, the diffusion varies an order of magnitude faster than the switching processes.

The rest of the chapter is arranged as follows. Section 12.2 is devoted to the case of fast switching. It begins with the precise formulation of the problem. Then we derive weak convergence results and demonstrate that the complicated problem can be "replaced" by a limit problem in which the system coefficients are averaged out with respect to the stationary measures of the switching process. In Section 12.3, we continue our study for the case of fast varying diffusions. Again, by means of weak convergence methods, we obtain a limit system. Section 12.4 gives remarks on specialization and generalization of the asymptotic results. Section 12.5 gives remarks on numerical approximation for switching-jump-diffusion processes. Section 12.6 concludes the chapter.

12.2   Fast-Varying Switching

12.2.1   Fast-Varying Markov Chain Model

Consider a continuous-time inhomogeneous Markov chain $\alpha(t)$ with generator $Q(t)$. Recall that the Markov chain $\alpha(t)$ or the generator $Q(t)$ is weakly irreducible if the system of equations

$$\begin{cases} \nu(t)Q(t) = 0, \\ \displaystyle\sum_{i=1}^{m_0} \nu_i(t) = 1 \end{cases}$$

has a unique nonnegative solution. The nonnegative solution (row-vector-valued function) $\nu(t) = (\nu_1(t),\ldots,\nu_{m_0}(t))$ is termed a quasi-stationary distribution.

For the fast-varying Markov chain model, by introducing a small parameter $\varepsilon > 0$ into the problem, suppose that $\alpha(t) = \alpha^\varepsilon(t)$ with the generator of the Markov chain given by

$$Q^\varepsilon(t) = \frac{1}{\varepsilon}\widetilde{Q}(t) + \widehat{Q}(t).$$

(12.4)

Both $\widetilde{Q}(t)$ and $\widehat{Q}(t)$ are generators, where $\widetilde{Q}(t)/\varepsilon$ represents the rapidly changing part and $\widehat{Q}(t)$ describes the slowly varying part. The slow and fast components are coupled through weak and strong interactions in the sense that the underlying Markov chain fluctuates rapidly within a single group $\mathcal{M}_k$ of states and jumps less frequently from group $\mathcal{M}_k$ to $\mathcal{M}_j$ for $k \neq j$. Suppose that the generator $\widetilde{Q}(t)$ has the form

$$\widetilde{Q}(t) = \text{diag}\left(\widetilde{Q}^1(t), \ldots, \widetilde{Q}^l(t)\right), \tag{12.5}$$

where for each $k = 1, \ldots, l$. $\widetilde{Q}^k(t)$ is a generator corresponding to the states in $\mathcal{M}_k = \{s_{k1}, \ldots, s_{km_k}\}$. Naturally, the state space can be decomposed to

$$\mathcal{M} = \mathcal{M}_1 \cup \cdots \cup \mathcal{M}_l = \{s_{11}, \ldots, s_{1m_1}\} \cup \cdots \cup \{s_{l1}, \ldots, s_{lm_l}\}. \tag{12.6}$$

The associated system can be written as

$$
X^\varepsilon(t) = x + \int_0^t f(s, X^\varepsilon(s), \alpha^\varepsilon(s))ds + \int_0^t \sigma(s, X^\varepsilon(s), \alpha^\varepsilon(s))dw \\
+ \int_0^t \int_\Gamma g(\gamma, X^\varepsilon(s^-), \alpha^\varepsilon(s^-))N(ds, d\gamma). \tag{12.7}
$$

Define the centered (or compensated) Poisson measure

$$\widetilde{N}(t, H) = N(t, H) - \lambda t \pi(H).$$

Then (12.7) may be rewritten as

$$
\begin{aligned}
X^\varepsilon(t) &= x + \int_0^t f(s, X^\varepsilon(s), \alpha^\varepsilon(s))ds \\
&+ \lambda \int_0^t \int_\Gamma g(\gamma, X^\varepsilon(s^-), \alpha^\varepsilon(s^-))\pi(d\gamma)ds \\
&+ \int_0^t \sigma(s, X^\varepsilon(s), \alpha^\varepsilon(s))dw \\
&+ \int_0^t \int_\Gamma g(\gamma, X^\varepsilon(s^-), \alpha^\varepsilon(s^-))\widetilde{N}(ds, d\gamma).
\end{aligned} \tag{12.8}
$$

Note that the last two terms are martingales. The operator of the regime-switching jump-diffusion process is given by

$$\mathcal{G}^\varepsilon F(t, x, \iota) = \left(\frac{\partial}{\partial t} + \mathcal{L}\right) F(t, x, \iota) + J(t, x, \iota) + Q^\varepsilon(t)F(t, x, \cdot)(\iota), \tag{12.9}$$

where

$$J(t, x, \iota) = \int_\Gamma \lambda[F(t, x + g(\gamma, x, \iota), \iota) - F(t, x, \iota)]\pi(d\gamma).$$

Remark 12.1. The use of the two-time-scale formulation with a small parameter $\varepsilon > 0$ stems from an effort to reduce complexity. Taking into consideration various factors, the number of states of the Markov chain is usually large; the system becomes complex; dealing with it is difficult. Nevertheless, we demonstrate by letting $\varepsilon \to 0$, a limit system can be obtained, which is an average of the original system with respect to a set of quasi-stationary measures.

To carry out the desired analysis, we postulate the following conditions.

(A12.1) $\mathbf{E}|x|^2 < \infty$, and the following conditions hold.

(a) The functions $f(\cdot)$ and $\sigma(\cdot)$ satisfy: For each $\alpha \in \mathcal{M}$, $f(\cdot, \cdot, \alpha)$ and $\sigma(\cdot, \cdot, \alpha)$ are defined and Borel measurable on $[0, T] \times \mathbb{R}$; for each $(t, x, \alpha) \in [0, T] \times \mathcal{M} \times \mathbb{R}$,

$$|f(t, x, \alpha)| \leq K(1 + |x|) \text{ and } |\sigma(t, x, \alpha)| \leq K(1 + |x|);$$

and for any $z, x \in \mathbb{R}$,

$$|f(t, z, \alpha) - f(t, x, \alpha)| \leq K|z - x| \text{ and}$$
$$|\sigma(t, z, \alpha) - \sigma(t, x, \alpha)| \leq K|z - x|.$$

(b) $0 < \lambda < \infty$. For each $\beta \in \mathcal{M}$, and $g(\cdot, \cdot, \alpha)$ is a bounded and continuous function, $g(0, x, \alpha) = 0$, and for each x, the value of γ can be determined uniquely by $g(\gamma, x, \alpha)$.

(A12.2) Both $\widetilde{Q}(t)$ and $\widehat{Q}(t)$ are generators that are bounded and Borel measurable such that for each $k = 1, \ldots, l$ and $t \in [0, T]$, $\widetilde{Q}^k(t)$ is weakly irreducible with the associated quasi-stationary distribution $\nu^k(t) = (\nu_1^k(t), \ldots, \nu_{m_k}^k(t)) \in \mathbb{R}^{1 \times m_k}$.

Remark 12.2. In the above and throughout the chapter, K is used as a generic positive constant, whose values may change for different usages, so the conventions $K + K = K$ and $KK = K$ are understood.

For each fixed $\alpha \in \mathcal{M}$, the conditions on $f(\cdot, \cdot, \alpha)$ and $g(\cdot, \cdot, \alpha)$ are the "Itô conditions" that ensure the existence and uniqueness of the solution of the stochastic differential equation. The assumption on $\widetilde{Q}(t)$ leads to the partition of $\mathcal{M}$, the state space of $\alpha^\varepsilon(\cdot)$ as in (12.6), which in turn leads to the natural definition of an aggregated process. That is, by aggregating the states s_{kj} in $\mathcal{M}_k$ as one state, we obtain an aggregated process $\overline{\alpha}^\varepsilon(\cdot)$ defined by

$$\overline{\alpha}^\varepsilon(t) = k \text{ if } \alpha^\varepsilon(t) \in \mathcal{M}_k. \tag{12.10}$$

Thus, lump all the states in each weakly irreducible class into one state resulting in a process with considerably smaller state space. Note that the process $\overline{\alpha}^\varepsilon(\cdot)$ is not necessarily Markovian.

Condition (A12.2) is concerned with the Markov chain $\alpha^\varepsilon(\cdot)$, which is modeled after that of [176, Chapter 7.5, p. 210] and allows the generators to be time-dependent. Note that no continuity of $\widetilde{Q}(\cdot)$ and $\widehat{Q}(\cdot)$ is required, but merely boundedness and measurability are assumed; see also [178] for the inclusion of transient states. For our applications, the only requirement is the partitioned form $\widetilde{Q}(\cdot)$ and the weak irreducibility of each $\widehat{Q}^k(\cdot)$.

The following idea uses an averaging approach by aggregating the states in each weakly irreducible class into a single state, and replacing the original complex system by its limit, an average with respect to the quasi-stationary distributions. Using certain probabilistic arguments, we have shown in [176, Section 7.5] (see also [178]) that

(i) $\overline{\alpha}^\varepsilon(\cdot)$ converges weakly to $\overline{\alpha}(\cdot)$ whose generator is given by

$$\overline{Q}(t) = \operatorname{diag}(\nu^1(t), \ldots, \nu^l(t))\widehat{Q}(t)\operatorname{diag}(\mathbb{1}_{m_1}, \ldots, \mathbb{1}_{m_l}), \qquad (12.11)$$

where $\nu^k(t)$ is the quasi-stationary distribution of $\widetilde{Q}^k(t)$, for $k = 1, \ldots, l$, and $\mathbb{1}_\ell = (1, \ldots, 1)' \in \mathbb{R}^\ell$ is an ℓ-dimensional column vector with all components being equal to 1;

(ii) for $k = 1, \ldots, l$ and $j = 1, \ldots, m_k$, as $\varepsilon \to 0$,

$$\sup_{0 \leq t \leq T} \mathbf{E} \left(\int_0^t \left(I_{\{\alpha^\varepsilon(s)=s_{kj}\}} - \nu_j^k I_{\{\overline{\alpha}^\varepsilon(s)=k\}} \right) ds \right)^2 \to 0, \qquad (12.12)$$

where I_A is the indicator function of the set A.

12.2.2 Limit System

Working with the pair $Y^\varepsilon(\cdot) = (X^\varepsilon(\cdot), \overline{\alpha}^\varepsilon(\cdot))$, we obtain the following theorem. It indicates that there is a limit system associated with the original process leading to a reduction of complexity.

Theorem 12.3. *Assuming* (A12.1) *and* (A12.2), $Y^\varepsilon(\cdot) = (X^\varepsilon(\cdot), \overline{\alpha}^\varepsilon(\cdot))$ *converges weakly to* $Y(\cdot) = (X(\cdot), \overline{\alpha}(\cdot))$, *which is a solution of the martingale problem with operator*

$$\overline{\mathcal{G}}F(t, x, i) = \left(\frac{\partial}{\partial t} + \overline{\mathcal{L}} \right) F(t, x, i) + \overline{J}(t, x, \iota) + \overline{Q}(t)F(t, x, \cdot)(i), \qquad (12.13)$$

for $i \in \overline{\mathcal{M}} = \{1, \ldots, l\}$, *where*

$$\overline{Q}(t)F(t, x, \cdot)(i) = \sum_{j=1}^{l} \overline{q}_{ij}(t)F(t, j, x) = \sum_{\substack{j \in \overline{\mathcal{M}} \\ j \neq i}} \overline{q}_{ij}(t)(F(t, x, j) - F(t, x, i)),$$

$$(12.14)$$

$\overline{\mathcal{L}}$ is a second-order differential operator given by (12.3) with $\sigma^2(t,x,\iota)$, $J(t,x,\iota)$, and $f(t,x,\iota)$ replaced by $\overline{\sigma}^2(t,x,i)$, $\overline{J}(t,x,\iota)$, and $\overline{f}(t,x,i)$, respectively, and

$$\overline{f}(t,x,i) = \sum_{j=1}^{m_i} \nu_j^i(t) f(t,x,s_{ij}),$$

$$\overline{J}(t,i,x) = \sum_{j=1}^{m_i} \nu_j^i(t) J(t,x,s_{ij}) \tag{12.15}$$

$$\overline{\sigma}^2(t,x,i) = \sum_{j=1}^{m_i} \nu_j^i(t) \sigma^2(t,x,s_{ij}).$$

Remark 12.4. Theorem 12.3 characterizes the limit as a solution of the associated martingale problem with operator $\overline{\mathcal{G}}$. It can also be described by the limit stochastic differential equation, given by

$$
\begin{aligned}
X(t) = {}& x + \int_0^t \overline{f}(s, X(s), \overline{\alpha}(s)) ds + \int_0^t \overline{\sigma}(s, X(s), \overline{\alpha}(s)) dw \\
&+ \int_0^t \int_\Gamma \overline{g}(\gamma, X(s^-), \overline{\alpha}(s^-)) N(ds, d\gamma),
\end{aligned}
\tag{12.16}
$$

where $\overline{\alpha}(\cdot)$ is a Markov chain generated by $\overline{Q}(\cdot)$. In particular, if the Markov chain corresponding to the generator $\widetilde{Q}(t)$ consists of only one weakly irreducible block, then the limit or the averaged system becomes a jump-diffusion process. We state the result as follows.

Corollary 12.5. *Under the conditions of Theorem 12.3 with the modification that $Q^\varepsilon(t) = Q(t)/\varepsilon + \widehat{Q}(t)$ such that $\mathcal{M} = \{1,\ldots,m\}$ and that $Q(t)$ is weakly irreducible for each $t \in [0,T]$ with the associated quasi-stationary distribution $\nu(t) = (\nu_1(t),\ldots,\nu_{m_0}(t))$, then $Y^\varepsilon(\cdot)$ converges weakly to $Y(\cdot)$ such that*

$$
\begin{aligned}
X(t) = {}& x + \int_0^t \sum_{\iota=1}^{m_0} f(s, X(s), \iota) \nu_\iota(s) ds \\
&+ \int_0^t \sqrt{\sum_{\iota=1}^{m_0} \sigma^2(s, X(s), \iota) \nu_\iota(s)} dw(s) \\
&+ \int_0^t \sum_{\iota=1}^{m_0} g(\gamma, X(s^-), \iota) \nu_\iota(s^-) N(ds, d\gamma).
\end{aligned}
\tag{12.17}
$$

Proof of Theorem 12.3. To proceed, the detailed proof is given by establishing a series of lemmas. We first show that an a priori bound holds.

Lemma 12.6. *Under the conditions of Theorem 12.3,*

$$\sup_{0 \le t \le T} \mathbf{E}|X^\varepsilon(t)|^2 = O(1).$$

Proof. It follows from (12.7),

$$
\mathbf{E}|X^{\varepsilon}(t)|^2 \le K \Bigg(\mathbf{E}|x|^2 + \mathbf{E}\bigg| \int_0^t f(u, X^{\varepsilon}(u), \alpha^{\varepsilon}(u))du \bigg|^2
$$
$$
+ \mathbf{E}\bigg| \int_0^t \sigma(u, X^{\varepsilon}(u), \alpha^{\varepsilon}(u))dw(u) \bigg|^2
$$
$$
+ \mathbf{E}\bigg| \int_0^t \int_{\Gamma} g(\gamma, x^{\varepsilon}(u^-), \alpha^{\varepsilon}(u^-))N(du, d\gamma) \bigg|^2 \Bigg).
$$

By virtue of the argument of [103, p.39],

$$
\mathbf{E}\bigg| \int_0^t \int_{\Gamma} g(\gamma, X^{\varepsilon}(u^-), \alpha^{\varepsilon}(u^-))N(du, d\gamma) \bigg|^2 = O(1),
$$

and the bound holds uniformly in $t \in [0, T]$. Using the linear growth of $f(t, x, \alpha)$ and $\sigma(t, x, \alpha)$ given in (A12.1) together with properties of stochastic integrals, we obtain

$$
\mathbf{E}|X^{\varepsilon}(t)|^2 \le K + K \int_0^t \mathbf{E}|X^{\varepsilon}(u)|^2 du.
$$

The well-known Gronwall inequality yields

$$
\sup_{t \in [0,T]} \mathbf{E}|X^{\varepsilon}(t)|^2 \le K \exp(KT) < \infty
$$

as desired.   □

Next, we derive the tightness of $Y^{\varepsilon}(\cdot)$.

Lemma 12.7. *Assume that the conditions of Theorem 12.3 are satisfied. Then $Y^{\varepsilon}(\cdot)$ is tight in $D([0, T] : \mathbb{R} \times \mathcal{M})$, the space of functions that are right-continuous, and have left limits endowed with the Skorohod topology.*

Proof. Because $\overline{\alpha}^{\varepsilon}(\cdot)$ converges weakly to $\overline{\alpha}(\cdot)$ [176, p. 172], $\{\overline{\alpha}^{\varepsilon}(\cdot)\}$ is tight. Therefore, to obtain the tightness of $\{Y^{\varepsilon}(\cdot)\}$, it suffices to derive the tightness of $\{X^{\varepsilon}(\cdot)\}$.

Note that for any $t > 0$, $s > 0$, and any $\delta > 0$ with $0 < s \le \delta$, we have

$$
\mathbf{E}_t^{\varepsilon}|X^{\varepsilon}(t+s) - X^{\varepsilon}(t)|^2
$$
$$
\le K\mathbf{E}_t^{\varepsilon} \left(\int_t^{t+s} |f(u, X^{\varepsilon}(u), \alpha^{\varepsilon}(u))|du \right)^2
$$
$$
+ \mathbf{E}_t^{\varepsilon}\bigg| \int_t^{t+s} \sigma(u, X^{\varepsilon}(u), \alpha^{\varepsilon}(u))dw(u) \bigg|^2 \qquad (12.18)
$$
$$
+ \mathbf{E}_t^{\varepsilon}\bigg| \int_t^{t+s} \int_{\Gamma} g(\gamma, X^{\varepsilon}(u^-), \alpha^{\varepsilon}(u^-))N(du, d\gamma) \bigg|^2,
$$

where $\mathbf{E}_t^\varepsilon$ denotes the conditional expectation with respect to the σ-algebra generated by $\{\alpha^\varepsilon(u), X^\varepsilon(u) : u \leq t\}$. Using the argument as in [103, p. 39], since $s \leq \delta$,

$$\mathbf{E}_t^\varepsilon \left| \int_t^{t+s} \int_\Gamma g(\gamma, X^\varepsilon(u^-), \alpha^\varepsilon(u^-))N(du, d\gamma) \right|^2 \leq Ks = O(\delta).$$

Taking the expectation in (12.18), and applying Lemma 12.6 lead to

$$\begin{aligned}
\mathbf{E}|X^\varepsilon(t+s) - X^\varepsilon(t)|^2 &\leq K \left(\int_t^{t+s} (1 + \mathbf{E}|X^\varepsilon(u)|)du \right)^2 \\
&\quad + K \int_t^{t+s} (1 + \mathbf{E}|X^\varepsilon(u)|^2)du + O(s) \\
&\leq K(s^2 + s) + O(\delta) = O(\delta).
\end{aligned} \tag{12.19}$$

As a result,

$$\lim_{\delta \to 0} \limsup_{\varepsilon \to 0} \mathbf{E}|X^\varepsilon(t+s) - X^\varepsilon(t)|^2 = 0.$$

By virtue of the tightness criterion (see Lemma A.28 of this book, and also [43, Section 3.8, p. 132] or [102, p. 47] or [16]), the tightness of $\{X^\varepsilon(\cdot)\}$ follows.   □

Lemma 12.8. *Assume the conditions of Theorem 12.3 are fulfilled. Suppose that $\zeta(t, z)$ defined on $[0, T] \times \mathbb{R}$ is a real-valued function that is Lipschitz continuous in both variables, and that for each $x \in \mathbb{R}^r$, $|\zeta(t, x)| \leq K(1 + |x|)$. Denote*

$$v_{ij}^\varepsilon(t) = v_{ij}(t, \alpha^\varepsilon(t)) \quad \text{with} \quad v_{ij}(t, \alpha) = I_{\{\alpha = s_{ij}\}} - \nu_j^i(t)I_{\{\alpha \in \mathcal{M}_i\}}.$$

Then for any $i = 1, \ldots, l$, $j = 1, \ldots, m_i$,

$$\sup_{0 < t \leq T} \mathbf{E} \left| \int_0^t \zeta(u, X^\varepsilon(u))v_{ij}^\varepsilon(u, \alpha^\varepsilon(u))du \right|^2 \to 0 \quad as \quad \varepsilon \to 0. \tag{12.20}$$

Proof. The proof of this lemma is similar to that of Lemma 7.14 in [176]. Pick out $0 < \Delta < 1$. For any $t \in [0, T]$, partition $[0, t]$ into subintervals of equal length $\varepsilon^{1-\Delta}$ (without loss of generality, assume that $\ell_0 = t/\varepsilon^{1-\Delta}$ is an integer, otherwise, we can always take its integer part). Denote the partition boundaries by $t_k = k\varepsilon^{1-\Delta}$ for $0 \leq k \leq \ell_0$. Define

$$\tilde{\zeta}^\varepsilon(u) = \zeta(t_k, X^\varepsilon(t_k)), \quad u \in [t_k, t_{k+1}), \ 0 \leq k \leq \ell_0 - 1. \tag{12.21}$$

In view of the process $N(\cdot)$, $\alpha^\varepsilon(\cdot)$, and $w(\cdot)$, the same argument as in the proof of the tightness yields

$$\begin{aligned}
&\mathbf{E}|X^\varepsilon(t) - X^\varepsilon(t_k)|^2 \\
&= O(t - t_k) = O(\varepsilon^{1-\Delta}) \to 0 \quad as \quad \varepsilon \to 0
\end{aligned} \tag{12.22}$$

for $t \in [t_k, t_{k+1}]$, $0 \le k \le \ell_0 - 1$. It follows that

$$
\mathbf{E} \left| \int_0^t \zeta(u, X^\varepsilon(u))) v_{ij}^\varepsilon(u, \alpha^\varepsilon(u)) du \right|^2
$$
$$
\le 2\mathbf{E} \left| \int_0^t [\zeta(u, X^\varepsilon(u)) - \widetilde{\zeta}^\varepsilon(u)] v_{ij}^\varepsilon(u, \alpha^\varepsilon(u)) du \right|^2 \qquad (12.23)
$$
$$
+ 2\mathbf{E} \left| \int_0^t \widetilde{\zeta}^\varepsilon(u) v_{ij}^\varepsilon(u, \alpha^\varepsilon(u)) du \right|^2 .
$$

We claim that the term on the second line of (12.23) goes to 0. To see this (recall that K is a generic positive constant), by using the Cauchy–Schwarz inequality and the Lipschitz continuity of $\zeta(\cdot)$ and (12.22),

$$
\mathbf{E} \left| \int_0^t [\zeta(u, X^\varepsilon(u)) - \widetilde{\zeta}^\varepsilon(u)] v_{ij}^\varepsilon(u, \alpha^\varepsilon(u)) du \right|^2
$$
$$
\le T \int_0^t \mathbf{E} |\zeta(u, X^\varepsilon(u)) - \widetilde{\zeta}^\varepsilon(u)|^2 du
$$
$$
\le K \sum_{k=0}^{\ell_0-1} \int_{t_k}^{t_{k+1}} \mathbf{E}[(u - t_k)^2 + |X^\varepsilon(u) - X^\varepsilon(t_k)|^2] du
$$
$$
\le K \sum_{k=0}^{\ell_0-1} \int_{t_k}^{t_{k+1}} O(\varepsilon^{1-\Delta}) du
$$
$$
\to 0 \quad \text{as } \varepsilon \to 0.
$$

To estimate the term on the last line of (12.23), for each $i = 1, \ldots, l$, and $j = 1, \ldots, m_i$, define

$$
\eta_{ij}^\varepsilon(t) = \mathbf{E} \left| \int_0^t \widetilde{\zeta}^\varepsilon(u) v_{ij}^\varepsilon(u, \alpha^\varepsilon(u)) du \right|^2 .
$$

Then similar to the derivation of [176, pp. 191–192],

$$
\frac{d}{dt} \eta_{ij}^\varepsilon(t) = O(\varepsilon^{1-\Delta}), \quad \eta^\varepsilon(0) = 0,
$$

as $\varepsilon \to 0$. Thus, solving the above initial value problem leads to

$$
\sup_{0 \le t \le T} \eta_{ij}^\varepsilon(t) = \sup_{0 \le t \le T} \int_0^t O(\varepsilon^{1-\Delta}) ds = O(\varepsilon^{1-\Delta}) \to 0 \quad \text{as } \varepsilon \to 0
$$

as desired. $\square$

To proceed, for each $i \in \overline{\mathcal{M}} = \{1, \ldots, l\}$ and each $F(\cdot, \cdot, i) \in C_0^{1,2}$ ($C_0^{1,2}$ represents the class of functions that have compact support and that are continuously differentiable with respect to t and twice continuously differentiable with respect to x), consider the operator defined in (12.13). The next lemma gives the characterization of the limit process as a solution of a martingale problem.

Lemma 12.9. *Under the conditions of Theorem 12.3, the limit process* $\{Y(\cdot)\}$ *is the solution of the martingale problem with operator* $\overline{\mathcal{G}}$ *given by* (12.13).

Proof. Using an argument similar to that of Lemma 7.18 in [176], it can be shown that the martingale problem with operator $\overline{\mathcal{G}}$ has a unique solution for each initial condition.

To obtain the desired result, it suffices to show that for each $i \in \overline{\mathcal{M}}$ and $F(\cdot, \cdot, i) \in C_0^{1,2}$,

$$F(t, X(t), \overline{\alpha}(t)) - F(0, x, \overline{\alpha}) - \int_0^t \overline{\mathcal{G}} F(u, X(u), \overline{\alpha}(u)) du$$

is a martingale. To this end, we show that for any positive integer n_0, any bounded and continuous functions $h_\ell(\cdot)$, $\ell \le n_0$, and any $t, s, t_\ell \ge 0$ with $t_\ell \le t < t + s \le T$,

$$\mathbf{E} \prod_{\ell=1}^{n_0} h_\ell(X(t_\ell), \overline{\alpha}(t_\ell)) \Bigg(F(t+s, X(t+s), \overline{\alpha}(t+s)) - F(t, X(t), \overline{\alpha}(t))$$
$$- \int_t^{t+s} \overline{\mathcal{G}} F(u, X(u), \overline{\alpha}(u)) du \Bigg) = 0.$$

$$(12.24)$$

Let us begin with the process $Y^\varepsilon(\cdot)$. Define

$$\widehat{F}(t, x, \alpha) = \sum_{i=1}^{l} F(t, x, i) I_{\{\alpha \in \mathcal{M}_i\}} \quad \text{for each} \ \alpha \in \mathcal{M}.$$

Clearly, $\widehat{F}(t, X^\varepsilon(t), \alpha^\varepsilon(t)) = F(t, X^\varepsilon(t), \overline{\alpha}^\varepsilon(t))$. Moreover, for each $\iota \in \mathcal{M}$, $\widehat{F}(\cdot, \cdot, \iota) \in C_0^{1,2}$. The function $\widehat{F}(\cdot)$ allows us to conveniently use the available $\alpha^\varepsilon(\cdot)$ process in lieu of the aggregated process $\overline{\alpha}^\varepsilon(\cdot)$. Consider the operator $\mathcal{G}^\varepsilon$ defined in (12.9). Because $Y^\varepsilon(\cdot)$ is a Markov process,

$$\widehat{F}(t, X^\varepsilon(t), \alpha^\varepsilon(t)) - \widehat{F}(0, x, \alpha^\varepsilon(0)) - \int_0^t \mathcal{G}^\varepsilon \widehat{F}(u, X^\varepsilon(u), \alpha^\varepsilon(u)) du$$

is a martingale. Consequently,

$$\mathbf{E} \prod_{\ell=1}^{n_0} h_\ell(X^\varepsilon(t_\ell), \overline{\alpha}^\varepsilon(t_\ell)) \Big[\widehat{F}(t+s, X^\varepsilon(t+s), \alpha^\varepsilon(t+s)) - \widehat{F}(t, X^\varepsilon(t), \alpha^\varepsilon(t))$$
$$- \int_t^{t+s} \mathcal{G}^\varepsilon \widehat{F}(u, X^\varepsilon(u), \alpha^\varepsilon(u)) du \Big] = 0.$$

$$(12.25)$$

We proceed to obtain the limit in (12.25) as $\varepsilon \to 0$.

First, by the weak convergence $Y^\varepsilon(\cdot)$ to $Y(\cdot)$, the definition of $\widehat{F}(\cdot)$, and the Skorohod representation, as $\varepsilon \to 0$,

$$\mathbf{E} \prod_{\ell=1}^{n_0} h_\ell(X^\varepsilon(t_\ell), \overline{\alpha}^\varepsilon(t_\ell))[\widehat{F}(t+s, X^\varepsilon(t+s), \alpha^\varepsilon(t+s)) - \widehat{F}(t, X^\varepsilon(u), \alpha^\varepsilon(t))]$$

$$\to \mathbf{E} \prod_{\ell=1}^{n_0} h_\ell(X(t_\ell), \overline{\alpha}(t_\ell))[F(t+s, X(t+s), \overline{\alpha}(t+s)) - F(t, X(t), \overline{\alpha}(t))].$$

$$(12.26)$$

The definition of $\mathcal{G}^\varepsilon$ leads to

$$\mathbf{E} \prod_{\ell=1}^{n_0} h_\ell(X^\varepsilon(t_\ell), \overline{\alpha}^\varepsilon(t_\ell)) \left[\int_t^{t+s} \mathcal{G}^\varepsilon \widehat{F}(u, X^\varepsilon(u), \alpha^\varepsilon(u)) du \right]$$

$$= \mathbf{E} \prod_{\ell=1}^{n_0} h_\ell(X^\varepsilon(t_\ell), \overline{\alpha}^\varepsilon(t_\ell)) \left[\int_t^{t+s} \frac{\partial}{\partial u} \widehat{F}(u, X^\varepsilon(u), \alpha^\varepsilon(u)) du \right.$$

$$+ \int_t^{t+s} \widehat{F}_x(u, X^\varepsilon(u), \alpha^\varepsilon(u)) f(u, X^\varepsilon(u), \alpha^\varepsilon(u)) du$$

$$+ \int_t^{t+s} [\widehat{F}_{xx}(u, X^\varepsilon(u), \alpha^\varepsilon(u)) \sigma^2(u, X^\varepsilon(u), \alpha^\varepsilon(u))] du$$

$$+ \int_t^{t+s} Q^\varepsilon(u) \widehat{F}(u, X^\varepsilon(u), \cdot)(\alpha^\varepsilon(u)) du$$

$$\left. + \int_t^{t+s} J(u, X^\varepsilon(u), \alpha^\varepsilon(u)) du \right].$$

Note that

$$\mathbf{E} \prod_{\ell=1}^{n_0} h_\ell(X^\varepsilon(t_\ell), \overline{\alpha}^\varepsilon(t_\ell)) \left[\int_t^{t+s} \widehat{F}_x(u, X^\varepsilon(u), \alpha^\varepsilon(u)) f(u, X^\varepsilon(u), \alpha^\varepsilon(u)) du \right]$$

$$= \sum_{i=1}^{l} \sum_{j=1}^{m_i} \mathbf{E} \prod_{\ell=1}^{n_0} h_\ell(X^\varepsilon(t_\ell), \overline{\alpha}^\varepsilon(t_\ell))$$

$$\times \left[\int_t^{t+s} \widehat{F}_x(u, X^\varepsilon(u), s_{ij}) f(u, X^\varepsilon(u), s_{ij}) I_{\{\alpha^\varepsilon(u) = s_{ij}\}} du \right]$$

$$= \sum_{i=1}^{l} \sum_{j=1}^{m_i} \mathbf{E} \prod_{\ell=1}^{n_0} h_\ell(X^\varepsilon(t_\ell), \overline{\alpha}^\varepsilon(t_\ell))$$

$$\times \left[\int_t^{t+s} \widehat{F}_x(u, X^\varepsilon(u), s_{ij}) f(u, X^\varepsilon(u), s_{ij}) \nu_j^i(u) I_{\{\overline{\alpha}^\varepsilon(u) = i\}} du \right]$$

$$+ \sum_{i=1}^{l} \sum_{j=1}^{m_i} \mathbf{E} \prod_{\ell=1}^{n_0} h_\ell(X^\varepsilon(t_\ell), \overline{\alpha}^\varepsilon(t_\ell))$$

$$\times \left[\int_t^{t+s} \widehat{F}_x(u, X^\varepsilon(u), s_{ij}) f(u, X^\varepsilon(u), s_{ij}) \right.$$

$$\left. \times \left(I_{\{\alpha^\varepsilon(u) = s_{ij}\}} - \nu_j^i(u) I_{\{\overline{\alpha}^\varepsilon(u) = i\}} \right) du \right].$$

By virtue of Lemma 12.8, the use of the Cauchy–Schwarz inequality, and noting the boundedness of $h_\ell(\cdot)$, for each $i = 1, \ldots, l$, $j = 1, \ldots, m_i$,

$$
\mathbf{E} \bigg| \prod_{\ell=1}^{n_0} h_\ell(X^\varepsilon(t_\ell), \overline{\alpha}^\varepsilon(t_\ell)) \int_t^{t+s} \widehat{F}_x(u, X^\varepsilon(u), s_{ij}) f(u, X^\varepsilon(u), s_{ij})
$$
$$
\times \left(I_{\{\alpha^\varepsilon(u) = s_{ij}\}} - \nu_j^i(u) I_{\{\overline{\alpha}^\varepsilon(u) = i\}} \right) du \bigg|^2
$$
$$
\leq K \mathbf{E} \bigg| \int_t^{t+s} \widehat{F}_x(u, X^\varepsilon(u), s_{ij}) f(u, X^\varepsilon(u), s_{ij})
$$
$$
\times \left(I_{\{\alpha^\varepsilon(u) = s_{ij}\}} - \nu_j^i(u) I_{\{\overline{\alpha}^\varepsilon(u) = i\}} \right) du \bigg|^2
$$
$$
\to 0 \quad \text{as } \varepsilon \to 0.
$$

In view of [176, Lemma 2.4] and similar to [176, Theorem 7.30], it can be shown that

$$
(I_{\{\overline{\alpha}^\varepsilon(\cdot)=1\}}, \ldots, I_{\{\overline{\alpha}^\varepsilon(\cdot)=l\}}) \quad \text{converges weakly to} \quad (I_{\{\overline{\alpha}(\cdot)=1\}}, \ldots, I_{\{\overline{\alpha}(\cdot)=l\}}).
$$

By means of Cramér–Wold's device [16, p. 48], for each $i \in \mathcal{M}$, $I_{\{\overline{\alpha}^\varepsilon(\cdot)=i\}}$ converges to $I_{\{\overline{\alpha}(\cdot)=i\}}$ weakly. By the Skorohod representation (without changing notation and with a slight abuse of notation), we may assume $I_{\{\overline{\alpha}^\varepsilon(\cdot)=i\}} \to I_{\{\overline{\alpha}(\cdot)=i\}}$ w.p.1. Consequently, the weak convergence of $Y^\varepsilon(\cdot)$ to $Y(\cdot)$, the Skorohod representation, and the convergence of $I_{\{\overline{\alpha}^\varepsilon(\cdot)=i\}}$ to $I_{\{\overline{\alpha}(\cdot)=i\}}$ imply that

$$
\mathbf{E} \prod_{\ell=1}^{n_0} h_\ell(X^\varepsilon(t_\ell), \overline{\alpha}^\varepsilon(t_\ell)) \left[\int_t^{t+s} \widehat{F}_x(u, X^\varepsilon(u), \alpha^\varepsilon(u)) f(u, X^\varepsilon(u), \alpha^\varepsilon(u)) du \right]
$$
$$
\xrightarrow{\varepsilon \to 0} \sum_{i=1}^{l} \sum_{j=1}^{m_i} \mathbf{E} \prod_{\ell=1}^{n_0} h_\ell(X(t_\ell), \overline{\alpha}(t_\ell))
$$
$$
\times \left[\int_t^{t+s} \widehat{F}_x(u, X(u), s_{ij}) f(u, X(u), s_{ij}) \nu_j^i(u) I_{\{\overline{\alpha}(u)=i\}} du \right]
$$
$$
= \sum_{i=1}^{l} \mathbf{E} \prod_{\ell=1}^{n_0} h_\ell(X(t_\ell), \overline{\alpha}(t_\ell))
$$
$$
\times \left[\int_t^{t+s} F_x'(u, X(u), i) \overline{f}(u, X(u), i) I_{\{\overline{\alpha}(u)=i\}} du \right]
$$
$$
= \mathbf{E} \prod_{\ell=1}^{n_0} h_\ell(X(t_\ell), \overline{\alpha}(t_\ell)) \left[\int_t^{t+s} F_x'(u, X(u), \overline{\alpha}(u)) \overline{f}(u, X(u), \overline{\alpha}(u)) du \right].
$$
$$
\tag{12.27}
$$

Exactly the same argument as in the derivation of (12.27) yields

$$\mathbf{E} \prod_{\ell=1}^{n_0} h_\ell(X^\varepsilon(t_\ell), \overline{\alpha}^\varepsilon(t_\ell))$$
$$\times \left[\int_t^{t+s} [\widehat{F}_{xx}(u, X^\varepsilon(u), \alpha^\varepsilon(u)) \sigma^2(u, X^\varepsilon(u), \alpha^\varepsilon(u))] du \right] \qquad (12.28)$$
$$\to \mathbf{E} \prod_{\ell=1}^{n_0} h_\ell(X(t_\ell), \overline{\alpha}(t_\ell))$$
$$\times \left[\int_t^{t+s} [F_{xx}(u, X(u), \overline{\alpha}(u)) \overline{\sigma}^2(u, X(u), \overline{\alpha}(u))] du \right],$$

as $\varepsilon \to 0$ and

$$\mathbf{E} \prod_{\ell=1}^{n_0} h_\ell(X^\varepsilon(t_\ell), \overline{\alpha}^\varepsilon(t_\ell)) \left[\int_t^{t+s} \frac{\partial}{\partial u} \widehat{F}(u, X^\varepsilon(u), \alpha^\varepsilon(u)) du \right]$$
$$\to \prod_{\ell=1}^{n_0} h_\ell(\overline{\alpha}(t_\ell), X(t_\ell)) \left[\int_t^{t+s} \frac{\partial}{\partial u} F(u, \overline{\alpha}(u), X(u), \overline{\alpha}(u)) du \right],$$
$$\qquad (12.29)$$

as $\varepsilon \to 0$.

Next, since

$$\widetilde{Q}^i(u) \mathbb{1}_{m_i} = 0 \quad \text{for each} \ \ i = 1, \dots, l,$$

the definition of $\widehat{F}(\cdot)$ yields

$$\widetilde{Q}(u) \widehat{F}(u, X^\varepsilon(u), \cdot)(\alpha^\varepsilon(u)) = 0.$$

Therefore, we have

$$\mathbf{E} \prod_{\ell=1}^{n_0} h_\ell(X^\varepsilon(t_\ell), \overline{\alpha}^\varepsilon(t_\ell)) \left[\int_t^{t+s} Q^\varepsilon(u) \widehat{F}(u, X^\varepsilon(u), \cdot)(\alpha^\varepsilon(u)) du \right]$$
$$= \mathbf{E} \prod_{\ell=1}^{n_0} h_\ell(X^\varepsilon(t_\ell), \overline{\alpha}^\varepsilon(t_\ell)) \left[\int_t^{t+s} \widehat{Q}(u) \widehat{F}(u, X^\varepsilon(u), \cdot)(\alpha^\varepsilon(u)) du \right]$$
$$= \sum_{i=1}^{l} \sum_{j=1}^{m_i} \mathbf{E} \prod_{\ell=1}^{n_0} h_\ell(X^\varepsilon(t_\ell), \overline{\alpha}^\varepsilon(t_\ell))$$
$$\times \left[\int_t^{t+s} \widehat{Q}(u) \widehat{F}(u, X^\varepsilon(u), \cdot)(s_{ij}) \nu_j^i(u) I_{\{\overline{\alpha}^\varepsilon(u) = i\}} du \right]$$
$$+ \sum_{i=1}^{l} \sum_{j=1}^{m_i} \mathbf{E} \prod_{\ell=1}^{n_0} h_\ell(X^\varepsilon(t_\ell), \overline{\alpha}^\varepsilon(t_\ell))$$
$$\times \left[\int_t^{t+s} \widehat{Q}(u) \widehat{F}(u, X^\varepsilon(u), \cdot)(s_{ij}) \right.$$
$$\left. \times [I_{\{\alpha^\varepsilon(u) = s_{ij}\}} - \nu_j^i(u) I_{\{\overline{\alpha}^\varepsilon(u) = i\}}] du \right].$$

By virtue of Lemma 12.8 again, the last term above goes to 0 as $\varepsilon \to 0$. For the next to the last term, we have

$$
\sum_{i=1}^{l} \sum_{j=1}^{m_i} \mathbf{E} \prod_{\ell=1}^{n_0} h_\ell(X^\varepsilon(t_\ell), \overline{\alpha}^\varepsilon(t_\ell))
$$
$$
\times \left[\int_t^{t+s} \widehat{Q}(u) \widehat{F}(u, X^\varepsilon(u), \cdot)(s_{ij}) \nu_j^i(u) I_{\{\overline{\alpha}^\varepsilon(u)=i\}} du \right]
$$
$$
= \sum_{i=1}^{l} \sum_{j=1}^{m_i} \mathbf{E} \prod_{\ell=1}^{n_0} h_\ell(X^\varepsilon(t_\ell), \overline{\alpha}^\varepsilon(t_\ell))
$$
$$
\times \left[\int_t^{t+s} \widehat{Q}(u) F(u, X^\varepsilon(u), \cdot)(i) I_{\{\overline{\alpha}^\varepsilon(u)=i\}} du \right]
$$
$$
\to \sum_{i=1}^{l} \mathbf{E} \prod_{\ell=1}^{n_0} h_\ell(X(t_\ell), \overline{\alpha}(t_\ell)) \left[\int_t^{t+s} \overline{Q}(u) F(u, X(u), \cdot)(i) I_{\{\overline{\alpha}(u)=i\}} du \right]
$$
$$
= \mathbf{E} \prod_{\ell=1}^{n_0} h_\ell(X(t_\ell), \overline{\alpha}(t_\ell)) \left[\int_t^{t+s} \overline{Q}(u) F(u, X(u), \cdot)(\overline{\alpha}(u)) du \right],
$$

as $\varepsilon \to 0$.

Arguing along the same line as in the above estimates, we obtain

$$
\int_t^{t+s} \lambda[\widehat{F}(u, X^\varepsilon(u^-) + g(\gamma, X^\varepsilon(u^-), \alpha^\varepsilon(u^-)), \alpha^\varepsilon(u^-))
$$
$$
- \widehat{F}(u, X^\varepsilon(u^-), \alpha^\varepsilon(u^-))] \pi(d\gamma)
$$
$$
= \sum_{i=1}^{l} \sum_{j=1}^{m_i} \int_t^{t+s} \lambda[\widehat{F}(u, X^\varepsilon(u^-) + g(\gamma, X^\varepsilon(u^-), s_{ij}), s_{ij})
$$
$$
- \widehat{F}(u, X^\varepsilon(u^-), s_{ij})] \pi(d\gamma) I_{\{\alpha^\varepsilon(u)=s_{ij}\}}
$$
$$
= \sum_{i=1}^{l} \sum_{j=1}^{m_i} \int_t^{t+s} \lambda[\widehat{F}(u, X^\varepsilon(u^-) + g(\gamma, X^\varepsilon(u^-), s_{ij}), s_{ij})
$$
$$
- \widehat{F}(u, X^\varepsilon(u^-), s_{ij})] \pi(d\gamma) \nu_j^i(u^-) I_{\{\alpha^\varepsilon(u^-)\in\mathcal{M}_i\}}
$$
$$
+ \sum_{i=1}^{l} \sum_{j=1}^{m_i} \int_t^{t+s} \lambda[\widehat{F}(u, X^\varepsilon(u^-) + g(\gamma, X^\varepsilon(u^-), s_{ij}), s_{ij})
$$
$$
- \widehat{F}(u, X^\varepsilon(u^-), s_{ij})] \pi(d\gamma)[I_{\{\alpha^\varepsilon(u^-)=s_{ij}\}} - \nu_j^i(u^-) I_{\{\alpha^\varepsilon(u^-)\in\mathcal{M}_i\}}]
$$
$$
= \sum_{i=1}^{l} \sum_{j=1}^{m_i} \int_t^{t+s} \lambda[\widehat{F}(u, X^\varepsilon(u^-) + g(\gamma, X^\varepsilon(u^-), s_{ij}), s_{ij})
$$
$$
- \widehat{F}(u, X^\varepsilon(u^-), s_{ij})] \pi(d\gamma) \nu_j^i(u^-) I_{\{\alpha^\varepsilon(u^-)\in\mathcal{M}_i\}} + o(1),
$$

where $o(1) \to 0$ in probability uniformly in t on any bounded set. By virtue of the weak convergence of $Y^\varepsilon(\cdot)$ to $Y(\cdot)$, the Skorohod representation, and the dominated convergence theorem, we have

$$
\mathbf{E} \prod_{\ell=1}^{n_0} h_\ell(X^\varepsilon(t_\ell), \overline{\alpha}^\varepsilon(t_\ell))
$$

$$
\times \left[\int_t^{t+s} \lambda[\widehat{F}(u, X^\varepsilon(u^-) + g(\gamma, X^\varepsilon(u^-), \alpha^\varepsilon(u^-)), \alpha^\varepsilon(u^-)) \right.
$$

$$
\left. - \widehat{F}(u, X^\varepsilon(u^-), \alpha^\varepsilon(u^-))]\pi(d\gamma) \right]
$$

$$
\to \mathbf{E} \prod_{\ell=1}^{n_0} h_\ell(X(t_\ell), \overline{\alpha}(t_\ell))
$$

$$
\times \left[\sum_{i=1}^l \sum_{j=1}^{m_i} \int_t^{t+s} \lambda[\widehat{F}(u, X(u^-) + g(\gamma, X(u^-), s_{ij}), s_{ij}) \right.
$$

$$
\left. - \widehat{F}(u, X(u^-), s_{ij})]\pi(d\gamma)\nu_j^i(u^-)I_{\{\overline{\alpha}(u^-)=i\}} \right],
$$

(12.30)

as $\varepsilon \to 0$. Combining (12.26)–(12.30), we obtain the desired result.   □

12.3   Fast-Varying Diffusion

This section presents another hybrid jump-diffusion model. Compared with the Markov modulated jump-diffusion model with fast switching discussed in the last section, there is an additional periodic fast-varying diffusion. From the motivation of an insurance point of view, the added periodic diffusion may be seen as a way of handling seasonal effects of uncertainty due to claims and premium incomes. This periodic diffusion varies at a faster pace compared with the other random effects.

Consider the system given by

$$
X^\varepsilon(t) = x + \int_0^t f(X^\varepsilon(s), \alpha(s), z^\varepsilon(s))ds + \int_0^t \sigma(X^\varepsilon(s), \alpha(s), z^\varepsilon(s))dw
$$

$$
+ \int_0^t \int_\Gamma g(\gamma, X^\varepsilon(s^-), \alpha(s^-), z^\varepsilon(s^-))N(ds, d\gamma),
$$

$$
z^\varepsilon(t) = z_0 + \frac{1}{\varepsilon} \int_0^t f_1(X^\varepsilon(s), z^\varepsilon(s))ds + \frac{1}{\sqrt{\varepsilon}} \int_0^t \sigma_1(X^\varepsilon(s), z^\varepsilon(s))dv,
$$

(12.31)

where $w(\cdot)$ and $v(\cdot)$ are independent standard Brownian motions. In the above, $z^\varepsilon(\cdot)$ represents a fast-varying diffusion. Relative to $z^\varepsilon(\cdot)$, $X^\varepsilon(\cdot)$ is a slowly varying jump-diffusion process, which is modulated by a continuous-time Markov chain. Due to its scaling, the process $z^\varepsilon(\cdot)$ does not blow up. Under suitable conditions, we show that $X^\varepsilon(\cdot)$ converges weakly to a jump-diffusion process modulated by the Markov chain $\alpha(\cdot)$, in which the system dynamics are averaged out with respect to the stationary measure of the fast process $z^\varepsilon(\cdot)$. For notational simplicity, we have chosen to treat the case where there is no explicit time dependence in the coefficients $f(\cdot)$,

$f_1(\cdot)$, $\sigma(\cdot)$, and $\sigma_1(\cdot)$. That is, both the diffusion process $z^\varepsilon(\cdot)$ and the jump diffusion $X^\varepsilon(\cdot)$ are time homogeneous. Treating z^ε as a parameter, the generator for $X^\varepsilon(\cdot)$ can be written as

$$
\mathcal{G}^\varepsilon F(x,\iota) = \mathcal{L}^\varepsilon F(x,\iota) + \int_\Gamma \lambda[F(x + g(\gamma, x, z^\varepsilon, \iota), \iota) - F(x,\iota)]\pi(d\gamma)
$$
$$
+ QF(x, \cdot)(\iota), \quad \text{for each } \iota \in \mathcal{M},
$$

(12.32)

where

$$
\mathcal{L}^\varepsilon F(x,\iota) = \frac{1}{2}\sigma^2(x,\iota,z^\varepsilon)\frac{\partial^2}{\partial x^2}F(x,\iota) + f(x,\iota,z^\varepsilon)\frac{\partial}{\partial x}F(x,\iota),
$$
$$
QF(x, \cdot)(\iota) = \sum_{\ell=1}^{m_0} a_{\iota\ell} F(x,\ell) = \sum_{\ell\neq\iota} a_{\iota\ell}[F(x,\ell) - F(x,\iota)].
$$

(12.33)

We need the following conditions.

(A12.3) $\mathbf{E}|x|^2 < \infty$ and $\mathbf{E}|z_0|^2 < \infty$. Moreover,

(a) for each $\alpha \in \mathcal{M}$, $f(\cdot, \alpha, \cdot)$ and $\sigma(\cdot, \alpha, \cdot)$ are continuous functions, $|f(x,\alpha,z)| \leq K(1 + |x|)$, and $|\sigma(x,\alpha,z)| \leq K(1+|x|)$ uniformly in z; for each $\alpha \in \mathcal{M}$, $x, z \in \mathbb{R}$, $|f(y,\alpha,z) - f(x,\alpha,z)| \leq K|y-x|$ and $|\sigma(y,\alpha,z) - \sigma(x,\alpha,z)| \leq K|y-x|$ uniformly in z. For each $\alpha \in \mathcal{M}$, $\lambda < \infty$, $g(\cdot, \cdot, \alpha, \cdot)$ is bounded and continuous, $g(0, x, \alpha, z) = 0$ for all x, z, and α; the value of γ can be determined uniquely by $g(\gamma, x, \alpha, z)$.

(b) Both $f_1(x, \cdot)$ and $\sigma_1(x, \cdot)$ are periodic functions with period 1 such that $\sigma_1(x, z) > 0$ for each x, z, $|f_1(x,z)| \leq K(1+|z|)$, $|\sigma_1(x,z)| \leq K(1 + |z|)$, $|f_1(x,z) - f_1(x,y)| \leq K|z-y|$, and $|\sigma_1(x,z) - \sigma_1(x,y)| \leq K|z-y|$ uniformly in x.

Remark 12.10. Part (b) yields that the fast-varying process $z^\varepsilon(\cdot)$ is a so-called periodic diffusion; see [10, 85] among others. This condition guarantees that there is an invariant density $\mu(x, z)$ (for a fixed x). In [85], under suitable conditions, it was proved that not only an invariant measure exists, but also asymptotic expansions of the transition density can be constructed. The choice of periodicity 1 is more or less for convenience; we could in fact use other positive constants as the periodicity. It seems to be more instructive to use simpler conditions as in the current setup.

Lemma 12.11. *Under* (A12.3), $\{X^\varepsilon(\cdot)\}$ *is tight in* $D([0,T];\mathbb{R})$.

Proof. The proof is somewhat similar to the previous case, therefore, we do not spell out all the details. Similar to Lemma 12.6, it can be shown that $\sup_{t \in [0,T]} \mathbf{E}|X^\varepsilon(t)|^2 \leq K$. For any $\delta > 0$, and any $t > 0$, $s > 0$ with $s \leq \delta$, we still have (12.18). Deriving a similar estimate as in that of Lemma 12.7

for the last term (12.18), using the estimate, the linear growth in x, for the first two terms on the right side of (12.18), and by virtue of Gronwall's inequality, we can show that

$$\lim_{\delta \to 0} \limsup_{\varepsilon \to 0} \mathbf{E}|X^\varepsilon(t+s) - X^\varepsilon(t)|^2 = 0.$$

Hence the tightness is obtained.   □

Theorem 12.12. *Assume* (A12.3). *Then* $(X^\varepsilon(\cdot), \alpha(\cdot))$ *converges weakly to* $(X(\cdot), \alpha(\cdot))$, *a jump-diffusion process modulated by the Markov chain* $\alpha(\cdot)$, *which is a solution of the equation*

$$
\begin{aligned}
X(t) = x &+ \int_0^t \widehat{f}(X(u), \alpha(u))du + \int_0^t \widehat{\sigma}(X(u), \alpha(u))dw \\
&+ \int_0^t \int_\Gamma \widehat{g}(\gamma, X(u), \alpha(u))N(du, d\gamma),
\end{aligned}
\tag{12.34}
$$

where

$$
\begin{aligned}
\widehat{f}(x, \alpha) &= \int_0^1 f(x, \alpha, z)\mu(x, z)dz, \\
\widehat{g}(\gamma, x, \alpha) &= \int_0^1 g(\gamma, x, \alpha, z)\mu(x, z)dz, \\
\widehat{\sigma}^2(x, \alpha) &= \int_0^1 \sigma^2(x, \alpha, z)\mu(x, z)dz, \\
\widehat{\sigma}(x, \alpha) &= \sqrt{\widehat{\sigma}^2(x, \alpha)}.
\end{aligned}
\tag{12.35}
$$

Remark 12.13. To proceed, a pertinent way of carrying out the averaging is to work with a truncated process (see [104, p. 248] for instance). The basic idea is: Let $M > 0$ be given and S_M be a sphere with radius M centered at the origin. Define $X^{\varepsilon,M}(t) = X^\varepsilon(t)$ up until the first exit from the M-sphere. Use a smooth truncation function $q_M(x)$ that is equal to 1 when it is inside the M-sphere and is 0 if it is outside the sphere with radius $M + 1$. Then rewrite the dynamics with the use of the M-truncated process and the truncation function. In this process, in lieu of $f(x, \alpha, z)$, $g(\gamma, x, \alpha, z)$, and $\sigma(x, \alpha, z)$, we use

$$
\begin{aligned}
f^M(x, \alpha, z) &\stackrel{\text{def}}{=} f(x, \alpha, z)q_M(x), \\
g^M(\gamma, x, \alpha, z) &\stackrel{\text{def}}{=} g(\gamma, x, \alpha, z)q_M(x), \\
\sigma^M(x, \alpha, z) &\stackrel{\text{def}}{=} \sigma(x, \alpha, z)q_M(x),
\end{aligned}
$$

respectively. Then we proceed to derive the weak convergence of $X^{\varepsilon,M}(\cdot)$ to $X^M(\cdot)$. Finally, using the uniqueness of the martingale problem letting $M \to \infty$, we conclude that $X^\varepsilon(\cdot)$ also converges to $X(\cdot)$. However, for notational simplicity, in what follows, we do not use the truncation notation, but simply assume that the process itself is bounded.

Proof. Consider the fast-varying process. To proceed, define

$$\widetilde{z}^\varepsilon(t) = z^\varepsilon(\varepsilon t), \ \widetilde{v}(t) = v(\varepsilon t)/\sqrt{\varepsilon}, \ \tau = t/\varepsilon,$$

where $v(\cdot)$ is the Brownian motion given in (12.31). Then the equation for $z^\varepsilon(t)$ may be written as

$$\widetilde{z}^\varepsilon(\tau) = z_0 + \int_0^\tau f_1(X^\varepsilon(\varepsilon u), \widetilde{z}^\varepsilon(u)) du + \int_0^\tau \sigma_1(X^\varepsilon(\varepsilon u), \widetilde{z}^\varepsilon(u)) d\widetilde{v}(u).$$
(12.36)

Note that as $\varepsilon \to 0$, $\tau \to \infty$. Note also that $X^\varepsilon(\varepsilon\tau)$ is slowly varying whereas $\widetilde{z}^\varepsilon(\tau)$ is fast changing. Intuitively, it tells us that we can treat $X^\varepsilon(\varepsilon\tau)$ as if it were a "constant" in a small interval. Taking x as a fixed parameter, in what follows, we consider the following fixed-x process (see [103] for an explanation of the fixed-x process),

$$\widetilde{z}^{\varepsilon,x}(\tau) = z_0 + \int_0^\tau f_1(x, \widetilde{z}^{\varepsilon,x}(u)) du + \int_0^\tau \sigma_1(x, \widetilde{z}^{\varepsilon,x}(u)) d\widetilde{v}(u),$$

and the associated generator is

$$\widetilde{L} = \frac{1}{2}\sigma_1^2(x, z)\frac{\partial^2}{\partial z^2} + f_1(x, z)\frac{\partial}{\partial z}.$$
(12.37)

Let $p(t, x, z_1, z)$ denote the transition density associated with the diffusion generated by $\widetilde{L}$. Then $p(t, x, z_1, z)$ satisfies the Kolmogorov forward equation

$$\frac{\partial p}{\partial t} = \widetilde{L}^* p,$$

$$\lim_{t \to 0^+} p(t, x, z_1, z) = \delta(z - z_1),$$

where $\widetilde{L}^*$ is the adjoint of $\widetilde{L}$. Because this diffusion is on a compact set, the classical theory of diffusion processes and related results on partial differential equations (see, e.g., Ikeda and Watanabe [72], and Agranovich [1]) yields that there is a unique invariant transition density $\mu(x, z)$ such that $p(t, x, z_1, z) \to \mu(x, z)$ as $t \to \infty$, where $\mu(x, z)$ is the unique solution of

$$\begin{cases} \widetilde{L}^* \mu(x, z) = 0, \\ \mu(x, z) \geq 0 \ \text{and} \ \int_0^1 \mu(x, z) dz = 1. \end{cases}$$
(12.38)

We note that $(X^\varepsilon(\cdot), \alpha(\cdot), z^\varepsilon(\cdot))$ is a Markov process and that $z^\varepsilon(\cdot)$ can be treated as a fast-varying noise process that will be averaged out in the limit process. The average is taken with respect to the stationary measure of $\widetilde{z}^\varepsilon(\cdot)$. Because $(X^\varepsilon(\cdot), \alpha(\cdot))$ is tight in $D([0, T]; \mathbb{R} \times \mathcal{M})$, by Prohorov's theorem we can extract a weakly convergent subsequence. Select such a subsequence and denote the limit by $(X(\cdot), \alpha(\cdot))$. For notational simplicity, still denote the subsequence by $(X^\varepsilon(\cdot), \alpha(\cdot))$. By virtue of the Skorohod representation, $(X^\varepsilon(\cdot), \alpha(\cdot))$ converges to $(X(\cdot), \alpha(\cdot))$ w.p.1 and the convergence is uniform on each bounded set. We proceed to show that the limit is a solution of the martingale problem with generator $\widehat{\mathcal{G}}$ given by

$$
\begin{aligned}
\widehat{\mathcal{G}}F(x, \alpha) &= \frac{1}{2}\widehat{\sigma}^2(x, \alpha)\frac{\partial^2 F(x, \alpha)}{\partial x^2} + \widehat{f}(x, \alpha)\frac{\partial F(x, \alpha)}{\partial x} + \widehat{J}(x, \alpha) \\
&\quad + QF(x, \cdot)(\alpha), \ \alpha \in \mathcal{M},
\end{aligned}
\tag{12.39}
$$

where

$$
\widehat{J}(x, \alpha) = \int_\Gamma \lambda[F(x + \widehat{g}(\gamma, x, \alpha), \alpha) - F(x, \alpha)]\pi(d\gamma),
$$

and $\widehat{\varphi}(\cdot)$ is an average of $\varphi(\cdot)$ as defined in (12.35) for $\varphi(\cdot)$ being $f(\cdot)$, $\sigma^2(\cdot)$, and $g(\cdot)$, respectively.

Using an argument as in [176, Lemma 7.18], it can be shown that the martingale problem with operator $\widehat{\mathcal{G}}$ has a unique solution. To show that $(X(\cdot), \alpha(\cdot))$ is indeed the solution of the martingale problem with operator $\widehat{\mathcal{G}}$, it suffices to verify that for each $\alpha \in \mathcal{M}$ and for any $F(\cdot, \alpha) \in C_0^2$,

$$
F(X(t), \alpha(t)) - F(x, \alpha) - \int_0^t \widehat{\mathcal{G}}F(X(u), \alpha(u))du
\tag{12.40}
$$

is a martingale. As in the previous section, we begin with the $X^\varepsilon(\cdot)$ process. For any positive integer n_0, and $0 \le t_\ell \le t$ and $\ell \le n_0$, and any bounded and continuous function $h_\ell(\cdot)$, the weak convergence and the Skorohod representation imply that

$$
\begin{aligned}
&\mathbf{E}\prod_{\ell=1}^{n_0} h_\ell(X^\varepsilon(t_\ell))[F(X^\varepsilon(t + s), \alpha(t + s)) - F(X^\varepsilon(t), \alpha(t))] \\
&\to \mathbf{E}\prod_{\ell=1}^{n_0} h_\ell(X(t_\ell))[F(X(t + s), \alpha(t + s)) - F(X(t), \alpha(t))] \quad \text{as } \varepsilon \to 0.
\end{aligned}
\tag{12.41}
$$

In addition,

$$
\begin{aligned}
&\mathbf{E}\prod_{\ell=1}^{n_0} h_\ell(X^\varepsilon(t_\ell))\left[F(X^\varepsilon(t + s), \alpha(t + s)) - F(X^\varepsilon(t), \alpha(t))\right. \\
&\quad \left. - \int_t^{t+s} \mathcal{G}^\varepsilon F(X^\varepsilon(u), \alpha(u))du\right] = 0.
\end{aligned}
$$

Thus in view of (12.41), it suffices to consider the limit of

$$
\mathbf{E} \prod_{\ell=1}^{n_0} h_\ell(X^\varepsilon(t_\ell)) \left[\int_t^{t+s} \mathcal{G}^\varepsilon F(X^\varepsilon(u), \alpha(u)) du \right]
$$

$$
= \mathbf{E} \prod_{\ell=1}^{n_0} h_\ell(X^\varepsilon(t_\ell)) \left[\int_t^{t+s} \left[f(X^\varepsilon(u), \alpha(u), z^\varepsilon(u)) \frac{\partial F(X^\varepsilon(u), \alpha(u))}{\partial x} \right. \right.
$$

$$
+ \frac{1}{2} \sigma^2(X^\varepsilon(u), \alpha(u), z^\varepsilon(u)) \frac{\partial^2 F(X^\varepsilon(u), \alpha(u), z^\varepsilon(u))}{\partial x^2}
$$

$$
+ \int_\Gamma \lambda[F(X^\varepsilon(u) + g(\gamma, X^\varepsilon(u), \alpha(u), z^\varepsilon(u)), \alpha(u))
$$

$$
- F(X^\varepsilon(u), \alpha(u))]\pi(d\gamma)
$$

$$
\left. \left. + QF(X^\varepsilon(u), \alpha(u)) \right] du \right].
$$

$$(12.42)$$

Consider the last term in (12.42). Using the weak convergence, the Skorohod representation, and the continuity of the function $F(\cdot, \alpha)$ for each $\alpha \in \mathcal{M}$, it can be shown that as $\varepsilon \to 0$,

$$
\mathbf{E} \prod_{\ell=1}^{n_0} h_\ell(X^\varepsilon(t_\ell)) \left[\int_t^{t+s} QF(X^\varepsilon(u), \alpha(u)) du \right]
$$

$$
\to \mathbf{E} \prod_{\ell=1}^{n_0} h_\ell(X(t_\ell)) \left[\int_t^{t+s} QF(X(u), \alpha(u)) du \right].
$$

$$(12.43)$$

Choose $0 \le \delta_\varepsilon$ such that $\delta_\varepsilon \to 0$ as $\varepsilon \to 0$ but $\delta_\varepsilon/\varepsilon \to \infty$. For any $t, s \ge 0$ with $0 \le t + s \le T$, by partitioning the interval $[t, t + s]$ into subintervals of length δ_ε, we can rewrite the terms (except the last one) in (12.42) as

$$
\int_t^{t+s} \widehat{H}^\varepsilon(u) du = \sum_{l=t/\delta_\varepsilon}^{(t+s)/\delta_\varepsilon - 1} \delta_\varepsilon \frac{1}{\delta_\varepsilon} \int_{l\delta_\varepsilon}^{l\delta_\varepsilon + \delta_\varepsilon} \widehat{H}^\varepsilon(u) du,
$$

$$(12.44)$$

where $\widehat{H}^\varepsilon(u)$ is a representation of any of the functions that appeared in the integrand of (12.42). We then work with each of the terms and find the corresponding limits.

Noting $t_\ell \le t$, $h_\ell(X^\varepsilon(t_\ell))$ is $\mathcal{F}^\varepsilon_{l\delta_\varepsilon}$-measurable, where

$$
\mathcal{F}^\varepsilon_{l\delta_\varepsilon} = \{ X^\varepsilon(u), \alpha(u), z^\varepsilon(u) : 0 \le u \le l\delta_\varepsilon \}.
$$

Making a change of variable from u to εu leads to

$$
\mathbf{E} \prod_{\ell=1}^{n_0} h_\ell(X^\varepsilon(t_\ell)) \left[\int_t^{t+s} f(X^\varepsilon(u), \alpha(u), z^\varepsilon(u)) \frac{\partial F(X^\varepsilon(u), \alpha(u))}{\partial x} du \right]
$$

$$
= \mathbf{E} \prod_{\ell=1}^{n_0} h_\ell(X^\varepsilon(t_\ell)) \left[\sum_{l=t/\delta_\varepsilon}^{(t+s)/\delta_\varepsilon - 1} \delta_\varepsilon \frac{\varepsilon}{\delta_\varepsilon} \int_{\frac{l\delta_\varepsilon}{\varepsilon}}^{\frac{l\delta_\varepsilon + \delta_\varepsilon}{\varepsilon}} \mathbf{E}_{l\delta_\varepsilon}^\varepsilon f(X^\varepsilon(\varepsilon u), \alpha(\varepsilon u), \tilde{z}^\varepsilon(u)) \right]
$$

$$
\times \frac{\partial F(X^\varepsilon(\varepsilon u), \alpha(\varepsilon u))}{\partial x} du \Bigg],
$$

where $\mathbf{E}_{l\delta_\varepsilon}^\varepsilon$ denotes the conditioning on the σ-algebra $\mathcal{F}_{l\delta_\varepsilon}^\varepsilon$. Denote $t_\varepsilon^l = l\delta_\varepsilon/\varepsilon$ and $T_\varepsilon = \delta_\varepsilon/\varepsilon$. Then as $\varepsilon \to 0$, $T_\varepsilon \to \infty$. By the Lipschitz continuity of $f(\cdot, \alpha, z)$ and the boundedness of $(\partial/\partial x)F(\cdot, \alpha)$,

$$
\mathbf{E} \left| \frac{\varepsilon}{\delta_\varepsilon} \int_{(l\delta_\varepsilon)/\varepsilon}^{(l\delta_\varepsilon + \delta_\varepsilon)/\varepsilon} \mathbf{E}_{l\delta_\varepsilon}^\varepsilon [f(X^\varepsilon(\varepsilon u), \alpha(\varepsilon u), \tilde{z}^\varepsilon(u)) - f(X^\varepsilon(l\delta_\varepsilon), \alpha(\varepsilon u), \tilde{z}^\varepsilon(u))] \right.
$$

$$
\left. \times \frac{\partial F(X^\varepsilon(\varepsilon u), \alpha(\varepsilon u))}{\partial x} du \right|
$$

$$
\leq K \frac{1}{T_\varepsilon} \int_{t_\varepsilon^l}^{t_\varepsilon^l + T_\varepsilon} \mathbf{E} |X^\varepsilon(\varepsilon u) - X^\varepsilon(l\delta_\varepsilon)| du
$$

$$
\leq K \sup_{l\delta_\varepsilon \leq \varepsilon u \leq l\delta_\varepsilon + \delta_\varepsilon} \mathbf{E} |X^\varepsilon(\varepsilon u) - X^\varepsilon(l\delta_\varepsilon)| \to 0 \quad \text{as} \quad \varepsilon \to 0.
$$

Similarly,

$$
\mathbf{E} \left| \frac{1}{T_\varepsilon} \int_{t_\varepsilon^l}^{t_\varepsilon^l + T_\varepsilon} \mathbf{E}_{l\delta_\varepsilon}^\varepsilon f(X^\varepsilon(l\delta_\varepsilon), \alpha(\varepsilon u), \tilde{z}^\varepsilon(u)) \right.
$$

$$
\left. \times \left[\frac{\partial F(X^\varepsilon(\varepsilon u), \alpha(\varepsilon u))}{\partial x} - \frac{\partial F(X^\varepsilon(l\delta_\varepsilon), \alpha(\varepsilon u))}{\partial x} \right] du \right|
$$

$$
\leq K \frac{1}{T_\varepsilon} \int_{t_\varepsilon^l}^{t_\varepsilon^l + T_\varepsilon} \mathbf{E} |X^\varepsilon(\varepsilon u) - X^\varepsilon(l\delta_\varepsilon)| du
$$

$$
\leq K \sup_{l\delta_\varepsilon \leq \varepsilon u \leq l\delta_\varepsilon + \delta_\varepsilon} \mathbf{E} |X^\varepsilon(\varepsilon u) - X^\varepsilon(l\delta_\varepsilon)| \to 0 \quad \text{as} \quad \varepsilon \to 0.
$$

Thus

$$
\frac{1}{\delta_\varepsilon} \int_{l\delta_\varepsilon}^{l\delta_\varepsilon + \delta_\varepsilon} \mathbf{E}_{l\delta_\varepsilon}^\varepsilon f(X^\varepsilon(u), \alpha(u), z^\varepsilon(u)) \frac{\partial F(X^\varepsilon(u), \alpha(u))}{\partial x} du
$$

$$
= \frac{1}{T_\varepsilon} \int_{t_\varepsilon^l}^{t_\varepsilon^l + T_\varepsilon} \mathbf{E}_{l\delta_\varepsilon}^\varepsilon f(X^\varepsilon(l\delta_\varepsilon), \alpha(\varepsilon u), \tilde{z}^\varepsilon(u)) \frac{\partial F(X^\varepsilon(l\delta_\varepsilon), \alpha(\varepsilon u))}{\partial x} du + o(1),
$$

where $o(1) \to 0$ in probability uniformly on any bounded t-set.

By virtue of the measurability of $(\partial/\partial x)F(X^\varepsilon(l\delta_\varepsilon), j)$ with respect to

$\mathcal{F}^\varepsilon_{l\delta_\varepsilon}$, and the independence of $\alpha(\cdot)$ and $z^\varepsilon(\cdot)$,

$$\frac{1}{T_\varepsilon} \int_{t^l_\varepsilon}^{t^l_\varepsilon + T_\varepsilon} \mathbf{E}^\varepsilon_{l\delta_\varepsilon} f(X^\varepsilon(l\delta_\varepsilon), \alpha(\varepsilon u), \widetilde{z}^\varepsilon(u)) \frac{\partial F(X^\varepsilon(l\delta_\varepsilon), \alpha(\varepsilon u))}{\partial x} du$$

$$= \sum_{j \in \mathcal{M}} \frac{1}{T_\varepsilon} \int_{t^l_\varepsilon}^{t^l_\varepsilon + T_\varepsilon} \mathbf{E}^\varepsilon_{l\delta_\varepsilon} f(X^\varepsilon(l\delta_\varepsilon), j, \widetilde{z}^\varepsilon(u)) \frac{\partial F(X^\varepsilon(l\delta_\varepsilon), j)}{\partial x} I_{\{\alpha(\varepsilon u) = j\}} du$$

$$= \sum_{i,j \in \mathcal{M}} \frac{1}{T_\varepsilon} \int_{t^l_\varepsilon}^{t^l_\varepsilon + T_\varepsilon} \mathbf{E}^\varepsilon_{l\delta_\varepsilon} f(X^\varepsilon(l\delta_\varepsilon), j, \widetilde{z}^\varepsilon(u)) \frac{\partial F(X^\varepsilon(l\delta_\varepsilon), j)}{\partial x}$$

$$\times \mathbf{P}(\alpha(\varepsilon u) = j | \alpha(l\delta_\varepsilon) = i) I_{\{\alpha(l\delta_\varepsilon) = i\}} du$$

$$= \sum_{i \in \mathcal{M}} \frac{1}{T_\varepsilon} \int_{t^l_\varepsilon}^{t^l_\varepsilon + T_\varepsilon} \mathbf{E}^\varepsilon_{l\delta_\varepsilon} f(X^\varepsilon(l\delta_\varepsilon), i, \widetilde{z}^\varepsilon(u)) \frac{\partial F(X^\varepsilon(l\delta_\varepsilon), i)}{\partial x}$$

$$\times I_{\{\alpha(l\delta_\varepsilon) = i\}} du + o(1),$$

where $o(1) \to 0$ in probability as $\varepsilon \to 0$ uniformly on any bounded t-set. The last step above follows from the well-known fact of continuous-time Markov chain: Because $\varepsilon u - l\delta_\varepsilon \to 0$ as $\varepsilon \to 0$,

$$\mathbf{P}(\alpha(\varepsilon u) = j | \alpha(l\delta_\varepsilon) = i) \to \delta_{ij} = \begin{cases} 1, & \text{if } i = j, \\ 0, & \text{otherwise.} \end{cases}$$

The above estimates indicate that we need only consider the term

$$\rho^\varepsilon = \frac{1}{T_\varepsilon} \int_{t^l_\varepsilon}^{t^l_\varepsilon + T_\varepsilon} \mathbf{E}^\varepsilon_{l\delta_\varepsilon} f(X^\varepsilon(l\delta_\varepsilon), i, \widetilde{z}^\varepsilon(u)) \frac{\partial F(X^\varepsilon(l\delta_\varepsilon), i)}{\partial x} I_{\{\alpha(l\delta_\varepsilon) = i\}} du$$

in the averaging.

We approximate the $X^\varepsilon(l\delta_\varepsilon)$ by a process taking finitely many values. To be more specific, as in [104, Section 6.1, p. 143 and Section 8.2, p. 227], for any $\Delta > 0$, let $\{B^\Delta_n : n \leq n_\Delta\}$ be a finite collection of disjoint sets that satisfies $\mathbf{P}(X^\varepsilon(l\delta_\varepsilon) \in \partial B^\Delta_n) = 0$ and that covers the range of $X^\varepsilon(l\delta_\varepsilon)$. Recall that we have assumed the boundedness of $X^\varepsilon(\cdot)$ for notational simplicity (see Remark 12.13). Select a point $X^\Delta_n \in B^\Delta_n$ and rewrite ρ^ε as

$$\rho^\varepsilon = \sum_{n=1}^{n_\Delta} I_{\{X^\varepsilon(l\delta_\varepsilon) \in B^\Delta_n\}} I_{\{\alpha(l\delta_\varepsilon) = i\}}$$

$$\times \frac{1}{T_\varepsilon} \int_{t^l_\varepsilon}^{t^l_\varepsilon + T_\varepsilon} \mathbf{E}^\varepsilon_{l\delta_\varepsilon} f(X^\Delta_n, i, \widetilde{z}^\varepsilon(u)) \frac{\partial F(X^\Delta_n, i)}{\partial x} du$$

$$+ \sum_{n=1}^{n_\Delta} I_{\{X^\varepsilon(l\delta_\varepsilon) \in B^\Delta_n\}} I_{\{\alpha(l\delta_\varepsilon) = i\}} \qquad (12.45)$$

$$\times \frac{1}{T_\varepsilon} \int_{t^l_\varepsilon}^{t^l_\varepsilon + T_\varepsilon} \mathbf{E}^\varepsilon_{l\delta_\varepsilon} \left[f(X^\varepsilon(l\delta_\varepsilon), i, \widetilde{z}^\varepsilon(u)) \frac{\partial F(X^\varepsilon(l\delta_\varepsilon), i)}{\partial x} \right.$$

$$\left. - f(X^\Delta_n, i, \widetilde{z}^\varepsilon(u)) \frac{\partial F(X^\Delta_n, i)}{\partial x} \right] du.$$

The term in the last three lines of (12.45) goes to 0 in probability as $\varepsilon \to 0$ and then $\Delta \to 0$ by the weak convergence, the Skorohod representation, the Lipschitz continuity of $f(\cdot, i, z)$, and the smoothness of $F(\cdot, i)$. We proceed to figure out the limit of the term in the first two lines of the right-hand side of (12.45).

Note that as $\varepsilon \to 0$,

$$I_{\{X^\varepsilon(l\delta_\varepsilon) \in B_n^\Delta\}} \to \begin{cases} 1, & \text{if } X^\varepsilon(l\delta_\varepsilon) - X_n^\Delta \to 0, \\ 0, & \text{otherwise.} \end{cases}$$

In view of (12.36) and the existence of the unique invariant density $\mu(x, \cdot)$ for each x, we have that

$$\sum_{i \in \mathcal{M}} \sum_{n=1}^{n_\Delta} I_{\{X^\varepsilon(l\delta_\varepsilon) \in B_n^\Delta\}} I_{\{\alpha(l\delta_\varepsilon)=i\}}$$

$$\times \frac{1}{T_\varepsilon} \int_{t_\varepsilon^l}^{t_\varepsilon^l + T_\varepsilon} \mathbf{E}_{l\delta_\varepsilon}^\varepsilon f(X_n^\Delta, i, \widetilde{z}^\varepsilon(u)) \frac{\partial F(X_n^\Delta, i)}{\partial x} du$$

$$= \sum_{i \in \mathcal{M}} \sum_{n=1}^{n_\Delta} I_{\{X^\varepsilon(l\delta_\varepsilon) \in B_n^\Delta\}} I_{\{\alpha(l\delta_\varepsilon)=i\}}$$

$$\times \frac{1}{T_\varepsilon} \int_{t_\varepsilon^l}^{t_\varepsilon^l + T_\varepsilon} \int_0^1 f(X_n^\Delta, i, z) \mathbf{E}_{l\delta_\varepsilon}^\varepsilon I_{\{\widetilde{z}^\varepsilon, X_n^\Delta(u) \in dz\}} \frac{\partial F(X_n^\Delta, i)}{\partial x} du$$

$$= \sum_{i \in \mathcal{M}} \sum_{n=1}^{n_\Delta} I_{\{X^\varepsilon(l\delta_\varepsilon) \in B_n^\Delta\}} I_{\{\alpha(l\delta_\varepsilon)=i\}}$$

$$\times \frac{1}{T_\varepsilon} \int_{t_\varepsilon^l}^{t_\varepsilon^l + T_\varepsilon} \int_0^1 f(X_n^\Delta, i, z) \mathbf{P}(\widetilde{z}^{\varepsilon, X_n^\Delta}(u) \in dz | \widetilde{z}^{\varepsilon, X_n^\Delta}(t_\varepsilon^l)) \frac{\partial F(X_n^\Delta, i)}{\partial x} du$$

$$\to \sum_{i \in \mathcal{M}} \int_0^1 I_{\{\alpha(u)=i\}} f(X(u), i, z) \mu(X(u), z) dz \frac{\partial F(X(u), i)}{\partial x}$$

$$= \widehat{f}(X(u), \alpha(u)) \frac{\partial F(X(u), \alpha(u))}{\partial x}.$$

Thus, as $\varepsilon \to 0$,

$$\mathbf{E} \prod_{\ell=1}^{n_0} h_\ell(X^\varepsilon(t_\ell)) \left[\int_t^{t+s} f(X^\varepsilon(u), \alpha(u), z^\varepsilon(u)) \frac{\partial F(X^\varepsilon(u), \alpha(u))}{\partial x} \right] du$$

$$\to \mathbf{E} \prod_{\ell=1}^{n_0} h_\ell(X(t_\ell)) \left[\int_t^{t+s} \widehat{f}(X(u), \alpha(u)) \frac{\partial F(X(u), \alpha(u))}{\partial x} \right] du.$$

Likewise, we obtain that as $\varepsilon \to 0$,

$$
\mathbf{E} \prod_{\ell=1}^{n_0} h_\ell(X^\varepsilon(t_\ell))
$$
$$
\times \left[\frac{1}{2} \int_t^{t+s} \sigma^2(\alpha(u), X^\varepsilon(u), z^\varepsilon(u)) \frac{\partial^2 F(X^\varepsilon(u), \alpha(u), z^\varepsilon(u))}{\partial x^2} \right] du
$$
$$
\to \mathbf{E} \prod_{\ell=1}^{n_0} h_\ell(X(t_\ell)) \left[\int_t^{t+s} \widehat{\sigma}^2(X(u), \alpha(u)) \frac{\partial^2 F(X(u), \alpha(u))}{\partial x^2} \right] du,
$$

and

$$
\mathbf{E} \prod_{\ell=1}^{n_0} h_\ell(X^\varepsilon(t_\ell)) \left[\int_t^{t+s} \int_\Gamma \lambda[F(X^\varepsilon(u) + g(\gamma, X^\varepsilon(u), \alpha(u), z^\varepsilon(u)), \alpha(u)) \right.
$$
$$
\left. - F(X^\varepsilon(u), \alpha(u))]\pi(d\gamma) \right]
$$
$$
\to \mathbf{E} \prod_{\ell=1}^{n_0} h_\ell(X(t_\ell)) \left[\int_t^{t+s} \int_\Gamma \lambda[F(X(u) + \widehat{g}(\gamma, X(u), \alpha(u)), \alpha(u)) \right.
$$
$$
\left. - F(X(u), \alpha(u))]\pi(d\gamma) \right].
$$

Combining these results, we obtain

$$
\mathbf{E} \prod_{\ell=1}^{n_0} h_\ell(X(t_\ell)) \left[F(X(t+s), \alpha(t+s)) - F(X(t), \alpha(t)) \right.
$$
$$
\left. - \int_t^{t+s} \mathcal{L}F(X(u), \alpha(u))du \right] = 0.
$$

Hence (12.40) holds. Consequently, the desired result follows.   □

12.4   Discussion and Remarks

We have studied two-time-scale hybrid jump diffusions. The motivational insurance risk models are more general than the classical compound Poisson models and compound Poisson models under diffusion perturbations. In our models, the rate at which the premiums are received and the rate of oscillation due to diffusion depend on the surplus level. Moreover, the models also take into consideration Markovian regime switching. Under suitable conditions, we have derived limit systems for fast switching and fast diffusion, respectively. The results obtained can be specialized to a number of cases.

Example 12.14. Consider (12.7) with $\sigma(t, x, \alpha) \equiv 0$. Under such a condition Theorem 12.3 still holds with $\overline{\sigma}^2(t, x, \alpha) \equiv 0$. Then the limit system becomes

$$X(t) = x + \int_0^t \overline{f}(s, X(s), \overline{\alpha}(s))ds + \int_0^t \int_\Gamma \overline{g}(\gamma, X(s^-), \overline{\alpha}(s^-))N(ds, d\gamma).$$

Under further specialized coefficients, it reduces to the surplus model involving regime switching [163].

Example 12.15. We consider the model in (12.7) again with $\widetilde{Q}(t) = Q(t)$ that is weakly irreducible. Then the limit system is given by Corollary 12.5. The limit does not involve the Markov switching process. Thus the complexity reduction is more pronounced. Further specification leads to the classical risk model.

Example 12.16. Consider the system given by (12.31) with $\sigma(t, x, \alpha) \equiv 0$. Then similar to Example 12.14, Theorem 12.3 still holds with the limit system being a Markovian modulated jump process.

12.5   Remarks on Numerical Solutions for Switching Jump Diffusions

This chapter focuses on switching jump diffusions. The central theme is the treatment of two-time-scale systems. Most of the jump-diffusion processes with regime switching are nonlinear. Even without two-time-scales, closed-form solutions are virtually impossible to obtain. As a viable alternative, one has to find feasible numerical schemes. In what follows, we suggest numerical algorithms for (12.1). The development is for systems without two-time-scales. The rationale is that if one has a system that involves fast and slow time scales, then one could first use the averaging ideals presented in the previous sections to reduce the amount of computation leading to a limit or reduced system. Thus, for numerical methods, we concentrate on the limit systems only. We present the algorithm and the corresponding convergence properties. The approach we are taking is again the martingale problem formulation, which brings out the profile and dynamic behavior of the process rather than dealing with the iterations directly. Because the detailed proof of convergence of numerical approximation algorithms has a certain similarity to that of Chapters 5 and 6, we omit the verbatim details, but provide a reference.

We use the notation as given in the appendix in A.6 of this book. Please consult that section for various definitions such as τ_n, ψ_n, $\psi(t)$ and the like. Consider the case where the coefficients have no explicit dependence on the time variable. For simplicity, we take $Q(t) = Q$, a constant matrix. We describe the algorithm as follows.

1. Choose $\Delta > 0$, a small parameter, as the step size.

2. Construct a discrete-time Markov chain α_n with transition probability matrix $P^\Delta = I + \Delta Q$, where Q is the generator of the continuous-time Markov chain $\alpha(t)$ given in (12.1), and I is an m_0-dimensional identity matrix.

3. To approximate the Brownian motion $w(\cdot)$, a usual practice in numerical solutions for stochastic differential equations is to use $\Delta w_n = w(\Delta(n+1)) - w(\Delta n)$ to approximate dw. Because $w(\cdot)$ has independent increments, $\{\Delta w_k\}$ is a sequence of independent and identically distributed random variables with mean 0 and covariance ΔI.

4. Let $\{\tau_n\}$ and $\{\psi_n\}$ be sequences of independent and identically distributed random variables such that τ_n has an exponential distribution with parameter λ for some $\lambda > 0$, and ψ_n is the impulse having distribution $\pi(\cdot)$ (see Section A.6 of this book in the appendix). Define

$$\tilde{\tau}_{n+1} = \tilde{\tau}_n + \tau_n, \quad \text{with } \tilde{\tau}_0 = 0.$$

It follows from the independence assumption, $\{\Delta w_k\}$, $\{\alpha_k\}$, $\{\tau_n\}$, and $\{\psi_n\}$ are also independent.

5. Construct the approximation algorithm

$$
\begin{aligned}
X_{n+1} = {} & X_0 + \Delta \sum_{k=0}^{n} f(X_k, \alpha_k) + \sum_{k=0}^{n} \sqrt{\Delta}\sigma(X_k, \alpha_k)\xi_k \\
& + \sum_{\tilde{\tau}_j \le \Delta n} g(\psi_j, X_{\lfloor \tilde{\tau}_j/\Delta \rfloor}, \alpha_{\lfloor \tilde{\tau}_j/\Delta \rfloor}),
\end{aligned}
\tag{12.46}
$$

where $\lfloor y \rfloor$ denotes the integer part of a real number y.

Remark 12.17. For construction of a discrete-time Markov chain, see [177, pp. 315–316]. The discrete-time Markov chain constructed is an approximation of a discretization obtained from $\alpha(t)$. In fact, we could define $\beta_n = \alpha(\Delta n)$ for any positive integer n. It is easily verified that the process so defined is a discrete-time Markov chain, whose transition probability matrix is given by $\exp(\Delta Q)$. This Markov chain has stationary transition probabilities or it is a time-homogeneous chain. The process β_n is known as a skeleton process in the literature [28]. One of the advantages of using a constant stepsize for the numerical procedure is that the skeleton process has stationary transition probabilities not depending on time, so it is easier to generate than that of a nonstationary process.

As in Chapter 5, in the algorithm, we have used another fold of approximation, namely, using $I + \Delta Q$ in lieu of $\exp(\Delta Q)$ for the transition matrix. This further simplifies the computation and reduces the complexity in calculating $\exp(\Delta Q)$. Intuitively, the discrete-time Markov chain we are constructing can be considered as one whose transition probability matrix is obtained from that of β_n by a truncated Taylor expansion. This approximation makes sense because we can invoke the results in [178] to show that an interpolated process of α_n with interpolation interval $[\Delta n, \Delta n + \Delta)$ converges weakly to $\alpha(\cdot)$ generated by Q.

To approximate the Brownian motion, in lieu of the usual approach as given above, we could generate a sequence of independent and identically distributed random variables $\{\xi_n\}$ such that $E\xi_n = 0$ and $E\xi_n = I$ to further simplify the computation. The functional central limit theorem ensures the approximation to the Brownian motion.

To proceed, define

$$J_n = \sum_{\tilde{\tau}_j \leq \Delta n} g(\psi_j, X_{\lfloor \tilde{\tau}_j/\Delta \rfloor}, \alpha_{\lfloor \tilde{\tau}_j/\Delta \rfloor}). \tag{12.47}$$

It is often desirable to write (12.46) recursively. This can be done as follows

$$X_{n+1} = X_n + \Delta f(X_n, \alpha_n) + \sqrt{\Delta}\sigma(X_n, \alpha_n)\xi_n + \Delta J_n, \tag{12.48}$$

where $\Delta J_n = J_n - J_{n-1}$.

Note that the sequences $\{\tau_n\} = \{\tilde{\tau}_{n+1} - \tilde{\tau}_n\}$ and $\{\psi_n\}$ are as in those discussed in the last section. The process $\tilde{\tau}_n$, in fact, represents the jump times of the underlying process.

To proceed, let us define the interpolated processes via piecewise constant interpolations as

$$\left.\begin{array}{l} X^\Delta(t) = X_n, \\[2mm] \alpha^\Delta(t) = \alpha_n, \\[2mm] w^\Delta(t) = \sqrt{\Delta} \sum_{k=0}^{t/\Delta-1} \xi_k, \\[2mm] J^\Delta(t) = J_n, \end{array}\right\} \quad \text{for } t \in [\Delta n, \Delta n + \Delta). \tag{12.49}$$

For convenience, with a slight abuse of notation, we omitted the floor function notation above. Henceforth, for instance, we use t/Δ to denote the integer part of t/Δ here. We state a result, whose proof is along the line of martingale averaging. Further details can be found in [173].

Theorem 12.18. *Under the conditions in* (A12.1), $(X^\Delta(\cdot), \alpha^\Delta(\cdot))$ *converges weakly to* $(X(\cdot), \alpha(\cdot))$ *as* $\Delta \to 0$ *such that* $(X(\cdot), \alpha(\cdot))$ *is the solution of the martingale problem with operator* $\mathcal{L}$.

As in Chapter 5, Theorem 12.18 implies that the algorithm we constructed is convergent. The limit is nothing but the solution of (12.1). As in Chapter 5, one of the variations of the algorithm is to use a sequence of decreasing step sizes. The modifications are as follows. Let $\{\Delta_n\}$ be a sequence of nonnegative real numbers such that $\Delta_n \to 0$ and $\sum_{n=0}^{\infty} \Delta_n = \infty$.

For example, we may take $\Delta_n = 1/n$, or $\Delta_n = 1/n^\gamma$ for some $0 < \gamma < 1$. Define

$$t_n = \sum_{l=0}^{n-1} \Delta_l, \quad \text{and} \quad \alpha_n = \alpha(t_n), \quad n \geq 0.$$

It then can be verified that $\{\alpha_n\}$ so defined is a discrete-time Markov chain with transition probability matrix

$$P^{n,n+1} = (p_{ij}^{n,n+1})_{m_0 \times m_0} = \exp((t_{n+1} - t_n)Q) = \exp(\Delta_n Q). \quad (12.50)$$

Using ideas in stochastic approximation [104], define

$$m(t) = \max\{n : t_n \leq t\}. \quad (12.51)$$

It is clear that now the Markov chain α_n is not time homogeneous. The approximate solution for the stochastic differential equation with jumps and regime switching (12.1) is given by

$$\begin{cases} X_{n+1} = X_0 + \displaystyle\sum_{k=0}^{n} f(X_k, \alpha_k)\Delta_k + \sum_{k=0}^{n} \sqrt{\Delta_k}\sigma(X_k, \alpha_k)\xi_k \\ \qquad + \displaystyle\sum_{\tilde{\tau}_j \leq n} g(\psi_j, X_{j-1}, \alpha_j), \\ X_0 = x, \quad \alpha_0 = \alpha. \end{cases} \quad (12.52)$$

With the algorithm proposed, we can then proceed to study its performance. The proof is along the same lines as that of the constant stepsize algorithm. The interested reader is referred to [173]; see also Chapters 5 and 6 of this book for further reading.

12.6   Notes

As a continuation of our study, based on the work Yin and Yang [175], this chapter has been devoted to two-time-scale jump diffusions with regime switching. Roughly, the fast-changing driving processes can be treated as a noise, whose stationary measure does exist. Under suitable conditions, the slow process is averaged out with respect to the stationary measure of the fast-varying process. Such an idea has been used in the literature. We refer the reader to Khasminskii [82], Papanicolaou, Stroock, and Varadhan [131], Khasminskii and Yin [86, 88, 89], Pardoux and Veretennikov [132], and references therein for related work in diffusions, and Yin [165] for that of switching diffusions.

The limit results obtained in this chapter can be useful for applications. For example, one may use such results in a subsequent study for obtaining bounds on ruin probability in risk management. It is conceivable that this approach will lead to approximations with tighter error bounds. Certain generalizations are possible. For example, the finite-state Markov chain may also include transient states. Treating (12.31), time-inhomogeneous systems may be considered. Further investigation may also include the replacement of the diffusion by a wideband noise [102] yielding more realistic systems. Finally, for notational simplicity, only scalar $X^\varepsilon(t)$ is treated in this chapter. The approach presented can be carried over to multidimensional cases.

Towards the end of the chapter, we also outlined numerical schemes for solutions of switching jump diffusions. For detailed development, the reader is referred to Yin, Song, and Zhang [173]. For simplicity, the switching process is assumed to be a continuous-time Markov chain independent of the Brownian motion. By combining the treatment of jump diffusions with that of the switching diffusions with x-dependent switching, x-dependent switching jump diffusions can also be treated.

Appendix A

Serving as a handy reference, this appendix collects a number of results that are used in the book. These results include Markov chains, martingales, diffusion processes, weak convergence, hybrid jump diffusions, and other miscellaneous results. In most of the cases, only results are presented. The detailed developments and discussions are omitted, but pointers are provided for further reading. We assume the knowledge of basic probability theory and stochastic processes. These can be found in standard textbooks for a course in probability, for example, Breiman [19], Chow and Teicher [27], among others.

A.1 Discrete-Time Markov Chains

The theory of stochastic processes is concerned with structures and properties of families of random variables X_t, where t is a parameter taken over a suitable index set $\mathbb{T}$. The index set $\mathbb{T}$ may be discrete (i.e., $\mathbb{T} = \{0, 1, \ldots, \}$), or continuous (i.e., an interval of the real line). Stochastic processes associated with these index sets are said to be discrete-time processes and continuous-time processes, respectively; see [80], for instance. The random variables X_t can be either scalars or vectors. For a continuous-time stochastic processes, we use the notation $X(t)$, whereas for a discrete process, we use X_k.

A stochastic process is wide-sense (or covariance) stationary, if it has finite second moments, a constant mean, and a covariance that depends only on the time difference. The ergodicity of a discrete-time stationary sequence $\{X_k\}$ refers to the convergence of the sequence $(X_1 + X_2 + \cdots + X_n)/n$ to its expectation in an appropriate sense; see for example, Karlin and Taylor [80, Theorem 5.6, p. 487] for a strong ergodic theorem of a stationary process. A stochastic process X_k is adapted to a filtration $\{\mathcal{F}_k\}$, if for each k, X_k is an $\mathcal{F}_k$-measurable random vector.

Suppose that α_k is a stochastic process taking values in $\mathcal{M}$, which is at most countable (i.e., it is either finite $\mathcal{M} = \{1, 2, \ldots, m_0\}$ or countable $\mathcal{M} = \{1, 2, \ldots\}$). We say that α_k is a Markov chain if it possesses the Markov property,

$$
\begin{aligned}
p_{ij}^{k,k+1} &= \mathbf{P}(\alpha_{k+1} = j | \alpha_k = i) \\
&= \mathbf{P}(\alpha_{k+1} = j | \alpha_0 = i_0, \ldots, \alpha_{k-1} = i_{k-1}, \alpha_k = i),
\end{aligned}
$$

for any $i_0, \ldots, i_{k-1}, i, j \in \mathcal{M}$.

Given i, j, if $p_{ij}^{k,k+1}$ is independent of time k, that is, $p_{ij}^{k,k+1} = p_{ij}$, α_k is said to have stationary transition probabilities. The corresponding Markov chain is said to be stationary or time-homogeneous or temporally homogeneous or simply homogeneous. In this case, denote the transition matrix by $P = (p_{ij})$. Denote the n-step transition matrix by $P^{(n)} = (p_{ij}^{(n)})$, with

$$
p_{ij}^{(n)} = \mathbf{P}(X_n = j | X_0 = i).
$$

Then $P^{(n)} = P^n$. That is, the n-step transition matrix is simply the matrix P to the nth power. Note that

(a) $p_{ij} \geq 0$, $\sum_j p_{ij} = 1$, and

(b) $(P)^{k_1+k_2} = (P)^{k_1}(P)^{k_2}$, for $k_1, k_2 = 1, 2, \ldots$ This identity is known as the Chapman–Kolmogorov equation.

Suppose that $P = (p_{ij}) \in \mathbb{R}^{m_0 \times m_0}$ is a transition matrix. Then the spectral radius of P satisfies $\rho(P) = 1$; see Section A.7 and also Karlin and Taylor [81, p. 3] for a definition of the spectral radius of a matrix. This implies that all eigenvalues of P are on or inside the unit circle.

For a Markov chain α_k, state j is said to be accessible from state i if $p_{ij}^{(k)} = \mathbf{P}(\alpha_k = j | \alpha_0 = i) > 0$ for some $k > 0$. Two states i and j, accessible from each other, are said to communicate with each other. A Markov chain is irreducible if all states communicate with each other. For $i \in \mathcal{M}$, let $d(i)$ denote the period of state i (i.e., the greatest common divisor of all $k \geq 1$ such that $\mathbf{P}(\alpha_{k+n} = i | \alpha_n = i) > 0$, define $d(i) = 0$ if $\mathbf{P}(\alpha_{k+n} = i | \alpha_n = i) = 0$ for all k). A Markov chain is aperiodic if each state has period one. In accordance with Kolmogorov's classification of states, a state i is recurrent if, starting from state i, the probability of returning to state i after some finite time is 1. A state is transient if it is not recurrent. Criteria on recurrence can be found in most standard textbooks of stochastic processes or Markov chains. In this book, we consider recurrence of switching diffusions, which as for its diffusion process counterpart, can be considered as a generalization of Kolmogorov's classifications.

Note that (see Karlin and Taylor [81, p. 4]) if P is a transition matrix for a finite-state Markov chain, the multiplicity of the eigenvalue 1 is equal to the number of recurrent classes associated with P. A row vector $\pi = (\pi_1, \ldots, \pi_{m_0})$ with each $\pi_i \geq 0$ is called a stationary distribution of α_k if it is the unique solution to the system of equations

$$\begin{cases} \pi P = \pi, \\ \sum_i \pi_i = 1. \end{cases}$$

As demonstrated in [81, p. 85], for i being in an aperiodic recurrent class, if $\pi_i > 0$, which is the limit of the probability of starting from state i and then entering state i at the nth transition as $n \to \infty$, then for all j in this class of i, $\pi_j > 0$, and the class is termed positive recurrent or strongly ergodic.

Theorem A.1. *Let $P = (p_{ij}) \in \mathbb{R}^{m_0 \times m_0}$ be the transition matrix of an irreducible aperiodic finite-state Markov chain. Then there exist constants $0 < \lambda < 1$ and $c_0 > 0$ such that*

$$\left| P^k - \overline{P} \right| \leq c_0 \lambda^k \quad for \quad k = 1, 2, \ldots,$$

where $\overline{P} = \mathbb{1}_{m_0}\pi$, $\mathbb{1}_{m_0} = (1, \ldots, 1)' \in \mathbb{R}^{m_0 \times 1}$, and $\pi = (\pi_1, \ldots, \pi_{m_0})$ is the stationary distribution of α_k. This implies, in particular,

$$\lim_{k \to \infty} P^k = \mathbb{1}_{m_0}\pi.$$

Suppose α_k is a discrete-time Markov chain with transition probability matrix P. One of the ergodicity conditions of Markov chains is Doeblin's condition (see Doob [33, Hypothesis D, p. 192]; see also Meyn and Tweedie [125, p. 391]). Suppose that there is a probability measure μ with the property that for some positive integer n, $0 < \delta < 1$, and $\Delta > 0$, $\mu(A) \leq \delta$ implies that $P^n(x, A) \leq 1 - \Delta$ for all $x \in A$. In the above, $P^n(x, A)$ denotes the transition probability starting from x reaches the set A in n steps. Note that if α_k is a finite-state Markov chain that is irreducible and aperiodic, then the Doeblin condition is satisfied.

Given an $m_0 \times m_0$ irreducible transition matrix P and a vector G, consider

$$F(P - I) = G, \tag{A.1}$$

where F is an unknown vector. Note that zero is an eigenvalue of the matrix $P - I$ and the null space of $P - I$ is spanned by $\mathbb{1}_{m_0}$. Then by the Fredholm alternative (see Lemma A.11), (A.1) has a solution if and only if $G\mathbb{1}_{m_0} = 0$, where $\mathbb{1}_{m_0} = (1, \ldots, 1)' \in \mathbb{R}^{m_0 \times 1}$.

Define $Q_c = (P - I : \mathbb{1}_{m_0}) \in \mathbb{R}^{m_0 \times (m_0+1)}$. Consider (A.1) together with the condition $F\mathbb{1}_{m_0} = \sum_{i=1}^{m_0} F_i = \widehat{F}$, which may be written as $FQ_c = G_c$ where $G_c = (G : \widehat{F})$. Because for each t, (A.9) has a unique solution, it follows that $Q_c(t)Q'_c(t)$ is a matrix with full rank; therefore, the equation

$$F[Q_cQ'_c] = G_cQ'_c \qquad (A.2)$$

has a unique solution, which is given by $F = G_cQ'_c[Q_cQ'_c]^{-1}$.

A.2 Continuous-Time Markov Chains

A right-continuous stochastic process with piecewise-constant sample paths is a jump process. Suppose that $\alpha(\cdot) = \{\alpha(t) : t \geq 0\}$ is a jump process defined on $(\Omega, \mathcal{F}, P)$ taking values in $\mathcal{M}$. Then $\{\alpha(t) : t \geq 0\}$ is a continuous-time Markov chain with state space $\mathcal{M}$, if

$$\mathbf{P}(\alpha(t) = i | \alpha(r) : r \leq s) = \mathbf{P}(\alpha(t) = i | \alpha(s)),$$

for all $0 \leq s \leq t$ and $i \in \mathcal{M}$, with $\mathcal{M}$ being either finite or countable.

For any $i, j \in \mathcal{M}$ and $t \geq s \geq 0$, let $p_{ij}(t, s)$ denote the transition probability $\mathbf{P}(\alpha(t) = j | \alpha(s) = i)$, and $P(t, s)$ the matrix $(p_{ij}(t, s))$. We name $P(t, s)$ the transition matrix of the Markov chain $\alpha(\cdot)$, and postulate that

$$\lim_{t \to s^+} p_{ij}(t, s) = \delta_{ij},$$

where $\delta_{ij} = 1$ if $i = j$ and 0 otherwise. It follows that for $0 \leq s \leq \varsigma \leq t$,

$$p_{ij}(t, s) \geq 0, \; i, j \in \mathcal{M},$$

$$\sum_{j \in \mathcal{M}} p_{ij}(t, s) = 1, \; i \in \mathcal{M},$$

$$p_{ij}(t, s) = \sum_{k \in \mathcal{M}} p_{ik}(\varsigma, s)p_{kj}(t, \varsigma), \; i, j \in \mathcal{M}.$$

The last identity is the Chapman–Kolmogorov equation as its discrete-time counterpart. If the transition probability $\mathbf{P}(\alpha(t) = j | \alpha(s) = i)$ depends only on $(t - s)$, then $\alpha(\cdot)$ is said to be time-homogeneous or it is said to have stationary transition probabilities. Otherwise, the process is nonhomogeneous or nonstationary. For time homogeneous Markov chains, we define $p_{ij}(h) := p_{ij}(s + h, s)$ for any $h \geq 0$.

Suppose that $\alpha(t)$ is a continuous-time Markov chain with stationary transition probability $P(t) = (p_{ij}(t))$. It then naturally induces a discrete-time Markov chain. For each $h > 0$, the transition matrix $(p_{ij}(h))$ is the transition matrix of the discrete-time Markov chain $\alpha_k = \alpha(kh)$, which is called an h-skeleton of the corresponding continuous-time Markov chain by Chung; see [28, p. 132].

Definition A.2 (q-Property). A matrix-valued function $Q(t) = (q_{ij}(t))$, for $t \geq 0$, satisfies the q-Property, if

(a) $q_{ij}(t)$ is Borel measurable for all $i, j \in \mathcal{M}$ and $t \geq 0$;

(b) $q_{ij}(t)$ is uniformly bounded. That is, there exists a constant K such that $|q_{ij}(t)| \leq K$, for all $i, j \in \mathcal{M}$ and $t \geq 0$;

(c) $q_{ij}(t) \geq 0$ for $j \neq i$ and $q_{ii}(t) = -\sum_{j \neq i} q_{ij}(t)$, $t \geq 0$.

For any real-valued function f on $\mathcal{M}$ and $i \in \mathcal{M}$, write

$$Q(t)f(\cdot)(i) = \sum_{j \in \mathcal{M}} q_{ij}(t)f(j) = \sum_{j \neq i} q_{ij}(t)(f(j) - f(i)).$$

Let us recall the definition of the generator of a Markov chain.

Definition A.3 (Generator). A matrix $Q(t)$, $t \geq 0$, is an infinitesimal generator (or in short, a generator) of $\alpha(\cdot)$ if it satisfies the q-property, and for any bounded real-valued function f defined on $\mathcal{M}$

$$f(\alpha(t)) - \int_0^t Q(\varsigma)f(\cdot)(\alpha(\varsigma))d\varsigma \tag{A.3}$$

is a martingale.

Remark A.4. Motivated by the applications we are interested in, a generator is defined as a matrix satisfying the q-property above, where an additional condition on the boundedness of the entries of the matrix is posed. It naturally connects the Markov chain and martingale problems. Definitions including other classes of matrices may be devised as in Chung [28]. To proceed, we give an equivalent condition for a finite-state Markov chain generated by $Q(\cdot)$.

Lemma A.5. Let $\mathcal{M} = \{1, \ldots, m_0\}$. Then $\alpha(t) \in \mathcal{M}$, $t \geq 0$, is a Markov chain generated by $Q(t)$ if and only if

$$\left(I_{\{\alpha(t)=1\}}, \ldots, I_{\{\alpha(t)=m_0\}} \right) - \int_0^t \left(I_{\{\alpha(\varsigma)=1\}}, \ldots, I_{\{\alpha(\varsigma)=m_0\}} \right) Q(\varsigma)d\varsigma \tag{A.4}$$

is a martingale.

Proof: For a proof, see Yin and Zhang [176, Lemma 2.4]. □

For any given $Q(t)$ satisfying the q-property, there exists a Markov chain $\alpha(\cdot)$ generated by $Q(t)$. If $Q(t) = Q$, a constant matrix, the idea of Ethier and Kurtz [43] can be utilized for the construction. For time-varying generator $Q(t)$, we need to use the piecewise-deterministic process approach as described in Davis [30], to define the Markov chain $\alpha(\cdot)$.

The discussion below is taken from that of Yin and Zhang [176], which was originated in the work of Davis [30]. Let $0 = \tau_0 < \tau_1 < \cdots < \tau_l < \cdots$ be a sequence of jump times of $\alpha(\cdot)$ such that the random variables $\tau_1, \tau_2 - \tau_1$, ..., $\tau_{k+1} - \tau_k$, ... are independent. Let $\alpha(0) = i \in \mathcal{M}$. Then $\alpha(t) = i$ on the interval $[\tau_0, \tau_1)$. The first jump time τ_1 has the probability distribution

$$\mathbf{P}(\tau_1 \in B) = \int_B \exp\left\{\int_0^t q_{ii}(s)ds\right\}(-q_{ii}(t))\,dt,$$

where $B \subset [0, \infty)$ is a Borel set. The post-jump location of $\alpha(t) = j$, $j \neq i$, is given by

$$\mathbf{P}(\alpha(\tau_1) = j|\tau_1) = \frac{q_{ij}(\tau_1)}{-q_{ii}(\tau_1)}.$$

If $q_{ii}(\tau_1)$ is 0, define $\mathbf{P}(\alpha(\tau_1) = j|\tau_1) = 0$, $j \neq i$. Then $\mathbf{P}(q_{ii}(\tau_1) = 0) = 0$. In fact, if $B_i = \{t : q_{ii}(t) = 0\}$, then

$$\begin{aligned}
\mathbf{P}(q_{ii}(\tau_1) = 0) &= \mathbf{P}(\tau_1 \in B_i) \\
&= \int_{B_i} \exp\left\{\int_0^t q_{ii}(s)ds\right\}(-q_{ii}(t))\,dt = 0.
\end{aligned}$$

In general, $\alpha(t) = \alpha(\tau_l)$ on the interval $[\tau_l, \tau_{l+1})$. The jump time τ_{l+1} has the conditional probability distribution

$$\begin{aligned}
&\mathbf{P}(\tau_{l+1} - \tau_l \in B_l|\tau_1,\ldots,\tau_l,\alpha(\tau_1),\ldots,\alpha(\tau_l)) \\
&= \int_{B_l} \exp\left\{\int_{\tau_l}^{t+\tau_l} q_{\alpha(\tau_l)\alpha(\tau_l)}(s)ds\right\}(-q_{\alpha(\tau_l)\alpha(\tau_l)}(t+\tau_l))\,dt.
\end{aligned}$$

The post-jump location of $\alpha(t) = j$, $j \neq \alpha(\tau_l)$ is given by

$$\mathbf{P}(\alpha(\tau_{l+1}) = j|\tau_1,\ldots,\tau_l,\tau_{l+1},\alpha(\tau_1),\ldots,\alpha(\tau_l)) = \frac{q_{\alpha(\tau_l)j}(\tau_{l+1})}{-q_{\alpha(\tau_l)\alpha(\tau_l)}(\tau_{l+1})}.$$

Theorem A.6. *Suppose that the matrix $Q(t)$ satisfies the q-property for $t \geq 0$. Then the following statements hold.*

(a) *The process $\alpha(\cdot)$ constructed above is a Markov chain.*

(b) *The process*

$$f(\alpha(t)) - \int_0^t Q(\varsigma)f(\cdot)(\alpha(\varsigma))d\varsigma \tag{A.5}$$

is a martingale for any uniformly bounded function $f(\cdot)$ on $\mathcal{M}$. Thus $Q(t)$ is indeed the generator of $\alpha(\cdot)$.

(c) *The transition matrix $P(t, s)$ satisfies the forward differential equation*

$$\begin{cases} \dfrac{\partial P(t,s)}{\partial t} = P(t,s)Q(t), \quad t \geq s, \\ P(s,s) = I, \end{cases} \tag{A.6}$$

where I is the identity matrix.

(d) *Assume further that $Q(t)$ is continuous in t. Then $P(t,s)$ also satisfies the backward differential equation*

$$
\begin{cases}
\dfrac{\partial P(t,s)}{\partial s} = -Q(s)P(t,s), & t \geq s, \\[2mm]
P(s,s) = I.
\end{cases}
\tag{A.7}
$$

Proof. For (a)–(c), see Yin and Zhang [176, Theorem 2.5]. As for (d), see [26, p. 402]. □

Note that frequently, working with $s \in [0,T]$, the backward equations are written slightly differently by using reversed time $\tau = T - s$. In this case, the minus sign in (A.7) disappears. Suppose that $\alpha(t)$, $t \geq 0$, is a Markov chain generated by an $m_0 \times m_0$ matrix $Q(t)$. The notions of irreducibility and quasi-stationary distribution are given next.

Definition A.7 (Irreducibility).

(a) A generator $Q(t)$ is said to be weakly irreducible if, for each fixed $t \geq 0$, the system of equations

$$
\begin{cases}
\nu(t)Q(t) = 0, \\[1mm]
\displaystyle\sum_{i=1}^{m_0} \nu_i(t) = 1
\end{cases}
\tag{A.8}
$$

has a unique solution $\nu(t) = (\nu_1(t), \ldots, \nu_{m_0}(t))$ and $\nu(t) \geq 0$.

(b) A generator $Q(t)$ is said to be irreducible, if for each fixed $t \geq 0$ the system of equations (A.8) has a unique solution $\nu(t)$ and $\nu(t) > 0$.

By $\nu(t) \geq 0$, we mean that for each $i \in \mathcal{M}$, $\nu_i(t) \geq 0$. A similar interpretation holds for $\nu(t) > 0$. It follows from the definitions above that irreducibility implies weak irreducibility. However, the converse is not true. For example, the generator

$$
Q = \begin{pmatrix} -1 & 1 \\ 0 & 0 \end{pmatrix}
$$

is weakly irreducible, but it is not irreducible because it contains an absorbing state corresponding to the second row in Q. A moment of reflection reveals that for a two-state Markov chain with generator

$$
Q = \begin{pmatrix} -\lambda(t) & \lambda(t) \\ \mu(t) & -\mu(t) \end{pmatrix}
$$

the weak irreducibility requires only $\lambda(t) + \mu(t) > 0$, whereas the irreducibility requires that both $\lambda(t)$ and $\mu(t)$ be positive. Such a definition is convenient for many applications (e.g., the manufacturing systems mentioned in Khasminskii, Yin, and Zhang [91, p. 292]).

Definition A.8 (Quasi-Stationary Distribution). For $t \geq 0$, $\nu(t)$ is termed a quasi-stationary distribution if it is the unique solution of (A.8) satisfying $\nu(t) \geq 0$.

Remark A.9. In the study of homogeneous Markov chains, stationary distributions play an important role. When we are interested in nonstationary (nonhomogeneous) Markov chains, stationary distributions are replaced by the corresponding quasi-stationary distributions, as defined above.

If $\nu(t) = \nu > 0$, it is a stationary distribution. In view of Definitions A.7 and A.8, if $Q(t)$ is weakly irreducible, then there is a quasi-stationary distribution. Note that the rank of a weakly irreducible $m_0 \times m_0$ matrix $Q(t)$ is $m_0 - 1$, for each $t \geq 0$. The definition above emphasizes the probabilistic interpretation. An equivalent definition using the algebraic properties of $Q(t)$ is provided next. One can verify their equivalence using the Fredholm alternative; see Lemma A.11.

Definition A.10. A generator $Q(t)$ is said to be weakly irreducible if, for each fixed $t \geq 0$, the system of equations

$$\begin{cases} f(t)Q(t) = 0, \\ \sum_{i=1}^{m_0} f_i(t) = 0 \end{cases} \tag{A.9}$$

has only the trivial (zero) solution.

A.3   Fredholm Alternative and Ramification

The Fredholm alternative, which provides a powerful method for establishing existence and uniqueness of solutions for various systems of equations, can be found, for example, in Hutson and Pym [71, p. 184].

Lemma A.11 (Fredholm Alternative). *Let $\mathbb{B}$ be a Banach space and A a linear compact operator defined on it. Let $I : \mathbb{B} \to \mathbb{B}$ be the identity operator. Assume $\gamma \neq 0$. Then one of the two alternatives holds:*

(a) *The homogeneous equation $(\gamma I - A)f = 0$ has only the zero solution, in which case $\gamma \in \rho(A)$, the resolvent set of A, $(\gamma I - A)^{-1}$ is bounded, and the inhomogeneous equation $(\gamma I - A)f = g$ also has one solution $f = (\gamma I - A)^{-1}g$, for each $g \in \mathbb{B}$.*

(b) *The homogeneous equation* $(\gamma I - A)f = 0$ *has a nonzero solution, in which case the inhomogeneous equation* $(\gamma I - A)f = g$ *has a solution if and only if* $\langle g, f^* \rangle = 0$ *for every solution* f^* *of the adjoint equation* $\gamma f^* = A^* f^*$.

Note that in (b) above, $\langle g, f^* \rangle$ is a pairing defined on $\mathbb{B} \times \mathbb{B}^*$ (with $\mathbb{B}^*$ denoting the dual of $\mathbb{B}$). This is also known as an "outer product" (see [71, p. 149]), whose purpose is similar to the inner product in a Hilbert space. If we work with a Hilbert space, this "outer product" is identical to the usual inner product. When one considers linear systems of algebraic equations, the lemma above can be rewritten in a simpler form.

Let B denote an $m_0 \times m_0$ matrix. For any $\gamma \neq 0$, define an operator $A : \mathbb{R}^{m_0 \times m_0} \to \mathbb{R}^{m_0 \times m_0}$ as

$$Ay = y(\gamma I - B).$$

Note that in this case, I is just the $m_0 \times m_0$ identity matrix I. Then the adjoint operator $A^* : \mathbb{R}^{m_0 \times m_0} \to \mathbb{R}^{m_0 \times m_0}$ is

$$A^* x = (\gamma I - B)x.$$

Suppose that b and $y \in \mathbb{R}^{1 \times m_0}$. Consider the system $yB = b$. If the adjoint system $Bx = 0$ where $x \in \mathbb{R}^{m_0 \times 1}$ has only the zero solution, then $yB = b$ has a unique solution given by $y = bB^{-1}$. If $Bx = 0$ has a nonzero solution x, then $yB = b$ has a solution if and only if $\langle b, x \rangle = 0$.

Suppose that the generator Q of a continuous-time Markov chain $\alpha_1(t)$ is a constant matrix and is irreducible. Then the rank of Q is $m_0 - 1$. Denote by $\mathcal{R}(Q)$ and $\mathcal{N}(Q)$ the range and the null space of Q, respectively. It follows that $\mathcal{N}(Q)$ is one-dimensional spanned by $\mathbb{1}$ (i.e., $\mathcal{N}(Q) = \text{span}\{\mathbb{1}\}$). As a consequence, the Markov chain $\alpha_1(t)$ with generator Q is ergodic. In what follows, denote the associated stationary distribution by $\nu = (\nu_1, \nu_2, \ldots, \nu_{m_0}) \in \mathbb{R}^{1 \times m_0}$. Consider a linear system of equations

$$Qc = \eta, \tag{A.10}$$

where c and $\eta \in \mathbb{R}^{m_0}$.

Lemma A.12. *The following assertions hold.*

(i) *Equation* (A.10) *has a solution if and only if* $\nu\eta = 0$.

(ii) *Suppose that* c_1 *and* c_2 *are two solutions of* (A.10)*. Then* $c_1 - c_2 = \gamma_0 \mathbb{1}$ *for some* $\gamma_0 \in \mathbb{R}$.

(iii) *Any solution of* (A.10) *can be written as*

$$c = \gamma_0 \mathbb{1} + h_0,$$

where $\gamma_0 \in \mathbb{R}$ *is an arbitrary constant,* $\mathbb{1} = (1, \ldots, 1)' \in \mathbb{R}^{m_0}$, *and* $h_0 \in \mathbb{R}^{m_0}$ *is the unique solution of* (A.10) *satisfying* $\nu h_0 = 0$.

Proof. We begin with a stochastic representation. First, if (A.10) has a solution c, then $\nu Q c = \nu \eta$. But $\nu Q = 0$, and hence $\nu \eta = 0$. Next, suppose that c_1 and c_2 are two solutions of (A.10). Define $\tilde{c} = c_1 - c_2$. Then we have $Q\tilde{c} = 0$. Thus $\tilde{c} \in \mathcal{N}(Q) = \text{span}\{\mathbb{1}\}$ and hence it follows that there exists a $\gamma_0 \in \mathbb{R}$ such that $\tilde{c} = \gamma_0 \mathbb{1}$.

Suppose that $\nu \eta = 0$. Using orthogonal decomposition, we can write η as $\eta = b + b_1$, where $b \in \mathcal{R}(Q')$ and $b_1 \in \mathcal{N}(Q)$. Thus, $b_1 = \beta \mathbb{1}$ for some $\beta \in \mathbb{R}$. Thus, η can be written as $\eta = b + \beta \mathbb{1}$. Since $\nu \eta = 0$, we obtain $\beta = -\nu b$. Then solving (A.10) is equivalent to that of

$$Qc = b - \nu b \mathbb{1}. \tag{A.11}$$

Denote by $\mathbf{E}_i$ the conditional expectation corresponding to the conditional probability $\mathbf{P}_i(\cdot) := \mathbf{P}(\cdot \mid \alpha_1(0) = i)$, and define a column vector $h = (h_1, h_2, \ldots, h_{m_0})' \in \mathbb{R}^{m_0}$ by

$$h_i = \mathbf{E}_i \int_0^\infty (\nu b - b_{\alpha_1(t)}) dt, \quad i \in \mathcal{M}. \tag{A.12}$$

It is readily seen that

$$\begin{aligned} \mathbf{E}_i b_{\alpha_1(t)} &= \sum_{j=1}^{m_0} b_j \mathbf{P}(\alpha_1(t) = j \mid \alpha_1(0) = i) \\ &= \sum_{j=1}^{m_0} p_{ij}(t) b_j \to \sum_{j=1}^{m_0} \nu_j b_j = \nu b \quad \text{as } t \to \infty. \end{aligned}$$

Moreover, because $\alpha_1(t)$ has a finite state space, the convergence above takes place exponentially fast; see [176, Appendix] and references therein. Thus, h is well defined. We proceed to show that h, in fact, is a solution of (A.11).

By direct calculation, it is seen that

$$\begin{aligned} Qh &= \int_0^\infty (Q\nu b \mathbb{1} - QP(t)b) dt = -\int_0^\infty QP(t)b \, dt \\ &= -\int_0^\infty \frac{dP(t)}{dt} b \, dt = -P(t)b \Big|_0^\infty = -\mathbb{1}\nu b + b, \end{aligned}$$

where $P(t) = (p_{ij}(t))$ is the transition matrix satisfying the Kolmogorov backward equation $(d/dt)P(t) = QP(t)$ with $P(0) = I$ and $\lim_{t\to\infty} P(t) = \mathbb{1}\nu$. Thus, the vector h satisfies equation (A.11) and hence (A.10). By using the result proved earlier, any solution of (A.10) can be represented by $c = h + \gamma \mathbb{1}$ for some $\gamma \in \mathbb{R}$.

Finally, we verify that any solution of (A.10) can be written as $c = h_0 + \alpha \mathbb{1}$ for $\alpha \in \mathbb{R}$, where h_0 is the unique solution of (A.10) satisfying $\nu h_0 = 0$. In fact, we have shown that h defined in (A.12) solves (A.10) and any solution of (A.10) can be represented by $c = h + \gamma \mathbb{1}$ for some $\gamma \in \mathbb{R}$. Now let $h_0 = h - \nu h \mathbb{1} \in \mathbb{R}^{m_0}$. Then we have

$$Qh_0 = Q(h - \nu h \mathbb{1}) = Qh - \nu h Q \mathbb{1} = Qh = \eta.$$

Also, we verify that

$$\nu h_0 = \nu(h - \nu h \mathbb{1}) = \nu h - (\nu h)(\nu \mathbb{1}) = 0.$$

Hence h_0 satisfies (A.10) and $\nu h_0 = 0$. Moreover, any solution of (A.10) can be represented by

$$c = h + \gamma \mathbb{1} = h_0 + (\gamma + \nu h)\mathbb{1} = h_0 + \gamma_0 \mathbb{1},$$

where $\gamma_0 = \gamma + \nu h \in \mathbb{R}$. It remains to show uniqueness. To this end, define an augmented matrix

$$Q_a = \begin{pmatrix} Q \\ \nu \end{pmatrix} \in \mathbb{R}^{(m_0+1) \times m_0}$$

and a new vector

$$\widehat{b} = \begin{pmatrix} b \\ 0 \end{pmatrix} \in \mathbb{R}^{m_0+1}.$$

Then (A.10) together with $\nu h = 0$ can be written as

$$Q_a h = \widehat{b}. \tag{A.13}$$

It can be shown that $Q_a' Q_a$ has full rank m_0 due to the irreducibility of Q, and the solution of (A.13) can be represented by

$$h = (Q_a' Q_a)^{-1} Q_a' \widehat{b} = h_0.$$

This leads to the desired uniqueness.   □

Remark A.13. In the literature, (A.10) is sometimes referred to as Poisson equation (see [14]), and the results in (i) and (ii) above are deemed to be well known. They are more or less a consequence of the Fredholm alternative. The irreducibility of Q used in the lemma can be relaxed to weak irreducibility. The result still holds.

Perhaps being interesting in its own right, we illustrate how the solutions may be obtained through a stochastic representation in the proof above. This angle of view has not been extensively exploited to our knowledge. Although classifying solutions of an algebraic system by the sum of homogenous part and a particular solution is a time honored concept, characterizing the unique particular solution using orthogonality with respect to the stationary distribution of Q is useful in many applications.

Similar to Lemma A.12, when studying stability of switching diffusions, we need to examine an equation

$$Qc = b - \frac{1}{2}\sigma^2 + \beta\mathbb{1}, \tag{A.14}$$

where $\beta \in \mathbb{R}$, $b \in \mathbb{R}^{m_0}$ is a constant vector, and $\sigma^2 = (\sigma_1^2, \ldots, \sigma_{m_0}^2)' \in \mathbb{R}^{m_0}$ with $\sigma_i^2 \geq 0$ for $i \in \mathcal{M}$. Then equation (A.14) has a solution if the Markov chain $\alpha_1(t)$ is ergodic. Let the associated stationary distribution be denoted by ν. Then β is given by

$$\beta = -\sum_{i=1}^{m_0} \nu_i \left(b_i - \frac{1}{2}\sigma_i^2 \right) = -\nu b + \frac{1}{2}\nu\sigma^2. \tag{A.15}$$

Moreover, let the column vector $h = (h_1, \ldots, h_{m_0})' \in \mathbb{R}^{m_0}$ be defined by

$$h_i = \mathbf{E}_i \int_0^\infty \left(\nu b - \frac{1}{2}\nu\sigma^2 - b_{\alpha_1(t)} + \frac{1}{2}\sigma_{\alpha_1(t)}^2 \right) dt, \quad i \in \mathcal{M}. \tag{A.16}$$

Then h is well defined and is a solution of (A.14) with β given by (A.15).

A.4   Martingales, Gaussian Processes, and Diffusions

This section briefly reviews several random processes including martingales, Gaussian processes, and diffusions.

A.4.1   Martingales

Many applications involving stochastic processes depend on the concept of the martingale. The definition and properties of discrete-time martingales can be found in Breiman [19, Chapter 5], Chung [28, Chapter 9], and Hall and Heyde [63] among others. This section provides a brief review.

Discrete-Time Martingales

Definition A.14. Suppose that $\{\mathcal{F}_n\}$ is a filtration, and $\{X_n\}$ is a sequence of random variables. The pair $\{X_n, \mathcal{F}_n\}$ is a martingale if for each n,

(a) X_n is $\mathcal{F}_n$-measurable;

(b) $\mathbf{E}|X_n| < \infty$;

(c) $\mathbf{E}(X_{n+1}|\mathcal{F}_n) = X_n$ a.s.

It is a supermartingale (resp., submartingale) if (a) and (b) in the above hold, and

$$\mathbf{E}(X_{n+1}|\mathcal{F}_n) \leq X_n \quad (\text{resp.,} \quad \mathbf{E}(X_{n+1}|\mathcal{F}_n) \geq X_n) \quad \text{a.s.}$$

In what follows if the sequence of σ-algebras is clear, we simply say that $\{X_n\}$ is a martingale.

Let $X_n = \sum_{j=1}^{n} Y_j$, where $\{Y_n\}$ is a sequence of independent and identically distributed (i.i.d.) random variables with zero mean. It is plain that

$$\mathbf{E}[X_{n+1}|Y_1, \ldots, Y_n] = \mathbf{E}[X_n + Y_{n+1}|Y_1, \ldots, Y_n]$$
$$= X_n + \mathbf{E}Y_{n+1} = X_n \quad \text{a.s.}$$

The above equation illustrates the defining relation of a martingale.

If $\{X_n\}$ is a martingale, we can define $Y_n = X_n - X_{n-1}$, which is known as a martingale difference sequence. Suppose that $\{X_n, \mathcal{F}_n\}$ is a martingale. Then the following properties hold.

(a) Suppose $\varphi(\cdot)$ is an increasing and convex function defined on $\mathbb{R}$. If for each positive integer n, $\mathbf{E}|\varphi(X_n)| < \infty$, then $\{\varphi(X_n), \mathcal{F}_n\}$ is a submartingale.

(b) Let τ be a stopping time with respect to $\mathcal{F}_n$ (i.e., an integer-valued random variable such that $\{\tau \leq n\}$ is $\mathcal{F}_n$-measurable for each n). Then $\{X_{\tau \wedge n}, \mathcal{F}_{\tau \wedge n}\}$ is also a martingale.

(c) The martingale inequality (see Kushner [102, p. 3]) states that for each $\lambda > 0$,

$$\mathbf{P}\left(\max_{1 \leq j \leq n} |X_j| \geq \lambda\right) \leq \frac{1}{\lambda}\mathbf{E}|X_n|,$$
$$\mathbf{E}\max_{1 \leq j \leq n} |X_j|^2 \leq 4\mathbf{E}|X_n|^2, \quad \text{if} \quad \mathbf{E}|X_n|^2 < \infty \quad \text{for each} \quad n. \tag{A.17}$$

(d) The Doob inequality (see Hall and Heyde [63, p.15]) states that for each $p > 1$,

$$\mathbf{E}^{1/p}|X_n|^p \leq \mathbf{E}^{1/p}\left(\max_{1 \leq j \leq n} |X_j|\right)^p \leq q\mathbf{E}^{1/p}|X_n|^p,$$

where $p^{-1} + q^{-1} = 1$.

(e) The Burkholder inequality (see Hall and Heyde [63, p.23]) is: For $1 < p < \infty$, there exist constants K_1 and K_2 such that

$$K_1\mathbf{E}\left|\sum_{j=1}^{n} y_j^2\right|^{p/2} \leq \mathbf{E}|X_n|^p \leq K_2\mathbf{E}\left|\sum_{i=j}^{n} y_j^2\right|^{p/2},$$

where $Y_n = X_n - X_{n-1}$.

Remark A.15. If $X(\cdot)$ is a right-continuous martingale in continuous time and $f(\cdot)$ is a nonnegative convex function, then

$$\mathbf{P}\Big[\sup_{s\leq t\leq T} f(X(t)) \geq \lambda \big| \mathcal{F}_s\Big] \leq \frac{\mathbf{E}\Big[f(X(t)\big|\mathcal{F}_s\Big]}{\lambda}, \qquad (\text{A.18})$$

$$\mathbf{E}\sup_{0\leq t\leq T} |X(t)|^2 \leq 4\mathbf{E}|X(T)|^2. \qquad (\text{A.19})$$

Consider a discrete-time Markov chain $\{\alpha_n\}$ with state space $\mathcal{M}$ (either finite or countable) and one-step transition probability matrix $P = (p_{ij})$. Recall that a sequence $\{f(i) : i \in \mathcal{M}\}$ is P-harmonic or right-regular (Karlin and Taylor [81, p. 48]), if (a) $f(\cdot)$ is a real-valued function such that $f(i) \geq 0$ for each $i \in \mathcal{M}$, and (b)

$$f(i) = \sum_{j\in\mathcal{M}} p_{ij} f(j) \ \text{ for each } \ i \in \mathcal{M}. \qquad (\text{A.20})$$

If the equality in (A.20) is replaced by $\geq$ (resp., $\leq$), $\{f(i) : i \in \mathcal{M}\}$ is said to be P-superharmonic or right superregular (resp., P-subharmonic or right subregular). Considering $f = (f(i) : i \in \mathcal{M})$ as a column vector, (A.20) can be written as $f = Pf$. Similarly, we can write $f \geq Pf$ for P-superharmonic (resp., $g \leq Pf$ for P-subharmonic). Likewise, $\{f(i) : i \in \mathcal{M}\}$ is said to be P left regular, if (b) above is replaced by

$$f(j) = \sum_{i\in\mathcal{M}} f(i)p_{ij} \ \text{ for each } \ j \in \mathcal{M}. \qquad (\text{A.21})$$

Similarly, left superregular and subregular functions can be defined.

There is a natural connection between a martingale and a discrete-time Markov chain; see Karlin and Taylor [80, p. 241]. Let $\{\alpha_n\}$ be a discrete-time Markov chain and $\{f(i) : i \in \mathcal{M}\}$ be a bounded P-harmonic sequence. Define $X_n = f(\alpha_n)$. Then $\mathbf{E}|X_n| < \infty$. Moreover, owing to the Markov property,

$$\begin{aligned}\mathbf{E}(X_{n+1}|\mathcal{F}_n) &= \mathbf{E}(f(\alpha_{n+1})|\alpha_n)) \\ &= \sum_{j\in\mathcal{M}} p_{\alpha_n,j} f(j) \\ &= f(\alpha_n) = X_n \ \text{ a.s.}\end{aligned}$$

Therefore, $\{X_n, \mathcal{F}_n\}$ is a martingale. Note that if $\mathcal{M}$ is finite, the boundedness of $\{f(i) : i \in \mathcal{M}\}$ is not needed.

As explained in Karlin and Taylor [80], one of the widely used ways of constructing martingales is through the utilization of eigenvalues and eigenvectors of a transition matrix. Again, let $\{\alpha_n\}$ be a discrete-time Markov chain with transition matrix P. Recall that a column vector f is a right eigenvector of P associated with an eigenvalue $\lambda \in \mathbb{C}$, if $Pf = \lambda f$. Let f be a right eigenvector of P satisfying $\mathbf{E}|f(\alpha_n)| < \infty$ for each n. For $\lambda \neq 0$, define $X_n = \lambda^{-n} f(\alpha_n)$. Then $\{X_n\}$ is a martingale.

Continuous-Time Martingales

Next, let us denote the space of $\mathbb{R}^r$-valued continuous functions on $[0,T]$ by $C([0,T];\mathbb{R}^r)$, and the space of functions that are right-continuous with left-hand limits endowed with the Skorohod topology by $D([0,T];\mathbb{R}^r)$; see Definition A.18. Consider $X(\cdot) = \{X(t) \in \mathbb{R}^r : t \geq 0\}$. If for each $t \geq 0$, $X(t)$ is an $\mathbb{R}^r$ random vector, we call $X(\cdot)$ a continuous-time stochastic process and write it as $X(t)$, $t \geq 0$, or simply $X(t)$ if there is no confusion.

A process $X(\cdot)$ is adapted to a filtration $\{\mathcal{F}_t\}$, if for each $t \geq 0$, $X(t)$ is an $\mathcal{F}_t$-measurable random variable; $X(\cdot)$ is progressively measurable if for each $t \geq 0$, the process restricted to $[0,t]$ is measurable with respect to the σ-algebra $\mathcal{B}[0,t] \times \mathcal{F}_t$ in $[0,t] \times \Omega$, where $\mathcal{B}[0,t]$ denotes the Borel sets of $[0,t]$. A progressively measurable process is measurable and adapted, whereas the converse is not generally true. However, any measurable and adapted process with right-continuous sample paths is progressively measurable.

Frequently, we need to work with a *stopping time* for applications. Consider $(\Omega, \mathcal{F}, P)$ with a filtration $\{\mathcal{F}_t\}$. A stopping time τ is a nonnegative random variable satisfying $\{\tau \leq t\} \in \mathcal{F}_t$ for all $t \geq 0$.

A stochastic process $\{X(t) : t \geq 0\}$ (real- or vector-valued) is a martingale on $(\Omega, \mathcal{F}, P)$ with respect to $\{\mathcal{F}_t\}$ if:

(a) For each $t \geq 0$, $X(t)$ is $\mathcal{F}_t$-measurable,

(b) $\mathbf{E}|X(t)| < \infty$, and

(c) $\mathbf{E}[X(t)|\mathcal{F}_s] = X(s)$ a.s for all $t \geq s$.

If $\mathcal{F}_t$ is the natural filtration $\sigma\{X(s) : s \leq t\}$, we often say that $X(\cdot)$ is a martingale without specifying the filtration $\mathcal{F}_t$. The process $X(\cdot)$ is a local martingale if there exists a sequence of stopping times $\{\tau_n\}$ such that $0 \leq \tau_1 \leq \tau_2 \leq \cdots \leq \tau_n \leq \tau_{n+1} \leq \cdots$, $\tau_n \to \infty$ a.s as $n \to \infty$, and $X^{(n)}(t) := X(t \wedge \tau_n)$ is a martingale.

A.4.2 Gaussian Processes and Diffusion Processes

A Gaussian random vector $X = (X_1, X_2, \ldots, X_r)$ is one whose characteristic function has the form

$$\phi(y) = \exp\left(\mathrm{i}y'\mu - \frac{1}{2}y'\Sigma y\right),$$

where $\mu \in \mathbb{R}^r$ is a constant vector, $y'\mu$ is the usual inner product, i denotes the pure imaginary number satisfying $\mathrm{i}^2 = -1$, and Σ is a symmetric nonnegative definite $r \times r$ matrix. In the above, μ and Σ are the mean vector and covariance matrix of X, respectively.

Consider a stochastic process $X(t)$, $t \geq 0$. It is a Gaussian process if for any $k = 1, 2, \ldots$ and $0 \leq t_1 < t_2 < \cdots < t_k$, $(X(t_1), X(t_2), \ldots, X(t_k))$ is a

Gaussian vector. A random process $X(\cdot)$ has *independent increments* if for any $k = 1, 2, \ldots$ and $0 \leq t_1 < t_2 < \cdots < t_k$,

$$(X(t_1) - X(0)),\ (X(t_2) - X(t_1)),\ \ldots,\ (X(t_k) - X(t_{k-1}))$$

are independent. A sufficient condition for a process to be Gaussian is given next, whose proof can be found in Skorohod [150, p. 7].

Lemma A.16. *Suppose that the process $X(\cdot)$ has independent increments and continuous sample paths almost surely. Then $X(\cdot)$ is a Gaussian process.*

Next, we consider the notion of Brownian motions. An $\mathbb{R}^r$-valued random process $w(t)$ for $t \geq 0$ is a Brownian motion, if

(a) $w(0) = 0$ almost surely;

(b) $w(\cdot)$ is a process with independent increments;

(c) $w(\cdot)$ has continuous sample paths almost surely;

(d) for all $t, s \geq 0$, the increments $w(t) - w(s)$ have Gaussian distribution with $\mathbf{E}(w(t) - w(s)) = 0$ and $\mathrm{Cov}(w(t), w(s)) = \Sigma |t - s|$ for some nonnegative definite $r \times r$ matrix Σ, where $\mathrm{Cov}(w(t), w(s))$ denotes the covariance.

A Brownian motion $w(\cdot)$ with $\Sigma = I$ is termed a standard Brownian motion. In view of Lemma A.16, a Brownian motion is necessarily a Gaussian process. For an $\mathbb{R}^r$-valued Brownian motion $w(t)$, let $\mathcal{F}_t = \sigma\{w(s) : s \leq t\}$. Let $h(\cdot)$ be an $\mathcal{F}_t$-measurable process taking values in $\mathbb{R}^{r \times r}$ such that $\int_0^t \mathbf{E}|h(s)|^2 ds < \infty$ for all $t \geq 0$. Using $w(\cdot)$ and $h(\cdot)$, we may define a stochastic integral $\int_0^t h(s) dw(s)$ such that it is a martingale with mean 0 and

$$\mathbf{E} \left| \int_0^t h(s) dw(s) \right|^2 = \int_0^t \mathbf{E}\Big[\mathrm{tr}(h(s)h'(s)) \Big] ds.$$

Suppose that $b(\cdot)$ and $\sigma(\cdot)$ are nonrandom Borel measurable functions. A process $X(\cdot)$ defined as

$$X(t) = X(0) + \int_0^t b(s, X(s)) ds + \int_0^t \sigma(s, X(s)) dw(s) \qquad (A.22)$$

is called a diffusion. Then $X(\cdot)$ defined in (A.22) is a Markov process in the sense that the Markov property

$$\mathbf{P}(X(t) \in A | \mathcal{F}_s) = \mathbf{P}(X(t) \in A | X(s))$$

holds for all $0 \leq s \leq t$ and for any Borel set A. A slightly more general definition allows $b(\cdot)$ and $\sigma(\cdot)$ to be $\mathcal{F}_t$-measurable processes. Nevertheless, the current definition is sufficient for our purpose.

Associated with the diffusion process, there is an operator $\mathcal{L}$, known as the generator of the diffusion $X(\cdot)$. Let $C^{1,2}$ be the class of real-valued functions on (a subset of) $\mathbb{R}^r \times [0, \infty)$ whose first-order partial derivative with respect to t and the second-order mixed partial derivatives with respect to x are continuous. Define an operator $\mathcal{L}$ on $C^{1,2}$ by

$$\mathcal{L}f(t,x) = \frac{\partial f(t,x)}{\partial t} + \sum_{i=1}^{r} b_i(t,x)\frac{\partial f(t,x)}{\partial x_i} + \frac{1}{2}\sum_{i,j=1}^{r} a_{ij}(t,x)\frac{\partial^2 f(t,x)}{\partial x_i \partial x_j},$$

(A.23)

where $A(t,x) = (a_{ij}(t,x)) = \sigma(t,x)\sigma'(t,x)$. The above may also be written in a more compact form as

$$\mathcal{L}f(t,x) = \frac{\partial f(t,x)}{\partial t} + b'(t,x)\nabla f(t,x) + \frac{1}{2}\text{tr}(\nabla^2 f(t,x)A(t,x)),$$

where ∇f and $\nabla^2 f$ denote the gradient and Hessian of f, respectively. Note that in this book, we use the notations $b'f$ and $\langle b, f \rangle$ interchangeably to represent an inner product.

The well-known Itô lemma (see Gihman and Skorohod [53], Ikeda and Watanabe [72], and Liptser and Shiryayev [110]) states that

$$df(t, X(t)) = \mathcal{L}f(t, X(t)) + \nabla f'(t, X(t))\sigma(t, X(t))dw(t),$$

or in its integral form

$$f(t, X(t)) - f(0, X(0))$$
$$= \int_0^t \mathcal{L}f(s, X(s))ds + \int_0^t \nabla f'(s, X(s))\sigma(s, X(s))dw(s).$$

By virtue of Itô's lemma,

$$M_f(t) = f(t, X(t)) - f(0, X(0)) - \int_0^t \mathcal{L}f(s, X(s))ds$$

is a square integrable $\mathcal{F}_t$-martingale. Conversely, suppose that $X(\cdot)$ is right continuous. Using the notation of martingale problems given by Stroock and Varadhan [153], $X(\cdot)$ is said to be a solution of the martingale problem with operator $\mathcal{L}$ if $M_f(\cdot)$ is a martingale for each $f(\cdot, \cdot) \in C_0^{1,2}$ (the class of $C^{1,2}$ functions with compact support).

A.5   Weak Convergence

The notion of weak convergence is a generalization of convergence in distribution in elementary probability theory. In what follows, we present definitions and results, including tightness, tightness criteria, the martingale problem, Skorohod representation, Prohorov's theorem, and so on.

Definition A.17 (Weak Convergence). Let $\mathbf{P}$ and $\mathbf{P}_k$, $k = 1, 2, \ldots$, be probability measures defined on a metric space $\mathbb{S}$. The sequence $\{\mathbf{P}_k\}$ converges weakly to $\mathbf{P}$ if

$$\int f d\mathbf{P}_k \to \int f d\mathbf{P}$$

for every bounded and continuous function $f(\cdot)$ on $\mathbb{S}$. Suppose that $\{X_k\}$ and X are random variables associated with $\mathbf{P}_k$ and $\mathbf{P}$, respectively. The sequence X_k converges to X weakly if for any bounded and continuous function $f(\cdot)$ on $\mathbb{S}$, $\mathbf{E}f(X_k) \to \mathbf{E}f(X)$ as $k \to \infty$.

Let $D([0, \infty); \mathbb{R}^r)$ be the space of $\mathbb{R}^r$-valued functions defined on $[0, \infty)$ that are right continuous and have left-hand limits; let $\mathbb{L}$ be a set of strictly increasing Lipschitz continuous functions $\zeta(\cdot) : [0, \infty) \mapsto [0, \infty)$ such that the mapping is surjective with $\zeta(0) = 0$, $\lim_{t \to \infty} \zeta(t) = \infty$, and

$$\gamma(\zeta) := \sup_{0 \leq t < s} \left| \log \left(\frac{\zeta(s) - \zeta(t)}{s - t} \right) \right| < \infty.$$

Similar to $D([0, \infty); \mathbb{R}^r)$, we also use the notation $D([0, T]; \mathbb{F})$ to denote the D-space of functions that take values in a metric space $\mathbb{F}$.

Definition A.18 (Skorohod Topology). For $\xi, \eta \in D([0, \infty); \mathbb{R}^r)$, the *Skorohod topology* $d(\cdot, \cdot)$ on $D([0, \infty); \mathbb{R}^r)$ is defined as

$$d(\xi, \eta) = \inf_{\zeta \in \mathbb{L}} \left\{ \gamma(\zeta) \vee \int_0^\infty e^{-s} \sup_{t \geq 0} \left(1 \wedge |\xi(t \wedge s) - \eta(\zeta(t) \wedge s)| \right) ds \right\}.$$

Analogous definitions and results are available for $D([0, T]; \mathbb{F})$, where $\mathbb{F}$ is a metric space; see Ethier and Kurtz [43] and Billingsley [16] for related references. Although we frequently work with $D([0, T]; \mathbb{R}^r)$ in this book, the following results are often stated with respect to the space $D([0, \infty); \mathbb{R}^r)$. This enables us to apply them to $t \in [0, T]$ for any $T > 0$.

Definition A.19 (Tightness). A family of probability measures $\mathcal{P}$ defined on a metric space $\mathbb{S}$ is *tight* if for each $\delta > 0$, there exists a compact set $K_\delta \subset \mathbb{S}$ such that

$$\inf_{\mathbf{P} \in \mathcal{P}} \mathbf{P}(K_\delta) \geq 1 - \delta.$$

The notion of tightness is closely related to compactness. The following theorem, known as Prohorov's theorem, gives such an implication. A complete proof can be found in Ethier and Kurtz [43].

Theorem A.20 (Prohorov's Theorem). *If $\mathcal{P}$ is tight, then $\mathcal{P}$ is relatively compact. That is, every sequence of elements in $\mathcal{P}$ contains a weakly convergent subsequence. If the underlying metric space is complete and separable, the tightness is equivalent to relative compactness.*

Although weak convergence techniques usually allow one to use weaker conditions and lead to a more general setup, it is often more convenient to work with probability one convergence for purely analytic reasons, however. The Skorohod representation provides us with such opportunities.

Theorem A.21 (The Skorohod representation (Ethier and Kurtz [43])). *Let X_k and X be random elements belonging to $D([0,\infty);\mathbb{R}^r)$ such that X_k converges weakly to X. Then there exists a probability space $(\widetilde{\Omega}, \widetilde{\mathcal{F}}, \widetilde{\mathbf{P}})$ on which are defined random elements $\widetilde{X}_k$, $k = 1, 2, \ldots$, and $\widetilde{X}$ in $D([0,\infty);\mathbb{R}^r)$ such that for any Borel set B and all $k < \infty$,*

$$\widetilde{\mathbf{P}}(\widetilde{X}_k \in B) = \mathbf{P}(X_k \in B), \quad and \quad \widetilde{\mathbf{P}}(\widetilde{X} \in B) = \mathbf{P}(X \in B)$$

satisfying

$$\lim_{k \to \infty} \widetilde{X}_k = \widetilde{X} \quad a.s.$$

Elsewhere in the book, when we use the Skorohod representation, with a slight abuse of notation, we often omit the tilde notation for convenience and notational simplicity.

Let $C([0,\infty);\mathbb{R}^r)$ be the space of $\mathbb{R}^r$-valued continuous functions equipped with the sup-norm topology, and C_0 be the set of real-valued continuous functions on $\mathbb{R}^r$ with compact support. Let C_0^l be the subset of C_0 functions that have continuous partial derivatives up to the order l.

Definition A.22. Let $\mathbb{S}$ be a metric space and $\mathcal{A}$ be a linear operator on $B(\mathbb{S})$ (the set of all Borel measurable functions defined on $\mathbb{S}$). Let $X(\cdot) = \{X(t) : t \geq 0\}$ be a right-continuous process with values in $\mathbb{S}$ such that for each $f(\cdot)$ in the domain of $\mathcal{A}$,

$$f(X(t)) - \int_0^t \mathcal{A}f(X(s))ds$$

is a martingale with respect to the filtration $\sigma\{X(s) : s \leq t\}$. Then $X(\cdot)$ is called a *solution of the martingale problem* with operator $\mathcal{A}$.

Theorem A.23 (Ethier and Kurtz [43, p. 174]). *A right-continuous process $X(t)$, $t \geq 0$, is a solution of the martingale problem for the operator $\mathcal{A}$ if and only if*

$$\mathbf{E}\left[\prod_{j=1}^{i} h_j(X(t_j)) \left[f(X(t_{i+1})) - f(X(t_i)) - \int_{t_i}^{t_{i+1}} \mathcal{A}f(X(s))ds \right]\right] = 0$$

whenever $0 \leq t_1 < t_2 < \cdots < t_{i+1}$, $f(\cdot)$ in the domain of $\mathcal{A}$, and $h_1, \ldots, h_i \in \mathcal{B}(\mathbb{S})$, the Borel field of $\mathbb{S}$.

Theorem A.24 (Uniqueness of Martingale Problems, Ethier and Kurtz [43, p. 184]). *Let $X(\cdot)$ and $Y(\cdot)$ be two stochastic processes whose paths are in $D([0,T];\mathbb{R}^r)$. Denote an infinitesimal generator by $\mathcal{A}$. If for any function $f \in \mathcal{D}(\mathcal{A})$ (the domain of $\mathcal{A}$),*

$$f(X(t)) - f(X(0)) - \int_0^t \mathcal{A}f(X(s))ds, \ t \geq 0,$$

and

$$f(Y(t)) - f(Y(0)) - \int_0^t \mathcal{A}f(Y(s))ds, \ t \geq 0$$

are martingales and $X(t)$ and $Y(t)$ have the same distribution for each $t \geq 0$, $X(\cdot)$ and $Y(\cdot)$ have the same distribution on $D([0,\infty);\mathbb{R}^r)$.

Theorem A.25. *Let $X^\varepsilon(\cdot)$ be a solution of the differential equation*

$$\frac{dX^\varepsilon(t)}{dt} = F^\varepsilon(t),$$

and for each $T < \infty$, $\{F^\varepsilon(t) : 0 \leq t \leq T\}$ be uniformly integrable. If the set of initial values $\{X^\varepsilon(0)\}$ is tight, then $\{X^\varepsilon(\cdot)\}$ is tight in $C([0,\infty);\mathbb{R}^r)$.

Proof: The proof is essentially in Billingsley [16, Theorem 8.2] (see also Kushner [102, p. 51, Lemma 7]). □

Define the notion of "p-lim" and an operator $\mathcal{A}^\varepsilon$ as in Ethier and Kurtz [43]. Suppose that $X^\varepsilon(\cdot)$ are defined on the same probability space. Let $\mathcal{F}_t^\varepsilon$ be the minimal σ-algebra over which $\{X^\varepsilon(s), \xi^\varepsilon(s) : s \leq t\}$ is measurable and let $\mathbf{E}_t^\varepsilon$ denote the conditional expectation given $\mathcal{F}_t^\varepsilon$. Denote

$$\overline{M}^\varepsilon = \big\{ f : f \text{ is real-valued with bounded support and is}$$
$$\text{progressively measurable w.r.t. } \{\mathcal{F}_t^\varepsilon\}, \ \sup_t \mathbf{E}|f(t)| < \infty \big\}.$$

Let $g(\cdot), f(\cdot), f^\delta(\cdot) \in \overline{M}^\varepsilon$. For each $\delta > 0$ and $t \leq T < \infty$, $f = p - \lim_\delta f^\delta$ if

$$\sup_{t,\delta} \mathbf{E}|f^\delta(t)| < \infty,$$

then

$$\lim_{\delta \to 0} \mathbf{E}|f(t) - f^\delta(t)| = 0 \ \text{ for each } \ t.$$

The function $f(\cdot)$ is said to be in the domain of $\mathcal{A}^\varepsilon$; that is, $f(\cdot) \in \mathcal{D}(\mathcal{A}^\varepsilon)$, and $\mathcal{A}^\varepsilon f = g$, if

$$p - \lim_{\delta \to 0} \left(\frac{\mathbf{E}_t^\varepsilon f(t+\delta) - f(t)}{\delta} - g(t) \right) = 0.$$

If $f(\cdot) \in \mathcal{D}(\mathcal{A}^\varepsilon)$, then Ethier and Kurtz [43] or Kushner [102, p. 39] implies that

$$f(t) - \int_0^t \mathcal{A}^\varepsilon f(u) du$$

is a martingale, and

$$\mathbf{E}_t^\varepsilon f(t+s) - f(t) = \int_t^{t+s} \mathbf{E}_t^\varepsilon \mathcal{A}^\varepsilon f(u) du \quad \text{a.s.}$$

In applications, ϕ-mixing processes frequently arise; see [43] and [102]. The assertion below presents a couple of inequalities for mixing processes. Further results on various mixing processes are in [43].

Lemma A.26 (Kushner [102, Lemma 4.4]). *Let $\xi(\cdot)$ be a ϕ-mixing process with mixing rate $\phi(\cdot)$ and let $h(\cdot)$ be $\mathcal{F}_t^\infty$-measurable and $|h| \le 1$. Then*

$$\left| \mathbf{E}(h(\xi(t+s))|\mathcal{F}_0^t) - \mathbf{E}h(\xi(t+s)) \right| \le 2\phi(s).$$

If $t < u < v$, and $\mathbf{E}h(\xi(s)) = 0$ for all s, then

$$\left| \mathbf{E}(h(\xi(u))h(\xi(v))|\mathcal{F}_0^t) - \mathbf{E}h(\xi(u))h(\xi(v)) \right| \le 4 \Big(\phi(v-u)\phi(u-t) \Big)^{1/2},$$

where $\mathcal{F}_\tau^t = \sigma\{\xi(s) : \tau \le s \le t\}$.

Example A.27. A useful example of a mixing process is a function of a stationary Markov chain with finite state space. Let α_k be such a Markov chain with state space $\mathcal{M} = \{1, \ldots, m_0\}$. Let $\xi_k = g(\alpha_k)$, where $g(\cdot)$ is a real-valued function defined on $\mathcal{M}$. Suppose the Markov chain or equivalently, its transition probability matrix, is irreducible and aperiodic. Then as proved in Billingsley [16, pp. 167–169], ξ_k is a mixing process with the mixing measure decaying to 0 exponentially fast.

A crucial step in obtaining many limit problems depends on the verification of tightness of the sequences of interest. A sufficient condition known as Kurtz's criterion appears to be rather handy to use.

Lemma A.28 (Kushner [102, Theorem 3, p. 47]). *Suppose that $\{Y^\varepsilon(\cdot)\}$ is a process with paths in $D([0,\infty); \mathbb{R}^r)$, and suppose that*

$$\lim_{K_1 \to \infty} \left\{ \limsup_{\varepsilon \to 0} \mathbf{P}\left(\sup_{0 \le t \le T} |Y^\varepsilon(t)| \ge K_1 \right) \right\} = 0 \quad \text{for each } T < \infty, \quad \text{(A.24)}$$

and for all $0 \le s \le \delta$, $t \le T$,

$$\mathbf{E}_t^\varepsilon \min \left(1, |Y^\varepsilon(t+s) - Y^\varepsilon(t)|^2 \right) \le \mathbf{E}_t^\varepsilon \gamma_\varepsilon(\delta),$$
$$\lim_{\delta \to 0} \limsup_{\varepsilon \to 0} \mathbf{E}\gamma_\varepsilon(\delta) = 0. \quad \text{(A.25)}$$

Then $\{Y^\varepsilon(\cdot)\}$ is tight in $D([0,\infty); \mathbb{R}^r)$.

Remark A.29. In lieu of (A.24), one may verify the following condition (see Kurtz [95, Theorem 2.7, p. 10]). Suppose that for each $\eta > 0$ and rational $t \geq 0$ there is a compact set $\Gamma_{t,\eta} \subset \mathbb{R}^r$ such that

$$\inf_{\varepsilon} \mathbf{P}\left(Y^{\varepsilon}(t) \in \Gamma_{t,\eta}\right) > 1 - \eta. \tag{A.26}$$

A.6 Hybrid Jump Diffusion

Let us recall the notion of a switching jump diffusion or hybrid jump diffusion. A hybrid jump diffusion is a jump diffusion modulated by an additional continuous-time switching process. In what follows, we confine ourselves to the case where the switching process is a continuous-time Markov chain. In this case, in lieu of one jump diffusion, we have a system of jump diffusions. The description of the jump-diffusion system below is a modification of that of [103] due to the appearance of the switching process. Suppose that $\alpha(\cdot)$ is a continuous-time Markov chain with state space $\mathcal{M} = \{1, \ldots, m\}$ and generator $Q(t) = (q_{ij}(t))$ [176, Sections 2.3–2.5]. Let $\{\tau_n\}$ be an increasing sequence of stopping times independent of $\alpha(t)$. Let $\{\psi_n\}$ be a sequence of random variables representing the "impulses," and $\psi(\cdot)$ be a random process defined by

$$\psi(t) = \begin{cases} \psi_n & \text{if } t = \tau_n, \\ 0, & \text{otherwise.} \end{cases}$$

The process is termed a point process if $\{\psi_n, \tau_n\}$ is a random sequence and $\{\tau_n\}$ has no finite accumulation point. In the above, the ψ_n are referred to as impulses.

Let Γ, a compact set not including the origin in some Euclidean space, be the range space of $\psi(\cdot)$. Denote the σ-algebra of Borel sets of Γ by $\mathcal{B}(\Gamma)$. Suppose that the impulse time $\tau_n \to \infty$ as $n \to \infty$. For each $H \in \mathcal{B}(\Gamma)$, and each $\iota \in \mathcal{M}$, define

$$N(t, H) = \{\# \text{ of impulses of } \psi(\cdot) \text{ on } [0, t] \text{ with values in } H\},$$

which is a counting process or counting measure. Suppose that $\mathbf{E}N(t, \Gamma) < \infty$, that $\mathcal{F}_t$ is a filtration such that $N(\cdot, H)$ is $\mathcal{F}_t$-adapted for each $H \in \mathcal{B}(\Gamma)$, and that $\psi(\cdot)$ is an $\mathcal{F}_t$-Poisson point process (i.e., $\{N(t + \cdot, H) - N(t, H) : H \in \mathcal{B}(\Gamma)\}$ is independent of $\mathcal{F}_t$) and $N(\cdot, \cdot)$ is an $\mathcal{F}_t$-Poisson measure. If $\psi(\cdot)$ is a Poisson process and the distribution of $\{N(t+s, H) - N(t, H) : H \in \mathcal{B}(\Gamma)\}$ is independent of t, then $\psi(\cdot)$ is a stationary Poisson point process. Then it is known that there exists a $\lambda > 0$ and probability measure $\pi(\cdot)$ on $\mathcal{B}(\Gamma)$ such that

$$\mathbf{E}[N(t + s, H) - N(t, H)|\mathcal{F}_t] = s\pi(H)\lambda,$$

where λ is known as the impulse rate of $\psi(\cdot)$ and/or the jump rate of $N(\cdot, \Gamma)$, and $\pi(H)$ is the jump distribution in the sense that

$$\mathbf{P}(\psi(t) \in H | \psi(t) \neq 0, \ \psi(u), \ u < t) = \pi(H).$$

The values and times of the impulses can be recovered from the integral

$$G(t) = \int_0^t \int_\Gamma \gamma N(ds, d\gamma) = \sum_{s \leq t} \psi(s).$$

With the setup above, the jump-diffusion process modulated by $\alpha(t)$ is given in

$$
\begin{aligned}
X(t) = x_0 &+ \int_0^t f(s, \alpha(s), X(s))ds + \int_0^t \sigma(s, \alpha(s), X(s))dw(s) \\
&+ \int_0^t \int_\Gamma g(\gamma, \alpha(s^-), X(s^-))N(ds, d\gamma),
\end{aligned}
\tag{A.27}
$$

where $w(\cdot)$ is a real-valued standard Brownian motion, and $N(\cdot, \cdot)$ is a Poisson measure.

A.7 Miscellany

Suppose that A is an $r \times r$ square matrix. Denote the collection of eigenvalues of A by Λ. Then the spectral radius of A, denoted by $\rho(A)$, is defined by $\rho(A) = \max_{\lambda \in \Lambda} |\lambda|$. Recall that a matrix with real entries is a positive matrix if it has at least one positive entry and no negative entries. If every entry of A is positive, we call the matrix strictly positive. Likewise, for a vector $x = (x_1, \ldots, x_r)$, by $x \geq 0$, we mean $x_i \geq 0$ for $i = 1, \ldots, r$; by $x > 0$, we mean all entries $x_i > 0$.

By a multi-index ζ, we mean a vector $\zeta = (\zeta_1, \ldots, \zeta_r)$ with nonnegative integer components with $|\zeta|$ defined as $|\zeta| = \zeta_1 + \cdots + \zeta_r$. For a multi-index ζ, D_x^ζ is defined to be

$$D_x^\zeta = \frac{\partial^\zeta}{\partial x^\zeta} = \frac{\partial^{|\zeta|}}{\partial x_1^{\zeta_1} \ldots \partial x_r^{\zeta_r}}; \tag{A.28}$$

see Friedman [47] or Gihman and Skorohod [54].

The following inequalities are widely used. The first of them is known as the Gronwall inequality and second one is the so-called generalized Gronwall inequality. Both of them can be found in [61, p. 36].

Lemma A.30. *If $\gamma \in \mathbb{R}$, $\beta(t) \geq 0$, and $\varphi(t)$ are continuous real-valued functions for $a \leq t \leq b$, which satisfy*

$$\varphi(t) \leq \gamma + \int_a^t \beta(s)\varphi(s)ds, \ t \in [a, b],$$

then

$$\varphi(t) \le \gamma \exp(\int_a^t \beta(s)ds), \ t \in [a,b].$$

Lemma A.31. *Suppose that $\varphi(\cdot)$ and $\gamma(\cdot)$ are real-valued continuous functions on $[a,b]$, that $\beta(t) \ge 0$ is integrable on $[a,b]$, and that*

$$\varphi(t) \le \gamma(t) + \int_a^t \beta(s)\varphi(s)ds, \ t \in [a,b].$$

Then

$$\varphi(t) \le \gamma(t) + \int_a^t \beta(s)\gamma(s)\exp(\int_s^t \beta(u)du)ds, \ t \in [a,b].$$

When we treat stochastic integrals, the Burkholder–Davis–Gundy inequality is used quite often. We state it below, a proof of which may be found in [120, p. 70].

Lemma A.32. *Let $\psi : [0,\infty) \mapsto \mathbb{R}^{r \times d}$ for some positive integers r and d, and suppose that $\mathbf{E}\int_0^\infty |\psi(s)|^2 < \infty$. Define*

$$Z(t) = \int_0^t \psi(s)dw(s) \quad and \quad a(t) = \int_0^t |\psi(s)|^2 ds,$$

where $w(t)$ is a d-dimensional standard Brownian motion. Then for any $p > 0$, there exist positive constants c_p and C_p such that

$$c_p \mathbf{E}|a(t)|^{p/2} \le \mathbf{E} \sup_{0 \le s \le t} |Z(s)|^p \le C_p \mathbf{E}|a(t)|^{p/2}, \quad for \ all \ t \ge 0.$$

A.8   Notes

One may find a nonmeasure-theoretic introduction to stochastic processes in Ross [141]. The two volumes by Karlin and Taylor [80, 81] provide an introduction to discrete-time and continuous-time Markov chains. Advanced treatments of Markov chains can be found in Chung [28] and Revuz [142]. A book that deals exclusively with finite-state Markov chains is by Iosifescu [76]. The book of Meyn and Tweedie [125] examines Markov chains and their stability. Doob's book [33] gives an introduction to stochastic processes. Gihman and Skorohod's three-volume work [55] provides a comprehensive introduction to stochastic processes, whereas Liptser and Shiryayev's book [110] presents further topics such as nonlinear filtering.

References

[1] M.S. Agranovich, Elliptic operators on closed manifolds, in *Current Problems in Mathematics: Fundamental Directions*, **63** (1994), 1–130.

[2] B.D.O. Anderson and J.B. Moore, *Optimal Control: Linear Quadratic Methods*, Prentice Hall, Englewood Cliffs, NJ, 1990.

[3] A. Arapostathis, M.K. Ghosh, and S.I. Marcus, Harnack's inequality for cooperative weakly coupled elliptic systems, *Comm. Partial Differential Eqs.*, **24** (1999), 1555–1571.

[4] G. Badowski and G. Yin, Stability of hybrid dynamic systems containing singularly perturbed random processes, *IEEE Trans. Automat. Control*, **47** (2002), 2021–2031.

[5] G. Barone-Adesi and R. Whaley, Efficient analytic approximation of American option values, *J. Finance*, **42** (1987), 301–320.

[6] G.K. Basak, A. Bisi, and M.K. Ghosh, Stability of a random diffusion with linear drift, *J. Math. Anal. Appl.*, **202** (1996), 604–622.

[7] G.K. Basak, A. Bisi, and M.K. Ghosh, Stability of degenerate diffusions with state-dependent switching, *J. Math. Anal. Appl.*, **240** (1999), 219–248.

[8] M.M. Benderskii and L.A. Pastur, The spectrum of the one-dimensional Schrodinger equation with random potential, *Mat. Sb.* **82** (1972), 273–284.

[9] M.M. Benderskii and L.A. Pastur, Asymptotic behavior of the solutions of a second order equation with random coefficients, *Teorija Funkeii i Functional'nyi Analiz*, **22** (1973), 3–14.

[10] A. Bensoussan, *Perturbation Methods in Optimal Control*, J. Wiley, Chichester, 1988.

[11] A. Bensoussan and P.L. Lions, Optimal control of random evolutions, *Stochastics*, **5** (1981), 169–190.

[12] A. Bensoussan and J.L. Menaldi, Hybrid control and dynamic programming, *Dynamics Continuous Disc. Impulsive Sys.*, **3** (1997), 395–442.

[13] B. Bercu, F. Dufour, and G. Yin, Almost sure stabilization for feedback controls of regime-switching linear systems with a hidden Markov chain, *IEEE Trans. Automat. Control*, **54** (2009).

[14] A. Benveniste, M. Metivier, and P. Priouret, *Adaptive Algorithms and Stochastic Approximations*, Springer-Verlag, Berlin, 1990.

[15] R.N. Bhattacharya, Criteria for recurrence and existence of invariant measures for multidimensional diffusions, *Ann. Probab.*, **6** (1978), 541–553.

[16] P. Billingsley, *Convergence of Probability Measures*, J. Wiley, New York, NY, 1968.

[17] T. Björk, Finite dimensional optimal filters for a class of Ito processes with jumping parameters, *Stochastics*, **4** (1980), 167–183.

[18] G.B. Blankenship and G.C. Papanicolaou, Stability and control of stochastic systems with wide band noise, *SIAM J. Appl. Math.*, **34** (1978), 437–476.

[19] L. Breiman, *Probability*, SIAM, Philadelphia, PA, 1992.

[20] J. Buffington and R.J. Elliott, American options with regime switching, *Int. J. Theoret. Appl. Finance*, **5** (2002), 497–514.

[21] P.E. Caines and H.-F. Chen, Optimal adaptive LQG control for systems with finite state process parameters, *IEEE Trans. Automat. Control*, **30** (1985), 185–189.

[22] P.E. Caines and J.-F. Zhang, On the adaptive control of jump parameter systems via nonlinear filtering, *SIAM J. Control Optim.*, **33** (1995), 1758–1777.

[23] M.-F. Chen, *From Markov Chains to Non-equilibrium Particle Systems*, 2nd ed., World Scientific, Singapore, 2004.

[24] Z.Q. Chen and Z. Zhao, Potential theory for elliptic systems, *Ann. Probab.*, **24** (1996), 293–319.

[25] Z.Q. Chen and Z. Zhao, Harnack inequality for weakly coupled elliptic systems, *J. Differential Eqs.*, **139** (1997), 261–282.

[26] C.L. Chiang, *An Introduction to Stochastic Processes and Their Applications*, Kreiger, Huntington, NY, 1980.

[27] Y.S. Chow and H. Teicher, *Probability Theory*, 3rd ed., Springer-Verlag, New York, NY, 1997.

[28] K.L. Chung, *Markov Chains with Stationary Transition Probabilities*, 2nd ed., Springer-Verlag, New York, NY, 1967.

[29] D.R. Cox and H.D. Miller, *The Theory of Stochastic Processes*, J. Wiley, New York, NY, 1965.

[30] M.H.A. Davis, *Markov Models and Optimization*, Chapman & Hall, London, UK, 1993.

[31] D.A. Dawson, Critical dynamics and fluctuations for a mean–field model of cooperative behavior, *J. Statist. Phys.*, 31 (1983), 29–85.

[32] G.B. Di Masi, Y.M. Kabanov, and W.J. Runggaldier, Mean variance hedging of options on stocks with Markov volatility, *Theory Probab. Appl.*, **39** (1994), 173–181.

[33] J.L. Doob, *Stochastic Processes*, Wiley Classic Library Edition, Wiley, New York, NY, 1990.

[34] N.H. Du and V.H. Sam, Dynamics of a stochastic Lotka–Volterra model perturbed by white noise, *J. Math. Anal. Appl.*, **324** (2006), 82–97.

[35] F. Dufresne and H.U. Gerber, Risk theory for the compound Poisson process that is perturbed by diffusion, *Insurance; Math. Economics*, **10** (1991), 51–59.

[36] F. Dufour and P. Bertrand, Stabilizing control law for hybrid modes, *IEEE Trans. Automat. Control*, **39** (1994), 2354–2357.

[37] N.H. Du, R. Kon, K. Sato, and Y. Takeuchi, Dynamical behavior of Lotka-Volterra competition systems: Non-autonomous bistable case and the effect of telegraph noise, *J. Comput. Appl. Math.*, **170** (2004), 399–422.

[38] E.B. Dynkin, *Markov Processes*, Vols. I and II, Springer-Verlag, Berlin, 1965.

[39] S.D. Eidelman, *Parabolic Systems*, North-Holland, New York, 1969.

[40] R.J. Elliott, *Stochastic Calculus and Applications*, Springer-Verlag, New York, NY, 1982.

[41] A. Eizenberg and M. Freidlin, On the Dirichlet problem for a class of second order PDE systems with small parameter. *Stochastics Stochastics Rep.*, **33** (1990), 111–148.

[42] A. Eizenberg and M. Freidlin, Averaging principle for perturbed random evolution equations and corresponding Dirichlet problems, *Probab. Theory Related Fields*, **94** (1993), 335–374.

[43] S.N. Ethier and T.G. Kurtz, *Markov Processes: Characterization and Convergence*, J. Wiley, New York, NY, 1986.

[44] W.H. Fleming and R.W. Rishel, *Deterministic and Stochastic Optimal Control*, Springer-Verlag, New York, NY, 1975.

[45] W.H. Fleming and H.M. Soner, *Controlled Markov Processes and Viscosity Solutions*, Springer-Verlag, New York, 1992.

[46] A. Friedman, *Partial Differential Equations of Parabolic Type*, Prentice-Hall, Englewood, Cliffs, NJ, 1967.

[47] A. Friedman, *Stochastic Differential Equations and Applications*, Vol. I and Vol. II, Academic Press, New York, NY, 1975.

[48] M.D. Fragoso and O.L.V. Costa, A unified approach for stochastic and mean square stability of continuous-time linear ystems with Markovian jumping parameters and additive disturbances, *SIAM J. Control Optim.*, **44** (2005), 1165–1191.

[49] J.P. Fouque, G. Papanicolaou, and R.K. Sircar, *Derivatives in Financial Markets with Stochastic Volatility*, Cambridge University Press, Cambridge, UK, 2000.

[50] J.P. Fouque, G. Papanicolaou, R.K. Sircar, and K. Solna, Singular perturbations in option pricing, *SIAM J. Appl. Math.*, **63** (2003), 1648–1665.

[51] J.P. Fouque, G. Papanicolaou, R.K. Sircar, and K. Solna, Multiscale stochastic volatility asymptotics, *Multiscale Modeling & Simulation*, **1** (2004) 22–42.

[52] M.K. Ghosh, A. Arapostathis, and S.I. Marcus, Ergodic control of switching diffusions, *SIAM J. Control Optim.*, **35** (1997), 1952–1988.

[53] I.I. Gihman and A.V. Skorohod, *Introduction to the Theory of Random Processes*, W.B. Saunders, Philadelphia, PA, 1969.

[54] I.I. Gihman and A.V. Skorohod, *Stochastic Differential Equations*, Springer-Verlag, Berlin, 1972.

[55] I.I. Gihman and A.V. Skorohod, *Theory of Stochastic Processes, I, II, III*, Springer-Verlag, Berlin, 1979.

[56] D. Gilbarg and N.S. Trudinger, *Elliptic Partial Differential Equations of Second Order*, Springer, Berlin, 2001.

[57] I.V. Girsanov, Strongly Feller processes I. General properties, *Theory Probab. Appl.*, **5** (1960), 5–24.

[58] K. Glover, All optimal Hankel norm approximations of linear multivariable systems and their L-error bounds, *Int. J. Control*, **39** (1984), 1145–1193.

[59] G.H. Golub and C.F. Van Loan, *Matrix Computations*, 2nd ed., Johns Hopkins University Press, Baltimore, MD, 1989.

[60] M.G. Garroni and J.L. Menaldi, *Green Functions for Parabolic Second Order Integro-Differential Equations*, Pitman Research Notes in Math. Series, No. 275, Longman, London, 1992.

[61] J.K. Hale, *Ordinary Differential Equations*, 2nd ed., R.E. Krieger, Malabar, FL, 1980.

[62] J.K. Hale and E.P. Infante, Extended dynamical systems and stability theory, *Proc. Nat. Acad. Sci.* **58** (1967), 405–409.

[63] P. Hall and C.C. Heyde, *Martingale Limit Theory and Its Application*, Academic Press, New York, NY, 1980.

[64] F.B. Hanson, *Applied Stochastic Processes and Control for Jump-diffusions: Modeling, Analysis, and Computation*, SIAM, Philadelphia, PA, 2007.

[65] Q. He and G. Yin, Invariant density, Liapunov exponent, and almost sure stability of Markovian-regime-switching linear systems, preprint, 2009.

[66] U. Helmke and J.B. Moore, *Optimization and Dynamical Systems*, Springer-Verlag, New York, NY, 1994.

[67] J.P. Hespanha, *Stochastic Hybrid Systems: Application to Communication Networks*, Springer, Berlin, 2004.

[68] J.P. Hespanha, A model for stochastic hybrid systems with application to communication networks, *Nonlinear Anal.*, **62** (2005), 1353–1383.

[69] J.C. Hull, *Options, Futures, and Other Derivatives*, 3rd ed., Prentice-Hall, Upper Saddle River, NJ, 1997.

[70] J.C. Hull and A. White, The pricing of options on assets with stochastic volatilities, *J. Finance*, **42** (1987), 281–300.

[71] V. Hutson and J.S. Pym, *Applications of Functional Analysis and Operator Theory*, Academic Press, London, UK, 1980.

[72] N. Ikeda and S. Watanabe, *Stochastic Differential Equations and Diffusion Processes*, North-Holland, Amsterdam, 1981.

[73] A.M. Il'in, R.Z. Khasminskii, and G. Yin, Singularly perturbed switching diffusions: rapid switchings and fast diffusions, *J. Optim. Theory Appl.*, **102** (1999), 555–591.

[74] A.M. Il'in, R.Z. Khasminskii, and G. Yin, Asymptotic expansions of solutions of integro-differential equations for transition densities of singularly perturbed switching diffusions, *J. Math. Anal. Appl.*, **238** (1999), 516–539.

[75] J. Imae, J.E. Perkins, and J.B. Moore Toward time-varying balanced realization via Riccati equations, *Math. Control Signals Syst.*, **5** (1992), 313–326.

[76] M. Iosifescu, *Finite Markov Processes and Their Applications*, Wiley, Chichester, 1980.

[77] J. Jacod and A.N. Shiryayev, *Limit Theorems for Stochastic Processes*, Springer-Verlag, New York, NY, 1980.

[78] Y. Ji and H.J. Chizeck, Controllability, stabilizability, and continuous-time Markovian jump linear quadratic control, *IEEE Trans. Automat. Control*, **35** (1990), 777–788.

[79] I.I. Kac and N.N. Krasovskii, On the stability of systems with random parameters, *J. Appl. Math. Mech.*, **24** (1960), 1225–1246.

[80] S. Karlin and H.M. Taylor, *A First Course in Stochastic Processes*, 2nd ed., Academic Press, New York, NY, 1975.

[81] S. Karlin and H.M. Taylor, *A Second Course in Stochastic Processes*, Academic Press, New York, NY, 1981.

[82] R.Z. Khasminskii, On an averaging principle for Ito stochastic differential equations, *Kybernetika*, **4** (1968), 260–279.

[83] R.Z. Khasminskii, *Stochastic Stability of Differential Equations*, Sijthoff and Noordhoff, Alphen aan den Rijn, Netherlands, 1980.

[84] R.Z. Khasminskii and F.C. Klebaner, Long term behavior of solutions of the Lotka-Volterra systems under small random perturbations, *Ann. Appl. Probab.*, **11** (2001), 952–963.

[85] R.Z. Khasminskii and G. Yin, Asymptotic series for singularly perturbed Kolmogorov-Fokker-Planck equations, *SIAM J. Appl. Math.* **56** (1996), 1766–1793.

[86] R.Z. Khasminskii and G. Yin, On transition densities of singularly perturbed diffusions with fast and slow components, *SIAM J. Appl. Math.*, **56** (1996), 1794–1819.

[87] R.Z. Khasminskii and G. Yin, Asymptotic behavior of parabolic equations arising from one-dimensional null-recurrent diffusions, *J. Differential Eqs.*, **161** (2000), 154–173.

[88] R.Z. Khasminskii and G. Yin, On averaging principles: An asymptotic expansion approach, *SIAM J. Math. Anal.*, **35** (2004), 1534–1560.

[89] R.Z. Khasminskii and G. Yin, Limit behavior of two-time-scale diffusions revisited, *J. Differential Eqs.*, **212** (2005) 85–113.

[90] R.Z. Khasminskii and G. Yin, Uniform asymptotic expansions for pricing European options, *Appl. Math. Optim.*, **52** (2005), 279–296.

[91] R.Z. Khasminskii, G. Yin, and Q. Zhang, Asymptotic expansions of singularly perturbed systems involving rapidly fluctuating Markov chains, *SIAM J. Appl. Math.*, **56** (1996), 277–293.

[92] R.Z. Khasminskii, C. Zhu, and G. Yin, Stability of regime-switching diffusions, *Stochastic Process. Appl.*, **117** (2007), 1037–1051.

[93] R.Z. Khasminskii, C. Zhu, and G. Yin, Asymptotic properties of parabolic systems for null-recurrent switching diffusions, *Acta Math. Appl. Sinica*, **43** (2007), 177–194.

[94] P.E. Kloeden and E. Platen, *Numerical Solution of Stochastic Differential Equations*, Springer-Verlag, New York, NY, 1992.

[95] T.G. Kurtz, *Approximation of Population Processes*, SIAM, Philadelphia, PA, 1981.

[96] N.V. Krylov, *Controlled Diffusion Processes*, Springer-Verlag, New York, NY, 1980.

[97] H.J. Kushner, *Stochastic Stability and Control*, Academic Press, New York, NY, 1967.

[98] H.J. Kushner, The concept of invariant set for stochastic dynamical systems and applications to stochastic stability, in *Stochastic Optimization and Control*, H.F. Karreman Ed., J. Wiley, New York, NY, 1968, 47–57.

[99] H.J. Kushner, On the stability of stochastic differential-difference equations, *J. Differential Eqs.*, **4** (1968), 424–443.

[100] H.J. Kushner, On the convergence of Lion's identification method with random inputs, *IEEE Trans. Automat. Control*, **15** (1970), 652–654.

[101] H.J. Kushner, Asymptotic distributions of solutions of ordinary differential equations with wide band noise inputs; approximate invariant measures, *Stochastics*, **6** (1982), 259–278.

[102] H.J. Kushner, *Approximation and Weak Convergence Methods for Random Processes, with Applications to Stochastic Systems Theory*, MIT Press, Cambridge, MA, 1984.

[103] H.J. Kushner, *Weak Convergence Methods and Singularly Perturbed Stochastic Control and Filtering Problems*, Birkhäuser, Boston, MA, 1990.

[104] H.J. Kushner and G. Yin, *Stochastic Approximation and Recursive Algorithms and Applications*, 2nd ed., Springer-Verlag, New York, NY, 2003.

[105] O.A. Ladyzenskaja, V.A. Solonnikov, and N.N. Ural'ceva, *Linear and Quasi-linear Equations of Parabolic Type*, Translations of Math. Monographs, Vol. 23, Amer. Math. Soc., Providence, RI, 1968.

[106] J.P. LaSalle, The extent of asymptotic stability, *Proc. Nat. Acad. Sci.*, **46** (1960), 365.

[107] J.P. LaSalle and S. Lefschetz, *The Stability by Liapunov Direct Method*, Academic Press, New York, NY, 1961.

[108] G.M. Lieberman, *Second-Order Parabolic Differential Equations*, World Scientific, Singapore, 1996.

[109] J.-J. Liou, Recurrence and transience of Gaussian diffusion processes, *Kodai Math. J.*, **13** (1990), 210–230.

[110] R.S. Liptser and A.N. Shiryayev, *Statistics of Random Processes I & II*, Springer-Verlag, New York, NY, 2001.

[111] Y.J. Liu, G. Yin, Q. Zhang, and J.B. Moore, Balanced realizations of regime-switching linear systems, *Math. Control, Signals, Sys.*, **19** (2007), 207–234.

[112] K.A. Loparo and G.L. Blankenship, Almost sure instability of a class of linear stochastic systems with jump process coefficients, in *Lyapunov Exponents*, Lecture Notes in Math., 1186, Springer, Berlin, 1986, 160–190.

[113] Q. Luo and X. Mao, Stochastic population dynamics under regime switching, *J. Math. Anal. Appl.*, **334** (2007), 69–84.

[114] X. Mao, *Stability of Stochastic Differential Equations with respect to Seminartingales*, Longman Sci. Tech., Harlow, Essex, UK, 1991.

[115] X. Mao, *Stochastic Differential Equations and Applications*, 2nd ed., Horwood, Chichester, UK, 2007.

[116] X. Mao, Stability of stochastic differential equations with Markovian switching, *Stochastic Process. Appl.*, **79** (1999), 45–67.

[117] X. Mao, A note on the LaSalle-type theorems for stochastic differential delay equations, *J. Math. Anal. Appl.*, **268** (2002), 125–142.

[118] X. Mao, S. Sabanis, and R. Renshaw, Asymptotic behavior of the stochastic Lotka-Volterra model, *J. Math. Anal. Appl.*, **287** (2003), 141–156.

[119] X. Mao, G. Yin, and C. Yuan, Stabilization and destabilization of hybrid systems of stochastic differential equations, *Automatica*, **43** (2007), 264–273.

[120] X. Mao and C. Yuan, *Stochastic Differential Equations with Markovian Switching*, Imperial College Press, London, UK, 2006.

[121] X.R. Mao, C. Yuan, and G. Yin, Numerical method for stationary distribution of stochastic differential equations with Markovian switching, *J. Comput. Appl. Math.*, **174** (2005), 1–27.

[122] X. Mao, C. Yuan, and G. Yin, Approximations of Euler-Maruyama type for stochastic differential equations with Markovian switching, under non-Lipschitz conditions, *J. Computational Appl. Math.*, **205** (2007), 936–948.

[123] M. Mariton, *Jump Linear Systems in Automatic Control*, Marcel Dekker, New York, NY, 1990.

[124] R.C. Merton, *Continuous-Time Finance*, Blackwell, Cambridge, MA, 1990.

[125] S.P. Meyn and R.L. Tweedie, *Markov Chains and Stochastic Stability*, Springer-Verlag, London, UK, 1993.

[126] G.N. Milstein, *Numerical Integration of Stochastic Differential Equations*, Kluwer, New York, NY, 1995.

[127] G.N. Milstein and M.V. Tretyakov, *Stochastic Numerics for Mathematical Physics*, Springer-Verlag, Berlin, 2004.

[128] C.M. Moller, Stochastic differential equations for ruin probability, *J. Appl. Probab.*, **32** (1995), 74–89.

[129] B.C. Moore, Principal component in linear systems: Controllability, observability, and model reduction, *IEEE Trans. Automat. Control*, **26** (1981), 17–31.

[130] B. Øksendal, *Stochastic Differential Equations, An Introduction with Applications*, 6th ed., Springer-Verlag, Berlin, 2003.

[131] G.C. Papanicolaou, D. Stroock, and S.R.S. Varadhan, Martingale approach to some limit theorems, in *Proc. 1976 Duke Univ. Conf. on Turbulence*, Durham, NC, 1976.

[132] E. Pardoux and A. Yu. Veretennikov, On Poisson equation and diffusion approximation 1, *Ann. Probab.* **29** (2001), 1061–1085.

[133] J.E. Perkins, U. Helmke, and J.B. Moore, Balanced realizations via gradient flow techniques, *Sys. Control Lett.*, **14** (1990), 369–379.

[134] L. Perko, *Differential Equations and Dynamical Systems*, 3rd ed., Springer, New York, NY, 2001.

[135] S. Peszat and J. Zabczyk, Strong Feller property and irreducibility for diffusions on Hilbert spaces, *Ann. Probab.*, **23** (1995), 157–172.

[136] K. Pichór and R. Rudnicki, Stability of Markov semigroups and applications to parabolic systems, *J. Math. Anal. Appl.*, **215** (1997), 56–74.

[137] M. Prandini, J. Hu, J. Lygeros, and S. Sastry, A probabilistic approach to aircraft conflict detection, *IEEE Trans. Intelligent Transport. Syst.*, **1** (2000), 199–220.

[138] M.H. Protter and H.F. Weinberger, *Maximum Principles in Differential Equations*, Prentice-Hall, Englewood Cliffs, NJ, 1967.

[139] H. Robbins and S. Monro, A stochastic approximation method, *Ann. Math. Statist.* **22** (1951), 400–407.

[140] T. Rolski, H. Schmidli, V. Schmidt, and J. Teugels, *Stochastic Processes for Insurance and Finance*, J. Wiley, New York, NY, 1999.

[141] S. Ross, *Stochastic Processes*, J. Wiley, New York, NY, 1983.

[142] D. Revuz, *Markov Chains*, 2nd ed., North-Holland, Amsterdam, 1984.

[143] W. Rudin, *Real and Complex Analysis*, 3rd ed., McGraw-Hill, New York, NY, 1987.

[144] H. Sandberg and A. Rantzer, Balanced truncation of linear time-varying systems, *IEEE Trans. Automat. Control*, **49** (2004), 217–229.

[145] E. Seneta, *Non-negative Matrices and Markov Chains*, Springer-Verlag, New York, NY, 1981.

[146] S.P. Sethi and Q. Zhang, *Hierarchical Decision Making in Stochastic Manufacturing Systems*, Birkhäuser, Boston, 1994.

[147] S.P. Sethi, H. Zhang, and Q. Zhang, *Average-cost Control of Stochastic Manufacturing Systems*, Springer, New York, NY, 2005.

[148] S. Shokoohi, L.M. Silverman, and P.M. Van Dooren, Linear time variable systems: Balancing and model reduction, *IEEE Trans. Automat. Control*, **28** (1983), 810–822.

[149] H.A. Simon and A. Ando, Aggregation of variables in dynamic systems, *Econometrica*, **29** (1961), 111–138.

[150] A.V. Skorohod, *Asymptotic Methods in the Theory of Stochastic Differential Equations*, Amer. Math. Soc., Providence, RI, 1989.

[151] M. Slatkin, The dynamics of a population in a Markovian environment, *Ecology*, **59** (1978), 249–256.

[152] Q.S. Song and G. Yin, Rates of convergence of numerical methods for controlled regime-switching diffusions with stopping times in the costs, *SIAM J. Control Optim.*, **48** (2009), 1831–1857.

[153] D.W. Stroock and S.R.S. Varadhan, *Multidimensional Diffusion Processes*, Springer-Verlag, Berlin, 1979.

[154] D.D. Sworder and J.E. Boyd, *Estimation Problems in Hybrid Systems*, Cambridge University Press, Cambridge, UK, 1999.

[155] D.D. Sworder and V.G. Robinson, Feedback regulators for jump parameters systems with state and control dependent transition rates, *IEEE Trans. Automat. Control*, **AC-18** (1973), 355–360.

[156] Y. Takeuchi, N.H. Du, N.T. Hieu, and K. Sato, Evolution of predator-prey systems decribed by a Lotka-Volterra equation under random environment, *J. Math. Anal. Appl.*, **323** (2006), 938–957.

[157] E.I. Verriest and T. Kailath, On generalized balanced realizations, *IEEE Trans. Automat. Control*, **28** (1983), 833–844.

[158] J.T. Wloka, B. Rowley, and B. Lawruk, *Boundary Value Problems for Elliptic Systems*, Cambridge University Press, Cambridge, UK, 1995.

[159] W.M. Wonham, Some applications of stochastic differential equations to optimal nonlinear filtering, *SIAM J. Control*, **2** (1965), 347–369.

[160] W.M. Wonham, Liapunov criteria for weak stochastic stability, *J. Differential Eqs.*, **2** (1966), 195–207.

[161] F. Xi, Feller property and exponential ergodicity of diffusion processes with state-dependent switching, *Sci. China Ser. A*, **51** (2008), 329–342.

[162] F. Xi and G. Yin, Asymptotic properties of a mean-field model with a continuous-state-dependent switching process, *J. Appl. Probab.*, **46** (2009), 221–243.

[163] H. Yang and G. Yin, Ruin probability for a model under Markovian switching regime, in *Probability, Finance and Insurance*, World Scientific, T.L. Lai, H. Yang, and S.P. Yung, Eds., 2004, 206–217.

[164] D.D. Yao, Q. Zhang and X.Y. Zhou, A regime-switching model for European options, in *Stochastic Processes, Optimization, and Control Theory Applications in Financial Engineering, Queueing Networks, and Manufacturing Systems*, H.M. Yan, G. Yin, and Q. Zhang, Eds., Springer, New York, NY, 2006, 281–300.

[165] G. Yin, On limit results for a class of singularly perturbed switching diffusions, *J. Theoret. Probab.*, **14** (2001), 673–697.

[166] G. Yin, Asymptotic expansions of option price under regime-switching diffusions with a fast-varying switching process, *Asymptotic Anal.*, **63** (2009).

[167] G. Yin and S. Dey, Weak convergence of hybrid filtering problems involving nearly completely decomposable hidden Markov chains, *SIAM J. Control Optim.*, **41** (2003), 1820–1842.

[168] G. Yin, V. Krishnamurthy, and C. Ion, Regime switching stochastic approximation algorithms with application to adaptive discrete stochastic optimization, *SIAM J. Optim.*, **14** (2004), 1187–1215.

[169] G. Yin and V. Krishnamurthy, Least mean square algorithms with Markov regime switching limit, *IEEE Trans. Automat. Control*, **50** (2005), 577–593.

[170] G. Yin, R.H. Liu, and Q. Zhang, Recursive algorithms for stock liquidation: A stochastic optimization approach, *SIAM J. Optim.*, **13** (2002), 240–263.

[171] G. Yin, X.R. Mao, and K. Yin, Numerical approximation of invariant measures for hybrid diffusion systems, *IEEE Trans. Automat. Control*, **50** (2005), 577–593.

[172] G. Yin, X. Mao, C. Yuan, and D. Cao, Approximation methods for hybrid diffusion systems with state-dependent switching diffusion processes: Numerical alogorithms and existence and uniqueness of solutions, preprint, 2007.

[173] G. Yin, Q.S. Song, and Z. Zhang, Numerical solutions for jump-diffusions with regime switching, *Stochastics*, **77** (2005), 61–79.

[174] G. Yin, H.M. Yan, and X.C. Lou, On a class of stochastic optimization algorithms with applications to manufacturing models, in *Model-Oriented Data Analysis*, W.G. Müller, H.P. Wynn and A.A. Zhigljavsky, Eds., Physica-Verlag, Heidelberg, 213–226, 1993.

[175] G. Yin and H.L. Yang, Two-time-scale jump-diffusion models with Markovian switching regimes, *Stochastics Stochastics Rep.*, **76** (2004), 77–99.

[176] G. Yin and Q. Zhang, *Continuous-time Markov Chains and Applications: A Singular Perturbations Approach*, Springer-Verlag, New York, NY, 1998.

[177] G. Yin and Q. Zhang, *Discrete-time Markov Chains: Two-time-scale Methods and Applications*, Springer, New York, NY, 2005.

[178] G. Yin, Q. Zhang, and G. Badowski, Asymptotic properties of a singularly perturbed Markov chain with inclusion of transient states, *Ann. Appl. Probab.*, **10** (2000), 549–572.

[179] G. Yin and C. Zhu, On the notion of weak stability and related issues of hybrid diffusion systems, *Nonlinear Anal.: Hybrid System*, **1** (2007), 173–187.

[180] G. Yin and C. Zhu, Regularity and recurrence of switching diffusions, *J. Syst. Sci. Complexity*, **20** (2007), 273–283.

[181] J. Yong and X.Y. Zhou, *Stochastic Controls: Hamiltonian Systems and HJB Equations*, Springer-Verlag, New York, 1999.

[182] C. Yuan and J. Lygeros, Stabilization of a class of stochastic differential equations with Markovian switching, *Syst. Control Lett.*, **54** (2005), 819–833.

[183] C. Yuan and X. Mao, Asymptotic stability in distribution of stochastic differential equations with Markovian switching, *Stochastic Process Appl.*, **103** (2003), 277–291.

[184] Q. Zhang, Stock trading: An optimal selling rule, *SIAM J. Control Optim.*, **40** (2001), 64–87.

[185] Q. Zhang and G. Yin, On nearly optimal controls of hybrid LQG problems, *IEEE Trans. Automat. Control*, **44** (1999) 2271–2282.

[186] X.Y. Zhou and G. Yin, Markowitz mean-variance portfolio selection with regime switching: A continuous-time model, *SIAM J. Control Optim.*, **42** (2003), 1466–1482.

[187] C. Zhu and G. Yin, Asymptotic properties of hybrid diffusion systems, *SIAM J. Control Optim.*, **46** (2007), 1155–1179.

[188] C. Zhu and G. Yin, On strong Feller, recurrence, and weak stabilization of regime-switching diffusions, *SIAM J. Control Optim.*, **48** (2009), 2003–2031.

[189] C. Zhu and G. Yin, On competitive Lotka–Volterra model in random environments, *J. Math. Anal. Appl.*, **357** (2009), 154–170.

[190] C. Zhu, G. Yin, and Q.S. Song, Stability of random-switching systems of differential equations, *Quarterly Appl. Math.*, **67** (2009), 201–220.

Index